AF181797

METHODS AND PROTOCOLS IN FOOD SCIENCE

Series Editor
Anderson S. Sant'Ana
University of Campinas
Campinas, Brazil

For further volumes:
http://www.springer.com/series/16556

Methods and Protocols in Food Science series is devoted to the publication of research protocols and methodologies in all fields of food science.

Volumes and chapters will be organized by field and presented in such way that the readers will be able to reproduce the experiments in a step-by-step style. Each protocol will be characterized by a brief introductory section, followed by a short aims section, in which the precise purpose of the protocol will be clarified.

Food Waste Valorization

Green Techniques in Sustainable Management

Edited by

Tanmay Sarkar

Department of Food Processing Technology, Malda Polytechnic, West Bengal State Council Technical Education, Government of West Bengal, Malda, West Bengal, India

Humana Press

Editor
Tanmay Sarkar
Department of Food Processing
Technology
Malda Polytechnic
West Bengal State Council Technical Education
Government of West Bengal
Malda, West Bengal, India

ISSN 2662-950X ISSN 2662-9518 (electronic)
Methods and Protocols in Food Science
ISBN 978-1-0716-4645-8 ISBN 978-1-0716-4646-5 (eBook)
https://doi.org/10.1007/978-1-0716-4646-5

© The Editor(s) (if applicable) and The Author(s), under exclusive license to Springer Science+Business Media, LLC, part of Springer Nature 2025, Corrected Publication 2025

This work is subject to copyright. All rights are solely and exclusively licensed by the Publisher, whether the whole or part of the material is concerned, specifically the rights of translation, reprinting, reuse of illustrations, recitation, broadcasting, reproduction on microfilms or in any other physical way, and transmission or information storage and retrieval, electronic adaptation, computer software, or by similar or dissimilar methodology now known or hereafter developed.
The use of general descriptive names, registered names, trademarks, service marks, etc. in this publication does not imply, even in the absence of a specific statement, that such names are exempt from the relevant protective laws and regulations and therefore free for general use.
The publisher, the authors and the editors are safe to assume that the advice and information in this book are believed to be true and accurate at the date of publication. Neither the publisher nor the authors or the editors give a warranty, expressed or implied, with respect to the material contained herein or for any errors or omissions that may have been made. The publisher remains neutral with regard to jurisdictional claims in published maps and institutional affiliations.

This Humana imprint is published by the registered company Springer Science+Business Media, LLC, part of Springer Nature.
The registered company address is: 1 New York Plaza, New York, NY 10004, U.S.A.

If disposing of this product, please recycle the paper.

Dedication

To my beloved mother, Arati Sarkar, and father, Bholanath Sarkar, and my sister, Tanusree Sarkar, whose unwavering love, support, and encouragement have shaped my journey.

And to God, for the strength and guidance throughout every step of the way.

Preface to the Series

Methods and Protocols in Food Science series is devoted to the publication of research protocols and methodologies in all fields of food science. The series is unique as it includes protocols developed, validated and used by food and related scientists as well as theoretical basis are provided for each protocol. Aspects related to improvements in the protocols, adaptations and further developments in the protocols may also be approached.

Methods and Protocols in Food Science series aims to bring the most recent developments in research protocols in the field as well as very well established methods. As such the series targets undergraduate, graduate and researchers in the field of food science and correlated areas. The protocols documented in the series will be highly useful for scientific inquiries in the field of food sciences, presented in such way that the readers will be able to reproduce the experiments in a step-by-step style.

Each protocol will be characterized by a brief introductory section, followed by a short aims section, in which the precise purpose of the protocol is clarified. Then, an in-depth list of materials and reagents required for employing the protocol is presented, followed by a comprehensive and step-by-step procedures on how to perform that experiment. The next section brings the do's and don'ts when carrying out the protocol, followed by the main pitfalls faced and how to troubleshoot them. Finally, template results will be presented and their meaning/conclusions addressed.

The *Methods and Protocols in Food Science* series will fill an important gap, addressing a common complain of food scientists, regarding the difficulties in repeating experiments detailed in scientific papers. With this, the series has a potential to become a reference material in food science laboratories of research centers and Universities throughout the world.

Campinas, Brazil ***Anderson S. Sant'Ana***

Preface

Food waste has now become a universal problem that influences environmental sustainability, economic stability, and food security. Millions of tons of food waste are created every year in the supply chain, from agricultural production to disposal by consumers. Whereas traditional waste management is disposed-oriented, food waste valorization offers a revolutionizing strategy-one that aims at recovering value from waste and reconstituting it into useful products like biofuels, bioplastics, nutraceuticals, and sustainable food ingredients.

Food Waste Valorization is an exhaustive anthology of state-of-the-art research and innovative practices that aim at reducing food waste and maximizing its potential. It is a useful resource for researchers, industry experts, policymakers, and students in food science, biotechnology, sustainability, and waste management. Focusing on environmentally friendly and technology-based solutions, the book investigates the potential of food waste to support a circular economy, mitigating the environmental impact while unlocking economic potential.

The book contains chapters discussing a wide variety of valorization methods. Anaerobic digestion and fermentation-based bioconversion to supercritical fluid extraction and enzymatic hydrolysis, the book sheds light on the most up-to-date developments in bioactive compound, protein, and essential nutrient recovery from food waste. Also, green solutions like waste decontamination with cold plasma technology, production of edible film and coating, and bioplastic development are discussed at length.

A standout feature of the book is the interdisciplinary nature. Input from stakeholders in various disciplines guarantees that the topic is touched upon from science, technology, and industry. In addition to this, regulatory frameworks, standards for safety, and economic viability are highlighted by the book as critical factors that must be used in the massive implementation of food waste valorization.

In light of increasing concerns related to climate change, natural resource depletion, and food security, the importance of food waste valorization cannot be emphasized more. The book intends to encourage researchers and industries to move away from conventional waste handling practices to sustainable valorization procedures, thereby supporting world sustainability initiatives.

I would like to express my heartfelt appreciation to all the contributors whose experience and commitment have turned this book into a reality. Their research, insights, and creative thinking have molded this volume into a worthwhile reference for both industry and academia. Special appreciation to the editorial and production staff for their steadfast support in seeing this project through.

I hope the book will prove to be a source of inspiration and a ready reference for everyone dedicated to improving food waste valorization and moving towards a sustainable future.

Malda, West Bengal, India ***Tanmay Sarkar***

Contents

Contributors

AMINU ABDULLAHI • *Department of Biotechnology, Modibbo Adama University, Yola, Nigeria; Department of Biotechnology, Vel Tech Rangarajan Dr. Sagunthala R&D Institute of Science and Technology, Chennai, Tamil Nadu, India; Centre for Biomaterials and Environmental Biotechnology, Vel Tech Rangarajan Dr. Sagunthala R&D Institute of Science and Technology, Chennai, Tamil Nadu, India*

P. AFRAH KOUNAIN • *Department of Biotechnology and Bioinformatics, JSS Academy of Higher Education Research, Mysore, Karnataka, India*

NAZIR AHMAD • *Department of Nutritional Sciences, Government College University Faisalabad, Faisalabad, Pakistan*

FAWAD ALI • *Institute of Biotechnology and Microbiology, Bacha Khan University, Charsadda, Peshawar, Pakistan*

FAIZA AMEEN • *British School of Leadership and Management, Glasgow Caledonian University (London Campus), Glasgow, Scotland, UK*

FAROOQ ANWAR • *Institute of Chemistry, University of Sargodha, Sargodha, Pakistan; Department of Food Science, Faculty of Food Science and Technology, University Putra Malaysia, Serdang, Selangor, Malaysia*

USAMA ANWAR • *Laboratory of Materials Science and Nanotechnology (LMNT), CR-INSTM, Department of Biomedical Sciences, University of Sassari, Sassari, Italy*

MAHENDERAN APPUKUTTY • *Sports Science Programme, Faculty of Sports Science and Recreation UniversitiTeknologi MARA (UiTM), Shah Alam, Selangor, Malaysia*

KHADGA RAJ ARAN • *Division of Neuroscience, Department of Pharmacology, I.S.F. College of Pharmacy, Moga, Punjab, India*

INSHA ARSHAD • *Department of Chemistry, University of Gujrat, Gujrat, Pakistan*

MUHAMMAD TAYYAB ARSHAD • *University Institute of Food Science and Technology, The University of Lahore, Lahore, Pakistan*

YUSUF ASAD • *Department of Biomedical Engineering, Rowan University, Glassboro, NJ, USA*

SUMEL ASHIQUE • *Department of Pharmaceutical Technology, Bharat Technology, Uluberia, West Bengal, India*

MUHAMMAD ADNAN AYUB • *Department of Chemistry, University of Sahiwal, Sahiwal, Pakistan*

ELHAM AZARPAZHOOH • *Agricultural Engineering Research Department, Khorasan Razavi Agricultural and Natural Resources Research and Education Center, AREEO, Mashhad, Iran*

NISARGA BABU • *Department of Biotechnology and Bioinformatics, JSS Academy of Higher Education Research, Mysore, Karnataka, India*

AYESHA BAIG • *Department de Quimica Analitica, Facultad de Quimica, Universidad de Sevilla, Sevilla, Spain*

SIBASHISH BAKSI • *Department of Biotechnology and Biosciences, University of Milan-Bicocca, Milan, Italy; Department of Biotechnology, Swami Vivekananda University, Barrackpore, India*

SIBOJYOTI BAKSI • *School of Biological Sciences, IIT Delhi, Delhi, India*

HOMA BEHMADI • *Agricultural Engineering Research Institute, Agricultural Research, Education and Extension Organization (AREEO), Karaj, Iran*
AMINA BIBI • *Department of Chemistry, University of Gujrat, Gujrat, Pakistan*
FULVIA BOVERA • *Department of Zoology, University of Naples Federico II, Naples, Italy*
AYAN CHATTERJEE • *School of Health Science and Technology, Medhavi Skills University, Singtam, Sikkim, India*
SUBHAJIT DAS • *Department of Allied Health, School of Health and Medical Sciences, Adamas University, Kolkata, West Bengal, India*
SUCHANDRA DATTA • *Department of Home Science, University of Calcutta, Kolkata, West Bengal, India*
BIPLAB DEBNATH • *Department of Pharmaceutical Technology, Bharat Technology, Uluberia, West Bengal, India*
PRAVIN S. DEHARKAR • *Department of Chemistry, Dr. Ambedkar College, Deekshabhoomi, Nagpur, India*
RAGUNATHAN DEVENDRAN • *Charles Tanford Protein Center, Martin Luther University, Halle (Saale), Germany*
SUMITHA ELAYAPERUMAL • *Department of Biotechnology and Bioinformatics, JSS Academy of Higher Education and Research, Mysore, Karnataka, India*
ZAHRA EMAM-DJOMEH • *Functional Food Research Core (FFRC), Transfer Phenomena Laboratory (TPL), College of Agriculture & Natural Resources, University of Tehran, Tehran, Iran*
MUJAHID FARID • *Department of Environmental Science, University of Gujrat, Gujrat, Pakistan*
SYEDA FATHIMA • *Department of Biotechnology and Bioinformatics, JSS Academy of Higher Education Research, Mysore, Karnataka, India*
GNANASELVAN GNANASEKARAN • *Department of Bioproducts and Biosystems Engineering, University of Minnesota, Minneapolis, MN, USA*
MOHAMMAD-TAGHI GOLMAKANI • *Department of Food Science and Technology, School of Agriculture, Shiraz University, Shiraz, Iran*
ANKITA GON • *Department of Allied Health, School of Health and Medical Sciences, Adamas University, Kolkata, West Bengal, India*
SEINUL ABDEEN FATHIMA HASANATH • *National Institute of Food Science and Technology, Faculty of Food Nutrition and Home Sciences, University of Agriculture, Faisalabad, Pakistan*
HADI HASHEMI • *Department of Food Science and Technology, School of Agriculture, Shiraz University, Shiraz, Iran*
HAIDER IJAZ • *Department of Chemistry, University of Gujrat, Gujrat, Pakistan*
RUKSHANA IRANI • *Department of Food & Nutrition, Raidighi College, Raidighi, West Bengal, India*
ANAS ISLAM • *Faculty of Pharmacy, Integral University, Lucknow, Uttar Pradesh, India*
BEENISH ISRAR • *Institute of Home Science, Faculty of Food Nutrition and Home Sciences, University of Agriculture, Faisalabad, Pakistan*
NIZWA ITRAT • *Institute of Home Science, Faculty of Food Nutrition and Home Sciences, University of Agriculture, Faisalabad, Pakistan*
DOUGLAS S. KALMAN • *Dr. Kiran C. Patel College of Osteopathic Medicine, Nova Southeastern University, Fort Lauderdale, FL, USA*
RAHUL B. KAMBLE • *Department of Botany, Dr. Ambedkar College, Deekshabhoomi, Nagpur, India*

WASEEM KHALID • *Department of Organic Chemistry, Faculty of Chemical Sciences and Technologies, University of Castilla La Mancha, Ciudad Real, Spain*
YAS KOHANSAL • *Department of Food Science and Technology, School of Agriculture, Shiraz University, Shiraz, Iran*
TARANGINI KORUMILLI • *Department of Biotechnology, Vel Tech Rangarajan Dr. Sagunthala R&D Institute of Science and Technology, Chennai, Tamil Nadu, India; Centre for Biomaterials and Environmental Biotechnology, Vel Tech Rangarajan Dr. Sagunthala R&D Institute of Science and Technology, Chennai, Tamil Nadu, India*
SHUN-YUAN LUO • *Department of Chemistry, National Chung Hsing University, Taichung, Taiwan, Republic of China*
ZAHRA MANSOURI KERMANSHAHI • *Department of Food Science and Technology, School of Agriculture, Shiraz University, Shiraz, Iran*
SAMMRA MAQSOOD • *National Institute of Food Science and Technology, University of Agriculture Faisalabad, Faisalabad, Pakistan*
PRAMILA MURUGESAN • *Department of Biotechnology, Faculty of Engineering and Technology, SRM Institute of Science and Technology, Tiruchirappalli, Tamil Nadu, India*
MONDHER MZOUGHI • *Department of Food Engineering, Faculty of Chemical and Metallurgical Engineering, Istanbul Technical University, Istanbul, Turkey*
BALAKRISHNAN NAVINA • *Department of Post-Harvest Technology, ETSEA-University of Lleida, Lleida, Spain*
SHAHA NOORAIN • *Department of Biotechnology and Bioinformatics, JSS Academy of Higher Education Research, Mysore, Karnataka, India*
PRITHA PAL • *Department of Microbiology, Swami Vivekananda University, Barrackpore, India*
TUHIN JAMES PAUL • *Department of Pharmacy Practice, I.S.F. College of Pharmacy, Moga, Punjab, India*
G. RAJESHWARI • *Department of Biotechnology, Faculty of Engineering and Technology, SRM Institute of Science and Technology, Tiruchirappalli, Tamil Nadu, India*
HOSAHALLI S. RAMASWAMY • *Department of Food Science and Agricultural Chemistry, Macdonald Campus of McGill University, Sainte-Anne-de-Bellevue, QC, Canada*
KÜBRA SAĞLAM • *Department of Food Processing, Food Technology Program, Istanbul Gelisim Vocational School, Istanbul Gelisim University, Istanbul, Turkey*
SAEED FARHANG SAHLEVAN • *Faculty of Engineering and Architecture, Arturo Prat University, Tarapacá, Chile*
RATNNADEEP C. SAWANT • *Department of Chemistry, Dr. Ambedkar College, Deekshabhoomi, Nagpur, India*
T. SELVA KUMAR • *Vel Tech Rangarajan Dr. Sagunthala R&D Institute of Science and Technology, Chennai, Tamil Nadu, India*
YUVARAJ SIVAMANI • *Crescent School of Pharmacy, B.S. Abdur Rahman Crescent Institute of Science and Technology, Vandalur, Chennai, Tamil Nadu, India*
SUBHASH R. SOMKUWAR • *Department of Botany, Dr. Ambedkar College, Deekshabhoomi, Nagpur, India*
SARANYADEVI SUBBURAJ • *Department of Biotechnology, Karpagam Academy of Higher Education, Coimbatore, Tamil Nadu, India*
SAJJAD HUSSAIN SUMRRA • *Department of Chemistry, University of Gujrat, Gujrat, Pakistan*
K. SURENDRA • *Department of Physics, Centurion University of Management and Technology, Vizianagaram, Andhra Pradesh, India*

C. M. Tisha Muthamma • *Department of Biotechnology and Bioinformatics, JSS Academy of Higher Education Research, Mysore, Karnataka, India*

Aman Ullah • *College of Natural and Applied Sciences (CNAS), Department of Agricultural, Food & Nutritional Science, University of Alberta, Edmonton, AB, Canada*

Mudassar Zafar • *Department of Biochemistry and Biotechnology, University of Gujrat, Gujrat, Pakistan*

Muhammad Nadeem Zafar • *Department of Chemistry, University of Gujrat, Gujrat, Punjab, Pakistan*

Muhammad Zubair • *Department of Chemistry, University of Gujrat, Gujrat, Pakistan*

Chapter 1

Anaerobic Digestion Protocols for Green Valorization of Food Waste

Shaha Noorain, Syeda Fathima, P. Afrah Kounain, C. M. Tisha Muthamma, Nisarga Babu, Ragunathan Devendran, and Sumitha Elayaperumal

Abstract

Anaerobic digestion represents a prominent sustainable technological process for converting food waste into bioenergy and digestate for sustaining environmental protection and resource recovery. The anaerobic digestion process occurs in the absence of air and is based on metabolization of organic substances into biogas by microbial consortia, consisting mainly of methane (CH_4) and carbon dioxide (CO_2). The efficiency of anaerobic digestion is highly dependent on a variety of factors such as feedstock characteristics, source of inoculum, temperature, pH, organic loading rate (OLR), and hydraulic retention time (HRT). This standard protocol outlines a complete process for optimizing anaerobic digestion parameters for food wastes, covering all aspects of the pre-treatment study, including sorting, size reduction, and dilution to enhance microbial activities. It also discusses reactor configurations and inoculation methods from the perspectives of essential operational parameters that maintain process stability and maximize methane yield. The study discusses biogas upgrading methods for better energy use and digestate management methods for its safe application as an organic soil amendment. Other major process challenges such as VFA accumulation, system acidification, and gas composition imbalance were discussed along with some workable solutions. This protocol outlines the path to promoting the circular bioeconomy by incorporating anaerobic digestion into waste management practices, thereby reducing greenhouse gas emissions and providing renewable energy. The guideline will be useful for researchers, industries, and policymakers in realizing new, efficient waste-to-energy solutions.

Key words Anaerobic digestion, Food waste valorization, Biogas production, Methane yield, Organic waste management, Sustainable energy, Circular bioeconomy, Digestate utilization, Microbial fermentation, Green technology

1 Feedstock (Food Waste and Co-substrates) in Anaerobic Digestion

The increasing growth of the world food waste problem and environmental concern is making sustainable waste management techniques much more important. Anaerobic digestion (AD) has proved to be one of the many methods developed for valuing organic waste as a highly efficient and sustainable alternative

Tanmay Sarkar (ed.), *Food Waste Valorization: Green Techniques in Sustainable Management*, Methods and Protocols in Food Science, https://doi.org/10.1007/978-1-0716-4646-5_1,
© The Author(s), under exclusive license to Springer Science+Business Media, LLC, part of Springer Nature 2025

[1]. This biological process has a double edge; it enhances the diversion of food waste from landfills while at the same time producing useful by-products, that is, biogas and nutrient-rich digestate, benefiting the circular economy and sustainable development goals. The systems of managing food waste can be supplemented with anaerobic digestion; therefore, they could enhance energy recovery from organic waste.

While at the same time reducing fossil energy reliance and minimizing environmental contamination. Food waste is a contextual issue around the world that has an impact on households, restaurants, food processing plants, and the agricultural sector with millions of tons generated every year. Approximately one-third of the global food production gets wasted. This amounts to around 1.3 billion tons a year in terms of wasted production, according to FAO, which the organization states is due to its non-market valuation. Such excessive waste aggravates food insecurity and causes severe environmental harm in the form of land degradation, water pollution, and even GHG emissions [2]. Belief has been attributed to many disposal methods, such as traditional landfill disposal, and incineration methods. All these suffer from drawbacks like high-operation costs, low-efficiency energy utilization, and adverse environmental effects. Compared to these, anaerobic digestion is a greater benefit for both the economics and the environment. The microbial degradation process occurs under anaerobic conditions in which complex organic compounds such as proteins, lipids, and carbohydrates can be biologically converted into simpler molecules. It consists of four basic biochemical stages: hydrolysis, acidogenesis, acetogenesis, and methanogenesis. Large organic molecules first undergo hydrolysis to produce their monomeric constituents (sugars, fatty acids, and amino acids). These monomers then enter the second-step process of acidogenesis to produce volatile fatty acids, alcohols, and gases including carbon dioxide and hydrogen [3]. The acetogenic product of these intermediates, finally, will be converted into acetic acid, hydrogen, and carbon dioxide, utilizing the methanogenic archaea for the production of biogas, a gas mainly consisting of methane (CH_4) and carbon dioxide (CO_2).

Anaerobic digestion, also referred to as green valorization, is the technique for converting food waste into useful products for a sustainable and circular economy. From this process, communities and industries can produce energy sustainably while reducing their waste and improving soil health. Anaerobic digestion forms an eco-cycle in which all organic poultry are recycled, not discarded, contrary to other waste management systems that operate on a linear economy; resources are consumed and thrown away after that. It is in line with the principles of resource efficiency, whereby food waste does not find its last resting place in landfills, but rather in economic and environmental sustainability [4]. Technological advances have now made anaerobic digestion much more practicable and efficient. Co-digestion, microbial optimization, and biogas upgrading are examples of innovations that have improved the functionality of anaerobic digestion in waste conversion into energy. Co-digestion improves methane production and stabilizes biological activity by simultaneous process treatment of different organic wastes, namely food waste, agricultural by-products, and wastewater. Additionally, raw biogas can now be purified to high-purity biomethane through contemporary biogas purification mechanisms such as membrane separation, pressure swing adsorption, and water scrubbing, and may serve as a green alternative to fossil fuels. With all these advancements, anaerobic digestion is increasingly accepted as a sustainable and scalable approach for converting organic waste into energy with the least possible adverse impacts on the environment [5].

Whereas dumping food waste is an economic and environmental matter related to overcrowding of landfills, depletion of resources, and increase in greenhouse gas emissions across the globe, on the other hand, anaerobic digestion (AD) represents the biological process.

Technique to convert food waste into useful resources in the process of sustainable disposal of such substrates includes anaerobic digestion, composting, vermicomposting, biochar production, and fermentation-based biorefineries that generate biofuels, bioplastics, and value-added chemicals. Further, this boosts a circular bioeconomy by converting food waste into beneficial by-products like energy in the form of biogas and soil nutrients in the form of digestate, besides reducing environmental pollution [6]. During anaerobic digestion, methane (CH_4), carbon dioxide (CO_2), and some traces of other gases are formed in the process conducted by the microbes under anaerobic conditions. The four major phases of anaerobic digestion are hydrolysis, acidogenesis, acetogenesis, and methanogenesis. All anaerobic digestion stages use different microbial consortia and are responsible for breaking down complex organic chemicals to generate biogas, an energy-rich by-product, through four key phases: hydrolysis, acidogenesis, acetogenesis, and methanogenesis—each mediated by specialized microorganisms that sequentially convert complex substrates into methane and carbon dioxide. To maximize the anaerobic digestion of food waste for achieving valorization, several criteria need to be put into consideration, namely characteristics of substrate, reactor designs, and operational parameters such as pH, temperature, and hydraulic retention time. For enhanced biogas yield and stability of the process, certain advanced methods of co-digestion, pre-treatment, and microbial enhancement have been examined. The anaerobic digestion of food waste for value addition helps accord with the global sustainable perspective, as it reduces landfill waste, promotes renewable energy generation, and lessens dependency on fossil fuels [7]. Furthermore, the digestate produced during AD could be utilized as an organic fertilizer, which would strengthen sustainable agricultural practices. This paper describes some of the standardized anaerobic digestion practices, advancements in technology, and possible hindrances associated with the green valorization of food waste. With this in mind, AD can be enhanced through the integration of appropriate innovations and optimization of existing AD conditions in favor of a sustainable management system [8].

The various processes of AD exploit batch vs. continuous systems, single-stage vs. multi stage digestion, and mesophilic vs. thermophilic settings to enhance the conversion of food waste. Each of these protocols has specific advantages and disadvantages impacting biogas generation, process stability, and sustainability in general. The review aims to discuss different anaerobic digestion approaches of food waste valorization, highlighting their advantages, disadvantages, and potential applications in waste management and production of green energy [9].

2 Materials

Materials selection is of utmost importance for optimizing anaerobic digestion (AD) processes for the sustainable valorization of food waste. Organic feedstock is majorly composed of food waste from households, agriculture, and food industries, providing major nutrients for microbial degradation. This is complemented by an inoculum source in the form of anaerobic sludge or manure, which is introduced to ensure there is an active population of hydrolytic, acidogenic, acetogenic, and methanogenic bacteria. Buffering agents, such as sodium bicarbonate and calcium carbonate, are included in the mixture to stabilize fluctuations in pH that may inhibit digestion [10]. Nitrogen-rich co-substrates such as lignocellulosic biomass are also employed to meet and optimize the carbon-to-nitrogen (C/N) ratio, thereby ensuring maximum microbial efficiency and biogas yield. The anaerobic digester system with temperature control, mixing, and collection of gas units is also very important in maintaining stable conditions for the process. In addition, analytical equipment like gas chromatography for biogas composition analysis and total solids measurement kits facilitate real-time process monitoring. These materials can be strategically selected and combined to promote biogas production, enhance digestate quality, and create another path toward sustainable waste management and energy recovery interventions.

2.1 Feedstock (Food Waste and Co-substrates) in Anaerobic Digestion

Choosing an optimal feedstock is the main consideration for efficient anaerobic digestion (AD) since it directly impacts the activity of microorganisms, biogas production, and the stability of the process [11]. Food waste from homes, restaurants, supermarkets, and food processing industries is one of the most important sources of organic feedstock mainly due to its organic nature and biodegradability. Carbohydrates (from fruits, vegetables, and grains), proteins (from meat, dairy, and legumes), and lipids (from oils and fats, as well as from dairy products) constitute the main part of food waste, and this can contribute to the production of volatile fatty acids (VFAs) and methane. But it is also seen that food waste frequently has an improper C/N balance which causes inefficiencies including acidification or ammonia inhibition. This problem can be solved by introducing co-substrates for balanced nutrient availability and improved microbial synergy. Lignocellulosic biomass, namely crops and forest residues, straw, sawdust, and other forestry wastes, can provide a constant carbon source while combating the fast acidification imposed by readily biodegradable waste food. Nitrogen-rich materials, such as animal manures, sewage sludge, and protein-rich waste, provide nutrients necessary for microbial growth and avert ammonia toxicity when employed in the appropriate ratios [12]. FOG wastes from the food industry and wastewater treatment plants are important co-substrates for the high-energy substrate propelling the methane yield by a large margin. On the other hand, excessive FOG leads to digester clogging and the buildup of long-chain fatty acids (LCFA), which subsequently inhibit microbial activity. The proper integration of food wastes with other co-substrates ensures balanced nutrient profiles, prevention of process disturbances, and enhancement of overall biogas output. Utilizing appropriate selection and proportions of feedstock components, anaerobic digestion may be optimally tuned to convert organic wastes efficiently into renewable energy, which underpins sustainable waste management and circular bioeconomy ideals. Largely, the anaerobic digestion (AD) scenario depends on how feedstock is selected, composition-wise, since they are the primary organic substrate for microbial degradation and biogas generation. Food wastes, the main feedstock available in AD, are generated from households, restaurants, supermarkets, and food processing industries [13]. Carbohydrates, proteins, and lipids are involved in the digestion process, each contributing in its way. Carbohydrates, present in fruits, vegetables, and grains, are rapidly broken down into volatile fatty acids (VFAs) and thus provide an immediate energy source for methanogens. Proteins contained in meat, dairy, and legumes undergo hydrolysis to amino acids and ammonia, in moderate concentrations showing favorable effects but proving toxic at higher concentrations. The highest methane potentials are assigned to lipids obtained from oil, butter, and greasy food residues that are highly energetic. However, a very high lipid accumulation impedes microbial activity by leading to long-chain fatty acid (LCFA) buildup and subsequently affects digesters' stability [14].

In practice, the digestion performance of the process is optimized by adding co-substrates that complement the carbon-to-nitrogen (C/N) ratio, promote microbial interactions, and boost methane yield. Lignocellulosic biomass in the form of agricultural residues, wood chips, straw, and paper wastes has proven to be a stable carbon source that mitigates rapid acidification and promotes digester stability. Nevertheless, due to the complex polymeric architecture (cellulose, hemicellulose, and lignin) of such materials, pre-treatment methods such as enzymatic hydrolysis, steam explosion, or acid treatment are often employed to enhance biodegradability [15]. Nitrogenous substrates such as animal manure, sewage sludge, and protein-rich industrial waste provide essential nutrients for the microbial communities, thus enhancing microbial growth and activity and reducing the potential for nutrient-based process failure. However, high nitrogen levels also lead to ammonia accumulation, which inhibits methanogenesis, requiring strategies to control pH and ammonium. Another important category of co-substrates is fat, oil, and grease (FOG) wastes from food industries, wastewater treatment plants, and grease traps. These substrates generally have good methane yield, often higher than that of carbohydrate- and protein-based materials. However, their hydrolysis is slow, and the accumulation of long-chain fatty acids necessitates controlled loading rates and co-digestion with other substrates to avoid overloading the digester. Fruit and vegetable residues are usually sugar- and organic acid-rich, very fast to degrade and enhance microbe's activity, although they need to be co-digested with fibrous or nitrogenous materials to balance the process [16]. Dairy wastes such as whey and cheese processing residues also constitute good estimable co-substrates concerning high organic content, but the protein and fat contents must remain under strict control to avoid inhibition of the process. Proper selection and proportioning of food waste and co-substrates become important for stabilizing the bioprocess, ensuring a balanced nutrient profile and optimum biogas production. Co-substrate use helps buffer against process limitations such as acidification, ammonia toxicity, and substrate inhibition, enhancing bioprocess stability and efficiency. With the smart combination of feedstock, AD is set to be able to provide sustainable valorization pathways for food waste toward renewable energy, waste reduction, and circular bioeconomy advancement [17].

2.2 Pre-treatment Materials for Anaerobic Digestion: Physical, Chemical, and Biological Processes

Much of the pre-treatment materials improve anaerobic digestion (AD) by enhancing the biodegradability of feedstock, promoting hydrolysis, and increasing biogas production. Food waste and lignocellulosic biomass, with their inherent complexities in composition, often need pre-treatment to first remove structural constraints such as lignin, cellulose, and hemicellulose from the pathways for microbial activity against organic matter. These pre-treatments may be grouped under physical, chemical, and biological categories that challenge different structural and compositional processes against feedstock processing [18]. Firstly, physical pre-treatments such as grinding, pulverization, and thermal processing break dry biomass down into smaller particles and increase surface area availability for microbial attack. Secondly, chemical pre-treatments such as strong acids, alkalis, oxidizing agents, and solvents help degrade complex organic compounds by solubilizing and disrupting complex polymers like lignin and proteins. Last but not least, biological pre-treatments involve microbial cultures or enzymatic formulations that are aimed at breaking down organic matter in biochemical pathways, greatly improving digestibility in an ecologically sustainable manner [19].

The appropriate choice of pre-treatment materials should be based on feedstock characteristics, process requirements, and economic feasibility since each pre-treatment method has its advantages and potential drawbacks. Applying an appropriate pre-treatment strategy is expected to facilitate the overall anaerobic digestion process, which eventually leads to increased methane yield and reduced retention times for better resource recovery, thus contributing to sustainable waste management and energy production [20].

Pre-treatment of feedstock thus plays a very significant role in enhancing anaerobic digestion (AD) efficiency by increasing the bioavailability of organic matter, degrading complex polymers, and reducing digestion time. Physical, chemical, and biological pre-treatment methods are employed to enhance the degradation of the substrate for optimization of biogas production [21]. The physical pre-treatment discusses an array of mechanical, thermal, or ultrasonic methods to reduce the particle size and surface area or break down structural barriers in the feedstock. Mechanically, pre-treatment such as cutting and crushing, grinding, shredding, or extruding may reduce the size of particles and with this make organic polysaccharides more susceptible to microbial degradation. Thermal pre-treatment was done between temperatures of 60 and 200 °C, which breaks down the lignocellulosic structure in plant-based waste and, therefore, increases solubilization of various organic compounds present. Steam explosion, a thermal method, applies very high pressure to heat and break down lignin and hemicellulose components in fibrous waste. In addition, ultrasonic pre-treatment causes cavitation by the use of high-frequency sound waves which rupture cell walls, thus promoting solubilization of organic matter, especially in sludge and complex food waste [22]. This is referred to as a chemical or solvent pre-treatment, which relies on the use of acids, alkalis, oxidizing substances, or solvents to break up or convert complex organic structures into simpler ones to improve the biodegradability of these compounds. Sulfuric acid (H_2SO_4) or hydrochloric acid (HCl) is most commonly employed in acidic pre-treatments. Acid hydrolysis provides hemicellulose hydrolysis and liberates carbohydrates from matrices, hence leading to easily available substrates for microbial digestion. However, higher concentrations of acid may also lead to the formation of some inhibitors like furfural and hydroxymethylfurfural, which are toxic to methanogenic bacteria. Alkaline pre-treatment uses sodium hydroxide (NaOH), calcium hydroxide ($Ca(OH)_2$), or ammonia (NH_3) and is said to give the greatest breakdown of lignin as well as improvements in digestibility for lignocellulosic biomass [23]. Ammonia fiber explosion (AFEX) is a specialized alkaline pre-treatment that prepares fibrous materials for enzymatic hydrolysis. The oxidative pre-treatment using hydrogen peroxide (H_2O_2) or ozone (O_3) is intended for breaking down complex organic matter while reducing pathogen load and odor. The other pre-treatment methods include solvent-based processes, wherein organic solvents like ethanol or methanol dissolve lignin; nevertheless, the application of this method has been really limited on account of cost and environmental concerns. On the other hand, biological pre-treatment breaks down organic matter into simpler components by microbial or enzymatic actions before anaerobic digestion [24]. Microbial pre-treatment employs fungi, bacteria, or other microbial consortia that excrete enzymes capable of degrading lignocellulosic and protein-rich substrates. The enzymes said to be involved in lignin degradation include laccase, manganese peroxidase, and lignin peroxidase produced by white-rot fungi such as Phanerochaete chrysosporium, thus improving the digestibility of the lignocellulosic biomass [25].

Enzymatic pre-treatment aims to accelerate biodegradation by breaking down cellulose, hemicellulose, protein, and lipid complex macromolecules into biodegradable building blocks. This technique can be applied to the treatment of fibrous, protein-rich food waste. Secondly, co-fermentation with hydrolytic bacteria like Clostridium species can help reduce lag phases in anaerobic digestion (AD) and increase solubilization of substrates. Each pretreatment has its considerations, which are determined from the characteristics of the feedstock, economic viability, and process efficiency perspective [26]. Compared with physical pre-treatment, which increases surface area and digestibility, chemical pre- treatment, in general, enhances hydrolysis through better solubilization but may introduce inhibitory compounds. Biological pre-treatments are environmentally benign but often take longer.

Therefore, combinations of these methods, known as integrated pre-treatment, are commonly implemented to optimize efficiency, improve biogas production, and further enhance sustainable valorization of organic waste through anaerobic digestion [27].

Hydrolysis is time-consuming step in anaerobic digestion (AD) processes involving food waste (FW) as the breakdown of complex molecules may take longer periods. Several physiochemical pre-treatment methods, such as thermal, chemical, ultrasonic, or hydrothermal techniques, attempt to hasten hydrolysis and consequently enhance biogas production. FW particle size is reduced during ultrasonic treatment, whereas, in thermal pre- treatment, chemical bonds are broken so that FW becomes more absorptive. Evidence shows that the performance of biogas yield multiplies exponentially from the synergetic effect between thermal and chemical treatment wherein thermochemical pre-treatment reaches as high as 615 mL/g VS. Hydrothermal and enzymatic pre-treatments have also been applied successfully to increase methane yield by up to 51%.

Food waste is characterized by high carbohydrate content which favors rapid biodegradation, but fatty acids of long carbon chains tend to slow down methane production temporarily [2]. With time, AD microbes would eventually adapt to the situation and break them down. Excessive thermal pre-treatment, however, would lead to the caramelization of sugars, causing an adverse impact on efficiency. Somewhere in between these two extremes lies the optimum condition. As far as AD system design studies go, it was determined that, indeed, a two-phase digester produces more methane than a single-phase digester, and that methane yield can be enhanced with the pre-treatment of the FW leachate. Under suitable conditions for pre-treatment, FW may be rendered readily biodegradable and converted into biogas useful as energy; thus, AD proves a feasible waste and energy management option [28].

2.3 Factors Affecting Anaerobic Digestion of Food Waste

The performance and stability of anaerobic digestion (AD) for food waste valorization can be affected by physicochemical, biological, and operational factors that influence microbial activity (Fig. 1), substrate degradation, and biogas production. Among the most critical parameters is the carbon-to-nitrogen (C/N) ratio, affecting the microbial nutrient availability. An optimal C/N ratio (usually between 20:1 and 30:1) allows the microorganisms to simultaneously utilize carbon for energy and nitrogen for growth to avoid either ammonia inhibition or organic acid accumulation [29]. pH and alkalinity are also crucial for the homeostasis of microbial systems, with the optimum range from 6.5 to 7.5 to keep the methanogenic activity up. Any disturbance of the pH due to high VFA accumulation or ammonia release will result in instability in the process, thereby affecting methane production. Temperature is yet another critical parameter, as anaerobic

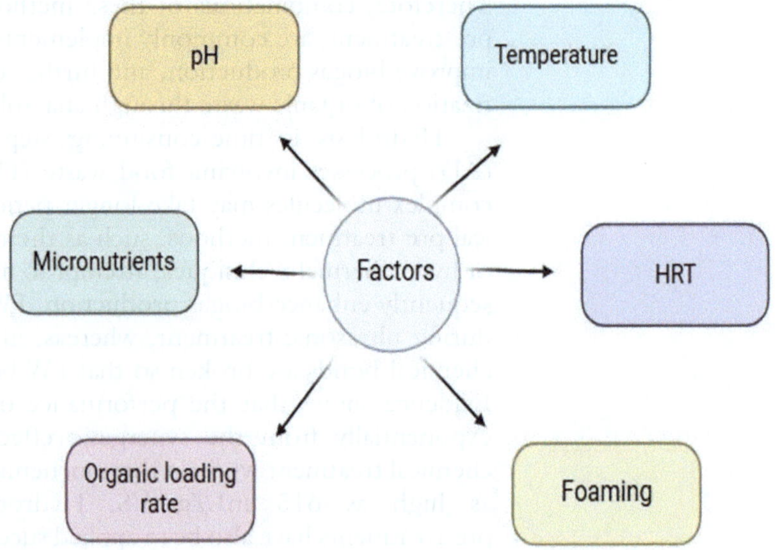

Fig. 1 Schematic representation of the factors that affect anaerobic digestion of food waste

digestion occurs at mesophilic (35–40 °C) or thermophilic (50–60 °C) conditions. Thermophilic digestion, in addition to being very advantageous in ensuring high reaction rates and pathogen removal, requires high energy input and is sensitive to process fluctuations. Hydraulic retention time (HRT) and organic loading rate (OLR) have a strong bearing on digestion performance; an optimal HRT (generally about 15–30 days) allows for adequate microbial digestion, whereas excessive OLR would then lead to acidification and digester failure. The composition of substrates, i.e., carbohydrates, proteins, and lipids, affects biodegradability as well as methane potential. Lipid-rich waste generates more energy, but there is a risk of accumulation of long-chain fatty acids (LCFAs) that will inhibit microbial activity. Inhibitors like ammonia (proceeding from protein degradation), sulfides (from sulfate-reducing bacteria), and heavy metals (industrial waste contamination) disrupt the microbial balance, thereby negating biogas production. Minimization of these effects can be done through substrate pre-treatment, co-digestion, or the addition of buffering agents. Digester mixing also promotes equal substrate distribution and stops stratification, which serves to enhance microbial contact with organic matter [30].

Digestion efficiency is controlled via microbial community dynamics, especially the balance between hydrolytic, acidogenic, acetogenic, and methanogenic bacteria. The inoculum source, such as anaerobic sludge or manure, has facilitated microbial acclimatization and the stabilization of anaerobic digestion processes. Optimization of these factors is of utmost importance in enhancing digestion efficiency, increasing methane yield, and promoting sustainable food waste valorization via anaerobic digestion [31]. Therefore, through the manipulation of these variables in a proper manner, AD shall be turned into a reliable renewable energy management option.

2.3.1 pH

pH stability that maintains anaerobic digestion efficiency is a neutral range between 7.2 and 8.5 for methanogenic microbes. Digressing from this range where the pH declines below 6.5 or rises over 8.2 usually causes an inhibitory effect on methane production often as a result of volatile fatty acid (VFA) accumulation. The optimum methane yield is best with a pH of 7; the process is then favored by a shift to hydrogen production via thermal pre-treatment, further lowering methane yield from 0.446 to 0.385 Nm^3/kg VS [32].

2.3.2 Temperature

Essentially right temperatures must be maintained to ensure anaerobic digestion (AD) through methanogen growth. Most operational conditions are in a mesophilic range (25–45 °C), with optimum temperatures between 30 and 40 °C. Too deviated conditions tend to slow down the digestion process since temperature deviations disturb those tiny organisms [33]. The thermophilic conditions would be excellent for better pathogen destruction and rapid waste disintegration; however, fewer thermophilic microbes are present, and therefore less popular. Also, AD, together with highly diluted wastewater, is usually deficient when cold.

2.3.3 Hydraulic Retention Time

Hydraulic retention time plays an important role in anaerobic digestion (AD), where long retention times stabilize the process, but may lead to depletion of substrate and consequent lower methane yield. Studies indicate that an HRT of approximately 12 days is optimal for the highest methane production, with shorter retention times leading to the washing out of methanogens and therefore less efficiency. On the flip side, excessive retention times may lead to the accumulation of VFA and ammonium concentration, which could impair the performance of the reactor. The suitable time will therefore depend on the substrate type, organic loading, and reactor design, hence requiring careful optimization [34].

2.3.4 Organic Loading Rate (OLR)

OLR is the amount of organic material provided to the digester daily per unit volume. Excessive OLR may cause the accumulation of fatty acids, inhibiting methanogenic bacteria and reducing methane production. Literature shows that the maximum OLR will vary with the type of digestion process; for mesophilic, it is 1.5 g VS/L/day, whereas thermophilic is 2.5 g VS/L/day [35]. Due to the washout of biomass, the methane yield is reduced by high OLR and short retention.

2.3.5 Micronutrients

Micronutrients that include nickel (Ni), iron (Fe), cobalt (Co), and selenium (Se) are important for anaerobic digestion for the growth of methanogens and enzyme activities. However, excess volatile fatty acids (VFA) and ammonia can hinder the process. Food waste (FW) lacks key micronutrients, which probably should be supplemented using manure or sludge to maintain balance in the digestion process and biogas generation.

Foaming in anaerobic digestion is generated mainly by surfactants, proteins, fatty acids, and CO_2, thereby trapping gas, lowering digester working volume, and blocking pipelines. Foaming can also be propelled by antibiotics, ammonia, or detergents [36]. Control measures to mitigate foam formation include using natural oils such as rapeseed oil, biodiesel, and silicones. Also, blending strategies, throttle feeding rates, and enzymatic hydrolysis can help reduce foaming.

2.4 Process Configuration

Methane production efficiencies of anaerobic digestion (AD) systems greatly depend on the way the system is configured. Many approaches are available, such as digestion and single-, two-, and multistage digestion. Each has its respective advantages and disadvantages, depending on the configuration and intended objectives [37].

2.4.1 Single-Stage Digestion: The Simple and Cost-Effective Option

Single-stage digestion is the most widely practiced method, especially in Europe, due to its straightforwardness and low-cost aspects. All the important biological processes—degradation of the waste, acid production, and acid conversion to methane—are performed in one operation within the same reactor. This setup is easy and neat and requires less capital investment [38]. Under this classification are "wet" digestion and "dry" digestion. Because wet digestion usually avoids volatile fatty acids (VFAs) inhibiting methane production, it is considered the more efficient option. On the other hand, since different microorganisms operate at different rates, single-stage systems can also tend to show stability problems. These problems are controlled through the management of organic loading rate (OLR) and retention time [39].

2.4.2 Two-Stage Digestion: Increased Efficiency but Increased Costs

The two-stage procedure involves two different reactors—one for hydrogen and acid production and the second for methane production. The separation enhances stability, improves methane yields, and boosts waste degradation. However, this comes at the cost of higher investment and maintenance. Studies indicate that two-stage digestion generally surpasses in performance of the single-stage counterpart, especially at shorter retention times. Though one school of thought proposes that single-stage thermophilic systems generate more energy, others emphasize the two-stage system's reliability in biogas generation and resistance to fluctuations [40].

2.4.3 Multistage Digestion: The Performance Champion

Further separation into different reactors for hydrolysis, acid production, and methane production is called multistage digestion. This configuration ensures that system failures are minimized while each stage is optimized for better efficiency. A classical type is "thermal- phased AD," where the first stage runs at high temperatures, accelerating the degradation of waste and killing pathogens, while the second stage runs at lower temperatures, favoring methane-producing microbes [41]. Hence, efficiency and energy recovery are enhanced. Multistage systems accommodate higher waste loads and remove more organic matter. Some studies have indicated that up to 50% additional methane and three times the solid waste elimination are achievable with these systems over the single-stage options. However, careful monitoring would be required to ensure energy is not lost [42].

2.4.4 Co-digestion: Bringing Different Waste Together for Higher Effects

Co-digestion is known as coordinating varied wastes; for example, food wastes with other wastes such as animal manure or sewage sludge. This mixing has been shown to contribute greatly to a robust methane output due to a stability improvement. Just food wastes are rather tough to digest on their own because of their low buffering capacity due to their very high C: N ratios. If food waste is mixed with manure or sludge, the process takes place more rapidly—from time to time, as much as more than **1.4 times faster** than with food waste alone. Therefore, although there are cost savings and efficiency improvement from specific co-digestion, it requires proper management of that mixture to avoid potential problems, such as foaming [43].

2.5 Resource Recovery

1. **Carbon Recovery**
   - Biogas production from digestate: Anaerobic digestion (AD) converts organic material to biogas (CH_4 and CO_2); incomplete degradation provides extra possibility for carbon recovery. Pre-treatment methods such as wet explosion and fungal treatment enhance the yield of biogas from fermentation residue [44].
   - Enhanced biogas by digestate-derived char addition: Pyrochar or hydrochar addition improves AD biomethane production with waste reuse.
   - Hydrothermal Carbonization (HTC) process water: HTC process water is very high in organic content and could be a potential raw material for biogas production from the feedstock, but methane yields from such sources are dependent on feedstock characteristics.
2. **Char Production Through Thermochemical Conversion**
   - Pyrolysis products: Pyrolysis digestate is syngas, bio-oil, and pyrochar. Pyrochar's composition and properties are

determined by temperature and feedstock. It's also for phosphorus recovery and reduction of contaminants [45].

- Hydrothermal carbonization (HTC) products: Hydrochar is produced under hydrothermal conditions and has a smaller surface area than pyrochar, but can be activated to improve adsorption properties. High process temperature severely affects categories of hydrochar properties.

3. **Nitrogen and Phosphorus Recovery**
   - Ammonia stripping and struvite precipitation: These methods for efficient NH_4-N and total phosphate (TP) recovery can recover up to 88% nitrogen and 96% phosphorus [46].
   - Pyrochar adsorption: Pyrochar from digestate can remove ammonium up to 66.64% efficiency from wastewater.
   - Hydrothermally processed for nutrient recovery: It goes up to 94.3% in the soil with the combined use of HTC and acid leaching to recover phosphorus making it more amenable as fertilizer.

2.6 Food Waste Anaerobic Digestion Reactors and Analytical Apparatus

The efficiency of anaerobic digestion (AD) for the treatment of food waste critically depends on the type of reactor configurations and the analytical apparatus used for monitoring process parameters. The choice of reactor plays a key role in the optimization of microbial activities of substrate degradation and biogas production, and analytical instruments ensure stability, efficiency, and operational control. Several configurations are used in AD depending on the feedstock type and the optimum condition of the process [47]. Batch reactors work by loading food wastes at the introduction and allowing digestion to occur for a particular time interval without adding substrate, thus being simple but less efficient for continuous biogas production. By contrast, continuously stirred tank reactors (CSTR) permit complete continuous feeding and removal of digestate, proper mixing, substrate suspension in digestate, and increases in methane yield [48].

PFRs are ideal for such high-solids feedstocks as food and agricultural wastes since the substrate moves through the reactor linearly, allowing for greater stability in digestion parameters and reducing the possibility of washout of the microorganisms. Advanced systems like UASB and EGSB use a dense microbial sludge blanket to degrade organic matter efficiently, yet allow very high recovery of biogas, thus being well suited to treat liquid food waste and wastewater [49]. Anaerobic Membrane Bioreactors (AnMBRs) are AnMBRs that divide membrane filtration for the separation of microbial biomass from the effluent, which allows for increased retention time of the digester and achieves higher quality digestate. Two- stage anaerobic digestion systems explicitly separate hydrolysis and acidogenesis from methanogenesis, thereby optimizing conditions for different microbial groups and thus enhancing process stability, especially while facing structurally complex food waste containing fats, proteins, and lignocellulosic material.

In the effective functioning of a reactor and averted failure in operations, various analytical tools have been employed continuously to monitor biological and physicochemical parameters. Alkalinity sensors and pH meters are required to maintain the ideal pH range between 6.5 and 7.5, because changes may inhibit microbial activity and reduce methane production. Gas chromatography (GC) explores methane (CH_4) and carbon dioxide (CO_2) levels in biogas, while trace gases that can prove harmful to methanogens such as hydrogen sulfide (H_2S) and ammonia (NH_3) are detected with infrared gases analyzers. The analyzers less commonly referred to as (TS) Total Solids or (VS) Volatile Solids calculate how much organic matter there is in feedstock and digestate, which gives information on the effectiveness of substrate degradation [50]. Therefore, analyzers for Chemical Oxygen Demand (COD) and Biochemical Oxygen Demand (BOD) are important for determining the organic load entering the digester, thereby evaluating the biodegradability of the food waste. A spectrophotometer or UV-Vis analyzer commonly used to quantify volatile fatty acids (VFA), ammonia, and major nutrients such as nitrogen and phosphorus ensures the balance in microbial metabolism. Scanning electron microscopy (SEM) helps visualize the microbial morphology and biofilm formation inside reactor systems, while Fourier Transform Infrared Spectroscopy (FTIR) is used to investigate structural changes in organic materials before and after their digestion [51]. The microbial population dynamics are analyzed by fluorescence microscopy and quantitative PCR methods, which ensure that essential methanogen cultures remain active and robust under variable process conditions. Respirometric systems measure respiration and degradation of substrates by microbes, hence optimizing substrate loading rates to prevent inhibition of the process. The integrated advanced reactor designs with real-time analytical monitoring are necessary to maximize biogas production, ensure process stability, and enhance the overall efficiency of food waste valorization. By optimizing anaerobic digestion as a sustainable approach for waste management, energy recovery, and circular bioeconomy applications through a carefully tailored combination of reactor technology and analytical control strategies, real-time monitoring can be integrated with different reactor technologies [52].

2.7 Safety Equipment and Monitoring Materials in Anaerobic Digestion

Anaerobic digestion (AD) is the decomposition of organic waste using microbial agents under oxygen-free conditions, resulting in biogas whose main components are methane (CH_4), carbon dioxide (CO_2), and some trace amounts of hazardous gases such as hydrogen sulfide (H_2S) and ammonia (NH_3). The above-mentioned requirements and other risks associated with pressure buildup, microbial contamination, flammability, toxicity, and corrosivity of gases require the installation of suitable safety equipment and monitoring systems for secure operations [53]. One of the most important safety measures in AD facilities is a gas detection system that continuously monitors levels of CH_4, CO_2, H_2S, NH_3, and O_2. These detectors, which may be fixed and portable variants, use electrochemical, infrared, or catalytic sensors for early detection of gas leakages preventing explosion and toxic burning.

A methane detector is to prevent fire and explosion hazards by detecting gas leaks in biogas storage and pipeline systems. Majorly, protection is offered from possible exposure and damage to workers as well as to equipment by detecting releases of H_2S gas through the installation of H_2S gas monitors. Oxygen sensors also ensure anaerobic conditions within digesters and monitor confined spaces where O_2 displacement by biogas could create asphyxiation risks [54].

In addition to gas monitoring, reactor stability is heavily reliant on temperature and pressure control systems. Burst discs and pressure relief valves (PRVs) arrest any excessive pressure buildup that could lead to an explosion or failure of the reactor. Temperature sensors and thermal monitoring equipment prevent microbial inhibition from heating or cooling by ensuring the digester operates within the ideal range of mesophilic (35–40 °C) or thermophilic (50–60 °C). Flame arrestors and explosion-proof electrical equipment in biogas storage facilities prevent fires or explosions that may be caused by ignition sources [55]. Biogas scrubbers and desulfurization units work to remove impurities from biogas, which contain moisture and other corrosive compounds. This prevents the corrosion and degradation of gas pipelines, engines, and storage facilities. The appropriate amount of personal protective equipment (PPE) is important for employee safety. Respirators, gas masks, gloves, chemical-resistant suits, and eye protection must all be used by operators in case of an emergency while handling digesters involving dangerous chemicals, digestate, and biogas leaks.

Entering maintenance or inspection into digester tanks requires equipment for confined space entry such as ventilation systems, harnesses, and self-contained breathing apparatus (SCBA), which are to be used for asphyxiation and harmful gas exposure prevention [56]. Responding and mitigating such k inds of events with gas leakages, overpressure, or any other dangerous condition have been made easier with real-time notifications from the automated alarm systems and emergency shutdown systems

(ESD). Fire suppression devices such as foam- based fire extinguishers or CO_2 would be put in strategic positions in the biogas facilities in anticipation of any possible fire dangers. Comprehensive monitoring systems and control systems would include real-time data acquisition—SCADA systems—that mounted gas composition, pressure, temperature, and flow rates to ensure optimal and safe operation of digesters. These digital monitoring solutions enable shorter response times in case of potential threats and increase the overall safety environment. It could dispose of anaerobic digestion at an efficient level but with the least hazards in compliance with the environment and protection of workers and infrastructures through advanced safety equipment, constant monitoring, and stringent operational protocols [57].

Anaerobic digestion (AD) refers to the microbiological break down of all organic wastes without the inclusion of oxygen, resulting in the product of biogas containing methane (CH_4) and carbon dioxide (CO_2), and small amounts of potentially dangerous gases such as hydrogen sulfide (H_2S) and ammonia (NH_3). Therefore, the presence of flammable, toxic, and corrosive gases, the risk of pressure buildup, and, of course, the risk of microbial contamination will result in the requirement of proper safety equipment and a monitoring system for secure operation. One of the most important safety measures within AD facilities includes a gas detection system that continuously measures any changes in levels of CH_4, CO_2, H_2S, NH_3, and O_2. A system of fixed and portable gas detectors (most commonly electrochemical, infrared, or catalytic) is used for the early detection of gas leakages which prevents explosions and toxic inhalation. Methane detectors specifically prevent fire and explosion hazards by directly identifying gas leaks in biogas storage and pipeline systems. Given the fact that hydrogen sulfide (H_2S) is very toxic and corrosive, H_2S gas monitors are additionally required for protection against exposure and worth of equipment damage [58]. Oxygen sensors are not just used to ensure anaerobic conditions inside digesters; they are also implemented in confined areas to analyze any risk of oxygen displacement by biogas leading to asphyxiation. Reactor stability, flowing from intense management of temperature and pressure, is left with gas monitoring as the control mechanism. Burst discs and pressure relief valves (PRVs) are used to protect against excessive pressure that may cause explosions or reactor failure. Temperature sensors and thermal monitoring allow the digester to operate in the desired mesophilic (35–40 °C) or thermophilic (50–60 °C) range, preventing microbial inhibition through potential warming or cooling. Flame arrestors and explosion-proof electrical equipment are affixed in biogas storage areas to stop fire or explosion ignitions. Biogas scrubbers and desulfurization units clean biogas from moisture and corrosive agents to prevent corrosion and deterioration of gas pipelines, engines, and storage places. Employee safety requires further implementation of proper personal protective equipment (PPE). Using respirators, gas masks, gloves, chemical-resistant suits, and eye protection to prevent exposure to harmful chemicals, biogas leakages, or digestate is paramount for digester operations [59]. To protect against asphyxiating conditions and harmful gases, entering digester tanks for maintenance or inspection ought to be done with proper confined space entry equipment, such as ventilation systems, harnesses, and self-contained breathing apparatuses (SCBA). Emergency shutdown systems (ESD) and automatic alarm systems provide immediate alerts and notifications in the event of gas leaks, overpressure, or any other condition posing a hazard that calls for immediate responses and mitigation processes. Fire safety equipment, including CO_2 or foam extinguishers, is strategically placed throughout the biogas facilities to respond to potential fire hazards. Hazardous gases, pressure, temperatures, and flow rates are measured and controlled by comprehensive monitoring and control systems by incorporating real-time data acquisition (SCADA systems) to ensure safe and efficient operation of the digester. Digital monitoring is enhancing the response time to emergencies and safety. Anaerobic digestion can be conducted efficiently when adopting advanced safety equipment, constant monitoring, and strict system operational rules while minimizing hazards and securing environmental compliance and workers and infrastructure [60].

3 Methods

3.1 Reactor Selection and Configuration

Reactor selection and configuration are imperative for optimizing food waste conversion into biogas and maintaining processing stability and efficiency. Reactor choice is influenced by the characteristics of the feedstock used, organic loading rate (OLR), hydraulic retention time (HRT), solids retention time (SRT), temperature regime (mesophilic or thermophilic), mechanisms of mixing, and operational complexity. Most common reactor types include batch reactors, continuously stirred tank reactors (CSTR), plug flow reactors (PFR), upflow anaerobic sludge blanket (UASB) reactors, anaerobic filter (AF) reactors, expanded granular sludge bed (EGSB) reactors, and two-stage digestion systems; each reactor type possesses its respective advantages and disadvantages on the consideration of waste composition and treatment goals. Batch reactors operate on fill-and-forget concepts with food waste being loaded, sealed, and kept for digestion for some 15–30 days [61]. They are, therefore, easy to operate and inexpensive for use but are unsuitable for continuous large-scale biogas production. On the other hand, CSTRs involve continuous feeding and mechanical mixing that guarantee homogeneous substrate distribution in the digester and steady biogas output; thus, they are considered well suited for high-solid-content food waste treatment but also demand high energy input and monitoring to maintain the process. Plug flow reactors (PFRs) are long tanks that are horizontal, and food waste will move through the tank in a unidirectional manner, and with little mixing. This makes them suitable for semi-solid waste treatment but prone to channeling for flow causing the digestion to become inefficient.

Advanced reactors like UASB and EGSB are designed for rapid digestion of liquid food waste, characterized by the formation of granular sludge and self-mixing by biogas bubbles, with short HRTs ranging from 6 to 10 days and compactness. However, these require well- adapted microbial consortia and do not perform equally well for solid-rich waste [62]. Anaerobic filter (AF) reactors, fitted with fixed media, favor microbial retention, giving them shock load resistance and making them applicable in treating low-solid high-organic content wastewater instead of needing pre-treatment against clogging. Two-stage anaerobic digester separates acetogenesis and methanogenesis into different reactors to enable the establishing of optimal conditions for microbial activity, thus leading to higher biogas yield and greater stability of the process, especially for wastes that have high lipid or carbohydrate content. It involves setting up an anaerobic digestion reactor with several important components and aspects. One of these is the gas-tight digester tank in which anaerobic conditions are maintained and oxygen intrusion is prevented such that methanogenic activity is ensured. The feedstock inlet and outlet system is designed depending on whether the reactor works in batch, semi-continuous, or continuous mode, considering solid-liquid separation to prevent clogging [63]. The biogas capturing and storing system, often composed of a floating drum, fixed dome, or gas bags, must be meticulously sealed in avoidance of leakage, to make its merits a cut above others. Temperature control is another very important parameter, as mesophilic ranges (35–40 °C) enable stability while thermophilic ranges (50–55 °C) augment degradation rates by increasing energy input requirements. Different reactors have different mixing mechanisms; for instance, in CSTR, mixing is achieved mechanically or hydraulically, whereas mixing in UASB and EGSB occurs through self-mixing gas recycling. Notably, this startup phase involves inoculating the digester with anaerobic sludge, such as either wastewater treatment plant sludge or livestock manure, with a progressive increase of organic load and monitoring various process parameters, including pH (6.5–7.5 being optimum), VFA accumulation, as well as gas composition (50–70% CH_4, 20–40% CO_2) to prevent process instability. If the reactor alkalinity, ammonia concentration, and organic loading rate are properly monitored, particular adjustments in acidity and methanogenic inhibition would not occur. Besides that, reactors should also be equipped with safety devices such as pressure relief valves to prohibit gas accumulation, and corresponding treatment environs in managing the effluents resulting from digestate, which can then be processed for biofertilizers. Anaerobic digestion reactor performance depends not only on the reactor selection but also on proper operational control, monitoring, and flexible adaptation to feedstock variability to valorize waste efficiently and generate biogas sustainably [64].

3.2 Types of Anaerobic Digesters

Anaerobic reactors are generally classified based on their flow characteristics, mixing mechanisms, and biomass retention strategies. These are:

1. Batch reactors
2. Continuous stirred tank reactors (CSTR)
3. Plug flow reactors (PFR)
4. Upflow anaerobic sludge blanket (UASB) reactors
5. Anaerobic filter (AF) reactors
6. Expanded granular sludge bed (EGSB) reactors
7. Two-stage anaerobic digesters.

3.3 Reactor Setup for Anaerobic Digestion of Food Waste

Setting up an anaerobic digestion (AD) reactor is another crucial step to ensure that microbial activity, biogas production, and process stabilization can be carried out efficiently. The design and installation of the reactor are intrinsically related to the type of reactor employed, the nature of food waste, required retention time, and operational conditions (mesophilic or thermophilic). In the setup, an initial step is to choose a reactor type appropriate to the percentage solid of food waste and biogas production desired [65]. Batch reactors do not require extensive infrastructure and are quick to setup, while continuous stirred tank reactors (CSTRs) and plug flow reactors (PFRs) would need a mechanical mixing system to ensure even substrate distribution. Upflow anaerobic sludge blanket (UASB) and expanded granular sludge bed (EGSB) reactors are more advanced and require a rigorous microbial consortium and an appropriate sludge retention mechanism. After the choice of reactor type, the construction of the digester tank is the next significant step.

Anaerobic conditions must be maintained in the digester, and hence the digester must be airtight. Methanogenic bacteria will be inhibited should oxygen get into the digester. Reinforced concrete, stainless steel, fiberglass, or high-density polyethylene (HDPE) can be used to construct the digester depending on the scale and cost considerations. The reactor should also be insulated when operating thermophilic conditions, particularly from 50 to 55 °C, to prevent heat loss. The size of the reactor is determined by its organic loading rates and the required hydraulic retention times, which normally range from 10 to 30 days depending on the food waste composition [66].

The feedstock input system will have to be designed in such a way as to allow trouble-free feeding without any clog formation. In the case of solid-rich food waste, the pre-treatment should include size reduction, homogenization, and the removal of non-biodegradable materials to ensure consistent particle size and texture, which facilitates smooth flow and prevents clog formation in the feedstock input system.

The requirement would include grinding such feed into slurry, thus enhancing the flat surface area for microbial degradation. In the case of feeding in CSTRs and PFRs, it would be much better if mechanization could be done to ensure continuous and even distribution of the substrate. On the other hand, UASB and EGSB reactors should all be fed with liquid mixes that should have low solid content (<5%) to avoid sludge bed disturbances. The mixing system is important as it serves to keep the conditions well mixed in the reactor. In these CSTRs, mechanical or hydraulic stirring ensures the homogeneity of the mixing of microbes and substrate. In the case of UASB and EGSB-type reactors, mixing is induced through gas recirculation wherein the ascending bubbles of biogas generate natural turbulence [67]. Generally, plug flow reactors do not require mixing, with the waste flow going in one direction. The biogas collection system is another critical component of the reactor arrangement. Biogas, mainly composed of methane (CH_4) and carbon dioxide (CO_2), is collected through a gas-tight storage system consisting of floating or fixed domes or gas bags. The gas must be purified from hydrogen sulfide (H_2S), moisture, and CO_2 to be used in the energy applications. Some setups are equipped with biogas upgrading systems to enhance methane purity for compressed natural gas (CNG) applications. An effluent management system treats the digested organic matter which can be in solid or liquid form and is produced after digestion. The solid digestate can further undergo dewatering and composting for use as an organic fertilizer while the liquid effluent might either be recycled back into the system or treated before disposal. An aerobic composting or filtration step post-treatment for digestate enhancement may be required. The last stage leading up to the reactor working phase is the startup phase, whereby the digester is inoculated with anaerobic sludge obtained from an existing digester or a wastewater treatment plant. The organic load is increased gradually over several weeks to allow the microbial communities to adapt and attain stable biogas production. There is continuous monitoring of key operational parameters like pH (6.5–7.5), volatile fatty acids (VFA), ammonia levels, and methane content (50–70%) to prevent failure of the systems through acidification or inhibition of methanogens. The successful setting up of the reactor necessitates well-thought planning study and monitoring to enable the efficient transformation of food waste to biogas and biofertilizers, thereby contributing to sustainable waste valorization and renewable energy production [68].

3.4 Inoculum Selection and Startup Phase in Anaerobic Digestion

Anaerobic digestion is primarily dependent upon the right selection of inoculum and startup phase since these aspects have a significant bearing on the nature of the microbial community during the organics' effective transformation into biogas. The inoculum is termed because it consists of anaerobic microorganisms expected to provide the initial microbial population to start up digestion, which ultimately determines biogas yield, process stability, and the time required to reach the steady state. The ideal inoculum should be rich in active methanogenic bacteria, balanced with an appropriate consortium of the microbial community, and well adapted to potential working conditions (mesophilic or thermophilic). Common inoculum sources are digestate from running anaerobic digesters, sludge from sewage treatment plants, cattle manure, or landfill leachate. It has to be diverse microbiologically to enhance the degradation of complex organic matter and be free from xenobiotic materials such as heavy metals, ammonia, or other inhibitors that could impede microbial activity. Startup is a critical time during which a community of microbes adapt and stabilize with the organic waste feedstock for an effective and stable digestion process. An appropriate startup strategy will mitigate the possibility of acids, foams, or inhibition of methanogenesis that can endanger the effectiveness of the process and failure of the system [69].

The startup phase begins with inoculum addition into the reactor at the recommended inoculation-to-substrate ratio of 1:1–3:1, depending on the type of reactor or characteristics of the feedstock. Here, the organic load is kept low for the initial stages (typically 0.5–1.5 g volatile solids per liter per day) to avoid the accumulation of volatile fatty acids (VFAs), which can drop pH and lead to instability of the system. Feeding rates are gradually increased over several weeks so that the growth and adaptation of hydrolytic, acidogenic, acetogenic, and methanogenic species can take place.

Several important parameters must be monitored throughout the initial period to allow for a smooth transition into full-scale operation [70]. The methanogens are extremely sensitive to acidic surroundings; therefore, maintaining a pH between 6.5 and 7.5 is of vital importance. The ratio of volatile fatty acids (VFAs) to alkalinity must be less than 0.4, indicating equilibrium between acid generation and methanogenesis. In addition, ammonia concentration must not go beyond 3000 mg/L, given that highly elevated concentrations may hinder microbial activity. Temperature stability is equally important, with mesophilic digesters requiring 35–40 °C and thermophilic setups 50–55 °C, to promote optimal bacterial performance. Therefore, a gradual temperature adjustment must be followed in the transition from mesophilic to thermophilic to avoid disturbing the microbial community. Another feature of the startup phase is biogas monitoring, in which methane concentrations should steadily increase to levels of 50–70% after microbial stabilization. Initially, biogas may contain higher amounts of carbon dioxide (CO_2) and hydrogen sulfide (H_2S); the concentration of CO_2 and H_2S levels should almost disappear as methanogenesis has been established and stabilized. If low methane production persists, some adjustments might be needed in the feeding rate, pH, or inoculum addition to stimulate microbial activity. Also, foam and scum formation should be assessed since excessive foaming may suggest an overloaded or unbalanced system. In high-rate digesters such as UASB and EGSB reactors, inoculum selection is of utmost importance, as these systems rely on the formation of granular sludge that enhances microbial retention and efficiency. Inoculum with well-formed anaerobic granules from mature digesters is a good choice to accelerate startup time and improve reactor stability [71]. CSTR and plug flow reactors must mix inoculum well to avoid settling and dead zones. The startup phase properly takes 4 to 8 weeks, depending on the type of reactor, organic loading rate, and temperature. Startup is confirmed when methane production stabilizes, VFA does not go out of control, and reactor performance is stable. Bioaugmentation strategies, which may result in the biological inoculation of natural bacteria or the addition of enzyme supplements, exist in some cases to allow faster establishment of microbial populations and better digestion efficiencies. Once stabilized, the reactor can be run continuously at conditions favorable to maximum biogas production with less organic waste and minimum process disruptions. The right choice of inoculum and sound management during the startup phase are essential components in ensuring an anaerobic digestion system for food waste valorization remains operational over the long-term and performs well [72].

3.5 Operating Conditions and Process Parameters in Anaerobic Digestion

Food waste valorization through anaerobic digestion (AD) processes depends on the efficiency and stability attained by maintaining the relevant operating conditions and process.

Parameters directly affect microbial activity, biogas production, and the overall performance of the system. These conditions include temperature, pH, organic loading rate (OLR), hydraulic retention time (HRT), concentration of volatile fatty acids (VFA), alkalinity, carbon-to-nitrogen (C/N) ratio, ammonia level, mixing, and biogas composition. Control of these parameters ensures proper functioning of the digester and inhibition of process failures, such as acidification, inhibition, and low methane yield. Temperature is probably one of the most important parameters affecting the metabolic rate of the microbial community and digestion efficiency [73]. AD operates within at least three temperature ranges: psychrophilic (<25 °C), mesophilic (35–40 °C), and thermophilic (50–55 °C). Mesophilic conditions are more commonly favored for their relatively stable microbial activity and lower energy requirements, whereas the advantages of thermophilic digestion include greater pathogen destruction, a faster digestion rate, and a high methane yield at the expense of being energy intensive and requiring thorough process control against microbial inhibition. Fluctuation in temperature will cause an unstable reaction environment, thus insulation and heating systems become imperative in maintaining desired conditions. The pH must therefore be stringently controlled as anaerobic microorganisms and methanogenic archaea, in particular, are severely impacted by fluctuations in pH. The optimal pH for hydrolysis and acidogenesis lies between 5.5 and 6.5, while for methanogenesis, it is between 6.8 and 7.5. A further drop in pH below 6.5 indicates an accumulation of volatile fatty acids (VFA), which can interfere with the activity of methanogens, reducing methane production. To avoid excessive pH fluctuations, alkalinity, acting as a buffer, must be maintained in a range of 2500–5000 mg $CaCO_3$/L. The ratio of VFA-to-alkalinity is an important indication of stability in the process and a value of greater than 0.4 implies possible acidification problems [74].

Organic loading rate (OLR) is the quantity of organic matter introduced into the reactor over a determined unit volume and per day, usually defined in g volatile solids (VS) per liter per day. An optimal OLR maintains the equilibrium regarding the substrate supply and the microbial degradation capacity concerned. On the one hand, if the digester is overloaded by food wastes for digestion, there may occur an accumulation of VFA, a drop in pH value, and even a reactor failure, while on the other hand, underloading would reduce biogas production. The OLR may also change with the type of reactor. For example, a CSTR (continuously stirred tank reactor) normally runs at 2–6 g VS/L/day, while high-rate reactors like an upflow anaerobic sludge blanket (UASB) and an expanded granular sludge bed (EGSB) can withstand OLRs of 10–30 g VS/L/day due to improved microbial retention. Hydraulically retention time (HRT) is another parameter that gets defined regarding the average time the substrate is retained in the digester. In addition, it varies from feedstock characteristics and reactor type; for example, solid-rich food waste would need longer HRT, which is around 20–30 days, to fulfill complete degradation, and for liquid-based systems, HRT suffices for 5–15 days. It results in insufficient digestion with aeration too short to generate maximum yields of biogas or a greater organic load in the effluent. This was because the carbon-to-nitrogen (C/N) ratio could stimulate microbial metabolism since carbon was regarded as an energy source, while nitrogen was used for protein synthesis. The ideal C/N ratio for anaerobic digestion will be between 20:1 and 30:1. A lower C/N ratio (<20:1) leads to an increase in ammonia concentration, which is toxic to the methanogens, while a higher C/N ratio (>30:1) results in slow growth of microbes with more inefficiency in digestion. Because of the high nitrogen content in food waste, usually low C/N ratios are found, implying that co-digestion with such wastes adds some substrates with a high carbon content like agricultural residues or paper waste, and thus improves process stability [75].

Table 1
Stages of anaerobic digestion

Methanogenesis: production of biogas, Stages	Major products	Substrate/input	Microbial groups involved	Key process conditions and factors
Hydrolysis	Simple sugars, amino acids, fatty acids	Complex organic matter (carbohydrates, proteins, lipids)	Hydrolytic bacteria	– **pH:** 5.5–6.5 – **Temperature:** Mesophilic (35–40 °C) or thermophilic (50–55 °C) – **Retention time:** 5–10 days
Acidogenesis	Volatile fatty acids (VFAs), alcohols, hydrogen (H_2), CO_2, ammonia (NH_3)	Hydrolysis products (simple sugars, amino acids, fatty acids)	Fermentative bacteria	– **pH:** 5.5–6.5 – **VFA** – **Accumulation:** Should be monitored to prevent acidification
Acetogenesis	Acetic acid (CH_3COOH), H_2, CO_2	VFAs (propionic acid, butyric Acid, lactic acid, ethanol)	Acetogenic bacteria	– **Hydrogen partial pressure:** Should be low to promote VFA conversion – **pH:** 6.3–7.0
Methanogenesis	Methane (CH_4), CO_2, trace H_2S	Acetic acid, hydrogen, CO_2	Methanogenic archaea	– **pH:** 6.8–7.5 – **Temperature stability:** Critical for methanogenic activity – **Ammonia levels:** Should be below 3000 mg/L to avoid inhibition

3.6 Stages of Anaerobic Digestion

1. The stages of the anaerobic digestion (Table 1) includes as follows Hydrolysis: breaking down complex organic matter.
2. Acidogenesis: generating organic acids and gases.
3. Acetogenesis: conversion of organic acids into acetic acids.

3.7 Biogas Production, Collection, and Monitoring Parameters

Biogas is produced and collected due to anaerobic digestion (AD) processes wherein microorganisms convert organic matter into methane-rich biogas without oxygen. Biogas production efficiency is influenced by many factors such as reactor design, operational conditions, microbial community, and substrate composition. They first hydrolyze these complex organic molecules (lipids, proteins, and carbohydrates) into simpler components.

Acidogenesis and acetogenesis then convert them into volatile fatty acids consisting of acetic acid, hydrogen, and carbon dioxide, instead of methanogenic archaea, which generate methane (CH_4) and sometimes carbon dioxide (CO_2), the principal components of biogas, together with some trace gases such as hydrogen sulfide (H_2S), ammonia (NH_3), and nitrogen (N_2), depending on the composition of the feedstock, in the final stage of the anaerobic digestion process [76].

One of the major outcomes of anaerobic digestion is the conversion of organic matter into methane-rich biogas. Several factors influence the efficiency of biogas formation, including the reactor con figuration, operating conditions, microbial community, and substrate composition. The first group of steps involves hydrolyzing complex organic molecules (lipids, proteins, and carbohydrates) into simpler compounds. Acidogenesis and acetogenesis further convert these compounds into volatile fatty acids, acetic acid, hydrogen, and carbon dioxide. Finally, methanogenic archaea convert hydrogen and acetic acid to methane (CH_4) and carbon dioxide (CO_2), which are the primary components of biogas. Depending on the feedstock composition, other minor gases such as hydrogen sulfide (H_2S), ammonia (NH_3), and nitrogen (N_2) can also be found [77].

3.8 System of Biogas Collection

Biogas collection is an integral part of the efficient use and prevention of gas loss. Different systems of collecting biogas are structured on the reactor type and scale of operation. Continuous stirred tank reactors (CSTRs) and plug flow reactors (PFRs) provide collection of biogas from the top of the digester with the help of a gas dome or gas holder. In contrast, UASB reactors collect gas from the top through a gas-liquid-solid separator, as buoyant biogas has been separated naturally in the reactor by its buoyancy and adjusts based on the volume of gas produced, floating drum digesters have a gas collection system that either rises or falls. Fixed-dome digesters, very widely used in rural biogas purifiers, store gas for pressured discharging through pipelines to the biogas plot. The collected biogas is then directed to storage tanks or balloon-type reservoirs to regulate the supply before being used. In large installations, biogas must undergo scrubbing and purification, because it is to be used at the end for generating electricity, heating, or as fuel for vehicles [78]. Safety valves and pressure regulators can also be included in storage systems to prevent gas leakage and ensure steady operation.

In general, the monitoring parameters for biogas production include anaerobic digestion (AD), which is a rather complicated biochemical process that is interrelated with numerous physical, chemical, and biological parameters, all of which must be controlled to maximize biogas production, process stability, and microbial health (Fig. 2). Temperature is one of the most important parameters that directly affect microbial activity and reaction

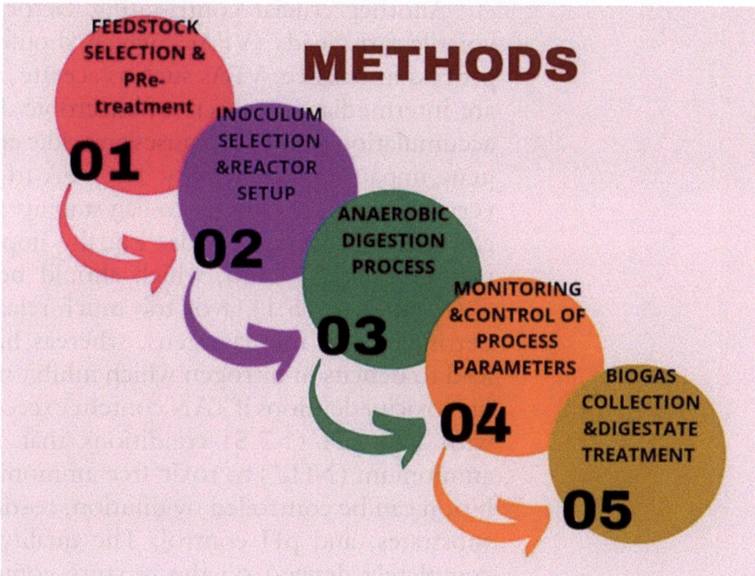

Fig. 2 Methods involved in anaerobic digestion protocols of food waste

rates. AD is carried out in three temperature ranges: psychrophilic 10–25 °C; mesophilic 35–40 °C; and thermophilic 50–55 °C. Among the three, mesophilic digestion is most widely used, as it provides stability for the microorganisms. Thermophilic digestion, on the other hand, has a greater yield of biogas and requires strict temperature control to avoid inhibition of microorganisms. Another important factor to always monitor is pH, which needs to be maintained in the range of 6.8–7.5 for methanogenic activity. Acidogenic bacteria produce VFAs, which can lower the pH to below 6.5, inhibiting methanogens and destabilizing the process. On the other hand, alkalinity (2000–5000 mg $CaCO_3$/L) must be provided to buffer against acid accumulation and, thus, avoid any pH fluctuations [79].

The OLR, organic loading rate, and HRT, hydraulic retention time, are important operational parameters to keep the equilibrium between biomass availability and microbial degradation. OLR, expressed in kg VS/m^3/day, determines the organic matter loading into a digester per unit volume. The system could lead to acid accumulation when OLR is higher than 6 kg VS/m^3/day, whereas lower values will result in less reactor utilization. HRT means the average time the substrate resides inside a digester and should be optimized depending on the reactor type. CSTRs are designed with 20–30 days of HRT, and UASB reactors function favorably with 5–15 days of HRT due to high microbial retention. In addition to this, SRT refers to the retention time for microbial biomass within the reactor. Sustained high SRT enhances microbial efficiency and prevents biomass washout, especially in high-rate digesters like UASB and expanded granular sludge bed (EGSB) reactors [80].

Another crucial contributing factor is the accumulation of volatile fatty acids (VFAs), which should be controlled to avoid process imbalance. VFAs such as acetate, propionate, and butyrate are intermediate products of anaerobic digestion. Their excessive accumulation (>2 g/L) causes an acidic environment for methanogens, impairing their activity. The VFA-to-alkalinity ratio (≤0.4) is a very important parameter to flag stability so that buffering capacity can neutralize excess acids. Equally important is the carbon-to-nitrogen (C/N) ratio, which should be kept at 20–30:1. Low C/N ratios (<15:1) favor too much release of ammonia, which is detrimental to methanogens, whereas high C/N ratios (>35:1) lead to deficits in nitrogen which inhibit microbial activity. Ammonia toxicity develops if TAN content exceeds 3000 mg/L, especially with high-pH (>7.5) conditions that favor the conversion of ammonium (NH_4^+) to toxic-free ammonia (NH_3). Ammonia inhibition can be controlled by dilution, co-digestion with carbon-rich substrates, and pH control. The quality and quantity of biogas completely depend on the mixture composition, yield, and flow rate. Methane content in biogas of 50–75% and 25–45% carbon dioxide (CO_2) is desirable. A concentration of H_2S higher than 500 ppm can corrode pipes and engines, and the gas should then be cleaned by iron salts ($FeCl_3$) or activated carbon. The gas flow rate should also be monitored for steady biogas production; sudden changes indicate disturbances in the process. Microbial health is essential since different groups of microorganisms carry out hydrolysis, acidogenesis, acetogenesis, and methanogenesis. Therefore, monitoring microbial diversity by 16S rRNA sequencing helps to maintain a balanced population in the digester and hence precludes possible process failures. Other factors found to inhibit microbial enzymes and decrease biogas yields are heavy metals (Pb, Cu, Zn, and Ni) above 50 mg/L and organic toxins such as pesticides and antibiotics [81].

3.9 Monitoring Parameters for Biogas Production

Anaerobic digestion is one of the most effective processes for the conversion of organic waste into biogas because it needs complete operational control to avoid potential disturbances that could otherwise lead to process failure (Table 2). Common problem areas arise when a pH imbalance takes place. pH imbalances are attributed to high volatile fatty acid (VFA) buildup that usually occurs during the hydrolysis and acidogenesis stages. A drop of pH below 6.5 could inhibit methanogenic activity, thus reducing methane production and leading to possible system failure. Adequate alkalinity levels, in the range of 2000–5000 mg $CaCO_3$/L, must be maintained to buffer against acid fluctuations. In the event of acid accumulation, the use of sodium bicarbonate, lime, or sodium hydroxide can combat such imbalances [82]. Another frequent challenge faced in ammonia inhibition comes from the breakdown of protein-rich feedstock and releases such high levels of total ammonia nitrogen (TAN). When the TAN concentration exceeds

Table 2
Key parameters for anaerobic digestion of food waste

Parameter	Optimal range	Importance
pH	6.8–7.5	Maintains microbial activity, prevents inhibition
Temperature	Mesophilic:35–40 °C Thermophilic: 50–55 °C	Affects microbial growth and biogas production
C/N ratio	20:1–30:1	Balances carbon and nitrogen for stable digestion
Solid content	10–15%	Ensures proper microbial activity and flow
Organic loading rate (OLR)	1–4 kg VS/m^3/day	Controls food waste input per digester volume
Hydraulic retention time (HRT)	Mesophilic: 20–30 days Thermophilic: 10–20 days	Determines digestion time and biogas yield
Methane content in biogas	55–70%	Indicates biogas quality
Volatile fatty acids (VFAs)	<4000 mg/L	Prevents acid accumulation and process failure
Ammonia concentration	<3000 mg/L	Prevents toxicity to methanogens
Mixing rate	Continuous or periodic	Enhances contact between microbes and substrate

3000 mg/L, free ammonia (NH3)-produced toxicity to methanogens especially when the pH is above 7.5 becomes critical. Control of ammonia toxicity may be achieved through different mechanisms, such as co-digestion with carbon-rich substrates, dilution of the reactors, and adjustment of the pH below 7.2. Besides the strategies mentioned, gradual acclimatization of the relevant microbial populations to progressively higher levels of ammonia may also lead to improved tolerance. Foaming and scum formation may additionally occur, with the presence of dissolved organic matter as proteins. Foaming leads to the loss of digester volume, gas collection blockage, and sometimes inefficient operations. Prevention and control of foaming may be conducted by the application of anti-foaming agents (e.g., silicone-based deformers), mechanical removal of foam, and optimizations of the organic loading rate. Proper mixing and the avoidance of sudden changes in feedstock would also help lessen foam formation [83].

A major issue causing process inefficiency is feedstock variation which might result in divergent substrate composition and hence process instability. Some waste materials such as fats, oils, and greases (FOG) slow down digestion and can lead to long-chain fatty acid (LCFA) accumulation causing inhibition by microorganisms. Addressing balanced carbon-to-nitrogen (C/N) ratio maintenance (20–30:1) and pre-treatment methods such as enzymatic hydrolysis or mechanical grinding improve substrate digestibility. Co-digestion with other compatible waste streams gives the treatment a balanced nutrient profile, thus making it stable. The two major operational challenges are hydraulic and organic overload. Overloading is an excess condition triggered when the organic loading rate (OLR) exceeds microbial capacity and results in acid accumulation and fluctuation of gas flow, and this tends to destabilize the reactor. Selim indicated that maintaining an optimum OLR (usually 2–6 kg VS/m^3/day, depending on the type of reactor) would prevent overloading. The avoidance of both organic and hydraulic overloads guarantees a proper HRT for substrate degradation. Slowly increase feed introduction and adjust flow rates to balance microbial activity and prevent system failure. Mixing inefficiencies lead to stratification, poor substrate–microbe contact, and dead zones in the digester. Inadequate mixing destroys any even temperature, pH, and nutrient distribution and causes a reduction in biogas yield. Mechanical agitators, biogas recirculation, or hydraulic mixing methods need to be used according to reactor design to maximize mixing [84]. However, under-mixing should be avoided since it will predispose the microbial populations to shear stress resulting in possible washout of the biomass in certain systems. Biogas leakage and its collection inefficiencies will also affect energy recovery and environmental performance significantly. Blocked gas collections might be from pipeline leaks or insufficient pressure control. Pressure monitoring, regular inspections, and installation of highly efficient gas collection systems with pressure relief valves contribute to maximizing gas capture and utilization. High levels of hydrogen sulfide (H_2S) (>500 ppm) can corrode engines and pipelines. This can be solved by the methods of desulfurization, which include iron-based adsorption, that is, by using $FeCl_3$, activated carbon filtration, or via biological scrubbers. To conclude, steady state operating conditions, adequately balanced feedstock composition, controlled loading rates, proper efficient mixing, and effective biogas handling maximize anaerobic digestion efficiency. Regular observation and proactive troubleshooting will minimize process failures, ensure sustained production of biogas, and contribute to the sustainability of anaerobic digestion systems [85].

3.10 Protocols for Treatment—An Anaerobic Digester

The pre-treatment of feedstock is most critical in this aspect of maximizing anaerobic digestion efficiency by improving the biodegradability of feedstock, increasing methane production, and shortening retention time. According to the type of substrate, varying pre-treatment methods are used. The pre-treatments are thermal, chemical, biological, and physical.

Physical pre-treatment techniques: Techniques such as mechanical shredding, milling, and ultrasound help reduce particle size and break down larger complex organic materials. This increases the surface area available for microbial action, enhancing anaerobic digestion's efficiency. Hydrolysis is the rate-limiting step improved by this technique, which is mostly done in anaerobic digestion [86].

Chemical pre-treatment techniques: Reduction of lignocellulosic and other recalcitrant materials using acids, alkalis, or oxidizing agents is chemical pre-treatment. While alkaline pre-treatments with sodium hydroxide or ammonia usually increase solubilization and digestibility, acid hydrolysis with sulfuric or hydrochloric acid helps to break down complex carbohydrates. By breaking apart cell structures, oxidative pre-treatments with hydrogen peroxide or ozone help increase the amount of fermentable sugars available.

A biological pre-treatment technique is where anaerobic digestion of complex organic materials is accomplished, by the action of microbes or enzymes. Enzymatic hydrolysis including cellulases and hemicellulases, enhances the breakdown of plant fibers, and

microbial hydrolysis by bacteria or fungus aids in breaking down lignocellulosic biomass. This way of biological pre-treatment is free of harsh chemicals, uses very little energy, and is therefore considered eco-friendly.

Thermal pre-treatment techniques apply heat to the feedstock to promote solubilization and enhance biogas yield. Pre-treatments like a steam explosion, autoclaving, and hydrothermal can break down lignocellulosic to enhance digestibility [87]. These techniques come in handy, particularly for treating substrates containing complex polymers resistant to biological degradation such as waste sludge and agro-wastes.

Anaerobic digestion comprises protocols that differ in their approaches based on the input material, the design of the vicinity within which this input material is destroyed, and subsequent variables. The anaerobic process applies both physical and chemical means to digest organic matter into the final products of digestate and biogas. Each method may vary slightly from one feedstock type to another; thus, it is important to outline the steps and best practices to achieve the eventual breakdown of organic material in an oxygen-free environment. This should involve anaerobic operation to digest organic material to digestate and biogas [88].

A. Feedstock Preparation.

- **Selection:** Anything from organic wastes such as agricultural residues, food, wastewater sludge, or manure could qualify.
- **Pre-treatment:** Shredding, pulping, or hydrothermal treatment might enhance digestibility.
- Optimization of C: N ratio: Maintain a constant, ideal carbon-by-nitrogen ratio (~20–30:1).
- Moisture content: More than 85% is to be maintained at this level for wet digestion.

B. Digester Design and Operation.

- **Inoculation:** The addition of digestate from an active digester would introduce necessary microbes.
- **Monitoring:** VFAs: Too high a concentration of VFA will indicate an imbalance in the process, alkalinity: needed to reduce negative impacts on rapid pH changes, ammonia: high ammonia concentration levels (>3000 mg/L) may inhibit methanogens.
- **Trace elements and nutrients:** Micronutrient supplementation is required to promote the health of the microorganisms.

Gas Production and Use

Biogas composition:

50–70% methane (CH_4)

30–50% carbon dioxide (CO_2) 0–3% hydrogen sulfide (H_2S).

Gas Cleaning:

H_2S removal: iron oxide/activated carbon or biological scrubbers. CO_2 removal: pressure swing adsorption or membrane separation.

Use for heating, electricity generation via gas engines, or upgrading to biomethane for grid injection and/or as vehicle fuel.

Digestate Management

Solid-liquid separation using a screw press or centrifuge. Nutrient recovery as organic fertilizer.

Pathogen reduction via thermophilic digestion or post-treatment.

Monitoring and Safety Protocols

Process monitoring:

Gas flow rates and composition, temperature, pH, and alkalinity. VFA-to-alkalinity ratio ($VFA/Alk < 0.4$ for stability).

Safety measures:

Detection of gas leakage and ventilation. Flame arresters on biogas lines.

Handling of anaerobically digested wastes appropriately to avoid contamination.

3.11 Procedures for Post-treatment

Anaerobic digestion (AD) post-treatment adopts digestate management, deviance from the environment, and resource recovery optimization. The major digestate that is left when anaerobic digestion is over is rich in essential nutrients but requires some pressing and processing to improve its usability and lessen its potential negative environmental impacts. Solid-liquid separation is probably the first step. All these can be mechanically by screw pressing, centrifugation, or gravity separation. Separation enables effective utilization of liquid and solid fractions for different purposes [89]. The liquid fraction is treated to lower nitrogen levels before discharge or reuse, while the solid fraction is rich in organic matter composted or dried to produce stable biofertilizers, which may include treatment through ammonia stripping, membrane filtration, or biological nitrification-denitrification.

3.11.1 Post-treatment of Digestate

The residue left after digestion is called digestate, which contains both liquid and solid components. Its value increases, and the effects it has on the environment are reduced with post-treatment.

A. Solid-Liquid Separation.

B. Solid Digestate Treatment.

C. Liquid Digestate Treatment.

Biogas Purification and Upgrading

A. Hydrogen sulfide (H_2S) removal

B. Carbon dioxide (CO_2) removal

C. Moisture and siloxane removal

D. Biogas utilization

Effluent Treatment

A. Ammonia stripping and capture

B. Electrodialysis

3.11.2 Environmental Compliance Audit and Monitoring

Post-treatment processes must comply with regulatory criteria for sustainability and legal operation.

A. Biogas digestive properties: Methane percentage, H_2S concentration, and CO_2 levels are to be monitored as quality indicators. Heavy metals, pathogens, and nutrients content (NPK levels). Quality of effluent: Total solids, pH, nitrogen, phosphorus, COD, BOD, carbon footprint, odor management, methane leakage, etc..

B. Compliance with local environmental standards regulations includes adhering to regulations on fertilizer application and limits on effluent discharge, while also preventing H_2S emissions and methane leakage under Air Quality Regulations. Health and Safety Regulations: Safety measures in biogas handling and reduction of pathogens in digestate are taken to improve. Although they get released from decomposing organic matter, methane, H_2S, and CO_2 concentrate more in anaerobic decompositions. High presence of digestive properties: heavy metals, pathogens, and nutrient content (NPK levels). Effluent qualities are total solids, pH, nitrogen, phosphorus, chemical oxygen demand (COD), and biological oxygen demand (BOD). Gas emission: carbon footprint, odor management, and methane leakage [90].

B. Compliance with the local environment standards regulations: Observance of the guidelines on the application of fertilizers and the limitations on effluent discharge. Air emission provisions: Prevent H_2S emissions and methane leakages. Health and Safety Regulations: Safety of biogas handling and reduction of pathogens in digestate. It gets released from decaying organic matter; methane, H_2S, and CO_2 concentrate more in anaerobic decomposition: heavy metal levels, pathogen, and nutrient contents (NPK levels).

Qualities of effluent: Fortified solids, seed-free, pH-sensitive for nitrogen and phosphorus, chemically demanding oxygen (COD), and biologically demanding oxygen (BOD).

B. Compliance with the Local Environment Standards regulations: Observance of the guidelines on the application of fertilizers and limits on effluent discharge. Air emission provisions: Prevalence of H_2S emissions and methane leakages. Health and Safety Regulations: Safety of biogas handling and reduction of pathogens in digestate.

4 Notes

Anaerobic digestion (AD), which provides a holistic and eco-friendly approach toward the green valorization of food waste, is further proceeded with the creation of biogas and nutrient-rich digestate. This indeed is a process wherein organic matter is degraded by microbial activity under anaerobic conditions, thus minimizing landfill disposal of organic.

Waste and greenhouse gas emissions while at the same time producing renewable energy. The efficiency of this AD process greatly depends on choosing suitable substrates such as food waste or a combination of food waste with agricultural residues and manure to make the nutrients nutritionally balanced [91]. Pre-treatment technologies increase biodegradability as well as the overall efficiency of the AD process and can be physical (such as shredding or dewatering), chemical (like pH adjustment or ozonation), or biological (such as enzymatic hydrolysis or microbial inoculation). The reactor can be selected as a batch, cold-operated continuous stirred tank, or UASB, depending on various circumstances, including characteristics of the waste, scale of operation, and desired retention time. Proper setting of the reactor involves inoculating with microbial consortia and optimizing operating conditions such as pH (6.5–7.5), temperature (mesophilic: 30–40 °C; thermophilic: 50–60 °C), and organic loading rate (OLR).

In the AD procedure, there are four principal microbial stages: hydrolysis—the transformation of complex organic molecules to simpler compounds; acidogenesis—the conversion of these compounds to volatile fatty acids; acetogenesis—the conversion of intermediate products to hydrogen and acetic acid; and methanogenesis—whereby the methanogens convert carbon dioxide along with methane as the final products. The efficiency of biogas production depends differently on varying factors, including the reactor stability, composition of substrates, and continuous monitoring of parameters such as gas production, levels of volatile fatty acids (VFA), alkalinity, and ammonia concentration [92].

A few of these purification techniques for biogas, such as H_2S removal, and CO_2 separation from moisture control, can also make digestate usable for electricity generation, heating, or upgrading it to biomethane for fuel in vehicles. However, post-treatment of digestate requires either composting or additional treatment for nutrient recovery to produce high-quality biofertilizers, for instance, solid-liquid separation. Effective monitoring, control procedures, safety measures control, pH regulation, and leak detection from gas sources are important variables for process stability and process efficiency [93]. Pre-treatment, digestion conditions, and treatment after digestion all need to be optimized to increase biogas production as well as nutrient recovery, which is what makes AD successful in the valorization of food waste. Although issues such as oscillation in reactor efficiency and shock feeding with varying feedstock types still abound, AD remains an efficient way of turning food waste into useful resources because of its sustainability. By improving operational procedures as well as developing advanced monitoring systems, anaerobic digestion might contribute significantly to renewable energy production and sustainable waste management, lessening its impact on the environment and promoting a circular economy [94].

References

1. Assis TI, Gonçalves RF (2022) Valorization of food waste by anaerobic digestion: a bibliometric and systematic review focusing on optimization. J Environ Manag 320:115763
2. Morales-Polo C, Cledera-Castro MD, Moratilla Soria BY (2018) Reviewing the anaerobic digestion of food waste: from waste generation and anaerobic process to its perspectives. Appl Sci 8(10):1804
3. Akturk AS, Demirer GN (2020) Improved food waste stabilization and valorization by anaerobic digestion through supplementation of conductive materials and trace elements. Sustain For 12(12):5222
4. Shewa WA, Hussain A, Chandra R, Lee J, Saha S, Lee HS (2020) Valorization of food waste and economical treatment: effect of inoculation methods. J Clean Prod 261:121170
5. Molino A, De Gisi S, Petta L, Franzese A, Casella P, Marino T, Notarnicola M (2019) Experimental and theoretical investigation on the recovery of green chemicals and energy from mixed agricultural wastes by coupling anaerobic digestion and supercritical water gasification. Chem Eng J 370:1101–1110
6. Sarker A, Ahmmed R, Ahsan SM, Rana J, Ghosh MK, Nandi R (2024) A comprehensive review of food waste valorization for the sustainable management of global food waste. Sustain Food Technol 2(1):48–69
7. Shukla KA, Sofian AD, Singh A, Chen WH, Show PL, Chan YJ (2024) Food waste management and sustainable waste to energy: current efforts, anaerobic digestion, incinerator and hydrothermal carbonization with a focus in Malaysia. Cleaner Prod 448:141457

8. Carucci A, Bortolussi A, Cappai G, Erby G, Tocco G, Milia S (2020) Application of anammox within an integrated approach to sustainable food waste management and valorization. New Biotechnol 58:1–9
9. Kannah RY, Merrylin J, Devi TP, Kavitha S, Sivashanmugam P, Kumar G, Banu JR (2020) Food waste valorization: biofuels and value added product recovery. Bioresour Technol Rep 11:100524
10. Degueurce A, Dabert P, Argence V, Blondel L, Le Bihan A, Lebreton M, Peu P, Sarrazin M, Picard S, Trémier A (2022) An innovative solid-state micro-anaerobic digestion process to valorize food waste: technical development constraints and consequences on biological performances. Waste Biomass Valor 13(1): 617–630
11. Lahboubi N, Habchi S, Hamraoui K, El Bari H (2024) New insight of combined hydrothermal and ultrasonic pretreatment for methane production from anaerobic digestion of food waste. Int J Green Energy 21(13):3101–3111
12. He K, Liu Y, Tian L, He W, Cheng Q (2024) Review in anaerobic digestion of food waste. Heliyon 10:e28200
13. Lelicińska-Serafin K, Rolewicz-Kalińska A, Manczarski P (2024) Challenges in the valorization of green waste in the Central European Region: case study of Warsaw. Energies 17(20): 5056
14. Chatterjee B, Mazumder D (2024) Valorization of fruit and vegetable waste in a novel three-stage hybrid anaerobic digester for enhanced biogas production: performance study and microbial community analysis. Biochem Eng J 209:109403
15. Zayen A, Gharsallah N, Jraou M, Loukil S, Nikolaou IE, Tsalis T, Sayadi S, Khoufi S (2025) Potential valorization of fruits and vegetables waste from the wholesale market in Sfax (Tunisia) via anaerobic digestion: long-term characterization and stakeholders' attitude. Biomass Convers Biorefinery:1–3
16. Shi Z, He P, Zhang H, Qiu J, Lü F (2024) Convert food waste into easily biodegradable liquid substrate: new insights into wet oxidation as a pretreatment for anaerobic digestion. J Environ Chem Eng 12(6):114316
17. Jia W, Li S, Wang J, Lee JT, Lin CS, Mašek O, Zhang H, Yuan X (2024) Sustainable valorisation of food waste into engineered biochars for CO2 capture towards a circular economy. Green Chem 26(4):1790–1805
18. Pal P, Singh AK, Srivastava RK, Rathore SS, Sahoo UK, Subudhi S, Sarangi PK, Prus P (2024) Circular bioeconomy in action: transforming food wastes into renewable food resources. Food Secur 13(18):3007
19. Castillo García P, Fernández-Rodríguez MJ, Borja R, Mancilla-Leytón JM, de la Lama CD (2024) Research Trends in the recovery of by-products from organic waste treated by anaerobic digestion: a 30-year bibliometric analysis. Fermentation 10(9):446
20. Kumari R, Singh A, Sharma R, Malaviya P (2024) Conversion of food waste into energy and value-added products: a review. Environ Chem Lett 22(4):1759–1790
21. Tsai WT, Lin YQ (2021) Analysis of promotion policies for the valorization of food waste from industrial sources in Taiwan. Fermentation 7(2):51
22. O'Connor J, Mickan BS, Rinklebe J, Song H, Siddique KH, Wang H, Kirkham MB, Bolan NS (2022) Environmental implications, potential value, and future of food-waste anaerobic digestate management: a review. J Environ Manag 318:115519
23. Alexis Parra-Orobio B, Soto-Paz J, Ricardo Oviedo-Ocaña E, Vali SA, Sánchez A (2023) Advances, trends and challenges in the use of biochar as an improvement strategy in the anaerobic digestion of organic waste: a systematic analysis. Bioengineered 14(1):2252191
24. Giduthuri AT, Ahring BK (2022) Current status and prospects of valorizing organic waste via arrested anaerobic digestion: production and separation of volatile fatty acids. Fermentation 9(1):13
25. Rodríguez-Jiménez LM, Pérez-Vidal A, Torres-Lozada P (2022) Research trends and strategies for the improvement of anaerobic digestion of food waste in psychrophilic temperatures conditions. Heliyon 8(10):e11174
26. Xiao Q, Hu J, Huang M, Shen F, Tian D, Zeng Y, Jang MK (2021) Valorizing the waste bottom ash for improving anaerobic digestion performances towards a "Win-Win" strategy between biomass power generation and biomethane production. J Clean Prod 295: 126508
27. Xu F, Li Y, Ge X, Yang L, Li Y (2018) Anaerobic digestion of food waste—challenges and opportunities. Bioresour Technol 247:1047–1058
28. Li L, Peng X, Wang X, Wu D (2018) Anaerobic digestion of food waste: a review focusing on process stability. Bioresour Technol 248:20–28
29. Srisowmeya G, Chakravarthy M, Devi GN (2020) Critical considerations in two-stage anaerobic digestion of food waste—a review. Renew Sust Energ Rev 119:109587

30. Ye M, Liu J, Ma C, Li YY, Zou L, Qian G, Xu ZP (2018) Improving the stability and efficiency of anaerobic digestion of food waste using additives: a critical review. J Clean Prod 192:316–326
31. Chew KR, Leong HY, Khoo KS, Vo DV, Anjum H, Chang CK, Show PL (2021) Effects of anaerobic digestion of food waste on biogas production and environmental impacts: a review. Environ Chem Lett 19(4):2921–2939
32. Ren Y, Yu M, Wu C, Wang Q, Gao M, Huang Q, Liu Y (2018) A comprehensive review on food waste anaerobic digestion: research updates and tendencies. Bioresour Technol 247:1069–1076
33. Bong CP, Lim LY, Lee CT, Klemeš JJ, Ho CS, Ho WS (2018) The characterisation and treatment of food waste for improvement of biogas production during anaerobic digestion—a review. J Clean Prod 172:1545–1558
34. Banks C, Heaven S, Zhang Y, Baier U (2018) Food waste digestion: anaerobic digestion of food waste for a circular economy. In: Murphy JD (ed) IEA bioenergy task, vol 37, p 12
35. Slorach PC, Jeswani HK, Cuéllar-Franca R, Azapagic A (2019) Environmental sustainability of anaerobic digestion of household food waste. J Environ Manag 236:798–814
36. Hunter SM, Blanco E, Borrion A (2021) Expanding the anaerobic digestion map: a review of intermediates in the digestion of food waste. Sci Total Environ 767:144265
37. Li W, Loh KC, Zhang J, Tong YW, Dai Y (2018) Two-stage anaerobic digestion of food waste and horticultural waste in high-solid system. Appl Energy 209:400–408
38. Kumar M, Dutta S, You S, Luo G, Zhang S, Show PL, Sawarkar AD, Singh L, Tsang DC (2021) A critical review on biochar for enhancing biogas production from anaerobic digestion of food waste and sludge. J Clean Prod 305:127143
39. Paranjpe A, Saxena S, Jain P (2023) A review on performance improvement of anaerobic digestion using co-digestion of food waste and sewage sludge. J Environ Manag 338: 117733
40. Tropea A (2022) Food waste valorization. Fermentation 8(4):168
41. Nayak A, Bhushan B (2019) An overview of the recent trends on the waste valorization techniques for food wastes. J Environ Manag 233:352–370
42. Plazzotta S, Manzocco L (2019) Food waste valorization. In: Saving food. Academic Press, pp 279–313
43. Socas-Rodríguez B, Álvarez-Rivera G, Valdés A, Ibáñez E, Cifuentes A (2021) Food by-products and food wastes: are they safe enough for their valorization? Trends Food Sci Technol 114:133–147
44. Cecilia JA, García-Sancho C, Maireles-Torres PJ, Luque R (2019) Industrial food waste valorization: a general overview. In: Biorefinery: integrated sustainable processes for biomass conversion to biomaterials, biofuels, and fertilizers, pp 253–277
45. Tsang YF, Kumar V, Samadar P, Yang Y, Lee J, Ok YS, Song H, Kim KH, Kwon EE, Jeon YJ (2019) Production of bioplastic through food waste valorization. Environ Int 127:625–644
46. Ong KL, Kaur G, Pensupa N, Uisan K, Lin CS (2018) Trends in food waste valorization for the production of chemicals, materials and fuels: case study South and Southeast Asia. Bioresour Technol 248:100–112
47. Talan A, Tiwari B, Yadav B, Tyagi RD, Wong JW, Drogui P (2021) Food waste valorization: energy production using novel integrated systems. Bioresour Technol 322:124538
48. Lytras G, Lytras C, Mathioudakis D, Papadopoulou K, Lyberatos G (2021) Food waste valorization based on anaerobic digestion. Waste Biomass Valor 12:1677–1697
49. Ma G, Ndegwa P, Harrison JH, Chen Y (2020) Methane yields during anaerobic co-digestion of animal manure with other feedstocks: a meta-analysis. Sci Total Environ 728:138224
50. Saady NM, Sivaraman S, Venkatachalam P, Zendehboudi S, Zhang Y, Palma RY, Shanmugam SR, Ruiz Espinoza JE (2024) Effect of veterinary antibiotics on methane yield from livestock manure anaerobic digestion: an analytical review of the evidence. Rev Environ Sci Biotechnol 23(1):133–161
51. Braguglia CM, Gallipoli A, Gianico A, Pagliaccia P (2018) Anaerobic bioconversion of food waste into energy: a critical review. Bioresour Technol 248:37–56
52. Mak TM, Xiong X, Tsang DC, Iris KM, Poon CS (2020) Sustainable food waste management towards circular bioeconomy: policy review, limitations and opportunities. Bioresour Technol 297:122497
53. Yuan Y, Hu X, Chen H, Zhou Y, Zhou Y, Wang D (2019) Advances in enhanced volatile fatty acid production from anaerobic fermentation of waste activated sludge. Sci Total Environ 694:133741
54. Banu JR, Kannah RY, Kumar MD, Gunasekaran M, Sivagurunathan P, Park JH, Kumar G (2018) Recent advances on

biogranules formation in dark hydrogen fermentation system: mechanism of formation and microbial characteristics. Bioresour Technol 268:787–796
55. Baleeiro FC, Kleinsteuber S, Neumann A, Sträuber H (2019) Syngas-aided anaerobic fermentation for medium-chain carboxylate and alcohol production: the case for microbial communities. Appl Microbiol Biotechnol 103(21):8689–8709
56. Cao W, Fang S, Wu Q, Wu Y, Feng L, Xie Z, Feng Q, Fang F, Xu Z, Luo J, Ma Y (2023) Choline chloride pretreatment on volatile fatty acids promotion from sludge anaerobic fermentation: in-situ deep eutectic solvents-like formation for EPS disintegration and associated microbial functional profiles upregulation. Chem Eng J 467:143556
57. Jiang Y, Liu Y, Zhang X, Gao H, Mou L, Wu M, Zhang W, Xin F, Jiang M (2021) Biofilm application in the microbial biochemicals production process. Biotechnol Adv 48: 107724
58. Buckel W (2021) Energy conservation in fermentations of anaerobic bacteria. Front Microbiol 12:703525
59. Lu J, Xu S (2021) Post-treatment of food waste digestate towards land application: a review. J Clean Prod 303:127033
60. Peng W, Pivato A (2019) Sustainable management of digestate from the organic fraction of municipal solid waste and food waste under the concepts of back to earth alternatives and circular economy. Waste Biomass Valor 10(2): 465–481
61. Grigatti M, Barbanti L, Hassan MU, Ciavatta C (2020) Fertilizing potential and CO2 emissions following the utilization of fresh and composted food-waste anaerobic digestates. Sci Total Environ 698:134198
62. Wang N, Huang D, Zhang C, Shao M, Chen Q, Liu J, Deng Z, Xu Q (2021) Long-term characterization and resource potential evaluation of the digestate from food waste anaerobic digestion plants. Sci Total Environ 794:148785
63. Bian X, Wang K, Gong H (2023) Biochar-enhanced agricultural application of liquid digestate from food waste anaerobic digestion for celery cultivation. Sci Total Environ 869: 161562
64. Zhu X, He M, Xu Z, Luo Z, Gao B, Ruan R, Wang CH, Wong KH, Tsang DC (2022) Combined acid pretreatment and co-hydrothermal carbonization to enhance energy recovery from food waste digestate. Energy Convers Manag 266:115855
65. Wystalska K, Kwarciak-Kozłowska A (2023) Utilization of digestate from agricultural and food waste for the production of biochar used to remove methylene blue. Sustain For 15(20): 14723
66. Chen T, Qiu X, Feng H, Yin J, Shen D (2021) Solid digestate disposal strategies to reduce the environmental impact and energy consumption of food waste-based biogas systems. Bioresour Technol 325:124706
67. Zhang C, Shao M, Wu H, Wang N, Chen Q, Xu Q (2021) Management and valorization of digestate from food waste via hydrothermal. Resour Conserv Recycl 171:105639
68. Fu SF, Wang DH, Xie Z, Zou H, Zheng Y (2022) Producing insect protein from food waste digestate via black soldier fly larvae cultivation: a promising choice for digestate disposal. Sci Total Environ 830:154654
69. Dahiya S, Kumar AN, Sravan JS, Chatterjee S, Sarkar O, Mohan SV (2018) Food waste biorefinery: sustainable strategy for circular bioeconomy. Bioresour Technol 248:2–12
70. Wu Y, Yao S, Narale BA, Shanmugam A, Mettu S, Ashokkumar M (2022) Ultrasonic processing of food waste to generate value-added products. Food Secur 11(14):2035
71. Panzella L, Moccia F, Nasti R, Marzorati S, Verotta L, Napolitano A (2020) Bioactive phenolic compounds from agri-food wastes: an update on green and sustainable extraction methodologies. Front Nutr 7:60
72. Martin-Rios C, Hofmann A, Mackenzie N (2020) Sustainability-oriented innovations in food waste management technology. Sustain For 13(1):210
73. Thakur M, Modi VK, Khedkar R, Singh K (2020) Sustainable food waste management
74. Mishra S, Singh PK, Mohanty P, Adhya TK, Sarangi PK, Srivastava RK, Jena J, Das T, Hota PK (2022) Green synthesis of biomethanol—managing food waste for carbon footprint and bioeconomy. Biomass Convers Biorefinery 12(5):1889–1909
75. Torres-Valenzuela LS, Ballesteros-Gómez A, Rubio S (2020) Green solvents for the extraction of high added-value compounds from agri-food waste. Food Eng Rev 12:83–100
76. Kosheleva RI, Mitropoulos AC, Kyzas GZ (2019) Synthesis of activated carbon from food waste. Environ Chem Lett 17:429–438
77. More PR, Jambrak AR, Arya SS (2022) Green, environment-friendly and sustainable techniques for extraction of food bioactive compounds and waste valorization. Trends Food Sci Technol 128:296–315

78. Vasyliev G, Vorobyova V (2020) Valorization of food waste to produce eco-friendly means of corrosion protection and "Green" synthesis of nanoparticles. Adv Mater Sci Eng 2020(1): 6615118
79. Jenkins SR, Morgan JM, Sawyer CL (1983) Measuring anaerobic sludge digestion and growth by a simple alkalimetric titration. Water Pollut Control Fed 55(5):448–453
80. Kapp H (1984) Schlammfaulung mit hohem Feststoffgehalt. Stuttgarter Berichte zur Siedlungswasserwirtschaft., Bd. 86,. Oldenbourg Verlag, München
81. Keitlinghaus H (2011) Gasmengen und –durchflussmessung—Konzepte und Eignung für Biogasanlagen. In: 6. Fachtagung Biogas 2011—Energieträger der Zukunft. 08.-09.06.2011, Braunschweig, Germany
82. Mahanty B, Zafar M, Han MJ, Park H-S (2014) Optimization of co-digestion of various industrial sludges for biogas production and sludge treatment: methane production potential experiments and modeling. Waste Manag 34:1018–1024
83. Mohamed SDY (2014) Influence of Oregano (Origanum vulgare L.), Fennel (Foeniculum vulgare L.) and Hop Cones (Humulus lupulus L.) on biogas and methane production. Universitat Giessen, Giessen, Germany
84. Hess J, Bernard O (2009) Advanced dynamical risk analysis for monitoring anaerobic digestion process. Biotechnol Prog 25:643–653
85. Barampouti EM, Mai S, Malamis D, Moustakas K, Loizidou M (2020) Exploring technological alternatives of nutrient recovery from digestate as a secondary resource. Renew Sust Energ Rev 134:110379. https://doi.org/10.1016/j.rser.2020.110379
86. Tait S, Harris PW, McCabe BK (2021) Biogas recovery by anaerobic digestion of Australian agro-industry waste: a review. J Clean Prod 299:126876. https://doi.org/10.1016/j.jclepro.2021.126876
87. European Parliamentary Research Service (2022) Future shocks 2022
88. Zhang C, Su H, Baeyens J, Tan T (2014) Reviewing the anaerobic digestion of food waste for biogas production. Renew Sust Energ Rev 38:383–392. https://doi.org/10.1016/j.rser.2014.05.038
89. Intergovernmental Panel on Climate Change (IPCC) (2023) Technical summary. In: Climate change 2022—impacts, adaptation and vulnerability. Cambridge University Press, pp 37–118. https://doi.org/10.1017/9781009325844.002
90. Haraldsen TK, Andersen U, Krogstad T, Sørheim R (2011) Liquid digestate from anaerobic treatment of source-separated household waste as fertilizer to barley. Waste Manag Res 29:1271–1276. https://doi.org/10.1177/0734242X11411975
91. Jin K, Pezzuolo A, Gouda SG, Jia S, Eraky M, Ran Y, Chen M, Ai P (2022) Valorization of bio-fertilizer from anaerobic digestate through ammonia stripping process: a practical and sustainable approach towards circular economy. Environ Technol Innov 27:102414. https://doi.org/10.1016/j.eti.2022.102414
92. Reuland G, Sigurnjak I, Dekker H, Michels E, Meers E (2021) The potential of digestate and the liquid fraction of digestate as chemical fertiliser substitutes under the RENURE Criteria. Agronomy 11:1–20. https://doi.org/10.3390/agronomy11071374
93. Elhaissoufi W, Ghoulam C, Barakat A, Zeroual Y, Bargaz A (2021) Phosphate bacterial solubilization: a key rhizosphere driving force enabling higher P use efficiency and crop productivity. J Adv Res 38:13–28. https://doi.org/10.1016/j.jare.2021.08.014
94. Alburquerque JA, de la Fuente C, Ferrer-Costa A, Carrasco L, Cegarra J, Abad M, Bernal MP (2012) Assessment of the fertiliser potential of digestates from farm and agroindustrial residues. Biomass Bioenergy 40:181–189. https://doi.org/10.1016/j.biombioe.2012.02.018

Chapter 2

Supercritical Fluid Extraction Techniques for Bioactive Recovery from Food Waste

Rukshana Irani, Suchandra Datta, Ayan Chatterjee, and Aminu Abdullahi

Abstract

Food waste management has been a colossal concern throughout the globe due to its perilous impact on the environment. It has led to the search of innovative extraction methods for the recovery of bioactives from wastes. An understanding of sustainability and the need for efficient waste valorization, especially in the food industry, is an area of paramount interest. Most of the wastes generated as a result of the processing of fruits, vegetables, grains, meat, fish, etc. are potential sources of bioactive ingredients but remain either underutilized or unused. Utilizing proper techniques for the extraction of compounds like polyphenols, flavonoids, essential oils, etc. can be highly beneficial in terms of health, economy, and environment. Among various methods, supercritical fluid extraction (SFE) has been a remarkable technique that offers a greener and efficient approach to recover bioactive compounds from food waste. SFE system consists of finely tuned sophisticated machinery that operates under multiple factors such as temperature, pressure, co-solvents, etc. The SFE process consists of some major steps, i.e., the selection and procurement of food wastes, pretreatment of the waste, interaction with the solvent, application of appropriate pressure and temperature, removal of solvent, and collection of the targeted bioactive components. High initial setup costs and the need for optimization of various parameters are the main challenges associated with SFE. Evaluating the effectiveness of SFE in comparison to other conventional methods and its relevance to sustainable practices in the food industry and other sectors is of great significance.

Key words Bioactives, Supercritical fluid extraction, Food wastes, Critical temperature, Sustainable

1 Introduction

1.1 Overview of Food Waste and Bioactive Compounds

The management of food waste is a significant environmental challenge throughout the globe. They are generated from various processes associated with the food production chain, including agricultural processing, industrial manufacturing, distribution, and consumption. An increase in food-related activities has caused an alarming rise in food waste. Statistics show that about 42% of waste is generated from household activities, 39% from food manufacturing, 14% from the food service sector, and 5% from distribution channels [1].

Tanmay Sarkar (ed.), *Food Waste Valorization: Green Techniques in Sustainable Management*,
Methods and Protocols in Food Science, https://doi.org/10.1007/978-1-0716-4646-5_2,
© The Author(s), under exclusive license to Springer Science+Business Media, LLC, part of Springer Nature 2025

Data from the Food and Agricultural Organization (FAO) in 2019 reported that about 14% of food worth $400 billion is wasted post-harvest until it reaches the retail stage every year worldwide. Immediate actions have to be taken to reduce food loss and waste, which can improve food security, sustainability, and economy. Addressing this issue can significantly help to achieve the United Nations Sustainable Development Goal (SDG) 12.3, which aims to halve per capita food waste by 2030 [2].

According to the United Nations Environment Programme (UNEP) Food Waste Index study (2021), an additional 17% of food (931 million tons of food) was wasted by consumers in the year 2019, which is primarily contributed by households and retail sectors [3, 4].

Poor food waste management and disposal can cause severe environmental hazards. An FAO study (2013) estimated that improper food waste handling releases 3.3 billion tons of CO_2 equivalent annually, contributing to 8% of total greenhouse gas emissions due to human activities [4, 5].

The Food and Agriculture Organization of the United Nations (FAO) report states that vegetable wastes have contributed substantially to a higher carbon footprint. At the same time, fruit wastage has been the cause of the major "blue water hotspot," primarily in the industrialized countries encompassing Europe and Asia [5].

The neglected by-product of food production can serve as a rich source of bioactive compounds and nutraceuticals, which can impart substantial potential in curing and preventing various health disorders and diseases. A variety of methods can extract secondary metabolites, vitamins, and minerals from food waste. These methods can serve as innovative solutions for the effective utilization of unused commodities and resources [6]. Table 1 describes the bioactives derived from food wastes.

In the past few decades, efforts have been made to discover methods and find innovative ways for the therapeutic usage of fruit and vegetable wastes. Conventionally, agro-industrial wastes have been widely used as livestock feed or organic fertilizer. However, current studies indicate the sustainable application of agro-industrial by-products for the making of high-value products such as cosmetics, foods, and medicines [7].

Bioactive compounds like polyphenols, flavonoids, flavonols, tannins, vitamins (A and E), fatty acids, essential minerals, volatiles, anthocyanins, pigments, etc., can be obtained from fruit and vegetable wastes or by-products. Bioactive peptides can be extracted from animal wastes. In the dairy industry, these bioactive peptides can be isolated from whey and colostrum. The marine sector obtains these bioactives from fish, shellfish, seaweeds, etc. The valorization of the wastes and by-products can lead to reduced

Table 1
Bioactives derived from food wastes [6]

Food wastes	Bioactive components
Fruit and vegetable waste: Peels, seeds, trimmings, stems	Polyphenols, carotenoids
Tomato by-products: Peels, seeds, pomace	Sterols, tocopherols, carotenes, terpenes, polyphenols, lycopene
Mango peel	Phenolic compounds, carotenoids, ascorbic acid, dietary fiber
Winery by-products (grape skins, seeds)	Polyphenols, tannins, anthocyanins, resveratrol, coumaric acid, ferulic acid, chlorogenic acid, cinnamic acid, gallic acid, proanthocyanidins, quercetin
Citrus processing waste (peels)	Flavonoids, hesperidin, naringin, eriocitrin, narirutin
Exhausted coffee waste	Polyphenols and tannins
Lees obtained from the Pisco making process	Polyphenols (gallic, protocatechuic, vanillic, syringic, ferulic derivatives, P-coumaric derivatives) Flavonoids (quercetin and its derivatives)
Bran of barley, mushroom by-products	β-Glucan
Rice bran	γ-Oryzanol, bran oil
Potato peel	Gallic acid, chlorogenic acid, caffeic acid, protocatechuic acid, syringic acid, *p*-hydroxyl benzoic acid, ferulic acid, coumaric acid
Pumpkin leaves	Caffeic acid, P-coumaric acid, ferulic acid, gentisic acid

generation of wastes. It can aid in fulfilling the well-known "zero waste concept" to meet the needs of the present day and the demands of the consumer and society [8].

A huge amount of waste is formed from fruit industries by pressing them to extract the juice and manufacturing value-added products. This waste can be extremely valuable. Studies indicate that the non-edible portions of fruits, such as peels or skin and twigs, often contain more bioactive components than the edible portion [9, 10].

The polyphenolic content of the peels of apples, citrus fruits, grapes, and seeds of avocado, jackfruit, and mango are 15% higher than the polyphenolic compounds of its pulp [9, 11].

Proposals have been made to add the pomace of apples and berries in bakery and dairy products to improve their antioxidants and dietary fiber content. The natural pigments and volatile compounds can also enhance the organoleptic properties of the finished products [8].

There are a variety of commonly used fruits, vegetables, seeds, and animal products whose waste can be reservoirs of many bioactive components.

Apple (*Malus domestica Borkh.*) is known for its delightful flavor and numerous health benefits. As per FAO statistics (2017), the global production of apples was more than 83 million tons. About 25% of the fresh fruit weight remains as apple pomace after the production of juices and ciders from them. Earlier, it was used as animal feed and fertilizer, but nowadays, apple pomace also serves as a chief source of pectin, which accounts for nearly 14% of the pectin supply throughout the world. This pectin is used in the food, pharmaceutical, and cosmetics industries as a thickening agent, gelling agent, and food stabilizing agent. The pectin also functions as dietary fiber and prebiotics that can lower blood cholesterol levels, improve post-prandial glycemic response, and promote satiety. Apple pomace contains health-promoting bioactive phytochemicals like phenolic acids, flavanols, flavonols, anthocyanins, and dihydrochalcones. The phenolics of apple pomace are found to impart radical scavenging activities. It also can inhibit protein glycation and tumor formation and progression. Phlorizin derived from apple by-products has anti-diabetic activities [8].

Bananas (*Musa L.* sp.) are tropical fruits whose production exceeded 113 million tons in 2017. The banana peel constitutes about 35% of the fruit's weight and is often thrown away as waste. It is a potential source of nutraceuticals due to the presence of higher contents of phenolic compounds, dietary fibers, and potassium. The quantity of polyphenols in banana peels is three times more than in the fruit flesh. The extract derived from banana peels has been found to exhibit strong antioxidant, antibacterial, antifungal, hypocholesterolemic, anti-diabetic, neuron-protecting, and anticarcinogenic activities. Berries are consumed fresh and are also used in the production of jams, juices, etc. The press residues of berries, after extracting the juice, are excellent sources of phenolic compounds. The unused stems and leaves of cauliflower are excellent sources of phenolic acids (ferulic acid, sinapic acid), flavonoids (kaempferol, quercetin), isothiocyanate, proteins, etc., that can impart antioxidant, antihypertensive, and anti-obesity properties. The pomace of beetroot contains ferulic acid, vanillic acid, caffeic acid, protocatechuic acid, *p*-hydroxy-benzoic acid, etc., having antioxidant and hepatoprotective activities. The oil obtained from seeds left due to berry processing is extremely valuable, having a distinctive fatty acid composition with higher amounts of lipid-soluble antioxidants, especially tocopherols. Mango seeds/ kernels are a major waste formed during mango processing. They contain numerous bioactive compounds such as polyphenols, flavonols, xanthones, alkyl-resorcinol, phytosterols (stigmasterol, campesterol), sitosterol, gallotannins, tocopherol, etc., having tremendous therapeutic potential. The seeds/ kernels serve as valuable sources of proteins and lipids. Proteins obtained from mango seeds are reported to have a high essential amino acids profile and excellent protein quality index. Whey is produced as a by-product in dairy industries during cheese making. Whey contains proteins that are good sources of bioactive peptides, which can exert therapeutic properties that involve antihypertensive and antioxidant potential. Meat processing industries also generate by-products like bones, blood, horns, fatty tissues, skin, and viscera. They contain bioactive peptides with antioxidant, antihypertensive, and antimicrobial activities both in vitro and in vivo. The fishing industry produces waste that is about 40–50% of the total fish weight. Both fish and shellfish by-products provide many bioactive compounds that can impart health-promoting activities. Shrimp are one of the most commonly consumed seafood. Its processing produces a huge quantity of waste. It contains chito-oligosaccharides, astaxanthin, polyunsaturated fatty acids, etc. [8].

Food waste impacts society, the economy, and the environment in a severely hazardous way. Food valorization rather than disposal can be an effective strategy in the management of food waste. Food waste can be converted into valuable resources that will not only reduce the harmful effects on the environment but will also support the economy. The recovery of bioactive compounds from food waste and by-products can be highly advantageous. Several new technologies have been developed for the extraction of bioactive compounds from food waste. The popular methods for the recovery of bioactive components from wastes include microwave-assisted extraction (MAE), pulse electric field extraction (PEFE), high hydrostatic pressure extraction (HHP), ultrasound-assisted extraction (UAE), enzyme-assisted extraction (EAE), supercritical fluid extraction (SFE), etc. [12].

Supercritical fluid extraction (SFE) has emerged as a promising green technology for obtaining bioactive compounds from various sources. SFE can be utilized as a sustainable and efficient technology for the transformation of food waste into valuable resources. The large amount of unused and unutilized bioactives from food waste can have multifarious benefits, favoring both the economy and the environment.

2 Materials

Supercritical fluid extraction (SFE) is a novel extraction technique characterized by its environmentally friendly nature. This technology is used for extracting bioactive compounds from plants and animal sources, like herbs, spices, aromatic and medicinal plants. SFE is highly efficient in the isolation and removal of bioactive substances from different sources.

The origins of this innovative technique can be traced back to 1822, when Baron Charles Cagniard de la Tour made an important observation regarding the supercritical phase.

He noticed changes in the behavior of the solvent at a particular temperature and pressure. Later, Thomas Andrews in 1869 made an insightful observation and coined the word "critical point." He used partially liquefied carbonic acid sealed in a glass tube to describe the critical point. Thomas Andrews defined the "critical point" as the point where the phase equilibrium curve ends, characterized by the critical temperature (Tc) and critical pressure (Pc), where the existence of the liquid and gas phase vanishes. In 1960, Hannay and Hogarth made significant contributions toward the application of supercritical fluids. It included utilizing CO_2 in the supercritical state, a key step in the development of SFE technology. The supercritical fluids were first used in Germany for decaffeinating green coffee beans. Later, it was used for the extraction of essential oils from hops in Australia. Around the 1980s, SFE technology was widely accepted in various industries of multiple countries due to its versatility and efficacy. Nowadays, this innovative technique is being extensively used throughout the globe to produce different kinds of products in the food, pharmaceutical, and natural product industries [13].

2.1 Selection of Food Waste Sources

The selection of food waste is a very important aspect of the supercritical fluid extraction (SFE) technique. Food waste is a rich source of valuable bioactive compounds having therapeutic and nutraceutical properties. Choosing the correct sources of food waste not only enhances economic viability but also promotes sustainability by decreasing the percentage of waste. Various plant materials, agro-industrial wastes, and by-products from the food processing industry can be selected for bioactive recovery using SFE techniques [14].

Peels, seeds, and pomace of fruits and vegetables can be used as they are excellent reservoirs of bioactive compounds. Agricultural residues and animal wastes like skin, bone, etc., can also be used for the recovery of these bioactives [6, 8].

The selection of food waste for supercritical fluid extraction involves evaluating the chemical composition of different types of wastes [15].

The collected food waste should be stored at a lower temperature (e.g., −20 °C, refrigeration, etc.) to minimize the degradation of bioactive compounds before extraction. Depending upon the waste type, it must be processed by washing, drying, grinding, etc., to prepare it for supercritical fluid extraction.

2.2 Targeted Bioactive Compounds

Bioactive compounds that can be targeted for extraction using SFE include polyphenols, flavonoids, carotenoids, vitamins, minerals, dietary fiber, essential oils, pigments, enzymes, etc. These are often more concentrated in the inedible parts of fruits, vegetables, and other plant residues. These bioactive components have been found to have remarkable anti-diabetic, antioxidant, anti-inflammatory, antimicrobial, anti-cancer, cardioprotective, and neuroprotective properties. The optimal extraction of these components can have wide applications in the food and pharmaceutical industries [16].

Wastes generated from milk, fish, meat, and marine industries can also be used to extract many bioactive components like bioactive peptides, proteins, fatty acids, pigments, minerals, etc. [8].

2.3 Solvents and Reagents Used in SFE

The efficiency of this extraction technique largely depends upon the reagents and solvents used in the process. The yield, purity, and quantity of extraction are directly associated with the solvents employed in the process.

Carbon dioxide has been an excellent solvent used in SFE due to its unique properties that favor the extraction of bioactive compounds. The lower viscosity and non-polar nature of CO_2 facilitate penetration into the matrix, which aids in rapid extraction. CO_2 with molecular weight, critical temperature, critical pressure, and critical density of 44.01 g/mol, 304.1 K, 7.38 MPa, and 0.469 g/cm^3, respectively, makes it suitable for SFE. The gaseous nature of supercritical CO_2 (SC-CO_2) efficiently dissolves less polar compounds and smaller molecules. Its liquid nature contributes to its high density and diffusion ability during extraction. The drawback of SC-CO_2 is its polarity, which makes it challenging to extract large and polar molecules effectively. Co-solvents like ethanol, methanol, or water are added to it to overcome this limitation. The addition of co-solvents augments the polarity of the extraction medium, which results in improved solubility of polar compounds, leading to greater overall yield. Ethanol, having a molecular weight of 46.07 g/mol and a critical temperature of 513.9 K, is often used as a good co-solvent due to its safety and efficient extraction abilities [13].

Many other compounds have also been regarded as supercritical fluids (SCFs). Examples include hydrocarbons like pentane, hexane, and butane; aromatic compounds such as benzene and toluene; various alcohols like methanol, ethanol, isopropanol, and n-butyl alcohol; as well as several gases such as ethylene and propane other than carbon dioxide. Water, methane, ethane, propane, ethylene, propylene, and acetone are used in the solvation process [13].

All solvents and reagents to be used must be of analytical grade or higher.

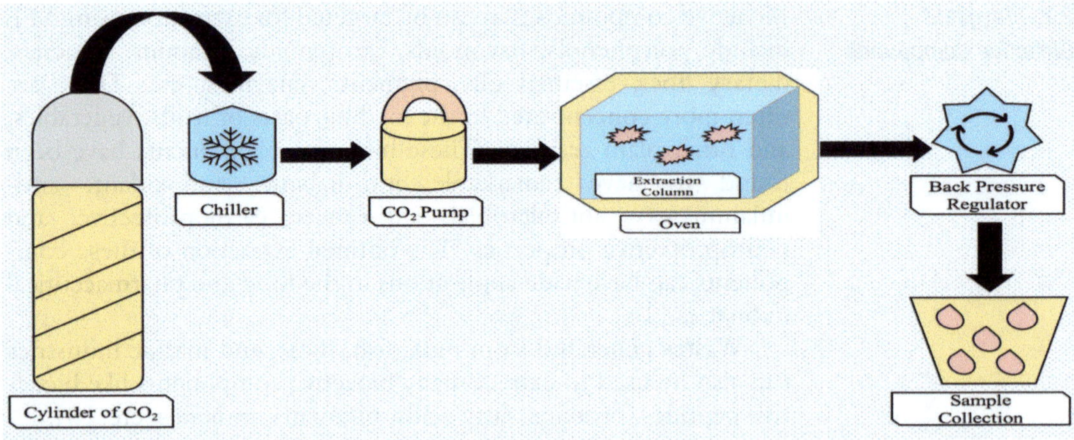

Fig. 1 Experimental setup of supercritical fluid extraction process [13, 17]

2.4 Equipment and Experimental Setup

The SFE process is carried out using a well-designed equipment setup that consists of some important components (Fig. 1).

2.4.1 Chiller

The chiller in the SFE system cools the solvent gas before it enters into the setup made for extraction. The cooling process is extremely important as it makes the solvent reach the optimal conditions necessary for the extraction process. Adequately chilled solvent maintains control over pressure and temperature, thus aiding in the creation of the supercritical conditions essential for extraction.

2.4.2 Solvent Pump

The solvent pump is another important component of an SFE system. It has been specifically designed to propel the supercritical fluid to the extraction apparatus. The solvent pump generates appropriate pressure to enable the conversion of the fluid into a supercritical state, which helps in penetrating the solid matrix effectively.

Whether the solvent pump would be a high-pressure CO_2 pump or a syringe pump depends on the scale of the extraction and operational requirements.

2.4.3 Extraction Column

The core extraction occurs in extraction columns. The column is packed with mostly pre-treated (dried and milled) solid samples. It is made to facilitate the contact between the supercritical fluid and the sample material. The extraction column's configuration can vary in diameter and length, depending upon the intended scale of extraction. It is typically made up of materials such as stainless steel, which is capable of withstanding the high pressures during the process.

2.4.4 Separators

Separators are an integral part of the SFE process. They are responsible for collecting the extracted components. Separators reduce the pressure of the fluid mixture and enable the separation of the dissolved compounds from the supercritical fluid. The multiple separators arranged in series can substantially improve the selectivity of the extraction process. It allows step-by-step separation based on differences in solubility.

2.4.5 Heat Exchangers

Heat exchangers of the SFE setup modify the temperature of various process materials in it. It helps in the adjustment of temperature, which is crucial for maintaining the conditions required for optimal extraction. It keeps the supercritical fluid in a state that is conducive to efficient extraction.

2.4.6 Oven

An oven keeps the temperature of the extraction column higher than the critical temperature of the supercritical solvent. This elevated temperature ensures that the extraction fluid sustains its supercritical phase during the entire extraction process. Ovens with programmable temperature controls can enhance the precision of the process.

2.4.7 Back Pressure Regulator

The back pressure regulator is essential for maintaining the overall pressure within the SFE system.

It ensures that the pressure is kept above the critical pressure of the supercritical fluid. It safeguards the integrity of the entire extraction process. Proper regulation of pressure is crucial for consistent and reliable extraction. It prevents unnecessary fluctuations that could adversely affect the quality of the extracted compounds [13].

3 Methods

3.1 Principles of Supercritical Fluid Extraction

The method of supercritical fluid extraction (SFE) uses supercritical fluids as solvents for extraction. Each supercritical fluid has a critical point, which is determined by its critical temperature and pressure. When the supercritical fluids exceed the critical temperature or are in a supercritical state, they cannot convert to liquid, yet their densities are comparable to that of liquids. Carbon dioxide (CO_2) is an ideal solvent for extraction because of its non-toxic, non-explosive nature. They are also readily available and can be easily removed from the extracted products. At room temperature and normal pressure, CO_2 exists as a gas, which facilitates the recovery of pure, solvent-free extracts or analytes [14].

Supercritical fluid extraction (SFE) is a separation technique where supercritical fluids extract desired compounds from a variety of matrices. SFE can be divided into two main steps, i.e., extraction and separation. Both solid and liquid samples can be processed using this technology. However, solid samples are usually employed. The supercritical fluid passes through the column containing the sample, under high pressure and temperature, which dissolves the extractable components from the solid matrix. These dissolved components are transported to the separator using diffusion, where they are separated from the solvent via pressure reduction, temperature increase, or both [13].

3.2 Optimization of Extraction Conditions

In supercritical fluid extraction (SFE), various parameters like extraction temperature, pressure, sample size, type and percentage of co-solvent, etc. must be optimized to enhance the yield of the desired compounds. These variables profoundly affect the solubility and mass transfer resistance of the raw sample materials. Increasing the solubility of extractable compounds is a key factor in determining the efficiency of the extraction process and the quality of the final extract. The solubility of solutes in the SFE process is controlled by temperature and pressure, as it influences the density of the fluid. SFE can easily be integrated with gas chromatography (GC) or supercritical fluid chromatography for analytical applications [14].

3.2.1 Temperature and Pressure Parameters

Temperature and pressure strongly regulate the extraction process. The density of the solvent decreases with an increase in temperature. An increase in pressure leads to higher yields due to improved solubility of the components within the matrix. However, higher pressure is not always recommended as potential interactions can occur between carbon dioxide (CO_2) and the solute. In supercritical fluid extraction (SFE), the saturation pressure of the solute increases with temperature, leading to enhanced solubility. Experimental studies indicate that higher pressure causes a greater yield of the extract. The density of supercritical CO_2 increases with its solubility. Overall, increased pressure greatly enhances extraction due to improved solubility within the matrix [14].

Temperature facilitates the rapid release of extracts from the sample matrix. Additionally, the pressure during the extraction plays a critical role in regulating the efficiency of extraction. Extraction yield increases as pressure rises at a fixed temperature. When the temperature reaches a threshold, the extraction process quickens even at lower pressure levels with increasing pressure. However, once high-pressure levels are reached, a slight decrease in the production of extracts has been observed [18].

An elevated temperature lowers the density of carbon dioxide, which in turn reduces the ability of the solvent to dissolve the solute. A temperature rise also increases the vapor pressure of the solutes, which results in an increased solubility during supercritical fluid extraction (SFE) with carbon dioxide. The solubility diminishes, stabilizes, or enhances with rising temperature at a constant pressure. It is determined based on whether the reduction in solvent density or the elevation in the vapor pressure of solute is dominant [14].

The temperature and pressure are two extremely crucial factors that determine the efficiency of supercritical fluid extraction (SFE) as they strongly regulate solubility. An increase in pressure at a given temperature improves the density of the solvent and thus the solubility of target compounds, which results in the requirement of smaller solvent volumes during extraction. Too much pressure can deteriorate antioxidant activity and also lower the diffusivity of the supercritical fluid (SCF), which results in decreasing the dissolution of solute. High pressure can lead to compact raw material, which negatively affects the yield of extraction. Higher temperatures reduce the density of the solvent and its solvating power while increasing the vapor pressure of the compounds. It improves solubility and yield but can also complicate the extraction process due to retrogradation, where solubility does not depend on solvent density. For heat-labile components, optimal extraction conditions are typically set between a temperature of 35 and 60 °C and a pressure of around 400 bar to avoid deterioration. These conditions of the extraction process vary depending upon the purpose of extraction and the nature of the compounds to be extracted [13].

3.2.2 Solvent Selection and Modifiers

A co-solvent can be referred to as an organic solvent that is mixed in varying proportions with carbon dioxide (CO_2). It can be dissolved alongside the supercritical fluid and maintain a significant solvent power for extracting the target compounds. CO_2 has been well recognized as the primary supercritical fluid in most of the industries; however, its lower polarity decreases its efficiency in extracting lipophilic and polar compounds. As a result, polar co-solvents such as ethanol, methanol, acetic acid, formic acid, water, and other polar or non-polar solvents can be used during the extraction process to avoid this limitation. This enhances the solvation power of supercritical CO_2, improves its affinity for poorly soluble solutes, enhances solubility, and ultimately improves the yield of extraction. In extraction, co-solvents such as methanol and ethanol are commonly utilized at concentrations lower than 10% of the CO_2 volume. The toxicity of ethanol is lower compared to methanol. The selection of the optimal co-solvent for a particular extraction process should be determined through several trials, taking into consideration the kind of sample, the compounds to be extracted, and the intended application. Research studies indicate that the choice of co-solvent affects not only the overall yield of extraction but also the properties of bioactivity in the extracts [13].

3.2.3 Effect of the Raw Matrix and Extraction Time on Supercritical Fluid Extraction

Factors such as the kind of raw material, moisture, size of the particle, shape, surface area, and porosity of the matrix can significantly impact the solubility and the mass transfer process in the extraction process. Correct selection of these factors makes the extraction of target compounds easy in a much-reduced time duration. Dried raw material must be used, as excess water can compete with the solute for interaction with the solvent, thereby decreasing the yield of extraction. However, sometimes, a moderate quantity of water may help in promoting solvent-solute interactions. Ideally, the moisture content should range from 4% to 14%. Additionally, the porosity and the particle size of solid samples directly regulate mass transfer rates during the extraction process. Smaller particle sizes enhance the efficiency of the extraction by shortening the path of the solvent's diffusion and increasing the surface area of contact, which in turn accelerates the extraction process. Very small particles can lead to increased internal mass transfer resistance, leading to channeling within the extraction column that, in turn, lowers the efficiency and yield of the process. Studies indicate that the mean optimal particle size for the supercritical fluid extraction of natural products typically ranges from 0.25 to 2.0 mm. Extraction time in supercritical fluid extraction (SFE) is very important as it can affect the composition of the extract. If the extraction time is very short, then it may result in incomplete or partial extraction. Too lengthy extraction times can cause wastage of both time and solvent. The degradation of bioactive compounds is also greater. The extraction time is directly associated with the flow rate of the solvent. A higher flow rate reduces the extraction time. Therefore, a preliminary trial can help to determine the optimal extraction time and fluid flow rate to obtain the highest yield of the extract [13].

3.3 Procedure of Extraction Process

The process of supercritical fluid extraction (SFE) has been described in a step-by-step manner in Fig. 2. All these steps need to be followed:

4 Notes

4.1 Comparative Analysis of SFE with Other Extraction Techniques

Supercritical fluid extraction (SFE) offers several advantages over traditional extraction methods such as solvent extraction, soxhlet extraction, and microwave-assisted extraction. SFE utilizes supercritical carbon dioxide (SC-CO_2), a non-toxic and easily removable solvent, minimizing the use of hazardous organic solvents and reducing environmental impact. The process is generally faster than conventional methods due to the high diffusivity of SC-CO_2, and the mild operating conditions preserve the quality and bioactivity of heat-sensitive compounds.

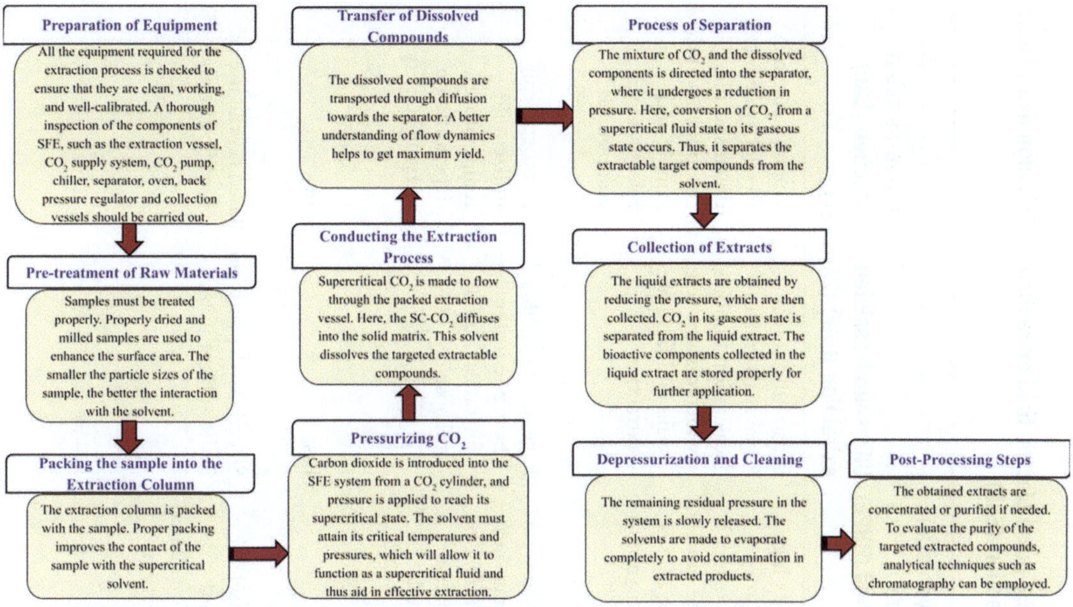

Fig. 2 Workflow of supercritical fluid extraction process [13, 14]

However, SFE also has limitations, including high initial investment costs and the need for specialized equipment. The selectivity of SC-CO_2 can be limited, particularly for polar compounds, although this can be addressed by using co-solvents.

A comparison of SFE with other popular extraction techniques, such as microwave-assisted extraction (MAE), pulse electric field extraction (PEFE), high hydrostatic pressure extraction (HHP), ultrasound-assisted extraction (UAE), enzyme-assisted extraction (EAE), and supercritical fluid extraction (SFE) is presented in Table 2. The key difference resides in the mechanism of action, heating process, selectivity, and relevancy for various types of compounds.

4.2 Industrial Applications and Commercial Viability

The recovery of bioactive compounds from food waste using SFE holds significant potential for industrial application. The extracted compounds can be used in various industries, including food, pharmaceuticals, cosmetics, and nutraceuticals. The demand for natural and sustainable ingredients drives the development of SFE technologies for food waste valorization [6].

A few industrial applications of SFE have been discussed below.

(i) **Application of SFE in the food industry**: The application of carbon dioxide in supercritical fluid extraction exhibited appealing properties such as non-toxic, inexpensive, odorless, colorless, and non-flammable. It also has low viscosity, high diffusivity, and near ambient critical temperature that made it

Table 2
A comparison among supercritical fluid extraction (SFE), microwave-assisted extraction (MAE), pulse electric field extraction (PEFE), high hydrostatic pressure extraction (HHP), ultrasound-assisted extraction (UAE), and enzyme-assisted extraction (EAE)

Properties	Supercritical fluid extraction (SFE) [19]	Microwave-assisted extraction (MAE) [20]	Pulse electric field extraction (PEF) [21]	High hydrostatic pressure extraction (HHPE) [22]	Ultrasound-assisted extraction (UAE) [23]	Enzyme-assisted extraction (EAE) [21]
Mechanism of action	It utilizes supercritical carbon dioxide to dissolve and extract target compounds by manipulating pressure and temperature	It uses microwave radiation to rapidly heat the sample and facilitate extraction by disrupting cell walls	It applies short pulses of high-voltage electric fields to disrupt cell membranes, which enhances extraction	It utilizes high pressure (hydrostatic) on the sample, which causes cell disruption	It uses ultrasonic waves to create cavitation bubbles, which disrupt cell walls and facilitate extraction	It employs enzymes to selectively break down specific components in the sample and releases targeted compounds
Heating process	This process can be performed at relatively low temperatures due to the use of supercritical CO_2	It generates significant heat within the sample due to microwave absorption	It is primarily a non-thermal process, although some heat generation may occur depending on the applied field	This process can generate heat depending on the pressure applied	It generates localized heat due to the cavitation bubble collapse	This process can utilize mild temperatures, depending on enzyme activity
Speed of the process	Moderate	Fast	Moderate to fast	Slow	Fast	Slow
Selectivity	Highly selective for non-polar compounds	In this process, both polar and non-polar compounds can be extracted, depending on the solvent system	Highly selective toward aqueous-based extraction	Highly selective toward aqueous-based extraction	Selectivity depends on the solvent usage	Based on the use of specific enzymes

Cost	High due to the usage of expensive equipment	Moderate	High	High due to the complex setup	Low to moderate	Low to moderate
Environmental impact	Low (use of CO_2)	Moderate	Low	Low	Moderate	Low
Applications	Widely used for extraction of essential oils, lipophilic compounds	Effective for extracting polar compounds such as phenol and flavonoids	Potential for extraction of sensitive bioactive compounds from plant tissues	Widely applied for extraction of antioxidants, bioactive compounds	Applicable to a broad range of compounds, including antioxidants and pigments	Suitable for extraction of specific components where enzymatic selectivity is advantageous such as proteins, sugars, and bioactive compounds

the preferred solvent for processing essential oils and food industry oils. The usage of carbon dioxide for extraction helps to preserve the aromatic ingredients in foods [13].

Studies have revealed that the SFE technology has been employed for the decaffeination of tea or coffee, extraction of flavors from herbs, extraction of fats and oils, extraction of sugars from bamboo, extraction of antioxidants such as vitamin E and C, extraction of aromas from different juices, de-acidification of oil, and extraction of tocopherols from vegetable oils [24]. It has also been applied for the extraction of phenolic compounds from freeze-dried apple pomace [25], and the release of extracts from apple seed oil under conditions such as appropriate pressure (300, 500, 750, 1000, and 1300 bar) and temperature (43, 53, and 63 °C) [26]. Similarly, the extraction of flavonoids from brown onion peels was done under optimized conditions (10 MPa, 40 °C and 85% ethanol), which showed a high content of quercetin and its derivatives. In the extraction of oil from green coffee beans, 0–5.7% ethanol as co-solvent under optimized conditions (200–400 bar and 40–60 °C) was used. SFE technology has also been utilized to separate oxygenated monoterpenes, α-humulene, viridiflorol, and manool from sage leaves under (100–300 bar and 40–60 °C) [27]. With the help of SFE, the phenolic compounds and flavonoids can be extracted from radish leaves by using pressure (300 and 400 bar), temperature (35, 40, and 50 °C), CO_2 flow rate (10 g/min), and co-solvent ethanol. In an experiment, the extraction of flavonoids from *Odontonema strictum* leaves was carried out under 200, 250 bar and 55–65 °C, CO_2 flow of (15 g/min) and 95% ethanol [28]. Tocopherol can also be extracted from quinoa under optimized conditions of pressure (200–400 bar) and temperature (40–60 °C) [29].

Several studies have been conducted on the extraction of β-carotene and lycopene from tomato processing waste. Experiment shows the effectiveness of SFE and the use of ethanol as a co-solvent for the recovery of phytochemicals with antioxidant capacity from soybean oil derived by-products. *Catharanthus roseus* is a rich source of alkaloids, and two dimeric alkaloids were extracted from it, which are extensively used as antineoplastic drugs (vinblastine and vincristine) in pharmacy [30]. A study was also carried out on rotten onion waste using SFE to evaluate the oleoresin extraction yield, sulfur content, and pyruvate content [31].

(ii) **Application of SFE in the pharmaceutical industry**: In the field of pharmaceuticals, advancements in the characteristics of active pharmaceutical compounds are typically preferred. Many pharmaceutical procedures for generating therapeutic particles and conducting pretreatments are still very basic, ineffective, and greatly limited. A major challenge in this regard is reducing particle size as it helps to enhance bioavailability. Therefore, the chief applications of supercritical fluids in the pharmaceutical industry include the formation of particles for drug delivery systems, crystal engineering, complexing cyclodextrins, coating, foaming, tissue engineering, impregnation, purification of pharmaceutical excipients, sterilization, and removal of solvents [32]. The most effective use of supercritical fluids in pharmaceuticals involves isolating bioactive compounds from a mixture (purification from reactions, quantification of an active enantiomer, and extraction from a natural matrix). Various plants have been utilized for the extraction of bioactive compounds to be used in pharmacy [33].

Artemisia annua L. exhibits strong antimalarial activity due to the presence of artemisinin and its derivatives (dihydroartemisinin). These compounds are especially effective against drug-resistant strains of *Plasmodium falciparum*. *Melocactus zehntneri* is a distinctive medicinal cactus that has unusual pharmacological properties as it contains isoquinoline and indole alkaloids, which are used in several drugs [34, 35].

(iii) **Application of SFE in the cosmetics industry**: Cosmetics are used as daily items that are presently available in different markets. These products are a complex mixture of different ingredients with different physical, chemical, and functional properties that are designed to achieve a product with the desired effectiveness and quality. Supercritical CO_2 ($SC\text{-}CO_2$) has proved to be a promising extraction solvent in this process and has the potential to substitute the toxic organic solvents. Supercritical fluid extraction can be easily incorporated into cosmetic formulations due to its distinct characteristics, appealing texture, absence of water or solvent traces, chemical stability, and inclusion of bioactive compounds [36]. Based on the operation parameters and the chemical composition, the obtained plant extracts can be utilized as common cosmetic ingredients such as thickeners, emollients, waxes, fragrances, and colorants. Additionally, $SC\text{-}CO_2$ also provides a significant concentration of bioactive compounds in the resulting extracts, which can be used in cosmetic formulations such as antioxidants, anti-aging, and photo-protective agents. It also provides antibacterial activity and inhibition of enzymes, such as *tyrosinase*, *lipoxygenase*, *collagenase*, and *hyaluronidase*. Studies have shown that $SC\text{-}CO_2$ can be used to isolate antioxidants and parabens from cosmetic products. In an investigation, $SC\text{-}CO_2$ has been used to obtain extracts from blackcurrant seeds, strawberry seeds, hop cones, and mint leaves for the production of shower gels and shampoo [37].

(iv) **Application of SFE in marine products**: The development of marine bioactive is a relatively recent field compared to those obtained from terrestrial sources. A significant diversity can be seen in the marine environments. In terms of chemical and biological perspectives, the ocean is an extraordinary source of valuable compounds. In this context, SFE has been extensively utilized for extracting bioactive constituents from algae, microalgae, and other marine-related organisms, such as crustaceans, fish, and their by-products. Marine origin, especially fish oil and fish by-products, serve as the chief and natural dietary source of omega-3 PUFAs (polyunsaturated fatty acids), mainly eicosapentaenoic acid (EPA) and docosahexaenoic acid (DHA). Studies have confirmed that the use of SFE with non-polar CO_2 is particularly very effective in extracting these kinds of nutrients [38].

An experiment was conducted to know the potentiality of supercritical fluid for fractionating triacylglycerols, chiefly (EPA and DHA) from fish oil under different pressure and temperature. The findings showed that SFE technology can efficiently fractionate triacylglycerols from fish oil [39]. The utilization of SFE technology in enhancing the value of waste from the fish industry is further illustrated by employing various fish by-products and certain marine invertebrates (sea urchins) as raw materials to extract omega-3 PUFAs. In an experiment, it has been found that the addition of ethanol significantly enhances the extraction yields of lipids and astaxanthin from red spotted shrimp waste compared to the extraction without ethanol as a co-solvent [40]. SFE has additionally been utilized to perform lipid extraction from microalgae, such as *Nannochloropsis oculata* and *Schizochytrium limacinum.* The application of this technology employs the extraction of antioxidant compounds, such as carotenoids, isoflavones, polyphenols, and flavonoids from algae and microalgae [41].

Supercritical fluid extraction (SFE) has the potential to be profitable for certain applications, particularly for the extraction of high-value natural products like nutraceuticals, food flavorings, and pharmaceuticals, due to its ability to produce high-quality extracts with minimal environmental impact. The commercial viability of SFE mainly depends on several factors, including the cost-effectiveness of the process, the scalability of the technology, and the market price of the extracted compounds. Considering these factors, SFE technology has numerous advantages, such as environment-friendly nature, high-quality product outcomes, and precise production control. Further research and development are needed to optimize SFE processes for different types of food waste and to reduce the cost of production.

4.3 Challenges and Future Perspectives

Currently, the demand for natural products is rapidly growing because of their extensive use in the area of functional food and the pharmaceutical industry. This need makes it necessary to establish suitable extraction technology which is capable of producing high-quality extracts from the respective materials. SFE has distinctive properties as it uses eco-friendly solvents for extraction. However, they also have some challenges that need to be addressed for effective application. A few major issues concerning the supercritical fluid extraction process include the following.

Restricted solubility: Supercritical fluids have limited solubility for polar and high molecular weight compounds. This can make it difficult when trying to isolate a compound whose material possesses high polarity or large molecular sizes [42].

High equipment and operating costs: The need for specialized equipment for handling high pressure and temperature is a huge capital expenditure. High operational costs associated with supercritical fluid extraction can limit the adoption of this technology in small-scale applications [43].

Process optimization and scalability: The optimization of extraction variables (e.g., pressure, temperature, flow rate) is very important for obtaining high extraction efficiency and selectivity. However, scaling up laboratory-based supercritical fluid extraction procedures to an industrial setting can be difficult because it requires accurate equipment design, effective heat transmission, and precise parameter control [44].

Safety issues: Supercritical fluids are used under high pressure, which can be hazardous if not handled correctly. Operating at high pressure and temperature requires measures to ensure safety and prevent accidents. Supercritical fluid extraction systems require safety protocols, training, and adherence to guidelines and regulations of operation [45].

Environmental impact: The supercritical fluids are generally found to be more environment-friendly than traditional ones. However, their production and disposal can still have negative effects on the environment. Therefore, solvent production must be managed carefully to decrease its hazardous impact on the environment [46].

The field of green solvents is still evolving, and further research can provide excellent opportunities. Several areas need to be explored for the development and use of green solvents. Future research may examine how to produce new green solvents which can be more selective, more efficient, and more stable than the present ones. Further efforts can be made to improve extraction techniques to enhance the sustainability and economic viability of green solvents. A detailed viewpoint on the application of supercritical fluid extraction in future is mentioned below.

Sources and availability of supercritical fluid:

- Assessment of different sources for the development of supercritical fluids that are easily accessible and sustainable.
- New techniques need to be developed to acquire and produce supercritical fluids from renewable resources.

Optimization of processes and cost reduction:

- Improving supercritical fluid extraction methods to enhance performance and lower energy consumption.
- Invention of economical methods for the creation and recovery of supercritical fluid [47].

Enhancement of solvent properties:

- Modification of characteristics of supercritical fluid by incorporating modifiers or co-solvents to enhance selectivity.
- Investigation of new solvents with customized characteristics for specific application in extraction.

Impact on environment and sustainability:

- Analysis of the ecological impact and carbon emission.
- Adaptation of sustainable practices in the production of supercritical fluid, such as using renewable energy sources and minimizing waste generation.

Expansion and industrial application:

- Exploration of scale-up approaches for supercritical fluid extraction methods to facilitate commercial production.
- The use of supercritical fluid in the extraction of natural products, pharmaceuticals, and food bioactives beside industrial use [14].

References

1. Mirabella N, Castellani V, Sala S (2014) Current options for the valorization of food manufacturing waste: a review. J Clean Prod 65:28–41. https://doi.org/10.1016/j.jclepro.2013.10.051
2. Food and Agriculture Organization of the United Nations (2019) The state of food and agriculture 2019: moving forward on food loss and waste reduction. FAO, Rome
3. United Nations Environment Programme. (2021). Food waste index report 2021. Nairobi
4. Siddique F, Ali Z, Arshad M, Mubeen K, Ghazala A (2024) Exploration of novel eco-friendly techniques to utilize bioactive compounds from household food waste: special reference to food applications. Front Food Sci Technol 4:1388461. https://doi.org/10.3389/frfst.2024.1388461
5. Food and Agriculture Organization of the United Nations (2013) Food wastage footprint: impacts on natural resources. FAO
6. Kumar K, Yadav AN, Kumar V, Vyas P, Dhaliwal HS (2017) Food waste: a potential bioresource for extraction of nutraceuticals and bioactive compounds. Bioresour Bioprocess 4(1):1–14. https://doi.org/10.1186/s40643-017-0148-6
7. Rudra SG, Nishad J, Jakhar N, Kaur C (2015) Food industry waste: mine of nutraceuticals. Int J Sci Environ Technol 4(1):205–229

8. Ben-Othman S, Jõudu I, Bhat R (2020) Bioactives from Agri-food wastes: present insights and future challenges. Molecules 25(3):510. https://doi.org/10.3390/molecules25030510
9. Gorinstein S, Martin-Belloso O, Park YS, Haruenkit R, Lojek A, Cız M, Caspi A, Libman I, Trakhtenberg S (2001) Comparison of some biochemical characteristics of different citrus fruits. Food Chem 74:309–315. https://doi.org/10.1016/S0308-8146(01)00157-1
10. Gorinstein S, Zachwieja Z, Folta M, Barton H, Piotrowicz J, Zemser M, Weisz M, Trakhtenberg S, Màrtín-Belloso O (2001) Comparative contents of dietary fiber, total phenolics, and minerals in persimmons and apples. J Agric Food Chem 49(2):952–957. https://doi.org/10.1021/jf000947k
11. Soong YY, Barlow PJ (2004) Antioxidant activity and phenolic content of selected fruit seeds. Food Chem 88:411–417. https://doi.org/10.1016/j.foodchem.2004.02.003
12. Chauhan S, Pandit NK, Mohanty A. Meena SS (2023). Resource recovery of bioactive compounds from food waste and their diverse industrial applications. Biomass Convers Biorefinery doi: https://doi.org/10.1007/s13399-023-04967-3
13. Uwineza PA, Waśkiewicz A (2020) Recent advances in supercritical fluid extraction of natural bioactive compounds from natural plant materials. Molecules 25(17):3847. https://doi.org/10.3390/molecules25173847
14. Fraguela-Meissimilly H, Bastías-Monte JM, Vergara C, Ortiz-Viedma J, Lemus-Mondaca-R, Flores M, Toledo-Merma P, Alcázar-Alay S, Gallón-Bedoya M (2023) New trends in supercritical fluid technology and pressurized liquids for the extraction and recovery of bioactive compounds from agro-industrial and marine food waste. Molecules 28(11):4421. https://doi.org/10.3390/molecules28114421
15. Aili Q, Cui D, Li Y, Zhige W, Yongping W, Minfen Y, Dongbin L, Xiao R, Qiang W (2024) Composing functional food from agro-forest wastes: selectively extracting bioactive compounds using supercritical fluid extraction. Food Chem 455:139848. https://doi.org/10.1016/j.foodchem.2024.139848
16. Ray A, Dubey KK, Marathe SJ, Singhal R (2023) Supercritical fluid extraction of bioactive from fruit waste and its therapeutic potential. Food Biosci 52:102418. https://doi.org/10.1016/j.fbio.2023.102418
17. Wei PC, May CY, Ngan MA, Hock CC (2005) Supercritical fluid extraction of palm carotenoid. Am J Environ Sci 1(4):264–269. https://doi.org/10.3844/ajessp.2005.264.269
18. De Melo MMR, Silvestre AJD, Silva CM (2014) Supercritical fluid extraction of vegetable matrices: applications, trends and future perspectives of a convincing green technology. J Supercrit Fluids 92:115–176. https://doi.org/10.1016/j.supflu.2014.04.007
19. Soquetta MB, Terra LDM, Bastos CP (2018) Green technologies for the extraction of bioactive compounds in fruits and vegetables. CyTA-J Food 16(1):400–412. https://doi.org/10.1080/19476337.2017.1411978
20. Panzella L, Moccia F, Nasti R, Marzorati S, Verotta L, Napolitano A (2020) Bioactive phenolic compounds from Agri-food wastes: an update on green and sustainable extraction methodologies. Front Nutr 7:1–27. https://doi.org/10.3389/fnut.2020.00060
21. Vernes L, Vian M, Chemat F (2020) Ultrasound and microwave as green tools for solid-liquid extraction. In: Liquid-phase extraction, pp 355–374. https://doi.org/10.1016/B978-0-12-816911-7.00012-8
22. Chan CH, Yusoff R, Ngoh GC, Kung FWL (2011) Microwave-assisted extractions of active ingredients from plants. J Chromatogr A 1218(37):6213–6225. https://doi.org/10.1016/j.chroma.2011.07.040
23. Drinic Z, Pljevljakusic D, Zivkovic J, Bigovic D, Savikin K (2020) Microwave-assisted extraction of O. Vulgare L. spp. hirtum essential oil: comparison with conventional hydro-distillation. Food Bioprod Process 120: 158–165. https://doi.org/10.1016/j.fbp.2020.01.011
24. Majid A, Phull AR, Khaskheli AH, Abbasi S, Sirohi MH, Ahmed I, Ujjan SH, Imdad Ali Jokhio IA, Waqar Ahmed W (2019) Applications and opportunities of supercritical fluid extraction in food processing technologies: a review. Int J Adv Appl Sci 6(7):99–103
25. Pando Bedriñana R, Rodríguez Madrera R, Loureiro Rodríguez MD, López-Benítez K, Picinelli Lobo A (2024) Green extraction of bioactive compounds from apple pomace from the cider industry. Antioxidants (Basel) 13(10):1230. https://doi.org/10.3390/antiox13101230
26. Montanes F, Catchpole OJ, Tallon S, Mitchell KA, Scott D, Webby RF (2018) Extraction of apple seed oil by supercritical carbon dioxide at pressures up to 1300 bar. J Supercrit Fluids 141:128–136. https://doi.org/10.1016/j.supflu.2018.02.002
27. Goyeneche R, Fanovich A, Rodrigues C, Nicolao MC, Di Scala K (2018) Supercritical CO2 extraction of bioactive compounds from radish leaves: yield, antioxidant capacity and cytotoxicity. J Supercrit Fluids 135:78–83. https://doi.org/10.1016/j.supflu.2018.01.004

28. Pimentel-Moral S, Borrás-Linares I, Lozano-Sánchez J, Arráez-Román D, Martínez-Férez A, Segura-Carretero A (2018) Supercritical CO2 extraction of bioactive compounds from Hibiscus sabdariffa. J Supercrit Fluids 147:213–221. https://doi.org/10.1016/j.supflu.2018.11.005
29. Benito-Román O, Rodríguez-Perrino M, Sanz MT, Melgosa R, Beltrán S (2018) Supercritical carbon dioxide extraction of quinoa oil: study of the influence of process parameters on the extraction yield and oil quality. J Supercrit Fluids 139:62–71. https://doi.org/10.1016/j.supflu.2018.05.009
30. Karimi M, Raofie F (2019) Micronization of vincristine extracted from Catharanthus roseus by expansion of supercritical fluid solution. J Supercrit Fluids 146:172–179. https://doi.org/10.1016/j.supflu.2019.01.021
31. Devani BM, Jani BL, Balani PC, Akbari SH (2020) Optimization of supercritical CO2 extraction process for oleoresin from rotten onion waste. Food Bioprod Process 119:287–295. https://doi.org/10.1016/j.fbp.2019.11.014
32. Adeoye O, Costa C, Casimiro T, Aguiar-Ricardo A, Cabral-Marques H (2018) Preparation of ibuprofen/hydroxypropyl-γ-cyclodextrin inclusion complexes using supercritical CO2-assisted spray drying. J Supercrit Fluids 133:479–485. https://doi.org/10.1016/j.supflu.2017.11.009
33. Falcão MA, Scopel R, Almeida RN, Espirito Santo AT, Franceschini G, Garcez JJ, Vargas RMF, Cassel E (2017) Supercritical fluid extraction of vinblastine from Catharanthus roseus. J Supercrit Fluids 129:9–15. https://doi.org/10.1016/j.supflu.2017.03.018
34. Baldino L, Reverchon E, Della Porta G (2017) An optimized process for SC-CO2 extraction of antimalarial compounds from Artemisia annua L. J Supercrit Fluids 128:89–93. https://doi.org/10.1016/j.supflu.2017.05.018
35. Bubalo CM, Vidović S, Redovniković IR, Jokić S (2018) New perspective in extraction of plant biologically active compounds by green solvents. Food Bioproduct Process 109:52–73. https://doi.org/10.1016/j.fbp.2018.03.001
36. Lee MR, Lin CY, Li ZG, Tsai TF (2006) Simultaneous analysis of antioxidants and preservatives in cosmetics by supercritical fluid extraction combined with liquid chromatography–mass spectrometry. J Chromatogr A 1120:244–251. https://doi.org/10.1016/j.chroma.2006.01.075
37. Vogt O, Sikora E, Ogonowski J (2014) The effect of selected supercritical CO_2 plant extract addition on user properties of shower gels. Pol J Chem Technol 16:51–54. https://doi.org/10.2478/pjct-2014-0069
38. Zhou J, Gullón B, Wang M, Gullón P, Lorenzo JM, Barba FJ (2021) The application of supercritical fluids technology to recover healthy valuable compounds from marine and agricultural food processing by-products: a review. Processes 9(2):357. https://doi.org/10.3390/pr9020357
39. De Melo MM, Sapatinha M, Pinheiro J, Lemos MFL, Bandarra NM, Batista I, Paulo MC, Coutinho J, Saraiva JA, Silva CM (2020) Supercritical CO_2 extraction of Aurantiochytrium sp . biomass for the enhanced recovery of omega-3 fatty acids and phenolic compounds. J CO2 Util 38:24–31. https://doi.org/10.1016/j.jcou.2020.01.014
40. Wan MC, Qin W, Lei C, Li QH, Meng M, Fang M, Song W, Chen J, Tay F, Niu L (2021) Biomaterials from the sea: future building blocks for biomedical applications. Bioactive Mater 6(12):4255–4285. https://doi.org/10.1016/j.bioactmat.2021.04.028
41. Sousa SC, Freitas AC, Gomes AM, Carvalho AP (2023) Extraction of *Nannochloropsis* fatty acids using different green technologies: the current path. Mar Drugs 21(6):365. https://doi.org/10.3390/md21060365
42. Khaw KY, Parat MO, Shaw PN, Falconer JR (2017) Solvent supercritical fluid technologies to extract bioactive compounds from natural sources: a review. Molecules 22(7):1186. https://doi.org/10.3390/molecules22071186
43. Lomba L, Ribate MP, Sangüesa E, Concha J, Garralaga MA, Errazquin D, García C, Giner B (2021) Deep eutectic solvents: are they safe? Appl Sci 11(21):10061. https://doi.org/10.3390/app112110061
44. Brunner G (2005) Supercritical dluids: technology and application to food processing. J Food Eng 67(1–2):21–33. https://doi.org/10.1016/j.jfoodeng.2004.05.060
45. Girotra P, Singh S, Nagpal K (2012) Supercritical fluid technology: a promising approach in pharmaceutical research. Pharm Dev Technol 18(1):22–38. https://doi.org/10.3109/10837450.2012.726998
46. Tang B, Zhang H, Row KH (2015) Application of deep eutectic solvents in the extraction and separation of target compounds from various samples. J Sep Sci 38(6):1053–1064. https://doi.org/10.1002/jssc.201401347
47. Si-Hung L, Bamba T (2022) Current state and future perspectives of supercritical fluid chromatography. TrAC Trends Anal Chem 149:116550. https://doi.org/10.1016/j.trac.2022.116550

Chapter 3

Green Enzymatic Hydrolysis Protocols for Protein Recovery from Food By-Products

Sumitha Elayaperumal, K. Surendra, Ragunathan Devendran, and Yuvaraj Sivamani

Abstract

Food by-products are waste produced during food production and processing that has a detrimental effect on the environment. To guarantee a sustainable and healthy diet for the world's population, reducing food loss and waste is an important component that has to be adopted. These by-products are rich in bioactive compounds and have a potential nutritional value. To meet sustainable development goals, there is an increasing concern about the availability of sufficient amounts of nutrients containing protein with suitable quality to maintain health. Therefore, much focus is given to the recovery of proteins from food by-products. Over time, green enzyme hydrolysis (GEH) has emerged as a suitable form of extracting high-value proteins from food processing waste. Based on the principles of circular bioeconomy and valorization of waste, it is considered an eco-friendly and very effective method. Specified proteases and mild conditions are used explicitly for biohazard protein-rich by-products that hydrolyze it with their qualities kept intact, benefitting its nutrition and functionality. Owing to the moderate severity under which the process is carried out, environmental pollution would be minimal. The process offers advantages such as high selectivity, tremendous energy reductions, and minimized by-product production. Fish waste, dairy residues, plant-based origins, and meat residues left from meat processing are some of the common food by-products usually used effectively. Such findings represent the wider potential uses of hydrolysates in providing food fortification, nutraceuticals, and biopolymers in association with more sustainable food systems. Enzyme type, pH, and temperature optimization are the main factors critical for the efficiency of enzymatic hydrolysis. Further advances of enzyme engineering and bioprocessing have strongly improved the efficiency of protein recovery and functional properties. The use of GEH technology with various latest technologies like hydrolysis assisted with ultrasound or a combination with membrane separation can make the process more efficient and purified protein. These findings are discussed in the current article, which advances protocols on GEH for protein recovery from food by-products and elucidates the significance of the commercial and environmental utility thereof. Further study would focus on enzyme formulations and scale-up applications.

Key words Enzymatic hydrolysis, Protein recovery, Food by-products, Green technology, Sustainable extraction, Circular economy, Protease, Bioactive peptides

Tanmay Sarkar (ed.), *Food Waste Valorization: Green Techniques in Sustainable Management*, Methods and Protocols in Food Science, https://doi.org/10.1007/978-1-0716-4646-5_3,
© The Author(s), under exclusive license to Springer Science+Business Media, LLC, part of Springer Nature 2025

1 Introduction

The current worldwide hunger and malnutrition crisis is enormous with 1.9 million people suffering from dreadful starvation [1]. But it also produces large amounts of waste and by-products, which present serious problems for the environment and the economy. As a result of a changing environment and increased awareness of the ethical use of resources and disposal practices, numerous ways to maximize the utilization of food waste have been developed [2]. In such framework where dietary waste is considered, the cleaning and enrichment of such food leftovers have open perspectives of their reuse. Of the several available methodologies that promote food waste minimization, enzymatic hydrolysis has proven to be the most applicable as a stage of integration of the biotechnological process, achieving economic viability [3].

Among the nutrients, protein is essential for building and repairing tissues, maintaining muscle mass, producing hormones and enzymes, and supporting immune function. The expanding population can only be fed by offering additive solutions, such as utilizing co-products, minimizing waste, and introducing new protein technology to support established industries. The idea is to extract proteins from food processing waste using enzymes, and this issue gives enough reason to review literature for more prominent explanations since the operation of enzymes is possible under milder conditions enhancing the recovery of quantity of proteins and destruction of the bright natural composition [4]. This approach is in consonance with the policy of the circular bioeconomy of a modern vision which militates against waste, provides effective and efficient utilization of natural resources, and advocates responsible manufacturer to promote sustainable practices in the industry [5]. By-products of food such as meat, dairy, marine, plant processing, and fermentation industries possess considerable characteristics such as proteins and bioactive compounds (Fig. 1). The

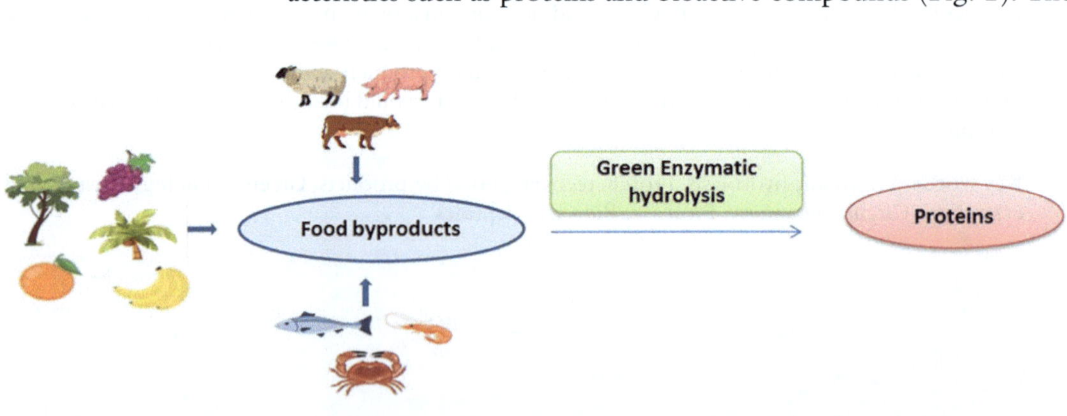

Fig. 1 Food by-products from meat, dairy, marine, and plant processing undergoing green enzymatic hydrolysis and converting to bioactive proteins

recovery and utilization of such food sources lie within the food, pharmaceutical, and nutraceutical domains [6]. Classic protein extraction methods generally involve many harsh chemicals, vast processing time, and exposure to high temperatures leading to denaturation and a decrease in bioavailability. Traditionally, enzymatic hydrolysis techniques enhance exceptional specificity, minimal energy use, and expanded functional properties of obtainable protein hydrolysates [6].

Industries must tap the potential of natural enzyme catalysts and economy technologies to transform their waste streams into valuable functionalities without destroying the environment [7]. The fundamental aspects of enzymatic hydrolysis will be alluded to in the present survey that will differentiate the process from general practice based on advantages and relevance to sustainable food production [8]. This chapter explores some salient protocols in enzymatic hydrolysis, the factors influencing the efficiency of the process, and some recent developments on the engineering and biological optimization of this bioprocess. In addition, enlightening an industry viewpoint, this also attempts to underscore how the application of specific technology will highlight the role that protein hydrolysates from food-sourced by-products may have in elevating food security and environmental health.

1.1 Sustainability and the Necessity for the Implementation of Enzymatic Treatment for the Backwards Recovery of Green Proteins

Over the past few years, combatting the increasing rates of protein deficiency, environmental pollution, and utilization of by-products during food processing has been an issue for the food industry [9]. For instance, the Food and Agricultural Organization (FAO) states that almost a third of fully produced food globally is thrown away alongside the pressure of degeneration of the earth's resources and its system. These waste products, however, when not properly handled, worsen global warming by emitting toxic gases or by killing useful constituents of the food [10]. Green enzymatic hydrolysis liquid extraction is one such technique that addresses the out-of-sight-out-of-mind years and recycles the high-value proteins integrated within the food wastes as protein hydrolysates, rather than allowing them to fester and get as heavy metals in the environment [11]. In terms of green protein recovery, the utilization value of enzymatic hydrolysis is high since hazardous solvents or thermal processes simply do not fit. The method also supports the production of bioactive peptides, which in turn have higher digestibility, solubility, and functional characteristics, thereby allowing proteins be used in new food and industrial applications [12]. In addition, this technique can coexist with the modern green technologies including ultrasound-assisted extraction, membrane filtration, and fermentation which also increases sustainability and the net benefit expected for environmental and health improvement.

Table 1
Types of proteases

Types of proteases depending on the mechanism	Functions
Endoproteases	Target internal peptide bonds and produce large peptide fragments
Exoproteases	Acts on terminal peptide bonds and release free amino acids

1.2 The Mechanistic and Key Parameters of Enzymatic Hydrolysis

In enzymatic hydrolysis, proteases break down complex structures of protein into smaller peptides and free amino acids with the help of enzyme catalysis. These proteases, better known as peptidases or proteinases, are a pioneer in hydrolyzing peptide bonds and producing protein hydrolysates, having specific bioactive characteristics [13]. It depends on enzyme type and substrate specificity, whether the hydrolysis takes place, and how long it takes for the reaction. Table 1 depicts the major types of proteases that are most commonly used in by-product hydrolysis in foods.

Specific proteases like trypsin, pepsin, alcalase, and papain, that each has its own substrate preferences, result in different hydrolysis patterns. The more significant process parameters include pH, temperature, enzyme concentrations, and reaction time, which affect the hydrolysis efficiency and functional properties of protein hydrolysates that would be obtained [14]. These parameters are the ones that ought to be a prime consideration in order to guarantee the strategies concerning the desired yield, bioactivity, and economic feasibility of large-scale processes in industry.

1.3 Food By-Products in Protein Sources for Sustainable Use

Several food by-products arising from processing (Table 2) have the potential to find use for protein recovery through enzymatic hydrolysis, e.g., meat and poultry. In the meat industry, the majority of waste is produced during the slaughtering process. These wastes include bones, tendons, skin, digestive contents, blood, and internal organs. By-products such as bones, skin, and cartilage contain a host of collagen, albumins, and globulins worth hydrolyzing into bioactive peptides [15]. Dairy industry residues: Whey protein, casein, and milk by-products—all these are excellent raw materials for enzymatic hydrolysis to produce peptides having functional and health benefits [16].

Discards from the seafood: Significant protein and collagen fibers are embedded in the fish heads, skin, and crustacean shells, making these wastes suitable for hydrolysis to produce foods and cosmetic items [17]. Vegetarian waste: Protein comprises a part of the most oilseed meal extracts, fruit skin materials, and various cereal brands. These proteins can be further metabolized using food application/enzymes or use as bio-materials [18].

Table 2
Sources of food by-products and their protein component

Animals	Food processed	By-products	Protein component
Cattles	Processed meat and poultry products	Blood, hides, hair, heads, Horns, hooves, meat trimmings, feathers, Feet, giblets	Keratin, collagen, albumin, globulins, gelatine
Fish	Canned fish, filleted fish, smoked fish, salted fish, processed crustaceans	Fish heads, scales, skin, and crustacean shells	Collagen, gelatine
Dairy	Milk, butter, cream, yoghurt, cheese, ice cream	Whey	Whey protein, casein
Plants			
Grains	Grain, flour, bread, Biscuits, cakes Starch, bakery goods	Straw, stems, leaves, husks, fiber, bran, germ	Tryptophan, tyrosine, cysteine
Fruits and vegetables	Tinned fruits and vegetables, juices, vegetable oils, starches, sugars	Rotten fruits and vegetables stem waste, pits, seeds, peels, pulp	Lysine

1.4 Advances in Green Enzymatic Hydrolysis Technology

In recent years, developments in green enzymatic hydrolysis technology have significantly increased the sustainability and efficiency of the enzyme hydrolysis process and allowed for the ultimate industrial application of protein recovery [19]. Immobilized enzyme systems are developed to improve enzymatic stability, reusability, and performance and also reduce the economics in processes. Ultrasound and microwave-induced hydrolysis also obviously improve enzyme–substrate interactions, leading to an increase in reaction rates, yield, as well as the functional aspects of products [20]. It has accordingly been improved by genetic modifications carried out in enzyme engineering over recent times in order to prepare proteases that will fit with a certain substrate with greater effectiveness and specificity.

In addition to that, there has been an integration of enzyme pre-treatment techniques with membrane separation and in particular pervaporation techniques, which has enhanced the quality and performance of protein hydrolysates [21]. Hydrolyzed products are segregated based on their molecular size which has been possible by the disruptive process of various molecular weight fractions through enzymatic centrifugal field flow fractionation [22]. The application of onsite biosystem with real-time control and ease of operation has allowed for processing to be done with much efficiency in that, the amount of enzyme required in the process most effective process parameters for the enzymes is coordinated [23]. Due to the progress in these systems especially in the recycling of proteins from food wastes, the costs of converting such wastes to proteins are extremely low; thereby, this technology fits well in modern requirements on reducing losses of food and developing green economy [24].

1.5 Protein Hydrolysates Applications of Food By-Products

The production of protein hydrolysates is regarded comprehensive by many people. Basically, these hydrolysates have had the potential to find ample application in various sectors. Majorly they serve to impart functionality in food products by altering the texture, solubility, and emulsification properties of dairy, meat, and bakery products [25]. In addition, bioactive peptides formant due to enzymatic hydrolysis have antioxidant, antimicrobial, and antihypertensive properties that recently concentrated to be appropriate to be integrated into nutraceuticals and functional foods [23]. These serve in the pharmaceutical industry as the starting materials in preparing drug formulations, wound healing materials, and dietary supplements rich in protein that boost recovery of muscle and immune health [25]. On the other hand, cosmetics have imbibed protein hydrolysates presently into skincare products because it possesses sequestered peptides from marine and animal by-products, which are good for anti-aging and also moisture-retaining formulations [26]. Most probably, hydrolyzed products are now seen in the research and development segment leading to better biodegradable packaging options and bio-based materials-aggregated sustainable alternatives to conventional plastics. The broad spectrum of applications of hydrolysates as well as possibilities of other related bioactivity show that this is something very much to do with how conceptually one can improve sustainability or resource efficiency in across the sectors [27].

2 Materials

2.1 Food By-Products

Dairy by-products (whey, cheese whey)

Wastes from meat and fish processing (fish heads, bones, skin)

Plant-based waste material (oil seed meals, fruit peels, cereal bran)

Malt and brewery wastes—spent grain residues and yeast residues

Enzymes used: Proteases—alcalase, bromelain, papain, trypsin, and pepsin

Commercial enzymes prepared for various applications

Buffers and reagents: PBS buffer maintains pH, sodium hydroxide and hydrochloric acid for the adjustment of pH

2.2 Preparing Analysis Solution with Deionized Water

Green enzyme hydrolytic protocols for food by-products are indeed believed to include eco-friendly processes through sustainable materials and methods to reduce waste, all to maximize protein recovery. Instead of using harsh chemical treatments, these approaches now focus on hydrolyzing with enzymes, thus rendering them safer and more sustainable [28]. Proteases such as alcalase, papain, bromelain, and pepsin are some of the commonly used enzymes that cleave complex protein macromolecules into bioavailable peptides and amino acids. The selection of one enzyme or the other depends on the substrate and the functionalities of the end product desired [29]. For example, alcalase, a serine protease, is used for its strong specific and high efficiency in hydrolyzing any protein coming from fish processing residues, meat industry by-products, and plant-based oils such as oilseed meals [30]. Bromelain and papain, derived from pineapple and papaya, respectively, use the same hydrolysis process to enhance the bioactivity and digestibility of animal and plant protein sources. With the enzymatic process, conditions are rather mild, considering optimal pH and temperature for enzyme activity, thereby causing low-energy consumption than in other methods [31]. Food-grade enzymes ensure the safety and applicability of recovered proteins as functional foods, nutraceuticals, and animal feed. Enzymatic hydrolysis is a remarkable step since it is capable of developing bioactive peptides with antioxidant, antihypertensive, antimicrobial, and immunomodulatory properties, thus enriching the value of food by-products [32]. Green protocols lay emphasis on renewable sources for use, for example, plant-derived or microbial enzymes, and minimize the reliance on synthetics or solvents. Further, techniques incorporating membrane filtration, spray drying, or freeze drying ensure effective protein concentration and stabilization while maintaining functionality [33]. Application of hydrolysis on food by-products such as fish viscera, meat trimmings, dairy waste, soybean residues, and brewery spent grains strengthens the circular bioeconomy as it converts waste streams into part of valuable proteins [34]. Raw materials, enzyme specificity, reaction conditions, and post-processing steps all play a crucial role in determining the quality, yield, and functional properties of the recovered proteins. In addition, integrating green extraction techniques such as ultrasound-assisted hydrolysis and subcritical water treatment can improve both enzymatic efficiency and protein recovery rates [35]. These modern techniques can promote mass transfer, which will shorten the time hydrolysis takes, and result in a more sustainable and scalable process. Enzyme recycling, enzyme-stable, and operational cost-effective approaches are being studied for prolonged enzymatic hydrolysis toward the goal of making it more eco-friendly [36]. Continuous enzymatic hydrolysis bioreactors integrated into the process are also promising for scaling the recovery of proteins from food without compromising the efficiency of the process. Another green issue in enzymatic hydrolysis is co-product valorization during protein recovery, for example, lipid fractions and residues rich in minerals that can further be diverted to food, pharmaceutical or bioenergy sectors [37]. Final life cycle assessments and sustainability analyses evaluate the environmental consideration of these protocols against circular economy principles. Green enzymatic hydrolysis, of course, is a sustainable approach to protein recovery from food waste, which is also in agreement with all sustainability aspirations to minimize waste and optimize use of resources as well as develop superior high-value proteins ingredients for diverse uses [38]. As research into the optimization of enzyme cocktails, processing conditions, and downstream processing strategies continues to advance, so too will yield and functionality improvements, leading to even more efficient and environmentally friendly methods of protein recovery in the food industry.

3 Methods

3.1 Pre-Treatment of By-Products

Solid waste drying and grinding to increase surface area, and defatting of oilseed meals and lipid-rich by-products using organic solvents [39] followed by heat treatment or sonication to increase enzyme accessibility.

3.2 Protocol for Hydrolysis by Enzymes

Preparation of enzymes: Dissolve the enzyme in an appropriate buffer, bringing it to the desired concentration, before using it for an experiment.

Hydrolysis conditions: Generally, for a particular enzyme, reactions are performed within an optimal temperature of about 40–60 ° C and pH 6–9 [40].

Enzyme-to-substrate ratio: They can be generally kept from a range of 1:50–1:10 (w/w) depending on the substrate.

Reaction time: For 2–24 h after which sample collection is performed.

Agitation: It should be kept stirring or shaking all the time to maintain uniform conditions in the reaction.

3.3 Post-Processing of Hydrolysis

Inactivation of enzymes: Heat treatment (85 °C for 10 min) or pH adjustment denature enzymes.

Separation of soluble proteins: Centrifugation (10,000 × *g*, 15 min) or membrane filtration.

Protein concentration: Freeze or spray dry hydrolysates for further applications [41].

3.4 Characterization of Hydrolysates

Analysis of Protein Content

Protein content assessment is crucial in hydrolysates characterization, which ensures efficiency of the yield and quality of the product. Two frequently used methods in determining protein concentration include Kjeldahl method and the bicinchoninic acid (BCA) assay.

3.4.1 Kjeldahl Method

This classical technique determines total nitrogen content, which is then converted into protein content using a nitrogen-to-protein conversion factor (commonly 6.25 for food proteins).

Digestion of the sample in concentrated sulfuric acid with catalyst followed by neutralization with strong base and titration for quantification of nitrogen completes the method.

It is very accurate, but needs hazardous chemicals and is also time consuming [42].

3.4.2 The BCA Assay

This colorimetric assay is based on the reduction of Cu^{2+} to Cu^{+} in an alkaline medium in which the Cu^{+} reacts with BCA forming a purple complex that is measurable at 562 nm.

The method is rapid, sensitive, and widely used in biochemical research and industrial applications.

It is more suitable for processed hydrolysates where free amino acids and peptides may interfere with other protein assays [42].

3.5 Degree of Hydrolysis (DH)

Degree of hydrolysis (DH) is a defining factor in assessing the extent of enzymatic hydrolysis. It provides a measure of the amount of peptide bond cleaved in relation to the total bonds available in the protein substrate.

3.5.1 Formol Titration

A classical way to assess free amino groups liberated during hydrolysis by titration against formaldehyde. It is a very simple method but may suffer from side reactions.

3.6 Quantification of Proteins

3.6.1 O-Phthaldialdehyde (OPA) Assay

The O-phthaldialdehyde (OPA) assay is an extremely sensitive spectrophotometric method that interacts with primary amines and forms a fluorescent complex which can be measured at 340 nm. The free amino groups are accurately quantified according to the degree of hydrolysis [43].

3.6.2 Trinitrobenzenesulfonic Acid (TNBS) Assay

Trinitrobenzenesulfonic acid (TNBS) assay is another popular assay; it reacts with free amino groups to form a yellow complex whose absorbance is recorded at 420 nm. It is reliable, but the assay is sensitive to pH changes. The functional characteristics of hydrolysates define their uses within food and industrial formulations. The properties dearest to functional hydrolysates include solubility, emulsification, and foaming capacity [43].

3.7 Solubility

Protein solubility is an important factor concerning the use of solubility in beverages, soups, and emulsions. Generally, hydrolysates, by virtue of relatively small peptide sizes and increased exposure of hydrophilic residues, show a higher degree of solubility than native proteins. Solubility measurements are usually quantified via dispersion of the hydrolysate into various pH conditions, followed by protein content quantification within the supernatant after centrifugation [43].

3.8 Emulsifying Properties

Hydrolysates are important for emulsifying and stabilizing emulsions, for example, dairy emulsions, meat products, and dressings. As previously indicated, emulsifying ability can be defined as blending the hydrolysate with oil and water, with then measuring the stability of the final emulsion. Emulsifying capacity depends on peptide length, charge distribution, and hydrophobicity.

3.9 Foaming Capacity and Stability

Foaming capacity is essential in bakery products and whipped toppings; in this case, foaming is measured using hydrolate solutions, as in aeration under ambient conditions. The stability of foaming can be evaluated by measuring foam collapse over time. Hydrolysis improves foaming by opening functional groups that stabilize air-water interfaces [43].

Protocols for green enzymatic hydrolysis for protein recovery from food by-products are a substitute for environmental-friendly ways to valorize waste. Specific proteolytic enzymes like alcalase, pepsin, papain, and bromelain are then used under optimum conditions aiming for efficient protein extraction while maximizing retention of bioactive properties [43]. Those properties of hydrolysates will depend on the choice of enzymes targeting substrate specificity, desired characteristics of the hydrolysate, and environmental influences like pH, temperature, and enzyme-to-substrate ratio. Enzymatic hydrolysis, on the whole, takes place in an aqueous environment and is often well buffered to eliminate extreme chemical environments and hazardous conditions typical of traditional methods for protein extraction [32]. Low-energy pre-treatments, such as milling, ultrasound-assisted extraction, and high-pressure processing enhance enzyme accessibility to protein substrates, thus promoting yield and efficiency. The ultrasound technology improves the enzyme–substrate interaction in cells by disrupting cell structure, thus reducing the duration of hydrolysis and improving recovery rates of the proteins [44]. High-pressure processing further denatures proteins to increase their susceptibility toward enzymatic degradation. Careful control of hydrolysis time and enzyme dosage allows an application of hydrolysis to obtain peptides of desired molecular weights; these have a direct influence on solubility, emulsifying properties, and antioxidant potential [45]. The post-hydrolysis processing method includes filtration, centrifugation, and membrane separation, such as ultrafiltration, to fractionate and purify protein hydrolysates. The residue removal and concentration of bioactive peptides having specific nutritional and functional properties can also follow this procedure [46]. Green enzymatic hydrolysis upholds sustainability principles in that it minimizes energy consumption and harmful solvents while ensuring maximum recovery of proteins with minimal waste production. Innovative solutions for economical, practical, and effective solutions such as immobilization of enzymes enhance the recycling of the process using more than one hydrolysis cycle [47]. The introduction of enzymatic hydrolysis into biorefinery techniques, such as microbial fermentation or supercritical fluid extraction, will further increase an efficient utilization of resources and reinforce the circular bioeconomy model [5]. New green approaches like deep eutectic solvents (DES) and ionic liquids (IL) are suggested as alternative media for the reactions, promising better stability and activity of enzymes while reducing environmental toxicity. Along with these advances, computational modeling and enzyme engineering approaches allow design of custom-made biocatalysts with enhanced specificity, thermal stability, and catalytic efficiency to further boost enzymatic hydrolysis [48]. Green enzymatic hydrolysis finds extended applicability across several food by-products, such as fish processing wastes, plant-based agricultural residues, dairy industry by-products, and meat processing waste, providing an alternative for upcycling protein-rich biomass into functional ingredients with applications in food, nutraceutical, and pharmaceutical industries [49]. For instance, fish viscera and scales, considered waste materials, can be treated with enzymes to yield bioactive peptides having antihypertensive and antioxidant activity.

Likewise, sustainable conversion of plant protein by-products such as soy and cereal bran into high-value hydrolysates for the enrichment of food formulations is possible. The new trend in the development of incorporating artificial intelligence and machine learning into optimizing hydrolysis parameters is to achieve fine-tuning in kinetics of reactions and maximization of protein recovery [50]. In addition, LCA studies also reinforce that the method of enzymatic hydrolysis is environmentally better than conventional methods of extraction because it has a much lower carbon footprint, energy consumption, and an overall improvement in the sustainability of processes as a whole [51]. On the other hand, limitations such as enzyme cost, process scalability, and hydrolysate stability need further research and development. Furthermore, enhanced economic amnesic and environmental impact of enzymatic hydrolysis for protein recovery can be achieved through advances in recombinant enzyme production, bioinformatics-driven enzyme screening, and integrated hydrolysis systems with real-time monitoring. Green enzymatic hydrolysis protocols are at the heart of what makes the extraction of high-value proteins from food waste sustainable, efficient, and scalable, principles aligned with those of a circular economy and that advance the development of functional protein ingredients with wide industrial applicability [52].

4 Industrial Applications and Commercial Viability

The green enzymatic hydrolysis operating procedures have grabbed much attention in industries for their capability of efficiently converting food wastes into high-value protein hydrolysates, bioactive peptides, and functional ingredients [53]. One of the most well-known industrial applications is the food and beverage industry, where hydrolyzed proteins are included in dietary supplements, sports nutrition products, and functional foods [54]. Hydrolyzed proteins are easily digestible and thus soluble in various mediums and bioactive products, perfect for fortifying infant formula, dairy products, and protein-enriched beverages. In addition, enzymatically hydrolyzed proteins exhibit better emulsifying, foaming, and gelling properties with great worth in food formulation and processing [55]. For example, hydrolysates obtained from fish processing by-products are generally added in protein-rich soups and sauces, while plant-based hydrolysates are used in plant-based meat substitutes for textural and nutritional enhancements.

Another major industrial application is in the nutraceutical and pharmaceutical fields, where bioactive peptides that possess health-promoting properties obtained through enzymatic hydrolysis comprise an important part. Antihypertensive, antioxidant, antimicrobial, and anti-inflammatory peptides have been incorporated into functional foods and dietary supplements to ameliorate conditions concerning cardiovascular diseases, metabolic disorders, and immune system modulation [56]. The increasing demand of consumers for natural and health-promoting ingredients has made enzymatically derived peptides commercially viable, leading to patents for therapeutic peptide preparations. The companies investing in nutraceuticals based on peptides are inclined to optimize the conditions for hydrolysis that will render the peptides stable, bioavailable, and effective toward the physiological effects intended [57].Enzymatic hydrolysis protocols are equally valuable in the cosmetic and personal care area since protein hydrolysates are among the major ingredients used in formulations for skin, hair, and anti-aging purposes. Hydrolysates of collagen and keratin extracted from biodegradable by-products of fish and animals are used in skin moisturizers, hair-strengthening treatments, and anti-wrinkle creams for further skin benefits such as hydration, elasticity, and regeneration [58]. On the contrary, with the increasing demand for sustainable and cruelty-free cosmetic ingredients, plant-based protein hydrolysates such as soy and rice peptides have found much favor for eco-conscious skincare formulations [59]. The emphasis on biodegradability and bioactivity of ingredients coincides with the tenets of green enzymatic hydrolysis in the cosmetic sphere. Thus, the potential high-value feed ingredient for the nutrition and digestibility enhancement of livestock feed is the protein hydrolysate obtained from food by-products [60]. Hydrolysates of fish protein are commonly used in aquaculture feeding to enhance growth performance and immune functionality in farmed fish. Similarly, plant protein hydrolysates incorporated into poultry feed and livestock rations improve absorption efficiency and gut health. Hydrolysation proteins by enzymes have the potential to make animal feeding fulfill the green agenda by lessening waste and supporting a circular economy approach using food industry by-products. Although promising in terms of industrial applications, no commercial viability of such applications would be achievable without cost effectiveness, scalability, and appropriate regulatory approvals for the application of enzymatic hydrolysis protocols [60]. The total cost of production is determined by enzyme costs, processing efficiencies, and purification methods for hydrolysates. Advances in enzyme immobilization techniques, recombinant enzyme production, and optimization of processes have contributed to the decreasing production costs and hence establishing economics of enzymatic hydrolysis. Any regulatory framework that pertains to food-grade enzymes, peptide safety, and labeling requirements will also ensure market acceptance and consumer trust [60]. Companies investing in enzymatic hydrolysis technologies deal with food safety aspects, clinical studies regarding bioactive peptides, and quality control processes that may lay a solid foundation for commercial endeavors involving protein hydrolysates [60].

Emerging trends in the industrial application of enzymatic hydrolysis bring in aspects of bioprocessing like microbial fermentation, high-pressure hydrolysis, and membrane filtration, which can serve to enhance protein recovery and functionality. AI and machine learning for process optimization will ensure that enzymatic reactions are controlled such that hydrolysate quality and yield are consistent. Furthermore, sustainability criteria-life cycle assessment, carbon footprint analysis, and waste valorization strategies will influence the long-term commercial viability of enzymatic hydrolysis protocols [61].

5 Challenges of Green Enzymatic Hydrolysis

5.1 Enzymes Are Costly, and Process Optimization Compounds the Cost

Among the most challenging problems in enzymatic hydrolysis is the high cost of proteolytic enzymes, for instance, pepsin, trypsin, alcalase, and papain, prohibitive in some cases. The costs are attributable to the microbial fermentation and purification steps involved in enzyme production. Besides, the optimization of hydrolysis conditions such as pH, temperature, enzyme-to-substrate ratio, and time of reaction must come through years of trials that add more to operational costs [62].

5.2 Enzyme Stability and Recyclability

Most industrial-scaled hydrolysis processes need enzymes that tolerate processing conditions like high temperature and extreme pH environments. However, many enzymes lose activity by way of denaturation of being activated or conserved through reasonable periods. Such enzymes require mix replenishment with fresh enzymes. Though these payoffs afford some approaches to recover the enzymes, in practice, their use ups the complexity of processes and require further research to make them suitable for scaling up [63].

5.3 Variability in Substrate Composition

Food by-products vary largely in composition due to species differences, processing methods, and seasonal variations. This heterogeneity affects the performance of the enzymes and the yield of protein. Unlike synthetic substrates, natural by-products contain fats, polysaccharides, and secondary metabolites, all of which can hinder enzymatic activity and often demand pre-treatment steps [63].

5.4 Scale-up Challenges and Process Integration

In enzymatic hydrolysis, everything works properly under lab conditions but not under industrial settings. Reason could be mass transfer limitations, diffusion issues with respect to substrate, or construction of large bioreactors leading to decreased enzyme efficiencies. Also, integration of enzymatic hydrolysis into conventional food processing lines (dairy and meat or plant bioprocesses) would involve infrastructural amendments and additional investment [61–63].

5.5 Environmental and Regulatory Constraints

Although enzymatic hydrolysis is a technology that could be classified under "green," it needs strict regulation approvals for application purposes. The use of genetically engineered microbial enzymes raises concerns regarding biosafety. Also, the effluent from the hydrolysis process, particularly when the effluents contain residual organics, needs to be treated as per environmental regulations to avoid pollution [63].

5.6 Recovery and Purification of Hydrolysates

Initially, following enzymatic hydrolysis, protein hydrolysates have to be separated, purified, and dried for application in food, feed, or pharmaceutical industries. The traditional separation methods such as ultrafiltration, chromatography, spray drying, etc., add extra processing steps and extra energy requirement that in turn influence the sustainability of the method [63].

5.7 Functional and Sensory Properties of Hydrolysates

Proteins recovered from enzymatic hydrolysis are often gauged on how much solubility, emulsification, and foaming have changed. Further, hydrolysis generates bitter peptides, restricting such peptides' possible usage in food products. Controlled hydrolysis, enzymatic debittering, and masking agents are in further research work to improve sensory acceptability [63].

6 Future Prospects in Green Enzymatic Hydrolysis Protocols

6.1 Development of Cheap Robust Enzymes

The advances in enzyme engineering including protein engineering and synthetic biology have opened the ways toward the design of much cheaper and much more robust enzymes; microbial strains such as Bacillus, Aspergillus, and Streptomyces are genetically tailored to yield highly productive and stable proteases. Furthermore, advancements in immobilization technique such as nanomaterials and bio-based carriers may increase the possibility of reusability and efficiency of the enzyme [64].

6.2 Novel Pre-Treatment Techniques

Pre-treatment treatments like ultrasound, microwave-assisted hydrolysis, and pulsed electric fields could also enhance the efficiency of enzymatic hydrolysis by disrupting the food matrices, thus increasing enzyme accessibility to its protein substrates. This reduces the processing time and enhances yield while minimizing enzyme consumption [65].

6.3 Circular Economy Integration

Green enzymatic hydrolysis fits the concept of a circular economy since it transforms food waste into high-value products. Hydrolysis can also be coupled with various green technologies, such as microbial fermentation, biorefinery techniques, and co-product recovery (e.g., bioactive peptides, lipids, and minerals), as a means of increasing sustainability and economic viability [5].

6.4 Process Optimization Via Computational Tools

The advent of various computational modeling systems, artificial intelligence (AI), and machine learning offers the possibility of the prediction and optimization of conditions for enzymatic hydrolysis. AI-based models for enzyme–substrate interactions can expedite the selection of optimal enzyme combinations and process parameters, thus minimizing the trial-and-error bench work involved [66].

6.5 Enzyme Production Biodegradable and Sustainable

The formulation of biodegradable enzymes through the use of plant and agro-waste-based materials as stabilizers are currently a hot topic. Furthermore, enzyme production using agricultural residues would offer another means to reduce the cost and environmental impact of hydrolysis processes [16].

6.6 Improved Techniques for Protein Hydrolysate Downstream Processing

The authors are exploring the application of membrane filtration, nanoencapsulation, and electrostatic precipitation, which are all efficient separation techniques, for enhanced protein recovery at a reduced cost. Retaining the bioactivity and functionality of hydrolysates can be enhanced by non-thermal drying methods such as freeze-drying and spray-freeze drying [67].

6.7 New Areas of Application

Research applications of the future focus on extending areas of protein hydrolysate application beyond food and feed. Hydrolysates with bioactive peptides could be incorporated into functional foods, nutraceuticals, cosmetics, and in applications such as wound healing and drug delivery, the biomedical field.

7 Notes

Enzymatic hydrolysis, therefore, could serve as a green and sustainable method of protein recovery from food waste by-products to further embrace principles of the circular economy and lessen environmental dumping. It involves employing some specific enzymes, particularly proteases, in breaking down complex protein structures into bioavailable peptides and amino acids under mild conditions while maintaining their functional and nutritional properties [68]. Various food waste matrixes, including fish waste, residuals from poultry processing, dairy wastes, residues of plant-based agriculture, and brewery spent grain materials, provide abundant sources for protein extraction. The pre-treatment processes include washing, grinding, and defatting in order to make the protein better accessible to enzymatic action [69].

Depending on the type of raw material under analysis, chemical treatments may include methods like lipid or non-protein impurity removal by ethanol extraction and acid hydrolysis, respectively. The enzymatic hydrolysis is typically performed, applying alcalase, papain, pepsin, or trypsin so selected that each one of them displays its highest activity under different sets of pH and temperature conditions [70]. Alcalase, for example, is being most active at a pH of about 8.0 and 50 °C, while pepsin prefers a much more acidic environment (pH 2.0, 37 °C). The enzyme-to-substrate ratio is normally kept at 1–5% (w/w) for 2–6 h; this period influences the degree of hydrolysis (DH%), which is usually checked using the OPA assay [70]. The pH and temperature conditions are further adjusted as the hydrolysis progresses to maintain the activity of the enzyme and enhance hydrolysis efficiency. After hydrolysis, the enzyme is destroyed by heat (80–90 °C for 5–10 min) or increase in pH for ensuring stability of recovered protein fractions [71].

After centrifugation (10,000 rpm for 15 min), protein recovery separates soluble peptides from insoluble residues, followed by membrane filtration to remove unwanted particulates. Depending on the level of purity desired, further fractionation techniques may be added, including isoelectric precipitation at pH 4.5–5.5 and ultrafiltration [63]. To improve storage stability and handling, the recovered protein extract is processed into a dry form by freeze drying or spray drying methods. It is also important to characterize the extracted protein samples to assess their quality and functionality by using analytical tools such as Kjeldahl or Bradford assays for protein quantification, SDS-PAGE or high-performance liquid chromatography (HPLC) for molecular weight distribution, and LC-MS for amino acid profiling [72]. Effectiveness in its potential application can also be assessed through bioactivity test methods such as antioxidant and antimicrobial assays (DPPH, ABTS). These protein hydrolysates exhibit enhanced solubility, emulsification, foaming, and gelling properties, enhancing their use in food formulations, functional foods, nutraceuticals, animal feed, and even for pharmaceutical applications [27, 73, 74]. The food industry employs these recovered proteins as emulsifiers, stabilizers, and nutritional supplements, while the pharmaceutical industry investigates their bioactive peptides for antimicrobial and antihypertensive functions. In addition, biodegradable protein-based films produced from enzymatic hydrolysis provide sustainable packaging alternatives [73, 75, 76]. Nevertheless, despite the aforementioned benefits, enzymatic hydrolysis encounters difficulties regarding enzyme costs, reaction scaling, and optimization of the process. For industrial-scale operations, enzyme efficiency can be improved by the use of immobilization techniques, cofactor supplementation, and enzyme recycling, which benefit process economics and feasibility [41, 77]. Enzyme hydrolysis can also be integrated with other biorefinery methods, such as microbial fermentation and membrane-based separations, to achieve maximum protein yield and valorization of secondary metabolites [78, 79]. Regulatory considerations regarding food-grade enzymes, safety assessments, as well as consumer acceptance are also factors significantly affecting the commercialization of enzymatically recovered proteins [80]. The green future for enzymatic hydrolysis lies in improving enzyme engineering, fully automating the process, and designing new biocatalysts directed at specific food by-products. With the ongoing shift of industries toward sustainable resource utilization, enzymatic hydrolysis can be a promising approach to minimize any food waste while generating high-value protein ingredients for various applications [81]. This green biotechnological approach not only promotes environmental sustainability but also impacts food security, economy, and the manufacture of functional protein-based products for global markets.

References

1. A Global Food Crisis | World Food Programme. [cited 2025 Mar 6]. Available from: https://www.wfp.org/global-hunger-crisis
2. Molina RA, Catan I (2021) Solid waste management awareness and practices among senior high school students in a state College in Zamboanga City, Philippines. Philipp Aquademia 2021(1):21001. https://doi.org/10.21601/aquademia/9579
3. Rahman T, Deb N, Alam MZ, Moniruzzaman M, Miah MS, Horaira MA et al (2024) Navigating the contemporary landscape of food waste management in developing countries: a comprehensive overview and prospective analysis. Heliyon 10(12):e33218

4. Franca-Oliveira G, Fornari T, Hernández-Ledesma B (2021) A review on the extraction and processing of natural source-derived proteins through eco-innovative approaches. Processes 9(9):1626. Available from: https://www.mdpi.com/2227-9717/9/9/1626/htm
5. Pal P, Singh AK, Srivastava RK, Rathore SS, Sahoo UK, Subudhi S et al (2024) Circular bioeconomy in action: transforming food wastes into renewable food resources. Food Secur 13:3007. Available from: https://www.mdpi.com/2304-8158/13/18/3007/htm
6. Siddiqui SA, Erol Z, Rugji J, Taşçı F, Kahraman HA, Toppi V et al (2023) An overview of fermentation in the food industry—looking back from a new perspective. Bioresour Bioprocess 10(1):85. Available from: https://pmc.ncbi.nlm.nih.gov/articles/PMC10991178/
7. Lee YY, Srinivaas M, Li IC, Keharika K, Pothu R, Boddula R et al (2024) Enhancing sustainable energy through cutting-edge waste biorefinery technologies. Reactions 5:1101–1147. Available from: https://www.mdpi.com/2624-781X/5/4/59/htm
8. Zarina R, Mezule L (2024) Enzymatic hydrolysis of waste streams originating from wastewater treatment plants. Biotechnol Biofuels Bioprod 17(1):1–11. https://doi.org/10.1186/s13068-024-02553-x
9. Safdar LB, Foulkes MJ, Kleiner FH, Searle IR, Bhosale RA, Fisk ID et al (2023) Challenges facing sustainable protein production: opportunities for cereals. Plant Commun 4(6):100716
10. Ishangulyyev R, Kim S, Lee SH (2019) Understanding food loss and waste—why are we losing and wasting food? Food Secur 8(8):297. Available from: https://pmc.ncbi.nlm.nih.gov/articles/PMC6723314/
11. Araujo J, Sica P, Costa C, Márquez MC (2021) Enzymatic hydrolysis of fish waste as an alternative to produce high value-added products. Waste Biomass Valor 12(2):847–855. https://doi.org/10.1007/s12649-020-01029-x
12. Mirzapour-Kouhdasht A, McClements DJ, Taghizadeh MS, Niazi A, Garcia-Vaquero M (2023) Strategies for oral delivery of bioactive peptides with focus on debittering and masking. NPJ Sci Food 7(1):1–20. Available from: https://www.nature.com/articles/s41538-023-00198-y
13. Alahmad K, Noman A, Xia W, Jiang Q, Xu Y (2023) Influence of the enzymatic hydrolysis using flavourzyme enzyme on functional, secondary structure, and antioxidant characteristics of protein hydrolysates produced from bighead carp (Hypophthalmichthys nobilis). Molecules 28(2):519. Available from: https://www.mdpi.com/1420-3049/28/2/519/htm
14. Barampouti EM, Mai S, Moustakas K, Malamis D, Loizidou M (2021) Status and perspectives of agricultural residues in a circular and resource-efficient context. Clean Energy Resour Recover Biomass Waste Based Biorefineries 1:49–102
15. Boboua SYB, Wen Q, Zhang L, Chen Y, Yu J, Chen P et al (2024) Valorization of animal waste proteins for agricultural, food production, and medicinal applications. Front Sustain Food Syst 8:1366333
16. Irazoqui JM, Santiago GM, Mainez ME, Amadio AF, Eberhardt MF (2024) Enzymes for production of whey protein hydrolysates and other value-added products. Appl Microbiol Biotechnol 108(1):354. Available from: https://pmc.ncbi.nlm.nih.gov/articles/PMC11142983/
17. Zhang J, Ahmmed MK, Regenstein JM, Wu H (2024) Recent advances of recycling proteins from seafood by-products: industrial applications, challenges, and breakthroughs. Trends Food Sci Technol 149:104533
18. Kamal H, Le CF, Salter AM, Ali A (2021) Extraction of protein from food waste: an overview of current status and opportunities. Compr Rev Food Sci Food Saf 20(3):2455–2475. https://doi.org/10.1111/1541-4337.12739
19. Liu N, Zhang X, Guo Q, Wu T, Wang Y (2022) 3D bioprinted scaffolds for tissue repair and regeneration. Front Mater 9:925321. Available from: www.frontiersin.org
20. Bhardwaj N, Kumar B, Verma P (2019) A detailed overview of xylanases: an emerging biomolecule for current and future prospective. Bioresour Bioprocess 6(1):1–36. https://doi.org/10.1186/s40643-019-0276-2
21. Osman AI, Chen Z, Elgarahy AM, Farghali M, Mohamed IMA, Priya AK et al (2024) Membrane technology for energy saving: principles, techniques, applications, challenges, and prospects. Adv Energy Sustain Res 5(5):2400011. https://doi.org/10.1002/aesr.202400011
22. Williams SKR, Caldwell KD (2012) Field-flow fractionation in biopolymer analysis. Springer Science & Business Media, pp 1–305
23. Weingart P, Joubert M, Connoway K (2021) Public engagement with science—origins, motives and impact in academic literature and science policy. PLoS One 16(7):e0254201
24. Sarangi PK, Singh AK, Sonkar S, Shadangi KP, Srivastava RK, Gupta VK et al (2023) Biorefinery solutions for food processing wastes: a sustainable bioeconomic perspective. Ind Crop Prod 205:117488

25. Patil U, Nilsuwan K, Benjakul S (2024) Functional ingredients from seafood processing wastes: protein hydrolysate and biocalcium. Turk J Fish Aquat Sci 24(4):25347. https://doi.org/10.4194/TRJFAS25347
26. Yang H, Zhang Q, Zhang B, Zhao Y, Wang N (2023) Potential active marine peptides as anti-aging drugs or drug candidates. Mar Drugs 21(3):144. Available from: https://pmc.ncbi.nlm.nih.gov/articles/PMC10053682/
27. Al Hajj W, Salla M, Krayem M, Khaled S, Hassan HF, El Khatib S (2024) Hydrolyzed collagen: exploring its applications in the food and beverage industries and assessing its impact on human health—a comprehensive review. Heliyon 10(16):e36433
28. Mujtaba M, Fernandes Fraceto L, Fazeli M, Mukherjee S, Savassa SM, Araujo de Medeiros G et al (2023) Lignocellulosic biomass from agricultural waste to the circular economy: a review with focus on biofuels, biocomposites and bioplastics. J Clean Prod 402:136815
29. Tacias-Pascacio VG, Castañeda-Valbuena D, Morellon-Sterling R, Tavano O, Berenguer-Murcia Á, Vela-Gutiérrez G et al (2021) Bioactive peptides from fisheries residues: a review of use of papain in proteolysis reactions. Int J Biol Macromol 184:415–428
30. David Troncoso F, Alberto Sánchez D, Luján FM (2022) Production of plant proteases and new biotechnological applications: an updated review. Chem Open 11(3):e202200017. https://doi.org/10.1002/open.202200017
31. Venetikidou M, Lykartsi E, Adamantidi T, Prokopiou V, Ofrydopoulou A, Letsiou S et al (2025) Proteolytic enzyme activities of bromelain, ficin, and papain from fruit by-products and potential applications in sustainable and functional cosmetics for skincare. Appl Sci 15(5):2637. Available from: https://www.mdpi.com/2076-3417/15/5/2637/htm
32. Cruz-Casas DE, Aguilar CN, Ascacio-Valdés JA, Rodríguez-Herrera R, Chávez-González ML, Flores-Gallegos AC (2021) Enzymatic hydrolysis and microbial fermentation: the most favorable biotechnological methods for the release of bioactive peptides. Food Chem Mol Sci 3:100047. Available from: https://pmc.ncbi.nlm.nih.gov/articles/PMC8991988/
33. Krakowska-Sieprawska A, Kiełbasa A, Rafińska K, Ligor M, Buszewski B (2022) Modern methods of pre-treatment of plant material for the extraction of bioactive compounds. Molecules 27(3):730. Available from: https://pmc.ncbi.nlm.nih.gov/articles/PMC8840492/
34. Ribeiro TB, Voss GB, Coelho MC, Pintado ME (2022) Food waste and by-product valorization as an integrated approach with zero waste: future challenges. Future Foods:569–596. https://doi.org/10.1016/b978-0-323-91001-9.00017-7
35. Bello U, Adamu H, Amran NA, Qamar M (2025) Green extraction technologies: process systems, techno-economic and lifecycle analyses. Clean Eng Technol:100881. Available from: https://linkinghub.elsevier.com/retrieve/pii/S2666790825000047
36. Abolore RS, Jaiswal S, Jaiswal AK (2024) Green and sustainable pretreatment methods for cellulose extraction from lignocellulosic biomass and its applications: a review. Carbohydr Polym Technol Appl 7:100396
37. Li SY, Ng IS, Chen PT, Chiang CJ, Chao YP (2018) Biorefining of protein waste for production of sustainable fuels and chemicals. Biotechnol Biofuels 11(1):1–15. https://doi.org/10.1186/s13068-018-1234-5
38. Chen TL, Kim H, Pan SY, Tseng PC, Lin YP, Chiang PC (2020) Implementation of green chemistry principles in circular economy system towards sustainable development goals: challenges and perspectives. Sci Total Environ 716:136998
39. Gil-Martín E, Forbes-Hernández T, Romero A, Cianciosi D, Giampieri F, Battino M (2022) Influence of the extraction method on the recovery of bioactive phenolic compounds from food industry by-products. Food Chem 378:131918
40. Amiri M, Hosseini SE, Asadi G, Khayambashi B (2024) Optimization of the alcalase and trypsin hydrolysis conditions of an isolated protein from Scenedesmus obliquus microalgae and characterization of its functional properties. LWT 210:116819. https://doi.org/10.1016/j.lwt.2024.116819
41. Ansorge-Schumacher MB (2024) Enzyme immobilization. In: Introduction to enzyme technology. Springer International Publishing, Cham, pp 219–241
42. Hayes M (2020) Measuring protein content in food: an overview of methods. Food Secur 9(10):1340. Available from: https://pmc.ncbi.nlm.nih.gov/articles/PMC7597951/
43. Mietsch F, Fehér J, Halász A (1989) Investigation of functional properties of partially hydrolyzed proteins. Food Nahrung 33(1):9–15
44. Ahmed T, Suzauddula M, Akter K, Hossen M, Islam MN (2024) Green technology for fungal protein extraction—a review. Separations 11(6):186. Available from: https://www.mdpi.com/2297-8739/11/6/186/htm

45. Fadimu GJ, Le TT, Gill H, Farahnaky A, Olatunde OO, Truong T (2022) Enhancing the biological activities of food protein-derived peptides using non-thermal technologies: a review. Food Secur 11(13):1823. Available from: https://pmc.ncbi.nlm.nih.gov/articles/PMC9265340/
46. Chen GQ, Qu Y, Gras SL, Kentish SE (2023) Separation technologies for whey protein fractionation. Food Eng Rev 15(3):438–465. https://doi.org/10.1007/s12393-022-09330-2
47. Osman AI, Fang B, Zhang Y, Liu Y, Yu J, Farghali M et al (2024) Life cycle assessment and techno-economic analysis of sustainable bioenergy production: a review. Environ Chem Lett 22(3):1115–1154. https://doi.org/10.1007/s10311-023-01694-z
48. Xu P, Zheng GW, Zong MH, Li N, Lou WY (2017) Recent progress on deep eutectic solvents in biocatalysis. Bioresour Bioprocess 4(1):1–18. https://doi.org/10.1186/s40643-017-0165-5
49. Ghinea C, Ungureanu-Comăniță ED, Țâbuleac RM, Oprea PS, Coșbuc ED, Gavrilescu M (2025) Cost-benefit analysis of enzymatic hydrolysis alternatives for food waste management. Food Secur 14(3):488. Available from: https://www.mdpi.com/2304-8158/14/3/488/htm
50. Bamidele MO, Bamikale MB, Cárdenas-Hernández E, Bamidele MA, Castillo-Olvera-G, Sandoval-Cortes J et al (2025) Bioengineering in solid-state fermentation for next sustainable food bioprocessing. Next Sustain 6:100105. https://doi.org/10.1016/j.nxsust.2025.100105
51. Alotaibi E, Arab MG, Abdallah M, Nassif N, Omar M (2022) Life cycle assessment of biocemented sands using enzyme induced carbonate precipitation (EICP) for soil stabilization applications. Sci Rep 12(1):6032. Available from: https://pmc.ncbi.nlm.nih.gov/articles/PMC9001663/
52. Aït-Kaddour A, Hassoun A, Tarchi I, Loudiyi M, Boukria O, Cahyana Y et al (2024) Transforming plant-based waste and by-products into valuable products using various "food industry 4.0" enabling technologies: a literature review. Sci Total Environ 955:176872
53. Yan F, Wang Q, Teng J, Wu F, He Z (2023) Preparation process optimization and evaluation of bioactive peptides from Carya cathayensis Sarg meal. Curr Res Food Sci 6:100408
54. Makhaye G, Mofokeng MM, Tesfay S, Aremu AO, Van Staden J, Amoo SO (2021) Influence of plant biostimulant application on seed germination. In: Biostimulants crop from seed germination to plant dev a pract approach, pp 109–135
55. Lükewille U, Uhlig HH (2007) Dairy products, probiotics and the health of infants and children. In: Functional dairy products, vol 2, 2nd edn. Woodhead Publishing, pp 46–62
56. Akbarian M, Khani A, Eghbalpour S, Uversky VN (2022) Bioactive peptides: synthesis, sources, applications, and proposed mechanisms of action. Int J Mol Sci 23(3):1445. Available from: https://pmc.ncbi.nlm.nih.gov/articles/PMC8836030/
57. Peighambardoust SH, Karami Z, Pateiro M, Lorenzo JM (2021) A review on health-promoting, biological, and functional aspects of bioactive peptides in food applications. Biomol Ther 11(5):631. Available from: https://pmc.ncbi.nlm.nih.gov/articles/PMC8145060/
58. Sheykhhasan M, Ahmadieh-Yazdi A, Heidari R, Chamanara M, Akbari M, Poondla N et al (2025) Revolutionizing cancer treatment: the power of dendritic cell-based vaccines in immunotherapy. Biomed Pharmacother 184:117858
59. Yang F, Hu Y, Wu M, Guo M, Wang H (2025) Biologically active components and skincare benefits of rice fermentation products: a review. Cosmet 12(1):29. Available from: https://www.mdpi.com/2079-9284/12/1/29/htm
60. Hou Y, Wu Z, Dai Z, Wang G, Wu G (2017) Protein hydrolysates in animal nutrition: industrial production, bioactive peptides, and functional significance. J Anim Sci Biotechnol 8(1):1–13. https://doi.org/10.1186/s40104-017-0153-9
61. Wang Q, Qi Z, Fu W, Pan M, Ren X, Zhang X et al (2024) Research and prospects of enzymatic hydrolysis and microbial fermentation technologies in protein raw materials for aquatic feed. Ferment 10(12):648. Available from: https://www.mdpi.com/2311-5637/10/12/648/htm
62. Jönsson LJ, Alriksson B, Nilvebrant NO (2013) Bioconversion of lignocellulose: inhibitors and detoxification. Biotechnol Biofuels 6(1):1–10. https://doi.org/10.1186/1754-6834-6-16
63. Vijayaraghavan P, Raj SRF, Vincent SGP (2016) Industrial enzymes: recovery and purification challenges. In: Agro-industrial wastes as feed enzym prod apply exploit emerg valuab use options waste biomass. Academic Press, pp 95–110
64. Ndochinwa OG, Wang QY, Amadi OC, Nwagu TN, Nnamchi CI, Okeke ES et al

(2024) Current status and emerging frontiers in enzyme engineering: an industrial perspective. Heliyon 10(11):e32673

65. Jain S, Kumar S (2024) Advances and challenges in pretreatment technologies for bioethanol production: a comprehensive review. Sustain Chem Clim Action 5:100053
66. Guarra F, Colombo G (2023) Computational methods in immunology and vaccinology: design and development of antibodies and immunogens. J Chem Theory Comput 19(16):5315–5333. https://doi.org/10.1021/acs.jctc.3c00513
67. Abduh MY, Prawitasari A, Fitrian UA, Firmansyah M (2023) Effects of enzymatic hydrolysis on the antioxidant activity of protein hydrolysate derived from the larvae of black soldier fly (Hermetia illucens L.). J Appl Biol Biotechnol 11(2):151–157. Available from: http://creativecommons.org/licenses/by-nc-sa/3.0/
68. Sapatinha M, Camacho C, Pais-Costa AJ, Fernando AL, Marques A, Pires C (2025) Enzymatic hydrolysis systems enhance the efficiency and biological properties of hydrolysates from frozen fish processing co-products. Mar Drugs 23(1):14. Available from: https://www.mdpi.com/1660-3397/23/1/14/htm
69. Liu Z, de Souza TSP, Holland B, Dunshea F, Barrow C, Suleria HAR (2023) Valorization of food waste to produce value-added products based on its bioactive compounds. Processes 11(3):840. Available from: https://www.mdpi.com/2227-9717/11/3/840/htm
70. Tang J, Yao D, Xia S, Cheong L, Tu M (2024) Recent progress in plant-based proteins: from extraction and modification methods to applications in the food industry. Food Chem X 23:101540
71. Vogelsang-O'dwyer M, Sahin AW, Arendt EK, Zannini E (2022) Enzymatic hydrolysis of pulse proteins as a tool to improve technofunctional properties. Food Secur 11(9):1307. Available from: https://pmc.ncbi.nlm.nih.gov/articles/PMC9104109/
72. Cayot P, Tainturier G (1997) The quantification of protein amino groups by the trinitrobenzenesulfonic acid method: a reexamination. Anal Biochem 249(2):184–200
73. Wang L, Wang N, Zhang W, Cheng X, Yan Z, Shao G et al (2022) Therapeutic peptides: current applications and future directions. Signal Transduct Target Ther 7(1):1–27. https://doi.org/10.1038/s41392-022-00904-4
74. Lu P, Ruan D, Huang M, Tian M, Zhu K, Gan Z et al (2024) Harnessing the potential of hydrogels for advanced therapeutic applications: current achievements and future directions. Signal Transduct Target Ther 9(1):166. https://doi.org/10.1038/s41392-024-01852-x
75. Gokhale AS, Satyanarayanajois S (2014) Peptides and peptidomimetics as immunomodulators. Immunotherapy 6(6):755. Available from: https://pmc.ncbi.nlm.nih.gov/articles/PMC4201125/
76. Wu YH, Samuel C, Chen Y, Wu YHS, Chen YC (2022) Trends and applications of food protein-origin hydrolysates and bioactive peptides. J Food Drug Anal 30(2):172. Available from: https://pmc.ncbi.nlm.nih.gov/articles/PMC9635901/
77. Maghraby YR, El-Shabasy RM, Ibrahim AH, Azzazy HMES (2023) Enzyme immobilization technologies and industrial applications. ACS Omega 8(6):5184–5196. https://doi.org/10.1021/acsomega.2c07560
78. Roy A, Khan A, Ahmad I, Alghamdi S, Rajab BS, Babalghith AO et al (2022) Flavonoids a bioactive compound from medicinal plants and its therapeutic applications. Biomed Res Int 2022(1):5445291
79. Gurib-Fakim A (2006) Medicinal plants: traditions of yesterday and drugs of tomorrow. Mol Asp Med 27(1):1–93
80. Sutay Kocabaş D, Grumet R (2019) Evolving regulatory policies regarding food enzymes produced by recombinant microorganisms. GM Crops Food 10(4):191. Available from: https://pmc.ncbi.nlm.nih.gov/articles/PMC6927691/
81. Phintha A, Chaiyen P (2022) Rational and mechanistic approaches for improving biocatalyst performance. Chem Catal 2(10):2614–2643

Chapter 4

Advanced Fermentation Protocols for Sustainable Bioconversion of Food Waste

Nizwa Itrat, Beenish Israr, Nazir Ahmad, Seinul Abdeen Fathima Hasanath, Faiza Ameen, and Fawad Ali

Abstract

Bioconversion of food waste into valuable products using advanced fermentation techniques is an innovative approach to sustainable waste management and resource recovery. This chapter explores three notable fermentation techniques: solid-state fermentation (SSF), submerged fermentation (SmF), and enzyme-fermentation processes (EFP), and each has particular advantages and challenges. SSF is a required minimal water and high yield fermentation method with scalability challenge. SmF provides better control of fermentation conditions along with scalability, but requires more water and energy. EFP enhances efficiency by integrating enzymatic hydrolysis with fermentation, but requires specific conditions to be effective. This chapter analyzed the environmental impact, economic feasibility, scalability, and quality of the products they generate. The findings highlighted the potential of these advanced fermentation techniques to support a circular economy by transforming food waste into valuable products. However, to make these methods viable for industrial applications, challenges like process control, contamination risks, and high initial investment costs must be addressed.

Key words Food waste, Sustainable bioconversion, Advanced fermentation techniques, Microorganisms, Waste management, Resource recovery, Value-added products, Economy viability

1 Introduction

1.1 Introduction to Food Waste and its Environmental Impact

A large amount of food produced globally goes to waste. Approximately 25% of all food worldwide is discarded [1]. As shown in Fig. 1, food waste can be classified into a few types: organic crop residues, animal by-products, mixed food waste, and uneatable food waste. However, each type of waste presents unique opportunities for the recovery of bioactive compounds, enzymes, proteins, fibers, and other value-added materials. Food waste disposal is

The original version of the chapter has been revised. A correction to this chapter can be found at https://doi.org/10.1007/978-1-0716-4646-5_17

Tanmay Sarkar (ed.), *Food Waste Valorization: Green Techniques in Sustainable Management*, Methods and Protocols in Food Science, https://doi.org/10.1007/978-1-0716-4646-5_4,
© The Author(s), under exclusive license to Springer Science+Business Media, LLC, part of Springer Nature 2025, Corrected Publication 2025

Fig. 1 Classification of food waste

associated with many problems, such as rotten odor from sulfur and nitrogen compounds, non-uniform and biodegradable structure, and high-moisture content, the release of harmful gases such as methane (CH_4) and carbon dioxide (CO_2), the production of toxic liquid wastes, the requirement of large disposal areas, and the high costs of collection and transportation of waste [2].

Food waste is an environmental and economic problem. It was approaching 1.3 billion tons of food wasted globally each year. This waste not only results in the loss of valuable resources, but rather contributes to gas emissions that lead to energy losses and environmental degradation. Advancing the conversion of food waste with sustainable methods is important for ensuring economic circulation and reducing its environmental impact [3].

Management of food waste has become a global problem, leading to increased research into innovative and sustainable solutions. One promising method is advanced fermentation, which has become a promising approach for the bioconversion of food waste into useful products. This chapter reviews the different fermentation techniques, how they work, their applications, advantages, and challenges.

1.2 The Science of Fermentation: An Overview

Fermentation is a biological process that has been used for centuries in the production of a broad range of foods and beverages. This indicates the process by which microorganisms like bacteria, yeasts, and fungi can significantly modify the nutritional value and sensory properties of foods through their metabolic activities. Ultimately they are producing new compounds with higher value and functionality [4]. Figure 2 provides an overview of how food wastes can be converted into value-added products using fermentation techniques.

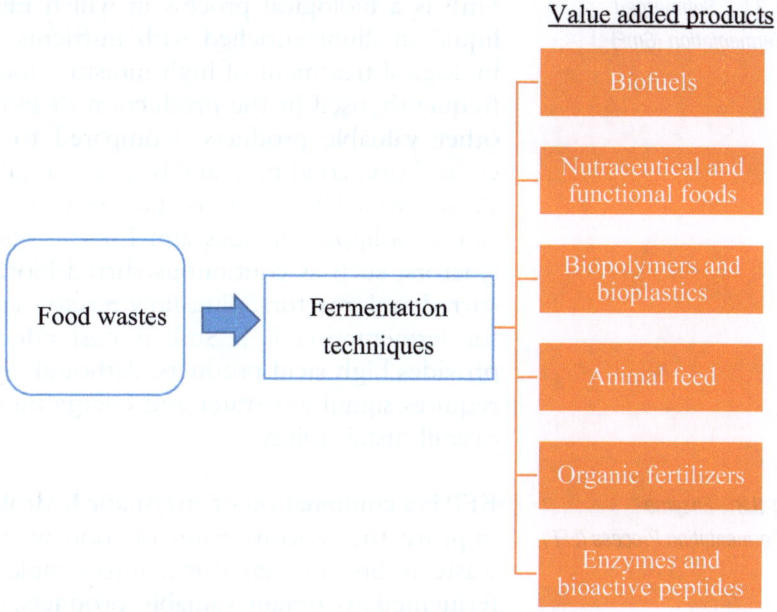

Fig. 2 Bioconversion of food waste into value added products

Bioconversion is a very effective method for waste reduction. To optimize the amount of food waste, advanced fermentation techniques have been developed, such as solid-state fermentation (SSF), submerged fermentation (SmF), and enzymatic fermentation processes (EFP).

1.3 Types of Fermentation Processes

1.3.1 Solid-State Fermentation (SSF)

SSF helps the microbial growth on solid materials without the availability of free-water. This method is especially suitable for treatment of food waste that has low moisture content. SSF is used in the enzymes production, organic acids and their derivatives from organic-based residues, such as food and agro-industrial by-products. For example, *Trichoderma asperellum* R has been used to convert fruit and vegetable wastes into valuable chemicals under abiotic conditions [5]. SSF technology usually has four setups: tray bioreactors, fluidized-bed bioreactors, rotating drum bioreactors, and packed-bed bioreactors. The selection of bioreactor depends on factors like the type of microorganism, available nutrients and substrate, desired end product, and resource availability. This process has advantages; the solid nature of the substrate and its less water requirements help reduce the risk of contamination. Then making it an environmentally friendly option. Also, it gives high product yields [6]. However, process control and scalability are needed to be focused on large-scale industrial applications.

1.3.2 Submerged Fermentation (SmF)

SmF is a biological process in which microorganisms grow in a liquid medium enriched with nutrients. This is suitable for the biological treatment of high-moisture food waste. This method is frequently used in the production of biofuels, organic acids, and other valuable products. Compared to SSF, SmF offers better control over conditions and is easier to scale. For example, modified *Pichia pastoris* has been used to convert restaurant food waste into isomaltooligosaccharides and L-lactic acid [3]. SmF uses various reactors, such as continuous stirred-bioreactors, continuous flow stirred tank reactors, plug-flow reactor, and fluidized-bed reactors for fermentation [7]. SmF is cost effective, easily scalable and provides high yield products. Although SmF has its advantages, it requires significant water and energy inputs, which may affect its overall sustainability.

1.3.3 Enzyme-Fermentation Process (EFP)

EFP is a combination of enzymatic hydrolysis and fermentation, to improve the bioconversion of food waste. In this process, food waste is first broken down into simpler components and then fermented to obtain valuable products. EFP is used to convert restaurant food waste into nutrient-rich animal feed containing isomaltooligosaccharides and L-lactic acid [3]. This approach improves the efficiency of combining enzymatic and fermentation steps, but requires tightly controlled commercial conditions and significant initial investment.

1.4 Microorganisms Involved in Fermentation

Microorganisms play a vital role in breaking down complex organic materials into simpler substances through enzymatic activity, making them essential in both solid-state fermentation and submerged fermentation. Microorganisms like filamentous fungi, yeast, and bacteria are commonly involved in fermentation processes. Fungi are particularly valuable for enzyme production and solid substrate degradation. Yeast strains contribute by producing detoxifiers, alcohol, biocontrol agents, and aroma compounds from agro-industrial waste, while bacteria help decompose substrates to generate enzymes and extract bioactive compounds. SSF also supports the growth of beneficial microbes like *Aspergillus*, *Rhizopus*, *Lactobacillus*, and *Bacillus*, which are widely used in fermented food production. Similarly, lactic acid bacteria (LAB) are widely used in the food industry for the SmF of food and beverages, as well as synthesizing exopolysaccharides (EPS), flavors, and bioactive compounds [8].

1.5 Benefits of Fermentation in Food Waste Management

1.5.1 Environmental and Economic Benefits

Advanced fermentation techniques offer significant environmental and economic benefits through the transformation of food waste. These methods support resource conservation and waste reduction, and greenhouse gas and landfill reduction, by transforming food waste into valuable products, which align with the principles of circular economy. Moreover, the ability to produce high-value products can be an economic incentive for the implementation of these technologies. However, to maximize their potential, challenges such as process optimization, contamination risks, and initial investment costs must be effectively managed [8].

1.5.2 Scalability and Industrial Application

EFF and SmF are more suitable for large-scale industrial applications than SSF, as they can be more easily integrated into existing production systems. However, due to difficulties in controlling process parameters and a higher risk of contamination, SSF faces scalability challenges [3].

1.5.3 Product Quality and Yield

Advanced fermentation technology ensures high quality and high yield of products. SSF can effectively produce high concentrations of bioactive products, while EFP can simultaneously produce excess valuable products. However, the diversity of food waste ingredients determines the quality and yield of the products, including the raw material standards required for process costs [8].

2 Materials

2.1 Selecting Suitable Food Waste for Fermentation

Selecting suitable substrates are crucial for recovery of specific value-added products. Generally, microorganisms require a substrate as a solid support and nutrient source. Plant-based and animal-based subtracts are mainly focused.

Agricultural wastes: straw, hulls, and other cellulose-rich materials.

By-products of the food industry: bagasse, hulls, corn cobs, molasses, seeds, pulp, pressed seed residue, fruits and vegetables, and tuber peels.

Synthetic materials: polyurethane foam, polystyrene, and carboxymethylcellulose.

Animal wastes: skin, feathers, scales, bones, fins, viscera.

Agro-industrial wastes are generally used because they are very cheap, environmentally friendly, and readily available. The wastes consist of cellulose, hemicellulose, lignin, starch, pectin, and proteins, which are nutritional sources of carbon and energy that support microbial growth. Similarly, animal wastes contain proteins and fat-rich animal refuse [9]. In fermentation, choosing the appropriate substrate is crucial as it directly impacts productivity, yield, quality, and efficiency. Moreover, the substrate type used can affect the behavior of microorganisms [8].

3 Methods

3.1 Pretreatment Methods for Food Waste

Pretreatment is a treatment method commonly used to increase the decomposition rate of food waste and improve the efficiency of the fermentation process. These pretreatment methods can yield higher by easily converting organic matter into biodegradable feedstocks [10]. Apart from improvements, it has disadvantages, such as higher costs, higher energy needs, and potential environmental impacts. In general, pretreatment technologies are used for different purposes. These include thermal treatment, physical or mechanical treatment, chemical treatment, biological treatment, and composite treatment [11]. Figure 3 presents the right pretreatment method depending on waste composition, cost effectiveness, necessary enzymes, and the microorganisms involved. Additionally, factors like resistance issues (toxic compound release and partial degradation), technical requirements, and the formation of harmful by-products should be considered [12].

3.1.1 Thermal Treatment

This pretreatment method is usually used for deformation of the food waste solid structure. In fact, the thermal treatment process of food waste was initially used to facilitate digestion as a conditioning process. In this process, the most important factor that increases the dissolution rate of food waste is temperature. It is usually generated by the heater, autoclave, or hydrothermal methods.

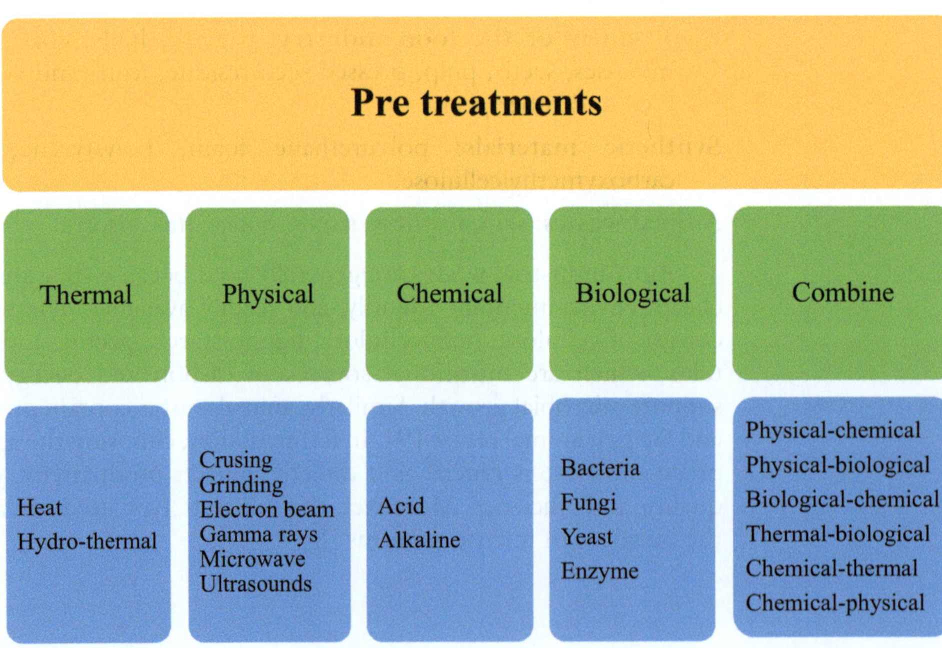

Fig. 3 Summary of pretreatment methods

3.1.2 Physical Treatment

Among the various pretreatment methods, the most common is the physical pretreatment method, which is used as the primary treatment method. This method improves the food waste properties by reducing the particle size and increasing the surface area, thus increasing the digestibility of the raw material. Physical pretreatment is also important in facilitating the transfer and feeding of food waste to the bioreactor. This includes grinding/mixing/crushing /chopping/milling, high-pressure homogenization, screw pressing, lysis centrifugation, and high-pressure and high-temperature steam explosion used to enhance the physical properties of food waste. Physical pretreatments such as grinding/chopping, which reduce particle size, are the most commonly used pretreatment steps to increase the hydrolysis of food waste. Other physical pretreatment of food waste such as microwaves, irradiation with gamma rays, electron beams, and ultrasonic irradiation requires high initial investment costs, requires more energy, and can produce a large amount of unwanted product.

3.1.3 Chemical Treatment

Chemical treatment aims to change the biological and physiochemical characteristics of the substrate through chemical reactions. There are two types. (1) alkaline and (2) acid. Alkaline pretreatment with sodium hydroxide, potassium hydroxide, calcium hydroxide, and liquid ammonia induces food waste swelling, which increases the internal surface area of raw materials and reduces the surface area of crystalline cellulose. Acid pretreatment uses hydrochloric acid, hydrogen peroxide, sulfuric acid, and acetic acid to hydrolyze hemicellulose and other biopolymers, making them more susceptible to microbial or enzymatic degradation. Acid pretreatment can also be used to solubilize complex biopolymers such as lignocellulosic biomass, thereby increasing the hydrolysis rate and improving enzyme accessibility. The results suggest that chemical pretreatment can enhance the degradation of complex biopolymers, thereby achieving more efficient COD removal and electricity generation. Chemical pretreatment systems require ongoing operating costs because they require continuous feeding. Although chemical pretreatment can increase the decomposition rate of food waste, it requires additional equipment due to potential corrosion caused by the use of chemicals.

3.1.4 Biological Treatment

Biological pretreatment does not require the use of chemical or toxic substances or solvents. However, these are usually time-consuming processes that require long hydraulic retention times. The biological pretreatment method can be divided into organic fermentation, fermentation, chemical fermentation, and anaerobic fermentation. The fermentation using bacteria and fungi to achieve the desired fermentation quality is time-consuming, targeted, and efficient. Although biological pretreatment has high capital costs, it is generally cheap, environmentally friendly, and energy-saving.

According to the literature, this technology involves enzymes, such as protease, amylase, viscozyme, flavourzyme, *Saccharomyces cerevisiae* KA4, and palatase. Also microorganisms such as *Aspergillus awamori, Aspergillus oryzae,* and *Monascus* are involved. The main purpose of the enzyme is to convert carbohydrates and proteins into their monomers glucose and amino acids, respectively [12].

3.1.5 Combined Pretreatment

The above pretreatments, each have their benefits and drawbacks. By combining these factors, we can eliminate the disadvantages and can gain additional advantages. Studies showed that the combination of chemical and physical pretreatment, physical and biological pretreatment increased the overall productivity while reducing economic and energy considerations [11].

3.2 Optimizing Fermentation Conditions

3.2.1 Condition of Reaction

For optimal performance of microbial activity and product yield, various conditions need to be optimized. These conditions include pH, temperature, humidity, and aeration. Maintaining ideal conditions during fermentation is important, because each microorganism functions best under specific conditions. Some processes, especially those involving aerobic microorganisms, may require controlled aeration to provide sufficient oxygen. The inoculation method and preculture used in fermentation can also influence the efficiency of the outcome. Microorganisms generally adapt to nutrient and environmental conditions by selecting an environment that imitates the substrate composition. Microorganisms can produce a wide range of enzymes, cellulases, and degradable by-products when applied to appropriate substrates [8].

pH

Changes in pH during fermentation can cause changes in the number of organisms and in their metabolic activity. The pH can be measured offline by mixing the sample with distilled water. Despite the availability of modern pH probes, it is difficult to accurately estimate pH. Therefore, a solid matrix with good buffering capacity is used. The pH is first adjusted to a certain range and then hydrated to the desired pH. The optimal pH for the production of glucoamylase from Sorghum bran by *Aspergillus awamori* was found to be 6.0.

Temperature

Temperature control is critical to optimize microbial activity and enzyme production during the fermentation process. Temperature monitoring methods include spot temperature measurement using thermocouples, non-contact monitoring of the sedimentation front using infrared sensors, and indirect measurement of bed temperature using oxygen uptake rate. To determine the optimal conditions for enzyme production and efficiency, packed-bed bioreactors were found to help maintain the temperature required for fructooligosaccharide production by *Aspergillus oryzae* IPT301 cells.

Humidity

Humidity and water temperature are also very important during fermentation. The increase in production depends largely on germination, fertilization, organisms, and even the production stage. On the other hand, low humidity can cause the substrate to be wet, increase water resistance, and encourage microbial growth. As microbes multiply, humidity naturally decreases due to increased bacterial and metabolic activity. Bioreactors measure water activity using potential and electrical sensors to provide optimum conditions and adjust accordingly, regulating humidity with extractors and solvents. In recent years, the feasibility of using *Saccharomyces cerevisiae, Bacillus subtilis, and Aspergillus niger* bacteria together in the recycling of tea waste has been investigated. The optimum humidity rate was found to be 54.14%, and this value increased the humidity rate while decreasing the sugar rate and mold activity and showed good results for the long-term processing of food waste.

Aeration

During the fermentation, aeration promotes growth, inhibits microbial growth, allows gas exchange, and introduces individual microorganisms into the biological environment. However, this is subject to fermentation by bacteria and fungi, and in the case of plants, the plant's digestive system degrades it, thus reducing the activity of antioxidant enzymes. For efficient enzyme production without damage to the proteins, the working ratio must be optimal. During this operation, they were continuously recorded by the sensors and the monitoring.

3.3 Fermentation Equipment and Technologies

Selection of suitable fermentation equipment and technologies for a specific fermentation depends on the desired product yields and management system for growing the desired microorganisms. Bioreactors are closed systems designed to carry out biological processes under carefully controlled conditions. Typically, cylindrical and made of stainless steel, these can range in volume from a few liters to several cubic meters. The main goal of an ideal bioreactor is to efficiently regulate biological reactions while maintaining optimum conditions. They allow cells or microorganisms to convert low-value compounds into more valuable products. When it comes to SSF, the nutrient composition plays a critical role in supporting microbial growth and activity.

3.3.1 Continuous Stirred Tank Bioreactors

Stirred tank bioreactors are the most commonly used aerobic bioreactors today, especially in biomass production facilities and FDA-approved equipment. This bioreactor has an optimized internal structure to create an efficient enzymatic and biochemical mixing loop design. This tank ratio varies based on the application, ranging from 1:1.5 for mammalian cell cultures to 1:3 for microbial fermentations. A baffle, usually one-tenth the tank diameter, is placed inside the tank to increase efficiency. The shape of the membrane depends on the process: marine membranes with a diameter of three and a half inches are used to mix suspensions in cell cultures, while some Rushton membranes help adsorb gases and improve mixing in microbial cultures.

3.3.2 Bubble Column Bioreactors

This is the simplest, tower-like fermenters which contain a glass or metal tube to which air is introduced through a perforated tube at the bottom and a microporous metal sheet or a diffuser for mixing and aeration process. The capacities of the fermentation tanks vary from 3 to 950 L, with a length-to-width ratio of up to 16:1.

3.3.3 Airlift Bioreactors

It is similar to a bubble column reactor, but contains a stack. There are two types of airlift tubes: internal tubes or external tubes. The tube improves oxygen circulation and exchange and balances the shear forces in the reactor. Air is usually directed through a distribution ring to the bottom of the central exhaust tube, which controls the circulation of air and the growing medium. The air rises through the tube, creating bubbles, and the exhaust gas is exhausted from the top of the tower. The degassed liquid then flows downward, and the product is discharged from the tank. Airlift bioreactors can be used on both a laboratory scale and a large production scale.

3.3.4 Trickle Bed Bioreactors

It is a cylindrical container filled with supporting materials such as wood chips, stone, or plastic structure. These supports have large open spaces for the circulation of liquids and gases in the support material and for the growth of microorganisms on the solid support. The liquid nutrient suspension is sprayed onto the substrate and dripped into the garden bed. Air can rise through the substrate in the opposite direction of the liquid flow. These fermenters are used in vinegar production and other processes. They are suitable for liquids with low viscosity and low suspended solids content [13].

3.3.5 Tray Bioreactor

Tray bioreactors consist of trays filled with a solid-liquid substrate. It provides a controlled aeration to maintain desired oxygen levels and regulate moisture content. This design offers a simple and efficient method for SSF, making it a valuable tool in bioactive compounds production. It supports a wide range of microbial processes. However, maintaining a uniform distribution of moisture and temperature within the trays is difficult. The bioreactor design and optimization are required to achieve consistent results and efficient mass and heat transfer.

3.3.6 Rotating Drum Bioreactor

Rotary drum bioreactor works by continuously rotating a drum filled with wet solid substrate. This rotation ensures that the substrate is thoroughly mixed and aerated, creating an ideal environment for microbial growth and metabolic activity. The moisture content of the substrate is regulated, and the rotary motion ensures homogeneous distribution, allowing efficient use of the solid substrate during SSF. Rotary drum bioreactors have the advantages of increased mass transfer and higher product yield. However, problems may arise in drum design, optimum rotation speed, and prevention of matrix agglomeration, and carefully designed solutions are needed for smooth and efficient operation. Rotary drum bioreactors with capacities up to 13,000 L have been evaluated in SSF.

3.3.7 Fluidized-Bed Bioreactor

The fluidized-bed bioreactor works on the principle that the bed is suspended in an ascending air flow. Effective mixing and contact between microorganisms and nutrients are provided by the air flow. To regulate the temperature, a cooling water control system can be installed. Fluidized bed bioreactor can provide high bioactive compound production. Because it improves aeration, humidity control, microbial growth, oxygen transfer, and nutrient utilization throughout the reactor. Although it has advantages, this type of bioreactor is very complex to design and operate and difficult to clean and maintain, and also it is limited for the flexible raw materials.

3.3.8 Packed-Bed Bioreactor

It is a tubular system filled with a moist solid substrate, which serves as a growth medium for microorganisms in SSF. Typically, the moisture content ranges from 12% to 70% depending on the application. Hence it plays a key role in imitating the natural microbial environment and regulating metabolic activity for the production of bioactive compounds. To support the growth of microorganisms, air with controlled humidity is provided to ensure enough oxygen supply. Controlling temperature and humidity also enhance microbial activity and create optimal conditions for fermentation. This type of bioreactor is particularly suited for the cultivation of filamentous fungi, yeasts, and to some extent bacteria. A major advantage of packed-bed bioreactors is the ability to improve both macroscopic and microscopic analysis of bioreactor behavior. This includes monitoring temperature, concentration gradients, and microbial growth. However, scaling up a packed-bed bioreactor has challenges. It requires careful design solutions to maintain optimal and sustained performance. Packed-bed bioreactors with capacities ranging from 100 to 130 L in SSF have their potential in bioprocess applications [6].

3.4 Case Studies: Successful Food Waste Fermentation Projects

3.4.1 Polyphenols from Grape Pomace

Many researchers have explored the recovery of polyphenols from grape pomace. These polyphenols are used in anti-aging cosmetics and functional foods. A study reviewed the importance of grape pomace and its wide range of properties: antioxidant, anti-inflammatory, anticancer, and antimicrobial effects. Moreover, the feasibility of using grape pomace as a food additive and cosmetic products is being analyzed. Also, application of grape pomace polyphenols in food products and skincare formulations has its potential benefits in both industries [14].

3.4.2 Chitosan Derived from Shrimp Shells

The importance of combining treatment methods using biological and chemical treatments to extract high-quality chitin and chitosan from shrimp shell waste was studied. It highlighted the importance of procedural sequence. The physicochemical properties of biopolymers obtained were excellent, suitable for applications in biomedical, industrial, and environmental fields [15].

3.4.3 Lycopene Extracted from Tomato Peels

In many studies, researchers reported that they successfully extracted lycopene from tomato peels and discussed its wide range of usage as a dietary supplement. A review from [16] explored the potential of tomato peels as a source of lycopene and carotenoids in cancer treatments. The results indicated the interesting therapeutic properties from the tomato peel extract.

3.4.4 Protein Hydrolysates Derived from Fish Waste

Many studies have reported on the synthesis of protein hydrolysates from fish waste for pharmaceutical applications. Fish protein hydrolysates are of particular interest due to their valuable biological activities and technological properties. Hence, they are used as ingredients for functional foods and dietary supplements. As reviewed in [17], enzymatic hydrolysates extracted from fish waste are important components for human nutrition. Also, it contributes to the development of fortified, enriched, and specialized dietary products, offering significant health benefits to consumers soon.

3.4.5 Essential Oils from Orange Peels

The valorization of citrus waste through fermentation processes showed an efficient, cost effective, and environmental friendly method of producing essential oils. It is applicable in the food, nutraceutical, cosmetic, and energy industries [18]. Orange peels contain phenolic compounds, particularly flavanones such as hesperidin and narirutin, which have potential health benefits. Research suggested that recovering phenolic compounds can increase their antioxidant activity and making them useful for food and beverage processing applications [19].

3.5 Nutrient Recovery from Fermented Food Waste

3.5.1 Organic Acids

Food wastes are rich in organic compounds, used in the production of organic acids. This process is driven by hydrolyzing bacteria, which break down complex compounds into simple sugars. These sugars are subsequently converted by acidogenic bacteria into various organic acids. This includes acetic, butyric, lactic, citric, succinic, and tartaric acids. Specific and different organic acids can be produced in two ways: (1) Fermentation using mixed bacterial cultures and (2) selectively fermented with pure cultures. Lactic acid is a key precursor for the production of biodegradable plastics, and it is an environmentally sustainable alternative. Hence organic acids have significant industrial relevance, serving as feed stocks for the synthesis of high-value products. Volatile fatty acids (VFAs) have gained attention as platform molecules for the synthesis of liquid fuels and chemicals. The co-production process enables the conversion of VFAs into promising gasoline alternatives. This includes the transformation of acetic acid into ethanol via methyl acetate and the conversion of butyric acid into methyl butyrate [20].

3.5.2 Proteins

Single-cell proteins (SCP), a group of microbial proteins, is a general term that includes proven proteins obtained from microorganisms such as bacteria, fungi, and algae. SCP production is expected to contribute to the evaluation of alternative rental and durable raw materials in agriculture and sustainable development, to the regulation of contracts made with source materials and to the increase of global nutritional security. SCP has a wide range of applications and can be used as an alternative protein source in animal nutrition, human nutrition, food, applications, animal nutrition, biostatistics, and nutrients. Among microbial producers, phosphate-soluble bacteria (PSB) appear to be a candidate for SCP production. PSB-based protein production offers significant advantages such as higher yield, reliable production rate, and reliable environmental impact compared to chemical methods. In addition, PSB has a unique capacity to simultaneously produce first-class bioproducts of high-value proteins such as polyhydroxybutyrate, alpha-lipoic acid, and coenzyme Q10. They prefer these microbial producers because of their capacity to synthesize a stable protein supplement. SCPs constitute 40–60% of the protein in cell membranes, with approximately 60% consisting of essential amino acids, and are therefore a very preferred protein source for food and dietary supplements [20].

3.5.3 Antioxidants

The ability of some chemicals to exhibit valuable functional properties makes them very important in industries such as food and cosmetics. Many studies have shown the potential of using different wastes as substrates for the production of antioxidant phenolic compounds in the SSF; the most widely used among these is olive oil production waste [21]. Phenolic compounds obtained from food waste have a protective effect against free radical damage, cancer, and cardiovascular diseases. In addition, fruit pomaces are a rich source of pectin, which acts as a natural thickener in various food products. This includes jams, confectionaries, and preserves. In addition, some tropical fruits contain proteolytic enzymes such as papain and bromelain, which are used as natural meat tenderizers and detergents. In addition, studies on mushroom biomass have revealed that tea waste can be a potential feed supplement for chickens and its importance in terms of sustainable animal nutrition [22].

3.6 Production of Biogas from Fermented Food Waste

Biogas is a renewable energy source consisting mainly of 55–65% methane (CH_4), 35–50% carbon dioxide (CO_2), hydrogen sulfide (H_2S), and water. Biogas is produced by anaerobic digestion of various organic wastes and offers a sustainable alternative to fossil fuels. They represent a significant problem globally among all food wastes, and their anaerobic digestion into biogas is a promising solution.

Pretreatment plays a key role in biogas production. Ultrasonic treatment is more efficient and yields 10.12%. Biogas production has also been increased by mixing food waste and biomass [23]. Biogas plays an important role in environmental protection. It helps in reducing and utilizing the wastes. Biogas production from renewable biomass serves a sustainable energy source. The ratio of volatile solids to total solids of the various raw materials, biogas generation for animal manure, fruit and vegetable waste, and canteen waste can reach 91.2%, 80.2%, and 78.3%, respectively.

According to a recent study [24], biogas production was significantly enhanced by adding rumen fluid and *S. cerevisiae isolate MUTJOF* as inoculum. The 6900.3 mL of biogas output was recorded as the highest volume obtained by mixing 100 mL of rumen fluid with 10 mL of *S. cerevisiae isolate MUTJOF* inoculum. This was suggested that biogas production can be optimized by microbial inoculation in the reactor. These results indicated the possibility of microbial-assisted anaerobic digestion to increase the yield of biogas and promote eco-friendly energy solutions.

3.7 Value-Added Products Derived from Fermented Waste

3.7.1 Organic Fertilizers

A study from [25] achieved high-quality organic fertilizers from locally available biomass waste. The fermentation process improved the soil nutrients such as nitrogen, phosphorus, potassium, calcium, and magnesium. It also improved pH and overall nutrient content compared to commercial fertilizer standards. This organic fertilizer improved soil health but reduced crop yields by 40%. They also independently follow the sustainable economy principles.

In fertilizer production, an anaerobic digestion of food waste known as the microbial fuel cells (MFCs) technology supported energy production. It also generated solid by-products as a result of the process. This process utilized fungal cells for MFC structure. in the synthesis of organic fertilizers, MFC enabled ultra-fast hydrolysis of organic waste, making the process highly efficient [26]. These improvements reveal the benefits of combining biotechnology with sustainable waste Management. This approach enhanced agricultural productivity while promoting environmental protection.

3.7.2 Bioactive Compounds

Pectin is a natural gelling agent commonly used in baked goods, confectionery, and meat products. It is an insoluble fiber abundantly found in orange peels. It is known for its functional benefits and contribution to gut microflora health. Likewise, blackberry waste contains high anthocyanin content (4.31 mg Cy3GlE/g) which is a great source of natural pigments and food supplements. Ultrasonic extraction was used to extract betalains from red beet waste to obtain a high amount of natural pigments. Researchers have used advanced extraction techniques to maximize the recovery of bioactive compounds from the food waste.

Carotenoids are extracted from pomegranate peel using organic solvents such as soybean oil and sunflower oil. A study found that 100 g of sunflower oil produced 0.6134 mg of carotenoids on dry skin. In comparison, soybean oil produced a slightly higher amount of carotenoids at 0.6715 mg [27]. SSF technique has proven to be an effective approach to increase the production of bioactive compounds. These include phenols, flavonoids, and peptides with antioxidant, anti-inflammatory, and antimicrobial properties. This process was done by peeling the rice hull. It is a sustainable method to convert agricultural by-products into valuable biomaterials. It is an efficient way to reuse waste and recover resources [8].

3.7.3 Biopesticides

Biopesticides are recognized as a sustainable substitute for chemical pesticides. They are biodegradable and selective and have a slight effect on the environment and food chain. Biological control agents are biopesticides derived from various microbial species that target insect and fungal pests. Research on SSF using food waste is still developing. It is because not all strains that produce biopesticides can survive in contaminated solid-state conditions. For example, *Lactobacillus amylovorus* DSM 20532 and *Rhodopseudomonas palustris* 42OL were used in a systematic process of photo-fermentative hydrogen (H_2) production and energy recovery. This process aligns with the bioremediation principles and supports the adoption of a circular economy [21].

3.7.4 Biofuels

Fish waste has been utilized as a valuable resource for biodiesel production. The high amount of fish oil waste makes it a valuable feedstock. Extraction of oil from fish waste is relatively easy and inexpensive. This crude oil can be converted into large quantities of stable biodiesel at a very low cost. Fish oils have been found to have a similar calorific value to petroleum oil. The conversion of fish oil into biodiesel requires a reaction with triglycerides and methanol to produce fatty acid methyl esters. The presence of compounds such as palmitic, palmitoleic, and oleic acids improves its potential. Researchers have made great progress in this area, by successfully obtaining high methanol from various raw materials [28].

Recent studies have explored the different substrates used for bioethanol production. For example, a study found that mixing white, pink, and sweet potatoes showed notable impacts and damages. Moreover, a blend of apple, banana, and papaya yielded the best results. The study also highlighted the need to examine energy utilization in bioethanol production [29].

3.7.5 Animal Feed

Some findings revealed that fermentation can increase the nutritional value of substrates and make them more available to animals. Microorganisms were used to break down complex nutrients, making them easier to digest. It required a proper management to produce animal feeds by using food waste. For example, plant waste often has high water content, which should be removed by drying. Additionally, due to the composition of the waste also varies, pet food manufacturers must modify their recipes accordingly. Recent advancements in recycling methods have made it possible to transform food waste into high-quality animal feed. Technologies like drying, pelleting, and extrusion help enhance its shelf life. The production of pelleted fish feed from protein-rich feed waste has grown in recent times. Researchers have also found that adding additives like pineapple and yeast can enhance the nutritional value of plant-based feeds. These additives also promote healthy growth in animals [30]. Conversion of food waste into animal feed is a promising approach. It can help to reduce waste, improve sustainability, and support animal health.

3.7.6 Enzymes

Transforming food waste into valuable products is crucial for sustainable development. The success of this process depends on two factors: (1) operating costs and (2) the cost of the final product. Using these fermentation technologies, it has become possible to convert carbohydrate-rich food waste into beneficial by-products like prebiotics. For example, researchers have used cocoa bean husk waste to extract xylooligosaccharides prebiotics (24.2% yield) [9]. Enzymes like α-amylase and phytase break down complex nutrients, making them more absorbable by the body. Fermentation further enhances mineral availability by reducing phytate and tannin levels. However, its effectiveness depends on the food waste composition. For example, high fiber content can limit the mineral absorption. Methods such as seed germination and enzymatic incubation with polyphenol oxidase or phytase can help overcome these limitations [4].

3.8 Quality Control in Fermentation Processes

For high-quality fermented products, it is essential to control parameters like temperature, acidity, and oxygen levels. However, the fermentation process can be affected by the quality of the raw materials and equipment used. To resolve these challenges, online monitoring systems and artificial intelligence algorithms help regulate parameters, ensuring food quality and safety. Despite their potential, these strategies are not widely adopted in industry. Fermentation optimization involves techniques like metabolic engineering and data analysis to enhance microbial pathways and regulate conditions such as temperature, pH, and nutrient levels. Advanced bioreactor design and automated systems ensure stable control during large-scale production [11].

3.9 Economic Viability of Food Waste Fermentation

Traditional methods of food waste disposal, such as landfilling, are expensive and unsustainable, requiring high collection, transportation, and disposal costs. However, recycling food waste into valuable products such as fertilizer, bioelectricity, and biofuel can be profitable. In Malaysia, the production of biofertilizer and biogas from waste supports the economy and the environment. Selling these products can be profitable. The study showed that the initial investment in a manufacturing plant was recouped within 4 years. Another study in China showed that combining food waste recycling with wastewater treatment created a profitable business. The energy produced from food waste required more water for wastewater treatment, generating excess electricity [26]. Overall, recycling food waste into valuable products contributes to a sustainable environment and economy.

3.10 Regulatory and Safety Considerations

The growing demand on fermented foods make it essential to address potential risks for consumer safety. International regulatory agencies such as the European Food Safety Authority and the U.S. Food and Drug Administration set guidelines on fermentation processes and permissible contaminant levels. These guidelines set microbial safety standards and apply Hazard Analysis and Critical Control Points (HACCP) to identify and reduce hazards during processing. However, regulations for fermented foods remain limited, especially for non-dairy products. This makes it difficult to regulate their production and safety. As novel fermented foods and production methods develop, regulatory agencies must adapt to balance innovation with consumer safety. To address this gap, initiatives promote a structured method to evaluating health risks. This includes integrating human studies, food analysis, and mechanistic data to develop a complete risk-benefit framework. Clear labeling is essential, especially regarding allergens, amine content, and storage guidelines.

Increasing public awareness and education is essential for reducing the risks associated with fermented foods. Consumers should be informed about proper handling, fermentation duration, and storage requirements. This reduces the risk of contamination or foodborne illness when making fermented foods at home. Enhancing supervision and consumer education is essential for maintaining food safety. This approach helps the industry to address the challenges arising from growing diversity of fermented foods. While regulatory requirements focus on safety, they also pose challenges for manufacturers. These include conducting validation activities, investing in specialized equipment and facilities to manage food safety risks [4]. The regulations for fermented foods are complex and constantly changing. For example, limits on histamine-producing lactic acid bacteria strains make it difficult to achieve desired flavors and textures. Ensuring safety and meeting regulatory standards for novel fermented foods and production methods require significant time and financial investment. Food safety authorities are working to improve standards for risk reduction, which is expected to lead to further risk reduction. Ultimately, managing the safety of fermented foods requires a collaborative effort between regulatory agencies, manufacturers, and consumers. Strengthening regulatory oversight, consumer education, and public awareness can ensure the safe production and consumption of fermented foods [31].

3.11 Challenges and Solutions in Food Waste Fermentation

Value-added fermented foods are known for their health benefits. They improve gut health and immune support. However, fermentation can cause risks if not properly controlled the process. Microbiological and chemical hazards can impact product safety. A balanced approach is crucial to protecting consumers and informing public health recommendations.

Microbial biotransformation can enhance food properties. However, several challenges must be addressed. One significant hurdle is the variability in microbial activity, which depends on factors like substrate type and fermentation conditions. This inconsistency affects product quality and complicates large-scale production. Additionally, spoilage microorganisms and pathogens continue to pose risks in fermented food production. They can cause unpleasant tastes, gas pockets, and discoloration during the aging process. Pathogens like *Listeria monocytogenes* and *Salmonella* spp. can contaminate raw ingredients or processing equipment and survive some fermentation controls. Implementing standard practices, such as GMPs and HACCP programs, ensures that risks to food safety are controlled.

Consumers tend to trust organic and transparent production processes. They expect ingredients to be listed on packaging with a clear indication of the methodology behind the production process. This has created niches for small-scale manufacturers of craft fermented foods, which are marketed as local and transparent. Maintaining a consistent product quality becomes increasingly difficult as microbial fermentation is scaled up. Maintaining uniform conditions, such as nutrient availability and temperature, becomes increasingly challenging at larger production scales. Effective heat and oxygen transfer demands a deep understanding of thermodynamics and precise process control [31].

Scaling of microbial fermentation methods requires a significant initial investment in production facilities, along with continuous costs of raw materials and maintenance. These financial challenges make it difficult for companies to expand production while managing expenses and maintaining sustainability. In addition to the constant cost of raw materials and upkeep, high volume microbial fermentation involves a massive initial investment in production facilities. The companies cannot keep costs down and ensure sustainable output as they increase levels of production due to these economic disappointments. When conducting large-scale operations, it is necessary to balance these factors, with an emphasis on concentrating fermentation operations and incorporating economic and environmental considerations [4].

3.12 Future Trends in Fermentation Technology

The bioconversion of food waste is rapidly evolving. Here are some key future trends:

3.12.1 Integration of Circular Bioeconomy Principles

Integrating food waste bioconversion into circular economy frameworks minimizes waste by transforming wastes to produce valuable products like biofuels, biofertilizers, and bioplastics.

3.12.2 Advanced Biotechnological Approaches

Using genetic engineering and synthetic biology enhances the efficiency of food waste bioconversion. This innovation can optimize the fermentation process and give high product yields.

3.12.3 Multiproduct Biorefineries

Developing biorefineries can produce multiple bioproducts from food waste. These include biofuels, enzymes, organic acids, and other valuable compounds.

3.12.4 Anaerobic Digestion and Biogas Generation

Converting food waste into biogas, a renewable energy source, using advanced anaerobic digestion. These advanced technologies improved reactor designs and microorganisms, and are expected to enhance efficient production of biogas [5].

3.12.5 Insect-Based Bioconversion

Another emerging trend is bioconversion of food waste by using insects like black soldier flies. They can transform food waste into protein-rich animal feeds. This is such a new technology to transform food waste into beneficial products.

3.12.6 Valorization of Food Waste into Value-Added Ingredients

Isolation of bioactive compounds, antioxidants, and prebiotics from food waste gives high-value products. These products are used as natural additives and supplements in the food and pharmaceutical industries.

3.12.7 Decentralized and Community-Based Systems

Implementing community-scale bioconversion systems to manage food waste locally is popular these days. These include community-scale anaerobic digesters and composting facilities.

3.12.8 Policy and Regulatory Support

Increasing policy and regulatory support for food waste bioconversion. Governments and international organizations are expected to implement incentives and regulations that promote sustainable waste management practices.

3.12.9 Digitalization and Smart Technologies

Utilizing AI (Artificial Intelligence) and machine learning to optimize fermentation parameters and improve product quality. These technologies can analyze real-time data to optimize and control fermentation parameters, predict process outcomes, and improve overall efficiency.

3.12.10 Public Awareness and Education

Raising public awareness and educating people are crucial for reducing food waste and promoting sustainable bioconversion methods. Campaigns and educational programs are effective ways to educate the public [3].

This effort will contribute to sustainable development, resource recovery, and environmental protection. Achieving this requires a collaborative effort between governments, industries, and consumers to promote sustainable waste management and reduce the environmental effects of food waste.

3.13 Community and Industrial Applications

According to estimates, the global precision fermentation market is expected to expand significantly. Annual growth rate of 46%, increasing from USD 3.2 billion in 2024 to over USD 104 billion by 2034. The growth is driven by consumer demand for sustainable goods and alternative protein sources, along with biotechnology's ability to efficiently produce food ingredients. The market is especially robust in North America and Europe, but it is growing quickly in Asia-Pacific because of shifting dietary preferences and economic expansion. Fermented foods such as probiotics and related foods are gaining popularity among the people for their benefits for gut health. Probiotic sales are expected to reach $79.4 billion in 2028, up from $52.1 billion in 2021. These include Kefir, kimchi, kombucha, yogurt, and other fermented foods. Probiotic foods give various benefits, such as weight control, digestion, immunity, and cognition. Customers are more assured about the organic component and prefer labels that list ingredients along with a brief explanation of the production process. As a result, there are opportunities for new local manufacturers of sourdoughs, ginger beers, pickles, sauerkraut, and artisanal kombuchas that are marketed as traceable and locally produced. Innovation in the traditional processes used to produce and enhance the quality of fermented foods [32]. One cutting-edge beauty trend that has no adverse effects on the environment is the use of bioferments. The cosmetics industry can replace problematic ingredients with countless alternatives. Fermentation makes it possible to obtain widely used ingredients like thickeners, preservatives, emollients, humectants, and biosurfactants and also cosmetic bioactive ingredients like ubiquinol, resveratrol, or hyaluronic acid. Fermentation produces active compounds with improved bioactivity, biocompatibility, and a favorable biodegradability profile [32].

4 Notes

Advanced fermentation techniques such as SSF, SmF, and EFP are a promising solution for the suitable bioconversion. They can offer significant benefits for the environment and economic stability. These techniques can convert waste materials into a wide range of valuable bioproducts. These include organic acids, biofuels, animal feeds, enzymes, proteins, antioxidants, biopesticides, fertilizers, bioplastics, and bioactive compounds.

This chapter provided an in-depth analysis focusing on (1) types of advanced fermentation techniques and equipment, (2) pretreatment methods of food waste, (3) conversion of food waste into different valuable products such as eco-friendly, (4) successful food waste fermentation projects and potential and limitations in the context of food waste management. Likewise different topics on bioconversion of food waste were discussed.

References

1. Cudjoe D, Zhu B (2025) Power generation from food waste bio-compressed natural gas: economic feasibility. Renew Sust Energ Rev 209:115110
2. Kim S, Lee Y, Lin KYA, Hong E, Kwon EE, Lee J (2020) The valorization of food waste via pyrolysis. J Clean Prod 259:120816
3. Bilal M, Niu D, Wang Z (2024) Novel enzyme-fermentation process for bioconversion of restaurant food waste into isomaltooligosaccharide-and L-lactic acid-enriched animal feed. Front Sustain Food Syst 8:1326304
4. Sawant SS, Park HY, Sim EY, Kim HS, Choi HS (2025) Microbial fermentation in food: impact on functional properties and nutritional enhancement—a review of recent developments. Fermentation 11(1):15
5. Bulgari D, Gobbi E, Cortesi P, Peron G (2024) Bioconversion of food and green waste into valuable compounds using solid-state fermentation in nonsterile conditions. Plan Theory 13(24):3494
6. Gómez-Ramos GA, Prado-Barragán LA, Rivera BA, Huerta-Ochoa S, Rivera VM, Couder-García M, Ángeles R, García-Hernández, Castillo-Araiza CO (2025) Bioreactor engineering for circular economy: bioactive compound production in solid-state fermentation. Chem Eng Technol 48(1):e202400289
7. Musoni M, Destain J, Thonart P, Bahama JB, Delvigne F (2015) Bioreactor design and implementation strategies for the cultivation of filamentous fungi and the production of fungal metabolites: from traditional methods to engineered systems. Biotechnol Agron Soc Environ 19(4):430–442
8. Bamidele MO, Bamikale MB, Cárdenas-Hernández E, Bamidele MA, Castillo-Olvera-G, Sandoval-Cortes J, Aguilar CN (2025) Bioengineering in solid-state fermentation for next sustainable food bioprocessing. Next Sustain 6: 100105
9. Nargotra P, Ortizo RGG, Wang JX, Tsai ML, Dong CD, Sun PP, Bajaj BK, Kuo CH, Sharma V (2024) Enzymes in the bioconversion of food waste into valuable bioproducts: a circular economy perspective. Syst Microbiol Biomanufact 4(3):850–868
10. Lytras G, Lytras C, Mathioudakis D, Papadopoulou K, Lyberatos G (2021) Food waste valorization based on anaerobic digestion. Waste Biomass Valor 12:1677–1697
11. Elgarahy AM, Eloffy MG, Alengebawy A, El-Sherif DM, Gaballah MS, Elwakeel KZ, El-Qelish M (2023) Sustainable management of food waste; pre-treatment strategies, techno-economic assessment, bibliometric analysis, and potential utilizations: a systematic review. Environ Res 225:115558
12. Zafar H, Peleato N, Roberts D (2022) A review of the role of pre-treatment on the treatment of food waste using microbial fuel cells. Environ Technol Rev 11(1):72–90

13. Kaur I, Sharma AD (2021) Bioreactor: design, functions and fermentation innovations. Res Rev Biotechnol Biosci 8:34–43
14. Prata C, Zalambani C, Rossi F, Rossello S, Cerchiara T, Cappadone C, Malucelli E (2025) Nutrients and nutraceuticals from Vitis vinifera L. pomace: biological activities, valorization, and potential applications. Nutrients 17(3):583
15. Vázquez-Aldana M, Sixto-Berrocal AM, Arcos-Casarrubias JA, Trujillo MAM, Cruz-Díaz MR (2025) Biological and chemical synergy in chitin and chitosan production: the role of process sequencing in shrimp shell waste treatment. Int J Biol Macromol 306:141247
16. Jiménez Bolaño DC, Insuasty D, Rodríguez Macías JD, Grande-Tovar CD (2024) Potential use of tomato peel, a rich source of lycopene, for cancer treatment. Molecules 29(13):3079
17. Honrado A, Miguel M, Ardila P, Beltrán JA, Calanche JB (2024) From waste to value: fish protein hydrolysates as a technological and functional ingredient in human nutrition. Food Secur 13(19):3120
18. Gervasi T, Mandalari G (2024) Valorization of agro-industrial orange peel by-products through fermentation strategies. Fermentation 10(5):224
19. Razola-Díaz MDC, De Montijo-Prieto S, Guerra-Hernández EJ, Jiménez-Valera M, Ruiz-Bravo A, Gómez-Caravaca AM, Verardo V (2024) Fermentation of orange peels by lactic acid bacteria: impact on phenolic composition and antioxidant activity. Food Secur 13(8):1212
20. Liang J, Zhang P, Zhang R, Chang J, Chen L, Zhang G, Wang A (2024) Bioconversion of volatile fatty acids from organic wastes to produce high-value products by photosynthetic bacteria: a review. Environ Res 242:117796
21. Sánchez A, Oiza N, Artola A, Font X, Barrena R, Moral-Vico J, Gea T (2024) Solid-state fermentation: a review of its opportunities and challenges in the framework of circular bioeconomy. Afinidad 81(601):51–57
22. Arshad RN, Abdul-Malek Z, Roobab U, Qureshi MI, Khan N, Ahmad MH, Liu ZW, Aadil RM (2021) Effective valorization of food wastes and by-products through pulsed electric field: a systematic review. J Food Process Eng 44(3):e13629
23. Sharma P, Bano A, Singh SP, Atkinson JD, Lam SS, Iqbal HM, Tong YW (2024) Biotransformation of food waste into biogas and hydrogen fuel—a review. Int J Hydrog Energy 52:46–60
24. Fentahun M, Kahsay B (2025) Enhancing biogas production from co-digestion of organic wastes mixed with cattle manure using rumen fluid and Saccharomyces cerevisiae isolate MUTJ0F. F1000Research 14:199
25. Thammayod A, Roschat W, Leelatam T, Phewphong S, Maneerat B, Leelatam N, Thangthong A, Moonsin P, Promarak V (2025) Development of high-quality organic fertilizer from biomass waste and the application of biochar in local areas of Sakon Nakhon Province, Northeastern Thailand. Trends Sci 22(2):9105–9105
26. Mahish PK, Verma DK, Ghritlahare A, Arora C, Otero P (2024) Microbial bioconversion of food waste to bio-fertilizers. Sustain Food Technol 2(3):689–708
27. Bangar SP, Chaudhary V, Kajla P, Balakrishnan G, Phimolsiripol Y (2024) Strategies for upcycling food waste in the food production and supply chain. Trends Food Sci Technol 143:104314
28. Umesh M, Kumar V, Priyanka K, Kathirvel P, Suresh S, Santhosh AS (2025) Fish waste valorisation through production of biodiesel and biopolymers for sustainable development: a mini review. Bioresour Technol Rep:102045
29. Dutra Fagundes V, Machado ÊL, de Souza Schneider RDC, Colla LM (2025) Life cycle of bioethanol production from blends of different food waste. Environ Technol 46(5): 731–751
30. Panda J, Amrit R, Mishra AK, Chakraborty A, Rustagi S, Nath PC, Sarabandi K, Sarma H, Wagh MS, Mohanta YK (2025) Sustainable valorization of fruit and vegetable waste for bioactive compounds: advancing functional food and wellness. Waste Biomass Valor:1–30
31. Praveen M, Brogi S (2025) Microbial fermentation in food and beverage industries: innovations, challenges, and opportunities. Food Secur 14(1):114
32. Krzyżostan M, Wawrzyńczak A, Nowak I (2024) Use of waste from the food industry and applications of the fermentation process to create sustainable cosmetic products: a review. Sustain For 16(7):2757

Chapter 5

Thermochemical Processes for Eco-Friendly Energy Recovery from Food Waste

G. Rajeshwari, Pramila Murugesan, Gnanaselvan Gnanasekaran, and Saeed Farhang Sahlevan

Abstract

Food waste, with approximately 931 million tons produced annually worldwide, poses a significant environmental and economic concern, with improper disposal contributing to greenhouse gas emissions (8–10%) and resource wastage (freshwater (25%) and agricultural land (28%)). Thermochemical processes offer a sustainable pathway for converting food waste into valuable energy products, aligning with the circular economy principles. This chapter explores the potential of thermochemical processes including pyrolysis, gasification, and hydrothermal carbonization recovery of energy from food wastes. The influencing parameters such as feedstock composition, feedstock concentration, operating conditions (temperature, steam flow rate, gas flow rate, and residence time) are critically analyzed. Moreover, it highlights the role of catalysts and the integration of the thermochemical process with other techniques (plasma process and microbial system) in enhancing energy recovery efficiency. Additionally, there has been discussion on the possible use of by-products (biochar and bio-oil) from gasification for efficient valorization. Overall, this chapter offers insights into scalable and environmentally friendly energy recovery applications for mitigating the food waste crisis.

Key words Food waste, Thermochemical conversion, Sustainable energy recovery, Simulation, Eco-friendly solutions

1 Introduction

Food waste, which has major ramifications for sustainability and resource management, has become a major environmental and social concern. Every year, an estimated 1.3 billion tons of food are lost worldwide in the supply chain, with an estimated USD 1 trillion in overall economic impact [1]. Due to the rising trends in food loss and food waste over the last 10 years, efficient waste reduction strategies are necessary. Almost 40% of municipal solid waste produced worldwide comprises food loss and food waste, which has important post-management and processing implications [2]. Food waste is a worldwide problem that poses significant

Tanmay Sarkar (ed.), *Food Waste Valorization: Green Techniques in Sustainable Management*,
Methods and Protocols in Food Science, https://doi.org/10.1007/978-1-0716-4646-5_5,
© The Author(s), under exclusive license to Springer Science+Business Media, LLC, part of Springer Nature 2025

social, economic, and environmental difficulties. Food waste, which mostly consists of organics (e.g., lipids, fats, proteins, and carbohydrates), is hereby converted into high-value-added products (e.g., gases, chemicals, biochar, and bioethanol) [3, 4]. Due to its low operating costs and ease of use, landfill technology has been used in the past to handle food waste; nevertheless, it takes up a lot of space and contaminates both the groundwater and the soil.

Recently, thermochemical conversion has been a highly effective route for converting organic waste into valuable energy products and is significantly utilized for energy generation. According to the International Energy Agency (IEA), biomass accounts for nearly 10% of the world's primary energy supply, with thermochemical processes playing a crucial role in bioenergy production [5]. Pyrolysis, gasification, and hydrothermal carbonization are the primary thermochemical methods (Fig. 1). The main products of pyrolysis, gasification, and hydrothermal carbonization are bio-oil (50–75%), syngas (85%), and char (60%), respectively [5]. Further, bio-oil is a promising route for the production of biofuels, while syngas could be adapted for synthetic fuel generation and power generation. In addition, the utilization of these thermochemical conversion technologies for food waste valorization could prevent carbon emission by up to 80% than fossil fuel-based energy generation processes. Caton et al. (2010) demonstrate that, compared to the wood used as comparison fuel, which had an energy content of 18.7 MJ/kg dry, food waste had an average energy content of 21.7 MJ/kg dry [6]. Moreover, the by-product (biochar) from these has been shown to improve the fertility of soil and carbon sequestration.

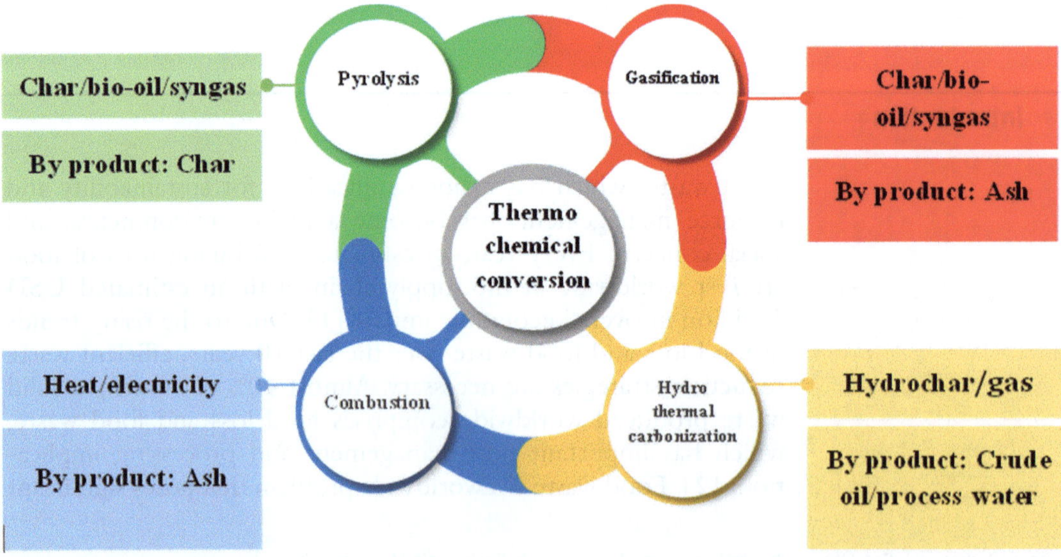

Fig. 1 Thermochemical conversion process

This chapter explores different thermochemical techniques, including pyrolysis, gasification, hydrothermal liquefaction, and combustion, detailing their underlying mechanisms and efficiency in generating energy-rich products (syngas, biochar, and bio-oil) and influencing factors. Additionally, it highlights the synergistic effect of thermochemical technologies with other techniques to enhance energy recovery from food waste. Moreover, it delves into the current research and future perspectives of these technologies.

2 Materials

The materials used in these processes can be categorized into food waste sources, catalysts, reactor type, and process additives. Various sources of food wastes include residential, commercial, and industrial food processing units. The waste was categorized into different fractions such as fruit and vegetable peels, dairy residues, meat by-products, and carbohydrate-rich leftovers. The excess moisture content of food waste was subjected to sorting and drying routes. It was then ground into fine particles using a mechanical grinder to achieve a uniform feedstock for thermochemical processing. The catalytic materials used in the thermochemical process (pyrolysis, gasification, hydrothermal carbonization, and combustion process) are metal oxides (zeolite, sodium hydroxide, and potassium hydroxide) and metals (nickel, zinc, and iron), which can play a key role in improving selectivity, minimizing tar formation, and enhancing the energy yield [7]. Moreover, catalysts play a vital role in reducing activation energy, increasing reaction rates, and enhancing the selectivity of desirable products. In addition, the process additives could improve the selectivity of product and reaction efficiency. Co-pyrolysis of food waste with other raw materials (biomass or plastic waste) approaches to enhance the product yield and quality [8, 9].

3 Methods

Thermochemical is a chemical process that breaks down and fractures long-chain, high molecular weight hydrocarbons into smaller, lower molecular weights. The thermochemical process can utilize a broad range of carbohydrate and lipid-content-rich feedstock biomass. The thermochemical processes are gasification, pyrolysis, hydrothermal carbonization, and combustion.

3.1 Gasification Process

The gasification process may transform the solid food waste into flammable syngas in a low-oxygen environment or oxygen-deficient, which can reduce the generation of dioxin. Solid carbonaceous resources are handled at high temperatures (700–1200 °C) in a gasifier using various gasifying media (carbon dioxide, oxygen, steam, and air) in the low-polluting thermochemical process known as gasification. The gasification process typically involves four steps: oxidation (combustion), reduction (gasification), pyrolysis (decomposition), and feedstock drying. The primary product of the gasification process is syngas (hydrogen and carbon monoxide). It can be utilized as substrate or fuel in fuel cells, boiler combustion, and the production of chemical compounds. The additional by-products of the gasification process include chars, oils, ash, and tars. Several influencing factors, including temperature, residence time, catalyst type, and precursor composition, have a substantial effect on the composition of the gasification products that are produced. Interestingly, a greater amount of carbon dioxide is generated when oxygen is employing a gasifying medium [10]. The hydrogen generation via the gasification process is represented in Fig. 2.

3.1.1 Classification of Gasification Process

Depending on the type of gasifying medium, the gasification process comprises two different types such as steam gasification and hydrothermal gasification (supercritical water gasification and subcritical water gasification). The gasification medium could alter the calorific value of gasses, concentration, and yield of hydrogen gas. The carbonaceous feedstock is transformed into hydrogen and carbon dioxide during the steam gasification route.

$$C_nH_m + 2nH_2O \leftrightarrow (2n + (m/2))H_2 + nCO_2$$

According to Valizadeh et al. (2021) reported that steam gasification produces more hydrogen from food waste than air gasification. In this process, the syngas properties are significantly influenced by

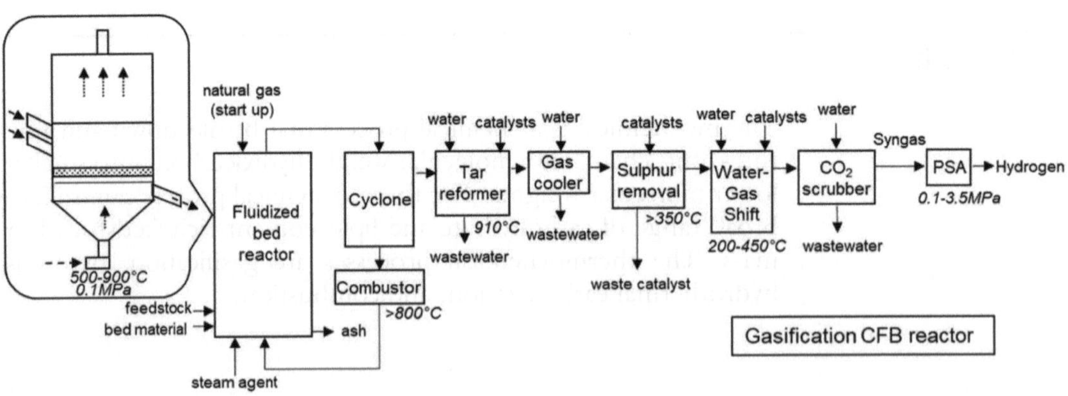

Fig. 2 Schematic representation of gasification process flow diagram for energy generation [11]

the temperature and the amount of steam or steam gas flow rate. Increases in flow rate of steam promote the water gas shift reaction (WGS) and methane reformation, improving syngas quality [12]. Recently, Singh et al. (2020) found that the steam gasification technique converts kitchen waste into hydrogen. The highest hydrogen yield was observed at a steam flow rate of 0.500 mL/min. Furthermore, the steam flow rate increase had no effect on the syngas characteristics above a certain threshold (0.625 mL/min) [13]. Similarly, Huang et al. (2019) reported that more hydrogen is produced when the temperature of the torrefied starchy food waste is continuously raised [14].

Hydrothermal gasification is a capable energy-efficient thermochemical technique for converting high-moisture raw material into hydrogen-rich syngas, which eliminates the need for the pre-drying process. It has the benefit of being highly efficient at comparatively low temperatures and requiring less time. The water gas shift reaction is encouraged by the high-moisture content of food waste and enhances the hydrogen yield. The hydrothermal gasification process typically uses either subcritical water (pressure < 22.1 MPa, temperature < 374 °C) or supercritical water (pressure > 22.1 MPa and temperature > 374 °C). Long-chain biomass molecules, including lignin, lipid, protein, and cellulose undergo hydrolysis in supercritical water to generate various intermediate products, including glucose, phenol, and fatty acids. Methanation ($CO + 3H_2 \leftrightarrow CH_4 + H_2O$), hydrogenation ($CO + 2H_2 \leftrightarrow CH_4 + 0.5O_2$), water gas shift (Eq. 1), and steam generation ($CH_XO_\Upsilon + (1 - \Upsilon)H_2O \rightarrow CO_2 + (1 - \Upsilon + \frac{X}{2})H_2$) are some of the chemical reactions that occurred in the creation of hydrogen by supercritical water gasification. The overall reaction for supercritical steam gasification reaction is as follows:

$$CH_XO_\Upsilon + (2 - \Upsilon)H_2O \rightarrow CO_2 + \left(2 - \Upsilon + \frac{X}{2}\right)H_2 \quad (1)$$

Recently, Yan et al. (2020) converted restaurant wastewater into electricity using both supercritical water and subcritical water. They discovered that supercritical water had a greater treatment efficiency than subcritical water [15].

3.1.2 Routes for Enhancing the Gasification Efficiency

More energy input, initial cost, and equipment manufacture requirements are necessary for gasification with high reaction pressure and temperature. Co-gasification and co-pyrolysis processes, which combine non-food waste, with food waste to create value-added products and reduce food waste accumulation, have recently been researched as ways to improve the effectiveness of the hydrothermal gasification process [16]. Adding an appropriate catalyst or catalyst to the hydrothermal gasification process is a cost-effective and energy-efficient method of increasing efficiency. Furthermore, catalysts may significantly increase the amount of liquid intermediates that are transformed into gaseous products, improving the gasification of food waste [17]. Carbon-based materials (activated carbon and char), transition metals (Ni, Pt, and Ru), and alkali (NaOH, CaO, $Ca(OH)_2$, $FeCl_3$, K_2CO_3, and Na_2CO_3) are some of the catalyst types that have been used in hydrothermal gasification to increase the production of hydrogen. The primary process by which the alkali metal catalyzes the gasification reaction is an intercalation reaction, which takes place in the interaction between the aromatic rings of organic biomass with alkali atom [18]. The response is as mentioned below:

$$M_2CO_3 + 2C \rightarrow 2M + 3CO$$

$$2M + 2H_2O \rightarrow 2MOH + H_2$$

$$2MOH + CO \rightarrow M_2CO_3 + H_2$$

$$2C + 2H_2O \rightarrow 2CO + 2H_2$$

According to the catalytic activity of Na_2CO_3, NaCl and $NaHCO_3$ on the supercritical water gasification of unsorted food waste, $NaHCO_3$ and Na_2CO_3 significantly impacted the H_2 yield more than NaCl. Either carbon dioxide removal process or partial oxidation of chemical compounds under super critical water gasification conditions occurs when $NaHCO_3$ and Na_2CO_3 catalysts are added [19].

3.1.3 Gasification Reactors for Food Waste Conversion

Different kinds of biomass gasifiers, including fixed beds, fluidized beds, supercritical water gasifiers, and entrained flow gasifiers, have been built based on moisture content and fuel availability.

Fixed-Bed Reactors

The gasification process frequently uses fixed-bed reactors because of their simple design and ease of use. The working pressure and temperature of fixed-bed gasifiers are 0–70 bar and 600–1000 °C. It has been used in three different configurations including cross-draft, downdraft, and updraft. Huang et al. (2019) utilized corn starch food waste for the steam gasification process. A tiny quartz tube batch reactor with 2.5 mm of inner diameter and 200 mm of length, operating at high temperatures (1000 °C) and high pressures (45 MPa), was recently employed for the supercritical food waste gasification. Additionally, food waste gasification studies have also been conducted utilizing fixed-bed downdraft reactors based on batch hoppers with feedstock auger feeding principles [14].

Fluidized Bed Gasifier

A fluidized bed gasifier minimizes feedstock agglomeration by distributing the system's temperature uniformly and ensuring adequate gas and material mixing. 0–70 bar and 923–1223 K are the operational pressure and temperature ranges for fluidized bed gasifiers, respectively. Either a tubular, fluidized bed is used to conduct the supercritical gasification reaction. [20] designed a fluidized bed reactor for the gasification process with operating conditions of up to 30 MPa and 923 K of temperature and pressure, respectively.

Entrained Flow Gasifier

An entrained flow gasifier is a continuous flow reactor in which the gasifying medium and feedstock are supplied into the gasifier simultaneously. From the bottom of the gasifier, the ash is collected while the syngas is removed from the top or bottom. For an entrained flow gasifier, the operating temperature and pressure are 1200–1600 °C and 2–70 bar, respectively. High flexibility of fuel, low residence time, homogenous distribution of temperature, and improved carbon conversion efficiency are all features of entrained flow gasifiers [21].

3.1.4 Influencing Factors for Energy Production from Gasification of Food Waste

The gas mixture composition after a gasification process depends on several influencing variables, such as physical characteristics, feedstock composition, residence time, feedstock concentration, and temperature. The gasification process operating temperature can vary between 350 and 1000 °C. As a result, at high temperatures, more hydrogen and carbon dioxide are formed, whereas at low temperatures, more carbon monoxide and methane are produced [22]. With a temperature adjustment to 500 °C from 420 °C, the hydrogen yield from mixed food waste improved from 3.92 mol/kg to 7.91 mol/kg under subcritical circumstances [15]. Furthermore, a reverse water gas shift reaction caused a minor diminishment in hydrogen concentration and an enhancement in carbon monoxide production when the temperature was raised above this threshold [23].

$$CO + H_2O \leftrightarrow CO_2 + H_2 \quad \Delta H^{\circ} = -41.2\ KJ/mol$$

A key element in water gas shift reactions and steam reforming reactions is feedstock concentration [24]. In the gasification process, a higher concentration of feedstock may result in lower hydrogen production. The hydrogen yield decreased as a result of the gasification process increased feedstock concentration and water concentration because water gas shift and steam reforming reactions are highly sensitive to reduced water concentrations [15]. The efficiency of the gasification reaction is also impacted by residence time. In general, more residence time improves the thermal cracking reaction and gives secondary reactions more opportunity to break down high molecular weight chemicals into syngas mixtures, boosting gasification efficiency. Additionally, it promotes the water gas shift process to generate better hydrogen. The biomass conversion and hydrogen yields are said to grow once the residency duration reaches a "residence time," after which there is no noticeable improvement. Several factors influence the "residence time," including the reactor vessel, the biomass type, and the reaction temperature [25].

The performance of gasification is affected by food waste properties such as size, moisture, ash content, volatile matter, lignin content, and mineral contents. For instance, the gasification reactivity can be greatly impacted by the amount of lignin present. Porosity enhanced the gasification process by facilitating carbon dioxide diffusion into the char particle. According to [26], inorganic content in food waste can also significantly impact the gasification process's reactivity. High-moisture content could improve the water gas shift reaction. The temperature in the gasifier reactor decreases as the fuel moisture content raised. The reactor reaches a critical point when the temperature falls too low, and the intensity of hydrogen generation decreases. The moisture content might lower the syngas heating value and carbon conversion effectiveness.

3.2 Pyrolysis

Pyrolysis is the thermal treatment of waste materials in an oxygen-starved atmosphere to yield gaseous, liquid, and solid fuels. The process of pyrolysis, which breaks down organic waste at temperatures ranging from 300 to 700 °C and pressures between 0.1 and 0.5 MPa without the presence of oxygen, yields valuable products with more heat value, such as bio-oil, biochar, and gases which can then be further processed to be used as transportation fuel. Dioxin production is lessened by the absence of air.

$$\text{Organic material} + \text{Heat} \rightarrow H_2 + CO + CH_4 + \text{Other products}$$

The two phases of pyrolysis are primary and secondary. During the primary pyrolysis, the volatile matter undergoes primary pyrolysis, they are broken down by decarboxylation, dehydrogenation, and dehydration. This leads to the condensation of bio-oil. The primary process by which heavyweight hydrocarbons break down and transform into char or gases (carbon dioxide, carbon monoxide, and methane) is secondary pyrolysis. The conversion process follows either thermal or catalytic conversion depending upon the conditions of pyrolysis, and it can be illustrated as follows

$$(C_6H_6O_6)n \rightarrow (H_2 + CO + CH_4 + \ldots + C_5H_{12}) + (H_2O + CH_3OH + CH_3COOH + \ldots) + C$$

The gas yield is represented by the first section of the product side, where various gases are created throughout the operation. The mixing of different liquid types is displayed in the second section of the product side, and the solid yield is the final term. One of the solid products, biochar, is a carbonaceous material resistant to decomposition and exhibits intriguing physicochemical properties, including high surface area and porosity. It has been used in many different applications, including the production of activated carbon and relevant compounds, energy production, carbon sequestration, catalysis, and soil amendments and adsorbents. To increase the

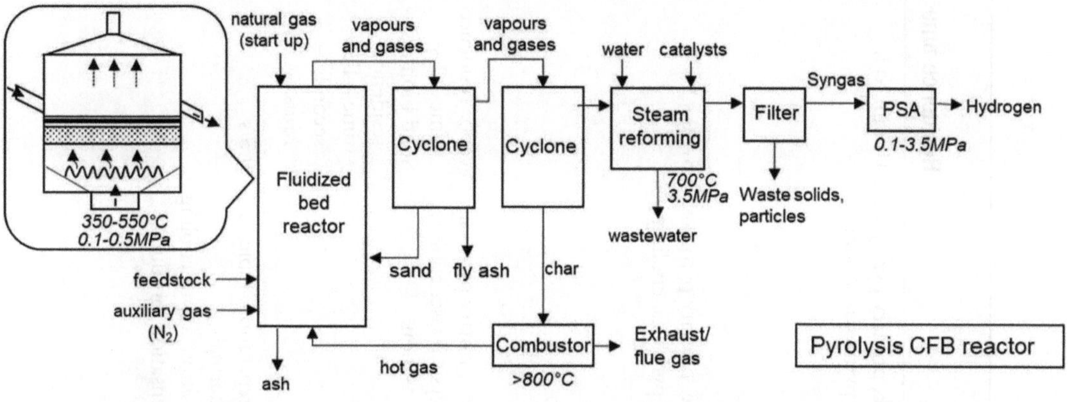

Fig. 3 Schematic illustration of pyrolysis process flow diagram for energy generation [11]

energy content of the gasification feedstock, pyrolysis can be carried out prior to gasification in order to produce biochar. To create extra hydrogen, the hydrogen-containing product gas can be changed using steam reforming and the water gas shift reaction. Figure 3 illustrates the energy generation via the pyrolysis process.

3.2.1 Types of Pyrolysis Process

Based on temperature, residence time, and heating rate, the pyrolysis process can be classified into seven types including slow pyrolysis, intermediate, fast, ultra-fast/ flash, hydropyrolysis, and vacuum as illustrated in Table 1.

Slow Pyrolysis

Slow pyrolysis is the traditional kind of pyrolysis, which is distinguished by a long residence period and a slow heating rate. The biomass was slowly pyrolyzed up to a temperature of 400–500 °C for 5–30 min with a heating rate of 0.1–1 °C/s. Char is more probable to form with slow pyrolysis, although small quantities of gaseous and liquid products are also produced. Reduced heating rate, extended vapor residence time, and delayed pyrolysis provide an appropriate environment and enough time for the secondary reactions to occur [27].

Intermediate Pyrolysis

Between slow and quick pyrolysis are the operating conditions for intermediate pyrolysis. Intermediate pyrolysis conditions prevent the generation of high molecular tars and yield high-quality bio-oil and dry char that can be used in agriculture or to produce energy. The temperature range for intermediate pyrolysis is 500–650 °C, with a residence duration of 300–1000 s and a heating rate of 0.1–10 °C/min.

Fast Pyrolysis

The food waste is heated to 850–1250 °C with a heating rate of 10–200 °C in a brief period ranging from 1 to 10 s during fast pyrolysis. Since the bio-oil yield from fast pyrolysis is dominated by the gaseous and char products, it is utilized to produce bio-oil.

Table 1
Different types of pyrolysis process

Pyrolysis process	Definition	Temperature	Heating rate	Steps	Residence time
Slow pyrolysis	Slow pyrolysis is a thermal decomposition process of organic material in an oxygen-limited environment, occurring at relatively low temperatures and slow heating rates	400–500 °C	0.1–1 °C/s	Drying phase, primary pyrolysis, secondary pyrolysis, cooling phase	10 min to h
Intermediate	Intermediate pyrolysis is a thermochemical process that decomposes organic materials at moderate temperatures and heating rates, producing a balanced mix of bio-oil, biochar, and gases	400–600 °C	10–200 °C/min	Drying phase, primary pyrolysis, secondary pyrolysis, cooling phase	5–30 s
Fast	Fast pyrolysis is a thermochemical process that rapidly decomposes organic materials in the absence of oxygen to produce bio-oil, gases, and char.	850–1250 °C	100–1000 °C	Drying phase, primary pyrolysis, main pyrolysis phase, cooling, and condensation	Vapor residence time—2 s Solid biomass residence time—Few seconds—Mins
Ultra-fast/ flash	Ultra-fast pyrolysis, also known as flash pyrolysis, is a thermochemical process that rapidly decomposes organic materials in the absence of oxygen to produce bio-oil, gases, and char	900–1200 °C	2500°C/s	Feedstock preparation, rapid heating, thermal decomposition, vapor cooling and condensation, product separation	0.5 s

Hydropyrolysis	Hydropyrolysis is a thermochemical process that involves the decomposition of organic materials in the presence of hydrogen and in the absence of oxygen	350–600 °C	10–300 °C/s	Feedstock preparation, heating in hydrogen atmosphere, thermal decomposition and hydrogenation, product collection	>15 s
Vacuum	Vacuum pyrolysis is a thermochemical decomposition process of organic materials conducted under reduced pressure and in the absence of oxygen	450 and 600 °C	0.1–1 °C	Feedstock preparation, application of vacuum, controlled heating, thermal decomposition, and product recovery	5–30 min
Microwave oven	Microwave pyrolysis is a thermochemical process that utilizes microwave energy to decompose organic materials in an oxygen-free environment	400–800 °C	Initial drying phase—120 °C, rapid heating phase—120–650 °C, completion phase—Above 650 °C	Feedstock preparation, microwave heating, thermal decomposition, and product recovery	0.5–10 s

Before it reacts to generate the undesirable char, the higher rate of heating required in the quick pyrolysis turns into a liquid product from input biomass.

Flash Pyrolysis

Flash pyrolysis is an improved and adapted version of fast pyrolysis. The temperature attained during flash pyrolysis ranges from 900 to 1200 °C, and the heat pulse that is generated for the biomass lasts for only 0.1–1 s. The quality and stability of the bio-oil are the issues with flash pyrolysis reactors since they are significantly impacted by the ash and char that are produced.

Vacuum Pyrolysis

The thermal degradation of biomass at less pressure and without oxygen is known as vacuum pyrolysis. During vacuum pyrolysis, the temperature is maintained between 450 and 600 °C, and the pressure range is typically 0.05–0.20 MPa. Vacuum pyrolysis uses less pressure and using a vacuum to extract the vapors rather than a gas purge. Quick extraction of organic gases produced through primary pyrolysis also considerably shortens the vapor residence period, which lowers the secondary reactions and ensures a high liquid product yield.

Hydropyrolysis

Along with the biomass, hydrogen and hydrogen-based compounds are supplied into the reactor in this process at pressures between 5 and 20 MPa, which are higher than the atmospheric pressure. Since hydrogen is known to be a reducing agent, its presence at high temperatures and pressures lowers the oxygen amount in the bio-oil that is produced and prevents the formation of char. Using a catalyst to extract water, oxygen, and other gases from the liquid product can be related to hydropyrolysis.

3.2.2 Influencing Factors

The type of temperature, heating rate, feedstock, catalyst, and process residence time all affect the product yield. These operating conditions have an impact on the pyrolysis products' quality in addition to controlling the char yield. A study demonstrated the slow pyrolysis of various food wastes, including walnut shells, pomegranate peels, and grape pomace. They found that heating rate, temperature, and residence time significantly impact biochar yield. For example, pyrolyzing walnut shells at 500 °C min a residence time of 60 min and with a heating rate of 15 °C/ resulted in approximately 30% biochar yield [28]. The biomass constituents are given enough time to respond when the vapor residence time is increased, which aids in the repolymerization process. However, the yield of biochar is decreased if the residence period is shorter because the biomass constituent's depolymerization is not finished [29]. The rapid transfer of heat to the cold end from the hot of the input biomass is reduced as the particle size rises because of the increased distance between the core and the surface of biomass. The char yield is favored by this temperature gradient [30]. For instance, Demirbas [143] examined the impact of particle size on the output of biochar from pyrolysis of tea waste, corncob, and olive husk at 677 °C. Olive husk biochar yield increased from 19.4% to 35.6% and corncob biochar yield increased from 5.7% to 16.6% when the particle size was raised from 0.5 mm to 2.2 mm [30]. A low heating rate also prevents biomass from thermally breaking, increasing the output of biochar. A high heating rate helps biomass fragmentation and enhancements liquid and gas production, reducing the probability of biochar formation [30]. Raising the temperature during pyrolysis has a detrimental effect on the yield of biochar because it permits the thermal cracking of heavy hydrocarbon components, which increases the amount of liquid and gaseous and lowers the yield of biochar [30]. It has been discovered that completing the pyrolysis under the influence of a pressure greater than the ambient pressure increases the production of biochar [31]. Increasing the flow rate lowers the yield of biochar, yet it has also been noted that the effect is not significant and that the flow rate only slightly lowers the yield of char. The vapors are forced out of the reaction zone by an increase in gas flow rate, which shortens the vapor residence time [32]. The pyrolysis process can be accelerated by utilizing inorganic salt-based catalysts including carbonates, chromates, and chloride. Because it should reduce and ease the burden of waste disposal, the use of co-pyrolysis of a coal mixture with organic wastes has recently gained attention in industrialized nations. However, pyrolysis requires the biomass to be pre-dried, which uses a lot of energy.

3.3 Hydrothermal Carbonization

Hydrothermal carbonization is a thermal conversion process that turns wet feedstocks into a value-added solid (also called hydrochar), liquid, and gaseous products at relatively low temperatures (<350 °C) and with relatively little energy input through the process of decarboxylation, hydrolysis, dehydration, and condensation (Fig. 4). This process eliminates the need to dry the feedstock before conversion. Hydrothermal carbonization is especially useful for processing high-moisture feedstocks that are difficult to treat with conventional dry thermal methods like pyrolysis and gasification, such as agricultural wastes, sewage sludge, and food waste. The resulting hydrochar has a greater energy density and can be utilized as a precursor to activated carbon, solid fuel, or soil amendment [34].

Hydrolysis has a lower activation energy than the other processes and starts biomass degradation primarily by cleaving ester and ether bonds. As a result, oligosaccharides and lignin fragments are further hydrolyzed or initiate other chemical pathways and products. After hydrolysis, the hydroxyl and carboxyl groups were removed by dehydration and decarboxylation, respectively. During the initial hydrothermal treatment, 5-hydroxymethylfurfural (HMF) is formed via dehydration reactions. The majority of the

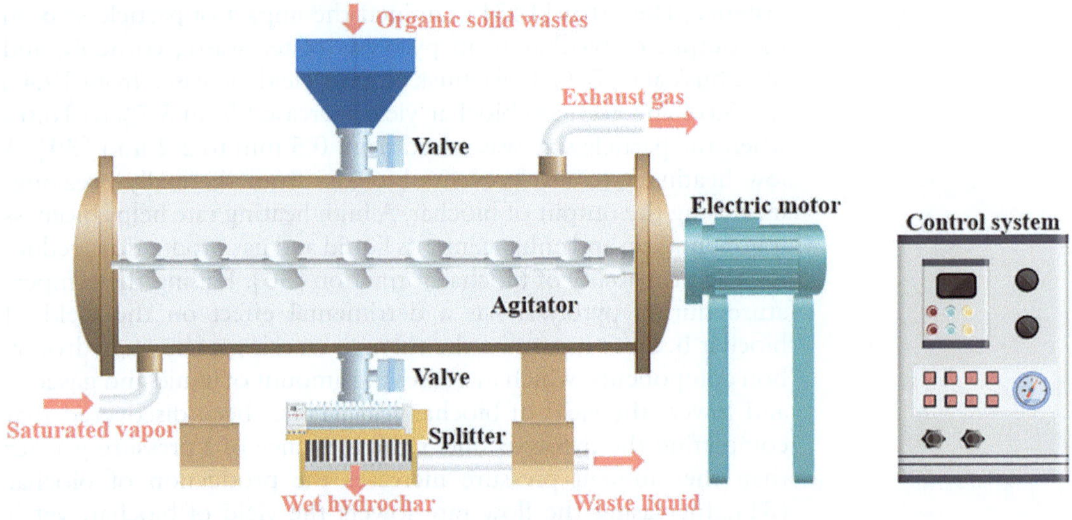

Fig. 4 Schematic representation of the hydrothermal carbonization process [33]

intermediates lead to the production of hydrochar, whereas additional degradation produces other compounds, such as organic acids and compounds. For example, formic acid, levulinic acid, and dihydroxyacetone can serve as catalysts to promote hydrolysis reactions, generating more HMF, which forms furanic compounds via a sequence of polymerization and condensation reactions. The hydrochar produced possesses higher dewaterability, drying performance, energy density, and porosity than the raw materials [35]. Subcritical water has a higher concentration of H^+ ions than ordinary water, which is a natural media that encourages acid-catalyzed chemical processes [36]. The use of subcritical water to food waste provides a number of beneficial effects.

- First, complex organic molecules were broken down into simpler ones like sugars, organic acids, and volatile organic compounds by the higher temperature and pressure of subcritical water.
- Second, subcritical water improves the sanitation and safety of the treated material by assisting in the efficient eradication of germs and bacteria present in the food waste.

3.3.1 Influencing Factors

Temperature, residence time, pressure, atmosphere reaction, and catalyst type utilized are among the process parameters that have a significant impact on the degree of carbonization and chemical structural evolution of food waste during the hydrothermal carbonization process.

Temperature and residence time have an impact on the structural evolution of hydrochar, as evidenced by the yield of food waste-derived hydrochar. Wang et al. (2018b) investigated the effect of temperature on hydrochar yield utilizing a 20 °C gradient and a residence time of 1 h, operating in the 180–260 °C range. As the temperature increased, the hydrochar production dropped from 53.8% to 35.0% [36]. Previous study illustrated that hydrothermal carbonization of food waste at temperatures lower than 200 °C can yield more hydrochar [37]. A progressive increase in hydrochar yield was noted as the residence period increased because of the continual polymerization of liquid-phase intermediates bound in the insoluble hydrochar [33]. The degree of carbonization carried on by the effects of temperature and residence time on the physicochemical characteristics of hydrochar is typically evaluated using the severity factor (log R_0).

$$R_O = t \times \exp\left(\frac{T - 100}{14.75}\right)$$

The residence duration and hydrothermal temperature have a major impact on the surface structure of hydrochar. A low heating rate may regulate the distribution of hydrothermal products and produce superior carbonization effects. In contrast, a high heating rate restricts the amount of heat and mass that can be transferred, which shortens the time it takes for secondary reactions involving intermediate products to occur. The self-generated pressure during hydrothermal carbonization of food waste is the primary focus of the current research, with little attention paid to the way pressure affects the process. According to Son Le et al. (2022), it was incorrect to state that pressure had no effect on the composition of hydrochar [38]. Interestingly, adding acidic catalysts can change the reaction path, speed up the rate of reaction, or strengthen the hydrogen ions. Food waste was impregnated with HCl by Zhu et al., which reduced the amount of ash and increased the thermal stability of hydrochar [39]. An increased hydrochar yield is a result of the rise of acidic conditions. Furthermore, a lower pH changes the pore structure of hydrochar and causes microspheres to develop on its surface [40].

3.4 Combustion

Combustion is a thermochemical conversion technology in which chemical energy stored in feedstock is transformed into heat, mechanical power, or electricity. When heating food waste biomass, its constituents start to hydrolyze, oxidize, dehydrate, and pyrolyze with increasing temperature, forming combustible volatiles, tarry substances, and highly reactive carbonaceous char. The reactions involved in the hydrothermal reactions are:

Volatile matter combustion:

$$C_xH_y + O_2 \rightarrow CO_2 + H_2O + \text{heat}$$

Char combustion:

$$C + O_2 \rightarrow CO_2 + \text{heat}$$

Table 2
Steps involved in the combustion process

Combustion process	Mechanism
Drying	The food waste comprises moisture that must be removed before combustion takes place. Radiation from flames and from the stored heat in the combustion body unit provides the heat for this process
Pyrolysis	The volatile gases are freed when the dry biomass temperature ranges from 200 °C to 350 °C. The products of pyrolysis include carbon monoxide (CO), carbon dioxide (CO_2), methane (CH_4), and tar. When these gases mix with airborne oxygen and combust, they produce a yellow flame. This mechanism is independent of utilizing gas heat generation to dry the fresh fuel and release extra volatile gases. Char is the leftover material after the cauterization of all the volatiles
Oxidation	The char decomposes at around 800 °C. Oxygen is needed again, for carbon oxidation, both in the fire bed and above the fire bed where it blends with carbon to form carbon dioxide that is released into the air
Complete combustion	This process could occur for the correct amount of time in a proper fuel-to-air ratio under suitable temperature and turbulence circumstances

Incomplete combustion (unwanted by-products):

$$C + \frac{1}{2}O_2 \rightarrow CO$$

Complete combustion entails the generation of heat due to oxidation of carbon- and hydrogen-rich biomass to carbon dioxide and water and with partial combustion, methane, carbon monoxide, and particulates are released. Fuel contaminants, including sulfur and nitrogen, are also linked to emissions of sulfur dioxide and nitrogen dioxide. The steps involved in the combustion process are drying, pyrolysis, oxidation, and complete combustion (Table 2).

Generally, fixed bed, fluidized bed, and suspension burners are commonly used for combustion technology. Zhou et al. demonstrate that food solid wastes (FSWs) exhibit superior combustion properties compared to coal and can be effectively utilized as a supplementary fuel in blast furnaces. The burnout capacity (ease of complete combustion) follows the order:

waste vegetables > waste steamed bread > waste rice > waste meat > coal [41].

3.5 Thermochemical Conversion Process with Other Techniques—Synergistic Method for Food Waste into Energy Conversion

Recently, increased hydrogen, methane, and nutrient-rich biochar generation has been achieved by the use of an integrated waste sorting system with gasification and anaerobic digestion. The pure food waste is broken down in the anaerobic digester, and the leftover food waste is then processed by gasification. The discarded FW has a higher moisture content, organic matter, and total nitrogen content, due to the high quantities of salts and proteins in the food waste. High-solid anaerobic digestate can be efficiently converted into syngas and methane-enriched biogas by combining the anaerobic digestion process with the gasification processes. Anaerobic digestion and gasification systems have only been the subject of a few number of investigations to date. Recently, it has been shown that incorporating biochar to anaerobic digestion can boost its buffering capability and make it easier to produce biogas and its methane percentage. As a result of its improved porous structure and surface area, the biochar aids in colonizing the microbial species, facilitating the removal of all organic carbon and accelerating the anaerobic digestion reaction rate [42]. Food waste and other types of waste may be treated using plasma gasification, a new gasification method. Trash is heated to a temperature between 2000 and 14,000 °C during the thermal process known as plasma gasification [43]. The following are the primary chemical reactions that occur in the plasma gasifier:

Devolatilization

$$\left(CH_xO_yN_zS_w \rightarrow \text{Char} + \text{Volatiles}\right)$$

Oxidation

$$C + 0.5\,O_2 \rightarrow CO \quad \Delta H^o = -268\ \text{kJ mol}^{-1}$$

Water gas reaction

$$C + O_2 \rightarrow CO_2 \quad \Delta H^o = -406\ \text{kJmol}^{-1}$$

Boudouard reaction

$$C + CO_2 \rightarrow 2CO \quad \Delta H^o = -172.6\ \text{kJmol}^{-1}$$

Nitrogenous species formation

$$CH_4 + H_2O \leftrightarrow CO + 3H_2O \quad \Delta H^o = -206\ \text{kJ mol}^{-1}$$

$$N_2 + 3H_2 \rightarrow 2NH_3$$

$$\text{Char} - \text{N H} \rightarrow HCN$$

$$HCN + H_2O \rightarrow NH_3 + CO$$

Sulfur species formation

$$H_2S + CO_2 \rightarrow COS + H_2O$$

$$H_2S + CO \rightarrow COS + H_2$$

Additionally, water gas shift reaction, methanation reaction, and steam methane reforming occur. Power plants and anaerobic digestion systems are two examples of systems that can be combined with hydrothermal technology to use the heat they provide to prepare hydrochar fuel, which is then utilized in a boiler. The wastewater produced by the hydrothermal treatment can be utilized again in anaerobic digestion and carbonization operations at the same time. When evaluating the economic viability of technological integration, drying hydrochar is a crucial component.

4 Notes

Thermochemical conversion processes such as gasification, pyrolysis, hydrothermal carbonization, and combustion could offer sustainable and effective pathways for the conversion of the environmental burden of food waste into renewable energy and valuable by-products. The potential energy carriers from the pyrolysis process are syngas, bio-oil, and biochar. While the gasification process produces the syngas for electricity generation and biofuel synthesis. Whereas hydrothermal processes convert the food waste into carbon-rich solid fuel. Gas flow rate, temperature, pressure, residence time, pretreatment of food waste, and addition of catalyst are the major influencing factors for the quantity and quality of products from the thermochemical conversion process. Hybrid systems that combine pyrolysis or gasification with carbon capture technologies can further reduce greenhouse gas emissions and enhance carbon utilization. Advancements in the incorporation of catalysts, reactor design, and pretreatment of food wastes can significantly enhance the thermochemical conversion process effectiveness and quality of products. Furthermore, integrating the thermochemical process with other waste treatment routes including plasma process and anaerobic digestion may offer potential solution waste to energy conversion. Adaptation of integration of artificial intelligence (AI) tools for process control and predictive modeling for minimizing the energy consumption and maximizing biofuel yield and quality. There is much potential for investigation into the use of the by-products produced by the food waste gasification routes.

References

1. Blakeney M (2019) Food loss and food waste: causes and solutions. Edward Elgar Publishing. https://doi.org/10.4337/9781788975391
2. Economou F, Voukkali I, Papamichael I, Phinikettou V, Loizia P, Naddeo V, Sospiro P, Liscio MC, Zoumides C, Țîrcă DM, Zorpas AA (2024) Turning food loss and food waste into Watts: a review of food waste as an energy source. Energies 17(13):3191. https://doi.org/10.3390/en17133191
3. Di Fraia S, Godvin Sharmila V, Banu JR, Massarotti N (2024) A comprehensive review on upcycling of food waste into value added products towards a circular economy: holistic approaches and life cycle assessments. Trends Food Sci Technol 143:104288. https://doi.org/10.1016/j.tifs.2023.104288
4. Yuan X, Shen Y, Withana PA, Mašek O, Lin CSK, You S, Tack FMG, Ok YS (2023) Thermochemical upcycling of food waste into engineered biochar for energy and environmental applications: a critical review. Chem Eng J 469:143783. https://doi.org/10.1016/j.cej.2023.143783
5. Schultz C, Rösch M, Bank L (2023) Renewables. In: Energy-related material flow simulation in production and logistics. Springer International Publishing, Cham, pp 179–203. https://doi.org/10.1007/978-3-031-34218-9_7
6. Caton PA, Carr MA, Kim SS, Beautyman MJ (2010) Energy recovery from waste food by combustion or gasification with the potential for regenerative dehydration: a case study. Energy Convers Manag 51:1157–1169. https://doi.org/10.1016/j.enconman.2009.12.025
7. Shao J, Yan R, Chen H, Haiping Y, Lee D (2010) Catalytic effect of metal oxides on pyrolysis of sewage sludge. Fuel Process

Technol 91:1113–1118. https://doi.org/10.1016/j.fuproc.2010.03.023
8. Amrullah A, Farobie O, Septarini S, Satrio JA (2022) Synergetic biofuel production from co-pyrolysis of food and plastic waste: reaction kinetics and product behavior. Heliyon 8: e10278. https://doi.org/10.1016/j.heliyon.2022.e10278
9. Kumar I, Tirlangi S, Kathiresan K, Sharma V, Madhu P, Sathish T, Ağbulut Ü, Murugan P (2024) Co-pyrolysis of furniture wood with mixed plastics and waste tyres: assessment of synergistic effect on biofuel yield and product characterization under different blend ratio. Sci Rep 14:24584. https://doi.org/10.1038/s41598-024-72809-x
10. Burra KG, Hussein MS, Amano RS, Gupta AK (2016) Syngas evolutionary behavior during chicken manure pyrolysis and air gasification. Appl Energy 181:408–415. https://doi.org/10.1016/j.apenergy.2016.08.095
11. Lui J, Chen W-H, Tsang DCW, You S (2020) A critical review on the principles, applications, and challenges of waste-to-hydrogen technologies. Renew Sust Energ Rev 134:110365. https://doi.org/10.1016/j.rser.2020.110365
12. Valizadeh S, Lam SS, Ko CH, Lee SH, Farooq A, Yu YJ, Jeon JK, Jung SC, Rhee GH, Park YK (2021) Biohydrogen production from catalytic conversion of food waste via steam and air gasification using eggshell- and homo-type Ni/Al2O3 catalysts. Bioresour Technol 320:124313. https://doi.org/10.1016/j.biortech.2020.124313
13. Singh D, Yadav S, Bharadwaj N, Verma R (2020) Low temperature steam gasification to produce hydrogen rich gas from kitchen food waste: influence of steam flow rate and temperature. Int J Hydrog Energy 45:20843–20850. https://doi.org/10.1016/j.ijhydene.2020.05.168
14. Huang J, Qiao Y, Wei X, Zhou J, Yu Y, Xu M (2019) Effect of torrefaction on steam gasification of starchy food waste. Fuel 253:1556–1564. https://doi.org/10.1016/j.fuel.2019.05.142
15. Yan M, Su H, Hantoko D, Kanchanatip E, Shahul Hamid FB, Zhang S, Wang G, Xu Z (2019) Experimental study on the energy conversion of food waste via supercritical water gasification: improvement of hydrogen production. Int J Hydrog Energy 44:4664–4673. https://doi.org/10.1016/j.ijhydene.2018.12.193
16. Yang Z, Koh SK, Ng WC, Lim RCJ, Tan HTW, Tong YW, Dai Y, Chong C, Wang CH (2016) Potential application of gasification to recycle food waste and rehabilitate acidic soil from secondary forests on degraded land in Southeast Asia. J Environ Manag 172:40–48. https://doi.org/10.1016/j.jenvman.2016.02.020
17. Cao W, Wei Y, Jin H, Liu S, Li L, Wei W, Guo L (2022) Characteristic of food waste gasification in supercritical water for hydrogen production. Biomass and Bioenergy. 163:106508. https://doi.org/10.1016/j.biombioe.2022.106508
18. Nzihou A, Stanmore B, Sharrock P (2013) A review of catalysts for the gasification of biomass char, with some reference to coal. Rev Energy. 58:305–317. https://doi.org/10.1016/j.energy.2013.05.057
19. Su W, Cai C, Liu P, Lin W, Liang B, Zhang H, Ma Z, Ma H, Xing Y, Liu W (2020) Supercritical water gasification of food waste: effect of parameters on hydrogen production. Int J Hydrog Energy 45:14744–14755. https://doi.org/10.1016/j.ijhydene.2020.03.190
20. Weiland F, Hedman H, Marklund M, Wiinikka H, Gebart R (2013) Pressurized oxygen blown entrained-flow gasification of wood powder. Energy Fuel 27:932–941
21. Miccio F, Picarelli A, Ruoppolo G (2016) Increasing tar and hydrocarbons conversion by catalysis in bubbling fluidized bed gasifiers. Fuel Process. Technol. 141: 31–37. https://doi.org/10.1016/j.fuproc.2015.06.007
22. Shen Y (2015) Chars as carbonaceous adsorbents/catalysts for tar elimination during biomass pyrolysis or gasification. Renew Sust Energ Rev 43:281–295. https://doi.org/10.1016/j.rser.2014.11.061
23. Xiong S, He J, Yang Z, Guo M, Yan Y, Ran J (2020) Thermodynamic analysis of CaO enhanced steam gasification process of food waste with high moisture and low moisture. Energy 194:116831. https://doi.org/10.1016/j.energy.2019.116831
24. Su H, Hantoko D, Yan M, Cai Y, Kanchanatip E, Liu J, Zhou X, Zhang S (2019) Evaluation of catalytic subcritical water gasification of food waste for hydrogen production: effect of process conditions and different types of catalyst loading. Int J Hydrog Energy 44:21451–21463. https://doi.org/10.1016/j.ijhydene.2019.06.203
25. Chen J, Fan Y, Jiaqiang E, Cao W, Zhang F, Gong J, Liu G, Xu W (2019) Effects analysis on the gasification kinetic characteristics of food waste in supercritical water. Fuel 241:94–104. https://doi.org/10.1016/j.fuel.2018.12.012
26. Lv D, Xu M, Liu X, Zhan Z, Li Z, Yao H (2010) Effect of cellulose, lignin, alkali and alkaline earth metallic species on biomass pyrolysis and gasification. Fuel Process Technol

91:903–909. https://doi.org/10.1016/j.fuproc.2009.09.014

27. Demirbaş A, Arin G (2002) An overview of biomass pyrolysis. Energy Sources 24:471–482. https://doi.org/10.1080/00908310252889979
28. Ismail IS, Othman MFH, Rashidi NA, Yusup S (2023) Recent progress on production technologies of food waste-based biochar and its fabrication method as electrode materials in energy storage application. Biomass Convers Biorefinery 1–17:1. https://doi.org/10.1007/s13399-023-03763-3
29. Park HJ, Park Y-K, Kim JS (2008) Influence of reaction conditions and the char separation system on the production of bio-oil from radiata pine sawdust by fast pyrolysis. Fuel Process Technol 89:797–802. https://doi.org/10.1016/j.fuproc.2008.01.003
30. Choi HS, Choi YS, Park HC (2012) Fast pyrolysis characteristics of lignocellulosic biomass with varying reaction conditions. Renew Energy 42:131–135. https://doi.org/10.1016/j.renene.2011.08.049
31. Jerry AM, Croiset E, Dai X, DeAlmeida C, Mok WS-L, Norberg N, Richard J-R, Al Majthoub M (1996) High-yield biomass charcoal. Energy Fuel 10:652–658. https://doi.org/10.1021/ef9501859
32. Demiral İ, Ayan EA (2011) Pyrolysis of grape bagasse: effect of pyrolysis conditions on the product yields and characterization of the liquid product. Bioresour Technol 102:3946–3951. https://doi.org/10.1016/j.biortech.2010.11.077
33. Wang Q, Wu S, Cui D, Zhou H, Wu D, Pan S, Xu F, Wang Z (2022) Co-hydrothermal carbonization of organic solid wastes to hydrochar as potential fuel: a review. Sci Total Environ 850:158034. https://doi.org/10.1016/j.scitotenv.2022.158034
34. Lanjekar PR, Panwar NL, Patel MR, Divyangkumar N (2024) Exploring sustainable energy: an overview of biochemical and thermochemical conversion of dairy and food waste. Environ Pollut Manag 1:152–166. https://doi.org/10.1016/j.epm.2024.09.003
35. Berenguer CV, Perestrelo R, Pereira JAM, Câmara JS (2023) Management of Agri-food waste based on thermochemical processes towards a circular bioeconomy concept: the case study of the Portuguese industry. PRO 11:2870. https://doi.org/10.3390/pr11102870
36. Wang T, Zhai Y, Zhu Y, Gan X, Zheng L, Peng C, Wang B, Li C, Zeng G (2018) Evaluation of the clean characteristics and combustion behavior of hydrochar derived from food waste towards solid biofuel production. Bioresour Technol 266:275–283. https://doi.org/10.1016/j.biortech.2018.06.093
37. Czerwińska K, Śliz M, Wilk M (2022) Hydrothermal carbonization process: fundamentals, main parameter characteristics and possible applications including an effective method of SARS-CoV-2 mitigation in sewage sludge: a review. Renew Sustain Energy Rev 154:111873. https://doi.org/10.1016/j.rser.2021.111873
38. Zhu X, He M, Xu Z, Luo Z, Gao B, Ruan R, Wang C-H, Wong K-H, Tsang DCW (2022) Combined acid pretreatment and co-hydrothermal carbonization to enhance energy recovery from food waste digestate. Energy Convers Manag 266:115855. https://doi.org/10.1016/j.enconman.2022.115855
39. Wang T, Zhai Y, Zhu Y, Li C, Zeng G (2018) A review of the hydrothermal carbonization of biomass waste for hydrochar formation: process conditions, fundamentals, and physicochemical properties. Renew Sust Energ Rev 90:223–247. https://doi.org/10.1016/j.rser.2018.03.071
40. Son Le H, Chen WH, Ahmed SF, Said Z, Rafa N, Le AT, Ağbulut U, Veza I, Nguyen XP, Duong XQ, Huang Z, Hoang AT (2022) Hydrothermal carbonization of food waste as a sustainable energy conversion path. Bioresource Technology 363:127958. https://doi.org/10.1016/j.biortech.2022.127958
41. Zhou D, Wei R, Long H, Li J, Qi L, Xu CC (2020) Combustion characteristics and kinetics of different food solid wastes treatment by blast furnace. Renew Energy 145:530–541. https://doi.org/10.1016/j.renene.2019.06.046
42. Zhang J, Cui Y, Zhang T, Hu Q, Peng Y, Tong YW, He Y, Dai Y (2021) Food waste treating by biochar-assisted high-solid anaerobic digestion coupled with steam gasification: enhanced bioenergy generation and porous biochar production. Bio Resour Technol 331:125051. https://doi.org/10.1016/j.biortech.2021.125051
43. Munir MT, Mardon I, Al-Zuhair S, Shawabkeh A, Saqib NU (2019) Plasma gasification of municipal solid waste for waste-to-value processing. Renew Sust Energ Rev 116:109461. https://doi.org/10.1016/j.rser.2019.109461

Chapter 6

Eco-Innovative Protocols for Developing Nutraceuticals from Food By-Product

Insha Arshad, Farooq Anwar, Fulvia Bovera, Ayesha Baig, Sajjad Hussain Sumrra, Mudassar Zafar, Muhammad Adnan Ayub, Haider Ijaz, and Muhammad Zubair

Abstract

With time, the growing awareness regarding food waste products and environmental sustainability resulted in the development of innovative green extraction techniques to obtain high-value nutraceuticals from food waste. These eco-friendly techniques employ sustainable methods for extraction, bioprocessing methods, and advanced technologies to extract high-value nutraceuticals from food wastes such as peels of fruits and vegetables, leftovers of vegetables, bran, and other agricultural by-products. Green and clean extraction technologies like supercritical fluid extraction, ultrasound-assisted extraction, and enzymatic extraction give advanced methods to recover nutraceuticals from food waste, which results in a reduction of environmental impacts. This chapter deals with the economic and ecological viability as well as regulatory considerations to produce high-value nutraceuticals from waste material. It resulted in the standing of a circular economy in reducing food waste, lowering the emission of greenhouse gases, and promoting resource-efficient industrial follows. The case studies highlighted the successful application and investigated how companies have changed waste into useful products that have great health impacts. Despite the considerable potential, it is crucial to tackle challenges related to scalability, standardization, regulatory adherence, and consumer acceptance.

Key words Extraction, Nutraceuticals, By-product, Bioactive, Eco-innovative

1 Introduction

Products made from food sources that provide health advantages beyond their fundamental nutritional value are known as nutraceuticals. The name was created by Dr. Stephen DeFelice in 1989 and is a mixture of the words "pharmaceutical" and "nutrition." Nutraceuticals are intended to enhance bodily functioning, prevent chronic diseases, postpone the aging process, and improve health. They fall under one of the following categories such as vitamins, minerals, herbs, amino acids, and enzymes that enhance the diet

Tanmay Sarkar (ed.), *Food Waste Valorization: Green Techniques in Sustainable Management*,
Methods and Protocols in Food Science, https://doi.org/10.1007/978-1-0716-4646-5_6,
© The Author(s), under exclusive license to Springer Science+Business Media, LLC, part of Springer Nature 2025

and are examples of dietary supplements [1]. Foods enhanced with nutrients or other healthful components are known as functional foods (e.g., probiotic yogurts and fortified cereals). Medicinal foods are specially prepared foods that are given for particular medical ailments under a doctor's supervision [2]. Pharmaceuticals are foods or crops that have undergone genetic modification and have therapeutic qualities. Nutraceutical and dietary supplement types include [3]: essential nutrients such as vitamins and minerals help maintain body processes and guard against deficits (e.g., vitamin D for bone health). Herbal supplements are plant-based substances with medicinal uses (e.g., ginseng for energy enhancement). Proteins (such as whey protein for athletes) and amino acids are the building blocks for muscle growth and repair. Probiotics and prebiotics, such as live-cultured yoghurt, help to enhance digestion and gut health. Supplemented foods are enhanced with extra nutrients (for instance, calcium-fortified orange juice). As they are rich in antioxidants, they fight against oxidative stress by eating flavonoid-rich berries [4]. They are formulated to address particular health conditions, such as metabolic diseases (e.g., low-glycemic foods for diabetic management). Chronic disorders, including obesity, diabetes, and cardiovascular disease, can be avoided with the help of nutraceuticals. Omega-3 fatty acids diminish the risk of heart disease and prevent inflammation. Antioxidants, such as vitamins C and E, lower the risk of cancer and guard against cellular damage.

The global nutraceutical market is projected to reach $750 billion by 2028, buoyed by increasing health consciousness. Progress in nutrigenomics (the interaction of genes with diet) paves the way for tailored functional foods. Consumers are increasingly inclined toward organic, non-GMO, and plant-based products [5]. The secondary products produced during the manufacturing and processing of food items are known as food by-products. Despite being regarded as garbage, these materials are rich in nutrients and bioactive substances. By decreasing food waste and improving resource efficiency, using food by-products can support sustainability [6]. Among the examples are peels from fruits and vegetables (such as potato skins and citrus peels), cereal husks and bran (such as rice husks and wheat bran), by-products of meat and fish, such as bones, skin, and scales, by-products of dairy (like whey), and waste products from distillation and brewing, such as discarded grains and grape pomace [7]. A lot of leftover materials are turned into livestock feed, which is an affordable and wholesome substitute for traditional feed [6]. Nutraceuticals, cosmetics, and pharmaceuticals use by-products that are rich in antioxidants, polyphenols, fibers, and essential oils [8]. Food ingredients and additives: They can be made into natural colorants (like beetroot pulp for coloring), thickeners, and stabilizers. By-products like rice husks and fruit peels can be converted into biofuels, biogas, and biochar. Composting and soil fertilizers: Organic waste is composted to produce nutrient-rich fertilizers, promoting sustainable agriculture. Packaging material: Agricultural leftovers are used to generate biodegradable packaging solutions [6].

The following are some environmental benefits of using food by-products as a source of nutraceuticals [9]. Nutraceuticals are products derived from food sources that offer health benefits, including the prevention and treatment of diseases. They encompass a wide range of supplements, functional foods, and pharmaceuticals that promote overall well-being [10]. Eco-innovative methods aim to reduce waste, pollution, and greenhouse gas emissions to minimize their effects on the environment. Through recycling, circular economy principles, and sustainable energy use, they seek to improve resource efficiency. Investing in green technologies and cutting operating expenses boost economic growth and competitiveness. Maintaining safety and enhancing public health promote social well-being. Lastly, they promote resilience against environmental hazards, stakeholder participation, and adherence to environmental legislation [11].

2 Materials

The terms nutraceuticals and functional foods are frequently used interchangeably, although they possess different definitions. Both categories emphasize health advantages that extend beyond basic nutrition, playing an integral role in preventive healthcare and the management of diseases. Nutraceuticals can be categorized into several types:

1. *Dietary supplements*: These consist of concentrated nutrients (like vitamins, minerals, amino acids, or bioactive compounds) available in forms such as capsules, tablets, powders, or liquids. Examples: Omega-3 fatty acid supplements for cardiovascular health, probiotic capsules for digestive health.
2. *Functional foods*: These are either whole or modified foods that offer additional health benefits in addition to basic nutrition. For example, fortified milk with vitamin D promotes bone health.
3. *Medicinal foods*: These are specifically formulated foods that are consumed under healthcare supervision for particular health issues. Example: Ketogenic diets are designed for managing epilepsy.
4. *Pharmaceuticals* (Pharma + Agriculture): These are genetically modified foods created to produce medicinal substances. Example: Golden Rice, which is fortified with β-carotene to address vitamin A deficiency [12].

Nutraceuticals have been extensively researched for their role in the prevention and management of various health issues. Omega-3 fatty acids, flavonoids, and phytosterols help to lower the risk of heart diseases. Calcium, vitamin D, and magnesium strengthen bones and help prevent osteoporosis. Probiotics and prebiotics enhance digestion and immune function. Cognitive function: Polyphenols and nootropics support brain health and may lower the risk of Alzheimer's disease. Metabolic disorders: Plant-based compounds aid in managing obesity and diabetes.

There are some challenges in the nutraceutical industry, such as regulatory issues. Nutraceuticals do not have standardized regulatory frameworks like pharmaceuticals, resulting in inconsistencies in quality and efficacy and bioavailability concerns. Certain bioactive compounds, like carotenoids and polyphenols, have low absorption rates, necessitating specialized formulations (e.g., nanotechnology-based delivery). Misleading claims about health benefits can confuse consumers.

Functional foods are whole, fortified, or enhanced foods that offer health benefits beyond fundamental nutrition. These foods are rich in bioactive compounds such as vitamins, minerals, fiber, and antioxidants that contribute to reducing disease risks [13]. Functional foods can be categorized into two groups: conventional functional foods are naturally abundant in compounds that promote health. Fruits and vegetables contain antioxidants (e.g., polyphenols found in berries and lycopene in tomatoes), and nuts and seeds are rich in omega-3 fatty acids and fiber (e.g., walnuts, flaxseeds). Modified (fortified) functional foods are foods that have been enhanced with additional nutrients to boost their health benefits. It includes fortified cereals, enriched with iron and folic acid to help prevent anemia. Vitamin D-fortified milk supports bone health and calcium absorption. Plant sterol-added margarine aids in lowering cholesterol levels [14]. Functional food helps to boost immunity, and probiotics, vitamins C and D, and zinc enhance the immune system. It also improves heart health as Omega-3 s, fiber, and polyphenols can help lower blood pressure and cholesterol. It aids digestion, regulates blood sugar, and supports skin health [5].

3 Methodology

Different techniques for the extraction of different components are used, and some conventional methods are applied, which have some disadvantages over the new, green, and clean extraction techniques.

Table 1
Comparison of conventional and green extraction techniques

Parameter	Conventional extraction	Green extraction
Extraction yield	Moderate too high but often requires longer times and higher temperatures	Higher yield in a shorter time due to enhanced mass transfer
Extraction yield	Moderate too high but often requires longer times and higher temperatures	Higher yield in a shorter time due to enhanced mass transfer
Efficiency	Lower efficiency due to solvent saturation and long diffusion time	More efficient extraction of bioactive compounds with less solvent
Time consumption	Hours to days (e.g., Soxhlet extraction can take up to 24 h)	Minutes to hours (e.g., UAE and MAE can be completed within 10–60 min)
Solvent usage	Large volumes of organic solvents (e.g., ethanol, methanol, hexane)	Lower solvent consumption, often using water or green solvents
Selectivity of compounds	Poor selectivity; extracts unwanted compounds (e.g., waxes, pigments)	Higher selectivity for target bioactives, reducing impurities
Environmental impact	High, due to solvent waste, energy-intensive processes, and toxic emissions	Lower, as it reduces energy consumption, toxic waste, and CO_2 footprint
Cost-effectiveness	Higher operational cost due to solvent and energy consumption	More cost-effective in the long run due to reduced solvent and energy needs

3.1 Green Extraction Technique

Sustainable extraction methods are designed to reduce ecological footprints, lower energy requirements, and promote the sustainability of extraction practices. These methods prioritize effectiveness, selectivity, and the utilization of non-toxic, renewable solvents [6]. A comparison of conventional and green extraction techniques is given in Table 1. Fundamental Principles of Green Extraction [7]. Alternative solvent usage: It emphasizes the use of supercritical liquids, water, deep eutectic solvents, and ionic solvents as the replacement of other harmful organic solvents. Energy consumption reduction: Use of techniques requiring low energy like microwave, ultrasound, or enzyme-assisted extraction. Waste minimization: Techniques that produce small amounts of by-products and can recover materials from waste. Process intensification: The efficiency of the extraction method can be increased by merging different methods. Renewable: Resource usage, focus on bio-based materials and sustainably sourced plant materials.

The comparison of the conventional techniques with green extraction is given in Table 1 [8, 9].

Supercritical fluid extraction (SFE) is an environmentally friendly extraction method that uses carbon dioxide (CO_2) to increase carotenoid production with great selectivity. Similarly, the extraction efficiency of ultrasound-assisted extraction (UAE) improved while reducing environmental impact. Processing time is reduced with the use of microwave-assisted extraction (MAE)

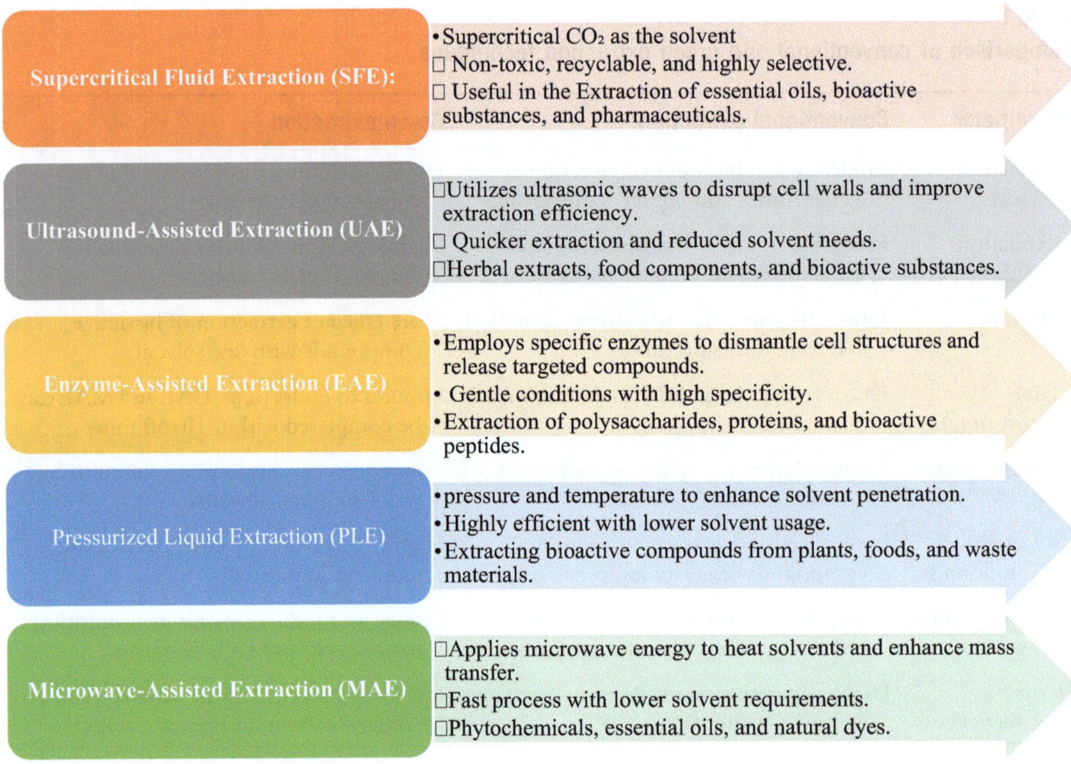

Fig. 1 Prominent sustainable extraction techniques [11]

and helps to preserve bioactive compounds. Enzyme-assisted extraction (EAE) utilizes specific enzymes to boost carotenoid yield under slight conditions. These techniques contribute to more efficient and sustainable extraction processes [10].

Certain nutrients play a vital role in promoting health and preventing disease. For instance, cancer prevention can be enhanced by reducing oxidative stress and inhibiting tumor growth. Cardiovascular health benefits from compounds like lycopene and β-carotene, which help prevent atherosclerosis and lower the risk of heart disease. Eye health can be supported by lutein and zeaxanthin, which protect against age-related macular degeneration. Additionally, astaxanthin aids in wound healing by supporting tissue repair and regulating glucose metabolism, contributing to overall metabolic health [11]. Different types of prominent sustainable extraction techniques are given in Fig. 1.

3.2 Supercritical Fluid Extraction (SFE)

Supercritical Fluid Extraction (SFE) is based on the unique properties of supercritical fluids (SCFs), which exhibit both liquid-like solvation power and gas-like diffusivity when subjected to specific temperature and pressure conditions. This allows SCFs to penetrate solid materials more effectively than liquids while selectively dissolving target compounds [6]. A comparison between the

Table 2
Advantages of SFE over conventional extraction

Parameter	Supercritical fluid extraction (SFE)	Conventional solvent extraction
Solvent use	CO_2 is non-toxic and recyclable and leaves no residue	Organic solvents are often toxic and require post-processing
Selectivity	High (adjustable via pressure and temperature)	Low (extracts many unwanted compounds)
Processing time	Short (high diffusivity of SCFs)	Long (requires extended solvent contact)
Extract purity	High (minimal contamination)	Often requires additional purification
Thermal stability	The low-temperature process preserves heat-sensitive compounds	High temperatures can degrade bioactive compounds
Environmental impact	Green and sustainable	High waste generation and toxicity

conventional extraction and SFE is given in Table 2. A fluid reaches the supercritical state when it is exposed to a temperature and pressure above its critical point (i.e., critical temperature (T_a) and critical pressure (P_a)). In this state, the fluid behaves as both a liquid and a gas, possessing: high density (similar to liquids), which increases the solubility of compounds; low viscosity (similar to gases) enhances mass transfer, allowing it to diffuse easily into solid materials. High diffusivity can penetrate porous matrices efficiently. Tunable solvating power: Adjusting pressure and temperature modifies its ability to dissolve different compounds. The most commonly used SCF in green extractions is carbon dioxide (CO_2) due to its low critical temperature (31.1 °C) and pressure (73.8 bar). It is non-toxic, non-flammable, and environmentally friendly nature. Easy removal by simple depressurization. The schematic diagram of the working of SFE is given in Fig. 2 and consists of a compressor/solvent pump and a modifier. The extraction material is placed in an extraction chamber, which is covered with a heat jacket. The extraction process consists of four main steps: Supercritical fluid contact with the sample: The sample (biomass) is loaded into an extraction vessel (e.g., CO_2 is pressurized above its critical point and introduced into the vessel). The SCF penetrates the solid matrix, reaching the solutes (target compounds). Solubilization of target compounds: The SCF dissolves specific compounds based on solubility principles (e.g., lipophilic or polar nature). Solubility depends on pressure, temperature, and the molecular interactions between the SCF and solutes. In some cases, co-solvents (e.g., ethanol) are added to enhance the extraction of polar compounds. Transport of extracted compounds: The SCF, now loaded with dissolved compounds, flows out of the

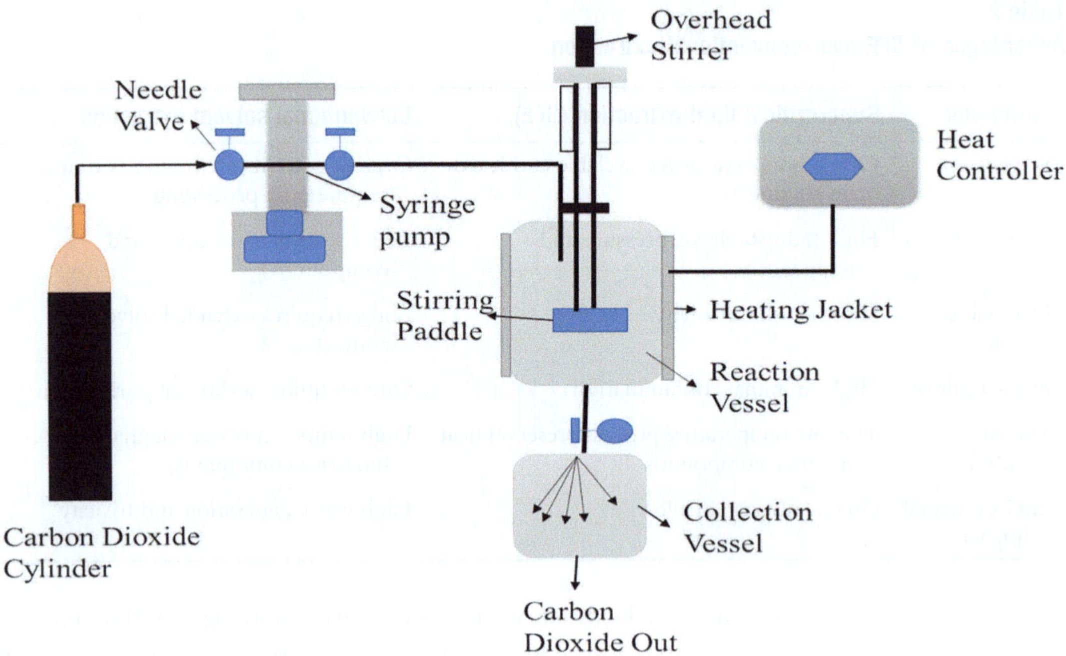

Fig. 2 Schematic diagram of working of SFE

extraction vessel. The extract-rich SCF is transported to a separation chamber. Depressurization and recovery: The SCF is depressurized, reducing its solubility and causing the dissolved compounds to precipitate out. The extract is collected in a separation vessel, while the CO_2 is either vented or recycled for further use.

Higher pressure increases the density of SCF, improving solubility and extraction efficiency. However, excessively high pressure can also dissolve unwanted compounds. Higher temperatures can improve solute diffusion but may reduce fluid density, affecting solubility. Some bioactive compounds degrade at high temperatures, requiring optimized conditions. Higher CO_2 flow rate improves mass transfer but may lead to inefficient solvent use. Lower flow rates enhance solute contact time but may slow the process. Pure CO_2 is non-polar, so co-solvents like ethanol, methanol, or water are used to extract polar compounds (e.g., polyphenols and alkaloids). Smaller particle sizes increase surface area, enhancing extraction. However, excessive grinding may lead to clogging and pressure drops in the system.

SFE is widely used in food, pharmaceuticals, cosmetics, and nutraceuticals, decaffeination of coffee and tea, extraction of essential oils (e.g., rosemary, thyme), removal of pesticides from food products, extraction of flavonoids, alkaloids, and polyphenols from medicinal plants, purification of omega-3 fatty acids from fish oil, extraction of antioxidants, pigments, and aroma compounds for

Table 3
Advantages of UAE over conventional extraction methods

Feature	Ultrasound-assisted extraction (UAE)	Conventional solvent extraction
Solvent consumption	Low (eco-friendly)	High (toxic organic solvents)
Energy efficiency	High (shorter extraction time)	Low (requires prolonged heating)
Extraction time	Fast (minutes to hours)	Slow (several hours to days)
Preservation of bioactive	Excellent (low temperatures)	Moderate (heat may degrade compounds)
Selectivity	High (adjustable via frequency and power)	Low (extracts unwanted compounds)

natural beauty products, and extraction of valuable bioactive compounds from agricultural and food waste (e.g., grape seed polyphenols, citrus flavonoids).

3.3 Ultrasound-Assisted Extraction (UAE)

It is a green extraction technique that utilizes high-frequency sound waves to enhance the recovery of bioactive compounds from plant materials. This method is based on the phenomenon of acoustic cavitation, which plays a crucial role in breaking down plant cell structures and improving mass transfer between the solvent and target compounds [6]. Table 3 describes the comparison between conventional extractions in the UAE. A schematic diagram of the working of UAE is given in Fig. 3 which shows that the key component of UAE is a transducer that generates high-frequency sound with the control of power supply, and the mechanism of UAE is as follows:

1. Formation of microbubbles: When ultrasound waves travel through a liquid, they create alternating compression (high pressure) and rarefaction (low pressure) cycles. During the rarefaction phase, tiny bubbles (vacuum cavities) form as the liquid molecules are pulled apart.
2. Bubble growth and coalescence: These microbubbles expand over multiple acoustic cycles, growing larger as they absorb more energy from the ultrasonic waves. Some bubbles merge to form larger cavitation bubbles, increasing their potential energy.
   - Collapse of cavitation bubbles: When the bubbles reach a critical size, they become unstable and collapse violently during the compression phase. This implosion generates localized hotspots with extreme conditions such as temperatures up to 5000 K (Kelvin), pressures exceeding 1000 atm (atmosphere), shockwaves, and high shear forces.

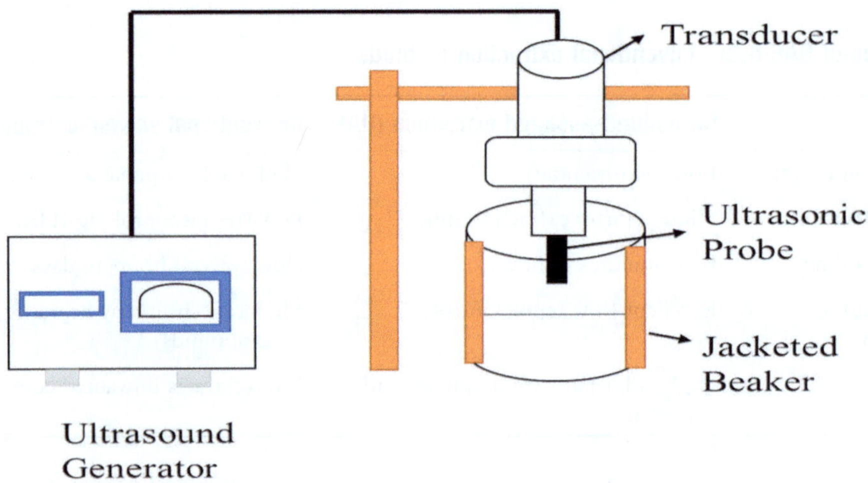

Fig. 3 Schematic diagram of working of UAE

3. Disruption of cell structures and enhanced mass transfer: The intense mechanical energy generated by collapsing bubbles ruptures cell walls, releasing bioactive compounds. The shockwaves and microjets improve the penetration of the solvent into the plant matrix, increasing the solubility and diffusion of target molecules. This reduces extraction time and enhances efficiency compared to conventional methods.

Extraction from the plant material can be done as the plant material is dried, ground, and mixed with the appropriate solvent. The mixture is placed in an ultrasonic extraction vessel. Ultrasound waves are applied using an ultrasonic probe or ultrasonic bath. Cavitation bubbles form, expand, and collapse, breaking down cell walls. The dissolution of the bioactive compound into the solvent is boosted by increased mass transfer. To increase the efficiency of the extraction, the solution is continuously mixed. The liquid extract is filtered to remove solid residues. The solvent may be evaporated or recovered, depending on the application. The bioactive compounds extracted are then subjected to further purification, like chromatographic techniques or membrane filtration [15].

3.4 Microwave-Assisted Extraction (MAE)

One of the eco-friendly extraction techniques is microwave-assisted extraction (MAE), in which the plant material is heated, and microwave energy is used for the extraction solvent. During this heating process, bioactive compounds are efficiently released. Faster heating and better mass transfer were observed in this method by letting microwaves interact with the polar molecules and ions of the sample [6]. The ionic conduction and dipole are resulted by the application of microwaves. Restricted heating results from this, raising the internal pressure in plant cells. The target compounds

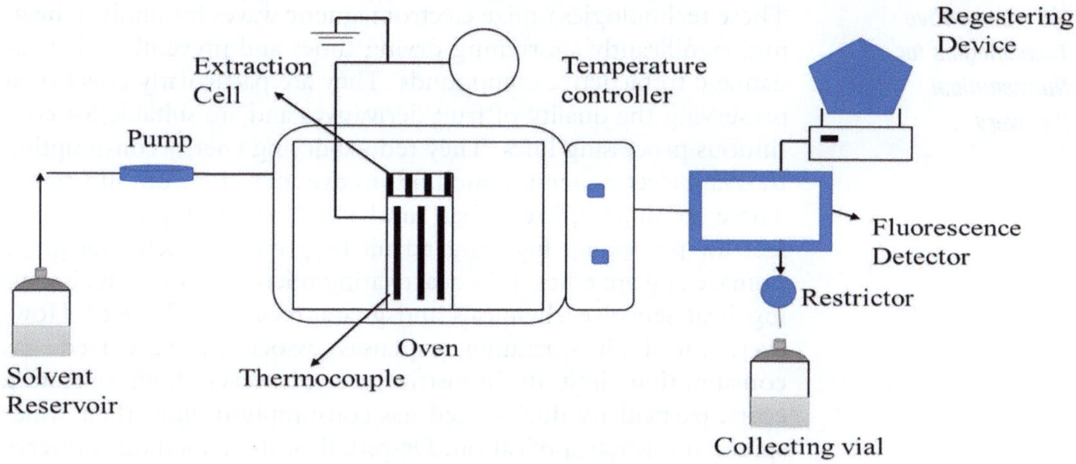

Fig. 4 Schematic diagram of working of MAE

are subjected to the solvent when the cell walls burst due to the increasing pressure. This mechanism improves extraction efficiency and shortens processing time by increasing the diffusion of bioactive components [16]. Figure 4 displays the instrumentation of MAE, the power control system adjusts the frequency of the waves, and the sample holder should be transparent.

It significantly reduces solvent consumption, minimizes energy usage, and shortens extraction time while preserving the integrity of heat-sensitive compounds. Additionally, the ability to control parameters such as microwave power, extraction time, temperature, and solvent type makes MAE a highly versatile technique suitable for various plant-derived bioactive compounds, including polyphenols, essential oils, flavonoids, and alkaloids [17].

Because of the demand for cleaner, greener processes and worries about solvent toxicity and environmental effects, solvent-free extraction techniques are becoming more and more popular in the nutraceutical sector. These techniques preserve bioactive chemicals while upholding safety and sustainability.

3.5 Supercritical Fluid Extraction (SFE)

SFE utilizes supercritical CO_2 (beyond its critical temperature and pressure) to extract bioactive compounds, functions as a solvent without leaving any hazardous residues, and allows for temperature and pressure control to modify the solubility of compounds. Benefits of CO_2 are environmentally friendly and reusable, allowing selective extraction by modifying conditions, and it operates at moderate temperature. It can be used for the extraction of omega-3 fatty acids that are extracted from fish oil and microalgae. Carotenoids, tocopherols (vitamin E), and phytosterols are separated from plant sources, and essential oils and flavors are extracted from herbs and spices.

3.6 Innovative Technologies for Nutraceutical Recovery

These technologies utilize electromagnetic waves for uniform heating, significantly shortening drying times and preventing thermal damage to bioactive compounds. They are particularly effective at preserving the quality of fruit derivatives and are suitable for continuous processing lines. They reduce drying energy consumption by using electrochemical methods to extract water from substrates. These are more efficient than mechanical dewatering and are perfect for processing high-sugar fruit by-products, including grape pomace and are efficient in deactivating microorganisms, maintaining heat-sensitive chemicals and guaranteeing food safety. However, the high operating expenses associated with feed gas consumption limit its industrial use. However, high operating costs, particularly due to feed gas consumption, limit their widespread industrial application. Despite this, these methods preserve the functional properties of bioactive substances by processing materials under high pressure without thermal damage. They are especially effective for enhancing the quality of protein-based products and for extracting antioxidants from grape by-products. Waves are used to break the structure of cells, which increases the release of bioactive compounds such as carotenoids and phenols. It requires less energy and time for extraction, so it has potential application on an industrial level for extraction. A high voltage electrical burst is used to destroy the cell structure and result in efficient bioactive extraction, as it requires less energy and gives high antioxidant output.

3.7 Technological Advancements in Nutraceutical Development

Due to the technological advancement in the drying process, the efficiency of the development and preservation of nutraceuticals has been increased. This advancement increased the shelf life of bioactive while preserving their activity. As the functional ability of this compound greatly depends upon the drying process, with the help of recent advancements the effectiveness, durability and quality have increased greatly. Advanced drying techniques for nutraceutical preservation include the following methods:

Freeze-Drying (Lyophilization): The chemical identity of the bioactive compound is greatly preserved by freeze-drying, so it is one of the popular methods to preserve heat-sensitive nutraceuticals. In this method, the product is first subjected to be frozen, and the water present there is sublimed. The bioactive factors, including probiotics, antioxidants, and vitamins, are preserved. The efficiency of the method has been increased by new advancements such as thin-film freeze drying (TFFD) which preserves biopharmaceuticals and inhalable bioactives. It is particularly for compounds that show sensitivity because of the improved size of particles to uniformity and less time required for drying.

Spray Drying and Microencapsulation: Spray drying is another technique to change liquid nutraceuticals to powder that is stable for spray drying. Microencapsulation, which requires coating bioactive chemicals in protective coatings to increase their stability and adjust their release, works efficiently. Spray drying is used to encapsulate essential nutrients, probiotics, and flavors, which ensures their bioavailability and protects them from environmental corrosion. Additionally, this process has been optimized by the incorporation of computational fluid dynamics (CFD) models, which allow enhanced control over particle size, shape, and thermal stability.

Heat Pump Drying (HPD): Heat pump drying is used to preserve compounds that are heat sensitive, it works at low temperatures as it is an energy-efficient method. It uses heat recovery devices to lessen the use of heat and control the drying environment. The phenolic content and antioxidants from functional food are successfully controlled by HPD. The level of humidity and temperature is controlled by this and results in the lowering of degradation of nutrients and better-quality products.

Infrared (IR) Drying: IR drying boosts the drying process by directly heating the surface of nutraceuticals with radiant radiations that remove the moisture content of the product fastly. In this method, the color, texture, and flavor of the product are retained along with the antioxidant quality. Recent studies have proved that foods like citrus peel, green vegetables, and garlic that are rich in antioxidants can be dried using IR with a small loss of bioactive.

Microwave-Assisted Drying: Electromagnetic waves are used in microwave-assisted drying to produce heat inside the product, which results in a fast and consistent drying procedure. With the reduction of drying time, it reduces the thermal deterioration and makes it useful to heat substances like vitamins and polyphenols that are heat sensitive. In the hybrid process, aligning low pressure and microwave heating is more beneficial than over Hoover drying assisted by microwave. As a result of this combination, heat-sensitive compounds are preserved better and quickly.

3.8 Emergence of Hybrid Drying Technologies

Hybrid drying technology is achieved by aligning different drying methods to overcome the drawbacks of that method. Infrared-assisted fluidized bed drying, spray freeze drying, and microwave-vacuum drying are examples of hybrid drying systems that are beneficial as less time required, less energy cost, and high output. Microwave-vacuum drying is used for functional food bioactive and also stops oxidation in the process. A combination of spray and freeze-drying methods is used to preserve delicate material.

3.9 Integration of Industry 4.0 Technologies in Drying Systems

The manufacture of nutraceuticals has seen a rebellion in drying techniques with the initiation of Industry 4.0. Real-time monitoring and drying condition optimization are made possible by smart drying systems that make use of sensors, the Internet of Things (IoT), artificial intelligence (AI), and machine learning (ML). Predictive maintenance is made possible by these cutting-edge technologies, which also improve the energy economy and guarantee constant product quality. To reduce energy usage and preserve bioactive components, machine learning algorithms can examine drying patterns and modify parameters.

3.10 Thermal Degradation Modeling for Process Optimization

The creation and preservation of nutraceuticals have been greatly impacted by developments in drying technologies. Innovations, including microencapsulation, hybrid drying systems, thin-film freeze drying (TFFD), and Industry 4.0 integration, have increased the stability, effectiveness, and quality of nutraceutical goods. Furthermore, these methods have been improved through the application of computational models for process optimization, guaranteeing the preservation of bioactive components during the drying process. Advanced computer models are used to forecast thermal deterioration during drying procedures in order to produce nutraceutical goods of the best quality. The effects of temperature on nutrient stability are simulated using models such as the Arrhenius equation and first-order reaction kinetics.

3.11 Nanotechnology in Encapsulation

Create nano-emulsions that enhance the bioavailability of nutraceuticals. Despite its benefits, scaling up nano-emulsion production remains costly and raises safety concerns regarding the impact of nanoparticles on biological systems. Table 4 summarizes several studies on encapsulation methods used to recover agricultural food wastes, emphasizing encapsulation techniques, wall materials employed, efficiency, and observed impacts [18].

4 Notes

This section covers the environmental and economic impacts of these eco-innovative methods for extraction. Growing concerns related to environmental safety have led to the development of eco-innovative protocols for converting food waste into useful byproducts such as high-value nutraceuticals. These procedures include green extraction techniques, bioprocessing strategies, and sustainable sourcing approaches to limit waste and preserve essential bioactive substances. Significant environmental and economic impacts are offered by eco-innovative protocols for developing nutraceuticals from food by-products. Waste materials are transformed to useful products, and it resulted in a reduction of food waste, lower environmental footprints, and presented. Industries

Table 4
Different encapsulation methods used to recover agricultural waste

Extract source	Encapsulation method	Wall material	Efficiency	Effect of encapsulation
Golden apple and red grape pomace (phenolic compounds)	Nano emulsification	Chitosan/soy protein	Up to 95%	Improvement of antioxidant activity of phenolic extracts
Sweet potato peels (carotenoids and phenolic compounds)	Spray drying/ freeze-drying	Whey protein	Up to 61.9%	Better quality of encapsulates based on various parameters; retention of compounds and prolonged shelf life under different storage conditions
Carotenoids from tomato peels	Electrospinning	Zein fiber	>90%	The nano-encapsulation method increases antioxidant properties up to 11 folds
Lemon byproducts (bioactive compounds)	Spray drying/ freeze-drying	Maltodextrin/soybean protein/ι-carrageenan	–	Highest retention of total phenolic content, total flavonoid content, and ferric ion-reducing antioxidant power with freeze-drying using maltodextrin and soybean protein
Beetroot pomace (bioactive compounds)	Freeze-drying	Soy protein	86.14%	Retention of polyphenols (76.67%), betalain pigments, betacyanins (17.77%), and betaxanthins (17.72%) during 3 months of storage at room temperature
Blueberry residue	Ionotropic gelation	Sodium alginate	–	67.01% retention of the phenolic compound

may considerably reduce their environmental impact and landfill trash by using food by-products such as citrus peels, olive pomace, and brewer's leftover grain. Additionally, because they need less energy and chemicals and are a successful substitute for conventional methods, eco-friendly extraction technologies like enzyme-assisted and ultrasound-assisted extraction are more affordable than solvent-based procedures. These advancements give a practical and sustainable method to the production of nutraceuticals, which support the global campaigns to the development of environmental preservation and a circular economy. The likely health benefits of nutraceuticals which include dietary supplements, functional foods, and natural health products have pinched a lot of interest. Additionally, affecting the healthcare industry, the increased demand for these stuffs had significant harmful effects on the economy and the environment. This section highlights the transition to green technology and looks at the economic growth associated with the nutraceutical business, as well as its importance in fostering environmental sustainability.

4.1 Reduction of Food Waste and Environmental Footprint

Reducing food waste is an essential way to lessen the environmental impact of rejected food, which contributes to the generation of greenhouse emissions and the depletion of supplies. In addition to reducing the quantity of waste dumped in landfills, using environmentally friendly methods to turn food waste into nutraceuticals also lessens the impact on the environment [19]. Recycling these leftovers could lower methane emissions and conserve energy, water, and arable land—all of which are utilized in the production of food. By turning potential trash into beneficial, health-promoting products, this method promotes a circular economy and is in line with sustainable development goals. With therapeutic properties ranging from antihypertensive to anti-inflammatory, these chemicals derived from waste food are perfect for use as functional meals for health. To put it briefly, addressing food waste by turning food waste into nutraceuticals from food by-products is a viable substitute for lowering environmental impact, preserving resources, and enhancing sustainability in the food industry.

4.2 Eco-Innovative Methods Compared to Conventional Approaches

This method increases efficiency and yield, such as PAE, which is used for the extraction of anthocyanin from red onion peel, reported 81.84%, as with MAE, it is 40.01% [20]. These methods use safe and sustainable solvents and utilize low energy as it reduces the extraction time, as MAE operates at low temperatures and less time is required. Enzyme-assisted extraction is an economical process because of the easy availability of enzymes [21]. Certain techniques can be incorporated as a part of any system to reduce the cost [22].

4.3 Market Expansion and Investment

The nutraceutical market has witnessed substantial growth driven by increasing consumer interest in health-promoting products. This industry is composed of functional foods, dietary supplements, and botanical formulations [23]. The nutraceutical sector has developed into a multi-billion-dollar industry, experiencing rapid growth attributed to increased consumer awareness and a growing focus on preventive healthcare. In 2016, the European nutraceutical market was estimated at USD 79.7 billion, with an anticipated compound annual growth rate (CAGR) of approximately 6.39% from 2018 to 2023.

4.4 Job Generation and Innovation

Nutraceutical investment promotes the creation of jobs in the domains of research, agriculture, processing, and marketing. Economic progress is being propelled by innovative formulation methods and biotechnology uses. Cost-effectiveness and medical spending: Through the prevention of chronic diseases like diabetes and heart disease, widespread usage of nutraceuticals could reduce healthcare expenses. This reduces the burden of disease, which benefits economies indirectly.

4.5 Regulatory Challenges and Ecological Advancement

Economic uncertainty is introduced by the lack of a standardized legal framework for nutraceuticals. While lax standards in certain areas create worries about quality, strong rules in others limit market entrance [24]. The manufacturing of nutraceuticals is being influenced by the shift to circular and sustainable economies. Utilizing eco-friendly resources such as microalgae, bacteria, and genetically modified crops has been shown to provide advantages.

Eco-friendly extraction and processing: Supercritical fluid extraction and ultrasonic-assisted extraction are two examples of environmentally friendly extraction methods that reduce ecological impact and solvent consumption. To increase effectiveness and environmental friendliness, the nutraceutical industry has embraced sustainable extraction techniques. Techniques like the use of natural deep eutectic solvents (NDESs) and ultrasound-assisted extraction (UAE) have shown promise in reducing energy consumption and chemical waste. These methods are consistent with international efforts to reduce industrial operations' carbon footprints while maintaining the potency of bioactive compounds [25].

Waste minimization and circular economy: The production of carotenoid-rich nutraceuticals from food and agricultural by-products, such as fruit and vegetable skins, is encouraging waste reduction initiatives. Sustainable production methods are being used to reduce environmental impacts as a result of the growing demand for nutraceuticals. The business's main focus is on the shift to a circular economy, which involves repurposing agricultural and food sector waste to extract bioactive ingredients. Utilizing microalgae and other renewable resources to produce nutraceutical components is a particularly encouraging development that reduces dependency on traditional agriculture and resource depletion [26].

Climate change considerations: Hydroponics and controlled salinity stress are two new farming methods that are increasing crop yields while using less water. Additionally, the manufacture of nutraceuticals is becoming more sustainable because of non-thermal preservation techniques like high-pressure processing. Climate-aware farming methods are becoming more and more popular in the nutraceutical sector as environmental concerns gain importance. These methods include water-efficient agricultural practices to reduce excessive water consumption. Sustainable cultivation of plant-based ingredients to reduce greenhouse gas emissions. Organic and non-GMO sourcing to promote environmental safety and enhance consumer trust. The economic and environmental dimensions of nutraceuticals are interconnected, underscoring the importance of sustainability and financial practicality in this expanding industry [10].

4.6 Case Study

Studies have been conducted on the application of defatted olive pomace powder, and its incorporation into baked goods as a functional ingredient. It was discovered that using this powder increased the total phenolic content of the products, improving the antioxidant properties. Such a method improves the nutritional profile of baked goods and offers an eco-friendly solution to the waste produced by the olive industry [27]. A researcher studied the potential fish growth capabilities with non-conventional feed ingredients, like plant by-products or processed animal protein, and its impact on gilthead seabream. These modified formulations indeed boosted fish growth performance, improved nutrient retention, and lessened environmental pollution compared to traditional diets based on fishmeal. This gives scope for developing cost-effective and eco-friendly aquaculture feed using food industry waste.

4.6.1 Regulatory and Safety Considerations

Nutraceuticals made from food by-products must meet high regulatory and safety criteria to assure customer safety and product efficacy. Key considerations include: everywhere across the world, nutraceuticals are regulated in varying ways. In the USA, the Food and Drug Administration regulates these categories of products per the Dietary Supplement Health and Education Act (DSHEA) of 1994, which has a definition of dietary supplements and a set of instructions concerning their labels. It is important to note that pre-market approval is not necessary under DSHEA, which means that the responsibility to make sure that the product is safe and effective lies with the manufacturers. Similarly, the European Union and Australia have separate regulatory systems, but there is no worldwide agreement on nutraceutical rules [28].

4.6.2 Quality Control and Safety Evaluation Protocols

Ensuring the quality and safety of nutraceuticals includes laborious protocols: manufacturing companies should follow GMP measures, which consist of people education, sanitation, apparatus authentication, and method justification [29]. Contaminants such as *E coli*, *Salmonella*, and *Staphylococcus aureus* must be tested regularly since they might be introduced at any point throughout the manufacturing process [30]. Using permitted analytical methods that guarantee that active components fulfil purity and effectiveness of standards [31]. Wide-ranging valuations, together with toxicological trials, are approved to confirm that both the raw constituents and the complete products are safe to use [32]. Clear labeling is vital for consumer trust and regulatory compliance: The name of the product should be labeled, along with a statement distinguishing it as a dietary supplement, the producer's information, a "Supplement Facts" panel, a list of additional components, and the net measure of contents [33]. To avoid devious customers, manufacturers should meet the regulatory limits, and any claims about health benefits must be supported by scientific evidence [34]. Addressing these regulatory and safety factors is essential for the effective development and promotion of nutraceuticals sourced from food by-products, thereby guaranteeing compliance and fostering consumer trust.

4.6.3 Applications and Market Potential

The creation of nutraceuticals from food by-products highlights environmentally sustainable methods that minimize waste while generating health-enhancing products. This strategy is in line with contemporary trends in the nutraceutical sector and addresses the increasing consumer preference for sustainable and eco-conscious products. Grape skins, which are a by-product of winemaking, are rich in resveratrol, a polyphenol celebrated for its antioxidant characteristics and possible health advantages, such as anti-inflammatory and cardioprotective effects [35]. Fish processing also produces a considerable quantity of waste, including heads, skins, and viscera, which are abundant in omega-3 fatty acids. These vital nutrients are linked to improved cardiovascular health and cognitive performance. By extracting omega-3 s from fish waste, it is possible to enhance value while simultaneously mitigating environmental impact [36]. Several customers are adopting plant-based products for health benefits along with ethical deliberation and the impacts of the environment. This has led to a rise in the research and growth of nutraceuticals obtained from plants. In addition to this, users are in search of products that should be obtained from sustainable sources and have been produced assuring ethically approved values. Which resulted in the market moving toward eco-friendly practices and high transparency in sourcing and production methods [37]. Current clients are more conscious about the environment and favor products that line up with their values. It is clear from studies that a huge portion of consumers are ready to pay the premium for articles that have sustainable and eco-friendly packaging and branding [2] with the rise in the expectation that the packaging material should be recycled, which reflects a great commitment to the sustainability of the environment [2].

5 Future Prospects

The production of nutraceuticals from food waste by utilizing eco-innovative approaches has great chances for sustainability and economic progress. However, many track races must be overwhelmed to rule these technologies and integrate them into typical manufacturing.

Food processing industries produce great amounts of by-products that are rich in bioactive ingredients, such as polyphenols, dietary fibers, proteins, and essential fatty acids. Revolutions in green extraction technologies, biotransformation, and nanoencapsulation are growing the potential of these underutilized resources. For instance, novel food processing means, such as ultrasonic, microwave, enzyme, supercritical fluid, and pressurized liquid extraction, are being explored to expand the bioavailability and functional characteristics of nutraceuticals gained from underutilized food sources and by-products. A recent study shows the possibility of upcycling grape stems, pomace, and skin by-products of the wine industry into nutraceutical components rich in bioactive substances, including phenolic acids and flavonoids.

Some technical, economic, and regulatory barriers are there for the reduction of nutraceuticals from food by-products. There is variability in the raw material because of sourcing and processing, which affects standardization. It also requires a high initial investment to set up this advanced technique, which has a large economic effect. A supportive governmental environment and cross-sector collaboration are essential for promoting the adoption of eco-innovative protocols. Collaboration between research institutes and the food industry may speed up technical breakthroughs and improve information transfer for better by-product utilization. Organizations like WHO highlight the lowering of food waste, so lining up these methods will give great regulatory support.

The future of eco-innovative nutraceuticals derived from food by-products looks promising. These products have the potential to reduce waste, improve nutrition, and encourage sustainable industrial practices. However, to fully realize this potential, it is essential to address challenges related to scaling, regulation, and consumer perception. This will require strategic investments, government support, and collaborative innovation. By tackling these issues, the nutraceutical industry could contribute to the development of a circular economy while promoting health and environmental sustainability.

6 Conclusion

Eco-innovative methods for creating nutraceuticals from food waste or by-products offer a sustainable and cost-effective method to lower food waste while increasing human health benefits. Green extraction techniques, such as supercritical fluid extraction, ultrasound-assisted extraction, and enzyme-based bioprocessing, can greatly recover high-value bioactive compounds with reduced environmental impact. These methods not only improve resource efficiency but also lower the belief in harmful chemical solvents, thereby reducing the overall environmental footprint of nutraceutical manufacturing.

The practice of eco-innovative measures in the expansion of nutraceuticals resulting from food by-products offers a game-changing approach to sustainability, economic viability, and human health. Food waste may be transformed into high-value bioactive nutraceuticals using green and clean extraction methods, bioprocessing advances, and improved formulation strategies, diminishing environmental influence while giving new economic opportunities. A multidisciplinary approach is essential for the successful implementation of these regulations, encompassing sustainable sourcing, eco-friendly processing, regulatory compliance, and consumer acceptance. The case studies conferred high-spot real-world applications where food waste has been proficiently converted into nutraceuticals, indicating both environmental and financial profits. Moreover, the economic potential of these strategies surpasses that of many conventional approaches, indicating a promising path forward for sustainable innovation.

However, significant complications persist, together with scalability, standardization, and regulatory limits. Overcoming these blocks demands policy support, cross-sector collaboration, and enduring innovation in food waste valorization. Government encouragement, industry-academia associations, and public awareness movements will be serious in rushing the adequacy and commercialization of sustainable nutraceuticals. The developments in AI-driven food waste optimization, innovative bioprocessing skills, and tailored nutrition show excessive potential for the area. With smart ventures and a commitment to eco-innovation, the future of nutraceuticals resulting from food by-products may donate to a circular economy, promising an improved world and people.

References

1. Paul-Chima UO, Ugwu CN, Alum EU (2024) Integrated approaches in nutraceutical delivery systems: optimizing ADME dynamics for enhanced therapeutic potency and clinical impact. RPS Pharm Pharmacol Rep 3(4): rqae024
2. Chopra AS et al (2022) The current use and evolving landscape of nutraceuticals. Pharmacol Res 175:106001
3. Rane BR, Kesarwani R (2020) Introduction to nutraceuticals and dietary supplements. In:

Nutraceuticals and dietary supplements. Apple Academic Press, pp 35–56
4. Gülcin I (2012) Antioxidant activity of food constituents: an overview. Arch Toxicol 86(3): 345–391
5. Daliri EB-M, Lee BH (2015) Current trends and future perspectives on functional foods and nutraceuticals. In: Beneficial microorganisms in food and nutraceuticals. Springer, pp 221–244
6. Arshad I et al (2024) Green and clean extraction technologies for novel nutraceuticals. In: Bioactive extraction and application in food and nutraceutical industries. Springer, pp 391–417
7. de Souza Mesquita LM et al (2024) Path2-Green: introducing 12 green extraction principles and a novel metric for assessing sustainability in biomass valorization. Green Chem 26(19):10087–10106
8. Zhang Z et al (2018) Comparison of selected clean and green extraction technologies for biomolecules from apple pomace. Electrophoresis 39:1934
9. Ozogul Y et al (2018) Comparison of green and conventional extraction methods on lipid yield and fatty acid profiles of fish species. Eur J Lipid Sci Technol 120(12):1800107
10. Galanakis CM (2013) Emerging technologies for the production of nutraceuticals from agricultural by-products: a viewpoint of opportunities and challenges. Food Bioprod Process 91(4):575–579
11. Bas TG (2024) Bioactivity and bioavailability of carotenoids applied in human health: technological advances and innovation. Int J Mol Sci 25(14):7603
12. Meléndez-Martínez AJ et al (2021) Carotenoids: considerations for their use in functional foods, nutraceuticals, nutricosmetics, supplements, botanicals, and novel foods in the context of sustainability, circular economy, and climate change. Annu Rev Food Sci Technol 12(1):433–460
13. Shao A (2017) Global market entry regulations for nutraceuticals, functional foods, dietary/food/health supplements. In: Developing new functional food and nutraceutical products. Elsevier, pp 279–290
14. Priyadarshini A, Tiwari BK, Rajauria G (2022) Assessing the environmental and economic sustainability of functional food ingredient production process. Processes 10(3):445
15. Kumar K, Srivastav S, Sharanagat VS (2021) Ultrasound assisted extraction (UAE) of bioactive compounds from fruit and vegetable processing by-products: a review. Ultrason Sonochem 70:105325
16. Destandau, E., Michel T (2022) Microwave-assisted extraction
17. Chemat F, Cravotto G (2012) Microwave-assisted extraction for bioactive compounds: theory and practice, vol 4. Springer Science & Business Media
18. Capanoglu E, Nemli E, Tomas-Barberan F (2022) Novel approaches in the valorization of agricultural wastes and their applications. J Agric Food Chem 70(23):6787–6804
19. Wibisono DAS et al (2025) Advancing industrial food byproduct management: strategies, technologies, and Progress in waste reduction. Processes 13(1):84
20. Mirzazadeh N et al (2024) Comparison of different green extraction methods used for the extraction of anthocyanin from red onion skin. Food Sci Nutr 12(10):7347–7357
21. Siddique F et al (2024) Exploration of novel eco-friendly techniques to utilize bioactive compounds from household food waste: special reference to food applications. Front Food Sci Technol 4:1388461
22. Lisboa HM et al (2024) Innovative and sustainable food preservation techniques: enhancing food quality, safety, and environmental sustainability. Sustain For 16(18):8223
23. Shinde N et al (2014) Nutraceuticals: a review on current status. Res J Pharmacy Technol 7(1):110–113
24. Santini A et al (2018) Nutraceuticals: opening the debate for a regulatory framework. Br J Clin Pharmacol 84(4):659–672
25. Shrivastav G et al (2024) Eco-friendly extraction: innovations, principles, and comparison with traditional methods. Separ Purif Rev:1–17
26. Kumar K et al (2017) Food waste: a potential bioresource for extraction of nutraceuticals and bioactive compounds. Bioresour Bioprocess 4: 1–14
27. Di Nunzio M et al (2020) Olive oil by-product as functional ingredient in bakery products. Influence of processing and evaluation of biological effects. Food Res Int 131:108940
28. Komala MG et al (2023) Investigating the regulatory process, safety, efficacy and product transparency for nutraceuticals in the USA, Europe and Australia. Food Secur 12(2):427
29. Bailey RL (2020) Current regulatory guidelines and resources to support research of dietary supplements in the United States. Crit Rev Food Sci Nutr 60(2):298–309
30. Mukherjee P et al (2017) Factors to consider in development of nutraceutical and dietary supplements. In: Pharmacognosy. Elsevier, pp 653–661

31. Brykman MC et al (2022) What should clinicians know about dietary supplement quality? AMA J Ethics 24(5):382–389
32. Sirois J et al (2024) Safety considerations for dietary supplement manufacturers in the United States. Regul Toxicol Pharmacol 147: 105544
33. Kołodziej G et al (2019) Knowledge concerning dietary supplements among general public. Biomed Res Int 2019(1):9629531
34. Puri V et al (2022) A comprehensive review on nutraceuticals: therapy support and formulation challenges. Nutrients 14(21):4637
35. Liu Z et al (2023) Valorization of food waste to produce value-added products based on its bioactive compounds. Processes 11(3):840
36. Borah MS et al (2023) Recent trends in valorization of food industry waste and by-products: encapsulation and in vitro release of bioactive compounds. Food Secur 12(20):3823
37. Santini C, Supino S, Bailetti L (2023) The nutraceutical industry: trends and dynamics. In: Case studies on the business of nutraceuticals, functional and super foods. Elsevier, pp 3–20

Chapter 7

Nutrient Recovery Techniques Using Eco-Safe Technologies for Agricultural Waste

Ratnnadeep C. Sawant, Rahul B. Kamble, Pravin S. Deharkar, Shun-Yuan Luo, and Subhash R. Somkuwar

Abstract

Nutrients like nitrogen, phosphate, and potassium are abundant in agricultural waste, which includes crop wastes, animal dung, and by-products of food manufacturing. However, poor handling of these wastes frequently results in resource loss, degradation of the environment, and a greater need for artificial fertilizers. Eco-safe nutrient recovery technologies have become well known as sustainable ways to recycle important nutrients back into agricultural systems in response to these issues. These eco-friendly techniques for turning trash into nutrient-rich fertilizers, improving soil health, and lowering greenhouse gas emissions include composting, anaerobic digestion, biochar synthesis, vermicomposting, and microbial-based recovery.

A comparison of these methods reveals their economic viability, environmental impact, and efficacy. Successful implementation and the advantages of implementing circular agriculture methods are illustrated through case studies and real-world applications. With an emphasis on the necessity of innovation and integrated approaches to sustainable nutrient management, the chapter ends with insights into the difficulties, policy issues, and hopes for nutrient recovery in the future. Agricultural systems can move toward a more resilient and circular model by putting eco-safe nutrient recovery technology into practice. This will reduce waste, enhance soil health, and lessen environmental concerns.

This chapter examines several environmentally friendly nutrient recovery methods, emphasizing their workings, advantages, and uses in sustainable farming. Farmers may increase crop productivity, reduce their negative effects on the environment, and stop the nutrient loop by implementing these techniques. Furthermore, incorporating these technologies into agricultural systems promotes long-term food security, resource conservation, and climate change mitigation. Cost, scalability, and farmer adoption are still obstacles in the way of their potential. In order to ensure a sustainable and resilient agricultural future, this study highlights the necessity of policy assistance, education, and innovation to encourage the broad adoption of eco-safe nutrient recovery technology.

Key words Nutrient recovery, Agricultural waste management, Eco-safe technologies, Sustainable agriculture, Biochar application

Tanmay Sarkar (ed.), *Food Waste Valorization: Green Techniques in Sustainable Management*, Methods and Protocols in Food Science, https://doi.org/10.1007/978-1-0716-4646-5_7,
© The Author(s), under exclusive license to Springer Science+Business Media, LLC, part of Springer Nature 2025

1 Introduction

The future of food production and the planet's general health depend on a balance of environmental preservation, economic viability, and social responsibility, which is what sustainable agriculture seeks to accomplish. The process of recovering nutrients from agricultural waste through environmentally friendly technologies is increasingly recognized as a sustainable approach. This practice not only aims to improve soil health but also seeks to minimize environmental pollution and foster circular economies. Agriculture [1] plays a crucial role in ensuring global food security, maintaining commercial stability, and supporting rural communities. Nevertheless, contemporary agricultural methods produce significant quantities of waste, which encompasses crop residues, livestock manure, by-products from agro-industrial activities, and waste generated from food processing [2].Inefficient management of these waste streams may result in environmental contamination, heightened greenhouse gas emissions, and the exhaustion of vital nutrients in the soil. The increasing demand for food, coupled with the continuous deterioration of soil quality [3], has intensified reliance on synthetic fertilizers. These fertilizers pose various risks, including water pollution, soil acidification, and increased production expenses. Agricultural waste [4] serves as a vital source of essential nutrients, such as phosphorus (P), nitrogen (N), and potassium (K), which are crucial for plant development. Nevertheless, improper disposal or management of this waste can lead to the loss of these nutrients through mechanisms like leaching, runoff, or volatilization. This mismanagement can consequently result in soil degradation, water pollution, greenhouse gas emissions, and financial incompetence [5]. The agricultural field invests billions of dollars in synthetic fertilizers, often overlooking the benefits of nutrient-dense waste. By recovering and repurposing nutrients from agricultural by-products, a closed-loop system can be created that enhances soil health, reduces environmental pollution, and lessens dependence on chemical inputs [6]. Agricultural waste [4], encompassing crop residues, livestock manure, and food processing by-products, represents a significant source of organic matter and vital nutrients, including nitrogen, phosphorus, and potassium. Ineffective management of these materials [7] frequently results in considerable environmental challenges, including soil degradation, water pollution, and heightened greenhouse gas emissions. Concurrently, the excessive dependence on synthetic fertilizers to fulfill the nutritional requirements of crops has led to resource depletion, deterioration in soil quality, and rising production costs. In this regard, the recovery of nutrients from agricultural waste [8] through eco-friendly technologies has emerged as a viable and sustainable approach to tackle these pressing concerns.

Sustainable agriculture and the reduction of the ecological impact of farming practices are significantly enhanced by eco-friendly nutrient recovery methods [9]. These approaches concentrate on the extraction of essential nutrients from organic waste and other sources, promoting their repurposing in agricultural production [10]. These technologies enhance soil fertility and boost crop production while contributing to climate change mitigation by lowering greenhouse gas emissions [11] and improving carbon sequestration. By adopting practices such as composting, anaerobic digestion, biochar production, and microbial recovery, farmers can transform waste into valuable resources, promoting a circular agricultural economy [12]. The technologies involve biological, chemical, and physical processes that effectively recover and recycle nutrients without harming ecosystems. Key strategies include composting and vermicomposting [13], anaerobic digestion [14], biochar application [15], algae-based recovery [16], and electrochemical and membrane-based recovery [17].

This methodology is consistent with the tenets of sustainable agriculture, which prioritize efficient resource use, environmental conservation, and enduring food security. As the world's population continues to grow and the demand for food increases, the incorporation of environmentally friendly nutrient recovery technologies becomes essential for ensuring the resilience and sustainability of agricultural practices. This paper explores various eco-conscious techniques for recovering nutrients from agricultural waste, emphasizing their advantages, challenges, and the potential prospects for broad adoption.

1.1 Significance of Nutrient Recovery from Agricultural Waste

The potential for solving important environmental, financial, and agricultural issues is what makes nutrient recovery from agricultural waste significant. This technique helps to protect the environment, manage resources sustainably, and increase agricultural output by removing vital nutrients like potassium (K), phosphorus (P), and nitrogen (N) from waste materials. The conservation of finite resources, such as phosphate rock, is facilitated by the recovery of nutrients from agricultural waste, which lessens reliance on synthetic fertilizers.

Environmental Protection: One important tactic for protecting the environment is the recovery of nutrients from agricultural waste. Crop leftovers, animal dung, and agro-industrial by-products are examples of agricultural waste that can cause serious environmental problems, such as soil erosion, water pollution, and greenhouse gas emissions, if it is not managed properly. These effects can be lessened, and a cleaner, healthier environment can be promoted by recovering and recycling nutrients like potassium (K), phosphorus (P), and nitrogen (N).

Improvement of Soil Health: Agriculture that is both productive and sustainable is based on healthy soil. Plant development, improved water infiltration and retention, carbon sequestration, and a variety of microbial communities are all supported by healthy soils. By reintroducing vital nutrients into the soil in an environmentally responsible way, nutrient recovery from agricultural waste is significant to enhancing soil health.

A revolutionary strategy for addressing a number of global issues, such as resource scarcity, food security, and environmental degradation, is nutrient recovery from agricultural waste. It encourages sustainable agriculture, lowers pollution, and boosts economic growth by turning waste into useful resources. Nutrient recovery will be essential to creating resilient and environmentally friendly agricultural systems as the globe shifts to more sustainable methods.

2 Materials

The effective execution of environmentally sustainable nutrient recovery methods depends on the accessibility of particular materials, equipment, and resources. Presented below is a comprehensive inventory of materials frequently utilized in diverse nutrient recovery procedures.

2.1 Composting [13]

A balanced mixture of organic materials, microbial inoculants, and ideal ambient conditions is necessary for successful composting. Together with necessary additives, the primary materials are divided into components that are nitrogen-rich (greens) and carbon-rich (browns).

- **Organic Waste**: Crop residues, livestock manure, food waste, and other biodegradable agricultural by-products.
- **Bulking Agents**: Dry materials like straw, sawdust, or wood chips to improve aeration.
- **Microbial Inoculants**: Optional additives to accelerate decomposition.
- **Water**: To maintain optimal moisture levels.
- **Compost Bins or Piles**: Structures to contain and manage the composting process.
- **Aeration Tools**: Pitchforks or mechanical turners to ensure oxygen supply.

2.2 Anaerobic Digestion [14]

Anaerobic digestion is a biological process that produces digestate, a nutrient-rich residue, and biogas, a mixture of carbon dioxide and methane, when organic materials are broken down without oxygen.

Waste management, the creation of renewable energy, and soil amendment are all common uses for this procedure. Anaerobic digestion is a sustainable and adaptable technique that helps achieve the objectives of the circular economy by turning waste into useful resources. For effective operation, materials must be chosen and managed carefully.

- **Organic Waste**: Manure, crop residues, and food waste.
- **Digester Tanks**: Sealed containers to facilitate anaerobic conditions.
- **Inoculum**: Anaerobic bacteria to kick-start the digestion process.
- **Biogas Storage**: Tanks or bags to store produced biogas.
- **Digestate Collection System**: To capture nutrient-rich slurry for use as fertilizer.

2.3 Biochar Production [15]

Pyrolysis, the heat breakdown of organic biomass in an oxygen-limited environment, produces biochar, a substance rich in carbon. It is frequently applied as a soil amendment to raise agricultural production, improve soil health, and sequester carbon. A sustainable method of managing organic waste, enhancing soil health, and slowing down climate change is the manufacture of biochar. The pyrolysis settings, feedstock selection, and post-processing techniques all have a big impact on the quality and use of biochar.

- **Biomass**: Agricultural residues like straw, husks, or wood chips.
- **Pyrolysis Unit**: A reactor or kiln for heating biomass in the absence of oxygen.
- **Cooling and Collection System**: To safely collect and store biochar.
- **Grinding Equipment**: To process biochar into a fine powder for soil application.

2.4 Phosphorus Recovery [18]

The process of removing phosphorus from waste streams, such as sewage sludge, wastewater, or agricultural runoff, in order to recycle it for use as fertilizer or in industrial settings, is known as phosphorus recovery. Since phosphorus is a non-renewable resource and an important nutrient for agriculture, recovery is crucial to sustainability. In order to minimize environmental contamination and lessen dependency on phosphate rock that is mined, phosphorus recovery is an essential part of sustainable resource management. The phosphorus source and the intended final product determine the recovery technique and materials to be used.

- **Source Material**: Manure, agricultural runoff, or wastewater.
- **Precipitation Agents**: Magnesium or calcium compounds for struvite formation.

- **Reactor Tanks**: For controlled precipitation processes.
- **Filtration Systems**: To separate recovered phosphorus solids.

2.5 Vermicomposting [13]

Vermicomposting, also referred to as worm castings, is the process of turning organic waste into nutrient-rich compost by employing earthworms. This process creates a useful soil amendment and is very effective and eco-friendly. Vermicomposting is an environmentally friendly method of managing organic waste, cutting down on landfill usage, and creating high-quality compost for farming and gardening. Successful vermicomposting depends on the selection of materials and appropriate system management.

- **Organic Waste**: Crop residues, manure and food scraps.
- **Earthworms**: Species like *Eisenia fetida* (red wigglers) for efficient decomposition.
- **Vermicompost Bins**: Containers with proper drainage and aeration.
- **Bedding Material**: Coconut coir, shredded paper, or straw for worm habitat.

2.6 Algal Biofertilizers [16]

Algal biofertilizers, which are used to improve soil fertility and encourage plant development, are fertilizers made from algae, mainly cyanobacteria (blue-green algae) and microalgae. These sustainable and environmentally friendly biofertilizers give plants vital nutrients and bioactive substances. Algal biofertilizers are a viable substitute for chemical fertilizers, providing an environmentally responsible and sustainable way to increase crop yields and soil fertility. The particular crop and soil conditions influence the selection of algae strains, production practices, and application strategies.

- **Agricultural Wastewater**: Nutrient-rich runoff or effluent.
- **Microalgae Strains**: Species like *Chlorella* or *Spirulina* for nutrient uptake.
- **Photobioreactors or Open Ponds**: For algae cultivation.
- **Harvesting Equipment**: Centrifuges or filters to collect algal biomass.

2.7 Nutrient Capture from Runoff [19]

Catching and eliminating surplus nutrients, such as phosphorus (P) and nitrogen (N), from industrial, urban, or agricultural runoff before it reaches water bodies are known as "nutrient capture from runoff." Ecosystem destruction, water pollution, and eutrophication are all prevented by this mechanism. One of the most important aspects of sustainable water management is the capture of nutrients from runoff. The particular nutrient sources, runoff properties, and environmental circumstances all influence the materials and techniques chosen.

- **Constructed Wetlands**: Plants, soil, and gravel to filter and absorb nutrients.
- **Bioreactors**: Woodchip or straw-filled trenches to treat runoff.
- **Filter Strips**: Vegetated areas to trap sediments and nutrients.

2.8 Liquid–Solid Separation [20]

The process of separating dissolved or suspended particles from liquids is known as liquid-solid separation. Numerous sectors, such as mining, food processing, wastewater treatment, and chemical manufacture, depend on this process. Many industries rely on liquid-solid separation as a crucial process, and the nature of the solids and liquids, the intended separation efficiency, and the particular application all influence the materials and equipment selection.

- **Manure or Slurry**: Raw agricultural waste.
- **Separators**: Mechanical screens, centrifuges, or screw presses.
- **Storage Tanks**: For separated liquid and solid fractions.

2.9 Microbial-Based Nutrient Recovery [21]

Using microorganisms to extract and recycle nutrients like phosphorous (P) and nitrogen (N) from waste streams like wastewater, agricultural runoff, or industrial effluents is known as microbial-based nutrient recovery. This method can yield useful by-products like biofertilizers or bioplastics and is environmentally benign and sustainable. One promising technique for long-term nutrient management is microbial-based nutrient recovery. The particular waste stream and intended final product determine which microbes, growth conditions, and recovery techniques are used.

- **Organic Waste**: Crop residues, manure, or food waste.
- **Microbial Inoculants**: Specific strains of bacteria or fungi for nutrient conversion.
- **Fermentation Tanks**: For controlled microbial activity.
- **Drying Equipment**: To produce stable biofertilizers.

2.10 Integrated Farming Systems [22]

In order to establish a sustainable and effective farming system, integrated farming systems, or IFS, combine a variety of farming techniques, including crop production, livestock husbandry, aquaculture, and agroforestry. Increasing agricultural productivity and profitability, minimizing environmental impact, and optimizing resource use efficiency are the objectives. A sustainable approach to contemporary agriculture, integrated farming systems support both economic viability and ecological equilibrium. Local factors, farmer preferences, and market demand all influence the materials and methods chosen.

- **Crop Residues**: Leftover plant material from harvest.
- **Livestock Manure**: Nutrient-rich waste from animals.

- **Recycling Infrastructure**: Composting pits, biogas digesters, or storage facilities.

2.10.1 Additional Supporting Materials

- **pH Meters and Sensors** [23]: To monitor and optimize conditions for nutrient recovery processes.
- **Nutrient Testing Kits** [24]: To analyze the nutrient content of recovered products.
- **Safety Gear** [25]: Gloves, masks, and protective clothing for handling waste and chemicals.
- **Transportation Equipment** [26]: Carts, trucks, or conveyors for moving waste and recovered materials.

The materials necessary for sustainable friendly nutrient recovery differ based on the particular technology employed and the scale of its application. Nonetheless, all approaches aim to convert agricultural waste into valuable resources while reducing environmental harm. By effectively utilizing these materials, farmers and agricultural stakeholders can embrace sustainable practices that improve soil health, decrease waste, and foster circular agriculture.

3 Methods

A range of environmentally friendly technologies can be utilized to extract essential nutrients, including phosphorus (P), nitrogen (N), and potassium (K) from agricultural waste. The process of nutrient recovery from agricultural waste encompasses several eco-safe technologies, each designed with particular methods aimed at enhancing both efficiency and sustainability. The following outlines the primary techniques employed for nutrient recovery, detailing their processes and advantages.

3.1 Composting Composting [27]

A balanced mixture of organic materials, microbial inoculants, and ideal ambient conditions is necessary for successful composting (Fig. 1). Together with necessary additives, the primary materials are divided into components that are nitrogen-rich (greens) and carbon-rich (browns).

Fig. 1 Composting process

- **Raw Material Collection**: Gather organic waste such as crop residues, food waste, and manure.
- **Pile Formation or Bin Setup**: Arrange waste in **windrows**, bins, or pits to facilitate microbial decomposition.
- **Carbon-to-Nitrogen (C/N) Ratio Optimization**: Maintain an optimal **C/N ratio (25–30:1)** by adding bulking agents like straw or sawdust.
- **Aeration and Moisture Control**: Turn the compost pile periodically and maintain **40–60% moisture**.
- **Maturation and Curing**: Allow decomposition to continue for **4–8 weeks (composting) or 6–12 weeks (vermicomposting)** until a stable, humus-like material is formed.
- **Final Screening and Storage**: Remove impurities and store compost for agricultural application.

3.2 Anaerobic Digestion and Biogas Production [28]

Anaerobic digestion (Fig. 2) is a biological mechanism that decomposes organic waste without the presence of oxygen, resulting in the production of biogas, which consists of methane and carbon dioxide, along with nutrient-dense digestate. This process offers a sustainable approach to handling agricultural waste, facilitating the generation of renewable energy and the recovery of nutrients for application as biofertilizers. The stages involved in anaerobic digestion and the production of biogas are outlined below.

- **Feedstock Selection**: Collect organic waste like livestock manure, food wastes and crop residues.
- **Loading into Digester**: Place waste into a sealed anaerobic digester.
- **Microbial Breakdown**: Microorganisms decompose waste into biogas (CH_4 and CO_2) and nutrient-rich digestate under oxygen-free conditions.
- **Biogas Collection and Utilization**: Capture biogas for energy production.

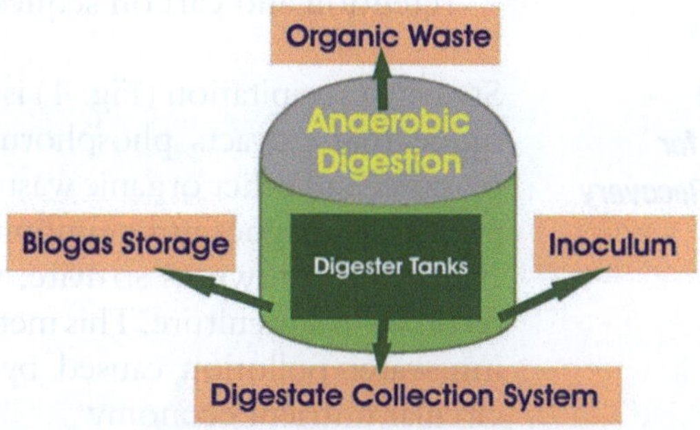

Fig. 2 Anaerobic digestion process

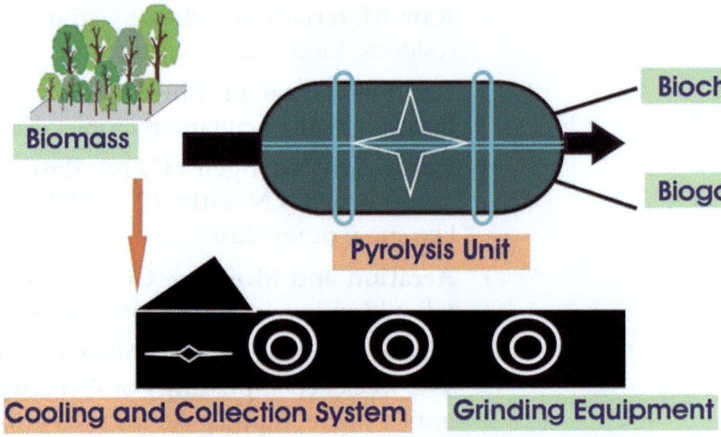

Fig. 3 Biochar production process

- **Digestate Treatment**: Separate liquid and solid fractions of digestate for nutrient recovery.
- **Application to Soil**: Use digestate as a fertilizer or soil conditioner.

3.3 Biochar Production and Application [29]

Biochar is a carbon-dense substance created via the pyrolysis of agricultural by-products in an oxygen-restricted environment (Fig. 3). It functions as a sustainable soil enhancer, augmenting soil fertility, facilitating carbon sequestration, and enhancing nutrient retention. Additionally, biochar contributes to waste valorization by transforming agricultural waste into a valuable resource for sustainable agricultural practices.

- **Feedstock Selection**: Use crop residues, wood chips, or manure as raw materials.
- **Pyrolysis Process**: Heat feedstock at 300–700 °C under low oxygen conditions in a pyrolysis reactor.
- **Biochar Collection and Processing**: Cool, grind, and sieve biochar into a fine, soil-applicable form.
- **Soil Application**: Mix biochar into soil to enhance nutrient retention and carbon sequestration.

3.4 Struvite Precipitation for Phosphorus Recovery [30]

Struvite precipitation (Fig. 4) is a sustainable nutrient recovery technique that extracts phosphorus (P) from agricultural wastewater, manure, and other organic waste streams. The process recovers phosphorus as magnesium ammonium phosphate ($MgNH_4PO_4{\cdot}6H_2O$), commonly known as struvite, which can be used as a slow-release fertilizer in agriculture. This method is particularly valuable for reducing water pollution caused by phosphorus runoff and creating a circular nutrient economy.

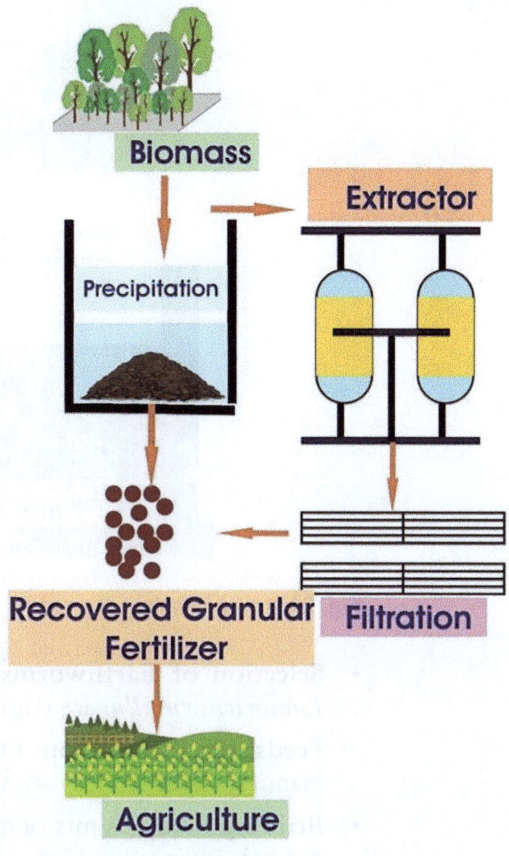

Fig. 4 Phosphorus recovery process

- **Collection of Phosphorus-Rich Wastewater**: Extract wastewater from manure lagoons, biogas digestate, or food processing industries.
- **Chemical Addition**: Add **magnesium (Mg^{2+}) and ammonium (NH_4^+) ions** to induce phosphorus precipitation.
- **Struvite Crystallization**: Maintain an optimal pH (8–9) to facilitate the formation of **struvite crystals**.
- **Solid Separation**: Filter and dry the recovered struvite.
- **Application in Agriculture**: Use struvite as a **slow-release phosphorus fertilizer**.

3.5 Vermicomposting [31]

Vermicomposting (Fig. 5) is a sustainable and eco-friendly method of converting organic agricultural waste into nutrient-rich compost using earthworms. This process not only recycles valuable nutrients but also enhances soil health, making it a key component of organic farming and sustainable agriculture. Vermicomposting is particularly advantageous for small-scale farmers due to its low cost, simplicity, and high-quality output.

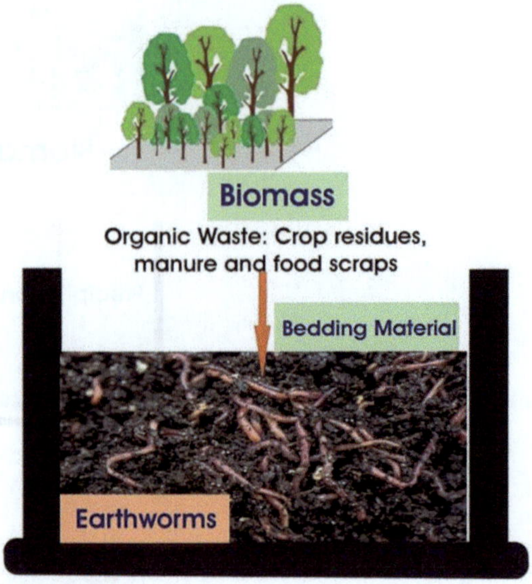

Fig. 5 Vermicomposting process

- **Selection of Earthworms**: *Eisenia fetida* (red wigglers) and *Lumbricus rubellus* are commonly used.
- **Feedstock Preparation**: Organic waste (e.g., vegetable scraps, manure, shredded leaves) is chopped for faster breakdown.
- **Bedding Setup**: A mix of moist soil, shredded paper, and coconut coir is placed in bins.
- **Earthworm Addition**: Worms are introduced and allowed to process the waste.
- **Moisture and Temperature Control**: Maintain 40–70% moisture and a temperature of 15–25 °C.
- **Harvesting Vermicompost**: After 6–12 weeks, nutrient-rich worm castings are collected.

3.6 Algae-Based Nutrient Recovery [32]

Algae-based nutrient recovery (Fig. 6) represents an environmentally sustainable biotechnological approach that employs both microalgae and macroalgae to extract and repurpose nutrients from agricultural waste. This technique contributes to wastewater management, the production of biofertilizers, and the sequestration of carbon, positioning it as a viable and sustainable option for nutrient recycling.

- **Wastewater Collection**: Extract nutrient-rich effluents from livestock farms, aquaculture, or food industries.
- **Algae Cultivation**: Grow algae in open ponds or photobioreactors with controlled light, CO_2, and nutrients.

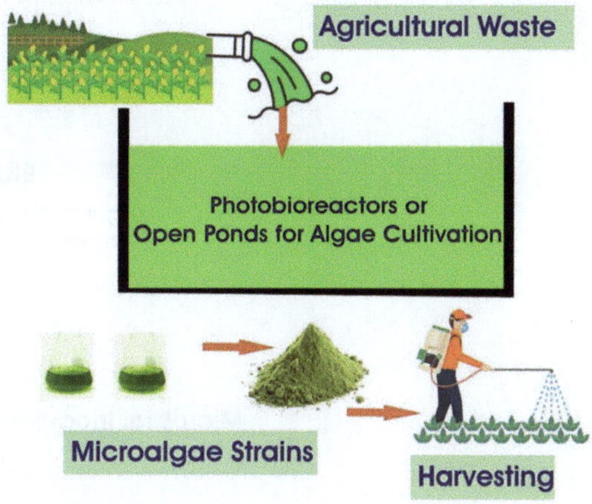

Fig. 6 Algal biofertilizers process

- **Nutrient Uptake: Algae absorb N, P, and other minerals from** wastewater.
- **Harvesting Algal Biomass**: Use filtration or centrifugation to extract algae.
- **Processing and Application**: Convert algae into biofertilizers, biostimulants, or animal feed supplements.

3.7 Bioelectrochemical Nutrient Recovery (Microbial Fuel Cells—MFCs) [33]

Bioelectrochemical nutrient recovery (Fig. 7) utilizes microbial fuel cells and microbial electrolysis cells to recover nitrogen (N), phosphorus (P), and other valuable nutrients from agricultural waste while simultaneously generating bioelectricity or hydrogen gas. This approach represents an innovative and sustainable solution for waste treatment, nutrient recycling, and renewable energy production.

- **Setup of Microbial Fuel Cells (MFCs)**: Install an anode and cathode in a wastewater-filled reactor.
- **Microbial Activity**: Microbes break down organic matter, releasing electrons.
- **Electricity Generation**: The electrons flow through an external circuit, generating electricity.
- **Nutrient Recovery**: Ammonium and phosphorus are extracted through **selective membranes.**
- **Reuse of Nutrients**: The recovered nutrients are processed into fertilizers.

3.8 Integrated Farming Systems [34]

Agroforestry, aquaculture, crop production, livestock rearing, and other related practices are all combined in an integrated farming system (IFS) (Fig. 8), a sustainable agricultural method that maximizes resource use, boosts output, and reduces waste. This

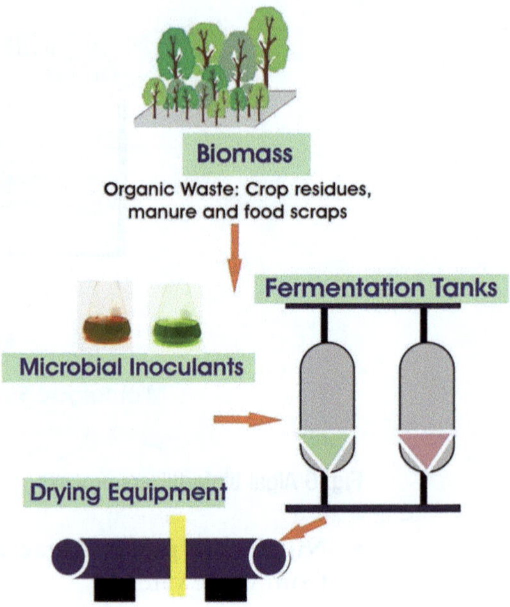

Fig. 7 Microbial-based nutrient recovery process

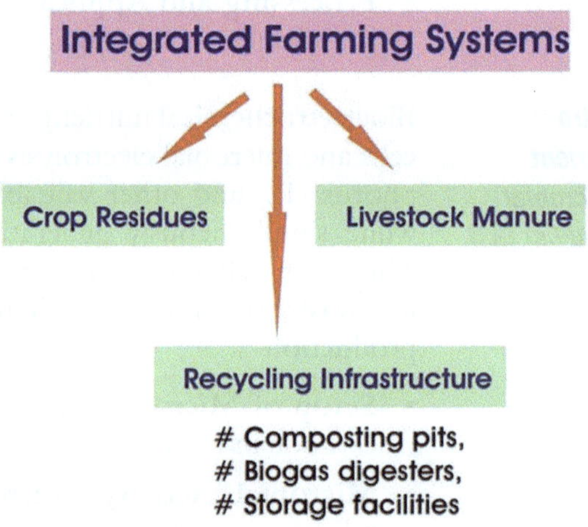

Fig. 8 Integrated farming systems

approach supports biodiversity preservation, nitrogen cycling, and farmer economic stability. A closed-loop system that integrates crop production, animal raising, and other farming practices is known as the integrated farming system (IFS). IFS improves nutrient recovery, reduces waste, and promotes resource efficiency by combining these elements. This technique reduces the demand for external inputs like synthetic fertilizers by recycling nutrients within the farming system.

3.9 Electrochemical and Membrane-Based Nutrient Recovery [31]

Electrochemical and membrane-based technologies present advanced and effective methods for the extraction of phosphorus (P), nitrogen (N), and potassium (K) from agricultural waste sources, including manure, wastewater, and runoff (Fig. 9). These approaches are especially beneficial for generating high-purity nutrient concentrates while simultaneously mitigating environmental pollution.

- **Wastewater Filtration**: Pre-treat agricultural wastewater to remove solids.
- **Membrane Separation**: Use ion-exchange membranes to capture dissolved phosphates and ammonium.
- **Electrochemical Process**: Apply a low-voltage electric field to selectively extract nutrients.
- **Nutrient Precipitation**: Convert recovered phosphorus into struvite ($MgNH_4PO_4$), a slow-release fertilizer.
- **Storage and Application**: Store the extracted nutrients for use in agriculture.

- **Wastewater Filtration** – Pre-treat agricultural wastewater to remove solids.
- **Membrane Separation** – Use **ion-exchange membranes** to capture dissolved **phosphates and ammonium**.
- **Electrochemical Process** – Apply a **low-voltage electric field** to selectively extract nutrients.
- **Nutrient Precipitation** – Convert recovered phosphorus into **struvite ($MgNH_4PO_4$)**, a slow-release fertilizer.
- **Storage and Application** – Store the extracted nutrients for use in agriculture.

Agricultural Wastewater
↓
[Wastewater Filtration]
↓
Filtered Wastewater (Dissolved Nutrients)
↓
[Membrane Separation]
↓
Concentrated Nutrient Solution (Phosphates, Ammonium)
↓
[Electrochemical Process]
↓
Separated Nutrients (Phosphate, Ammonium Ions)
↓
[Nutrient Precipitation]
↓
Struvite Crystals ($MgNH_4PO_4$)
↓
[Storage and Application]

Fig. 9 Flow chart of electrochemical and membrane-based nutrient recovery

3.10 General Steps for Implementing Nutrient Recovery Methods [35]

- **Waste Collection**: Gather agricultural waste from fields, livestock operations, or processing facilities.
- **Pre-Treatment**: Sort, shred, or mix waste to prepare it for processing.
- **Processing**: Apply the chosen nutrient recovery method (e.g., composting, anaerobic digestion).
- **Monitoring**: Track key parameters (e.g., temperature, pH, moisture) to optimize the process.
- **Harvesting and Application**: Collect the recovered nutrients and apply them to agricultural fields.

The methods for nutrient recovery using eco-safe technologies are diverse and adaptable to different types of agricultural waste and farming systems. By selecting the appropriate method and ensuring proper implementation, farmers can effectively recycle nutrients, reduce environmental impacts, and enhance agricultural sustainability.

4 Notes

A key tactic for advancing sustainability, lowering environmental pollution, and improving soil fertility is the recovery of nutrients from agricultural waste utilizing eco-safe methods. Nitrogen, phosphorus, and potassium are among the important nutrients found in agricultural waste, which is frequently thrown away or improperly handled. A circular agricultural system that optimizes resource efficiency can be achieved by implementing creative recovery techniques rather than letting these nutrients be lost through leaching, runoff, or greenhouse gas emissions.

Numerous environmentally safe nutrient recovery strategies have been covered in this chapter [10], such as sophisticated electrochemical and membrane-based techniques, composting, anaerobic digestion, biochar application, and algae-based nutrient capture. From enhancing soil structure and microbial activity to decreasing reliance on synthetic fertilizers and limiting ecological harm, each strategy offers distinct advantages. The viability of these strategies in actual agricultural contexts has been illustrated by case studies [36], underscoring their significance in sustainable farming methods.

Notwithstanding the benefits, there are also issues, such as gaps in policy, financial restrictions, and technology limitations. More funding for R&D, laws that encourage it, and the incorporation of nutrient recovery into larger agricultural sustainability plans are all necessary to overcome these obstacles. In order to persuade farmers and agribusinesses to implement eco-safe nutrient recovery systems, awareness-raising and capacity-building programs are also crucial. In the future, nutrient recovery procedures can be further optimized to become more effective and economical through developments in digital monitoring systems, precision agriculture, and biotechnology. Agriculture may move toward a more resilient and sustainable future by adopting the concepts of the circular economy and utilizing eco-friendly technologies, guaranteeing both environmental preservation and long-term food security.

References

1. Flannery KV (1973) The origins of agriculture. Annu Rev Anthropol:271–310
2. Sangha KK (2014) Modern agricultural practices and analysis of socio-economic and ecological impacts of development in agriculture sector, Punjab, India—a review. Indian J Agric Res 48(5):331–341
3. Gupta GS (2019) Land degradation and challenges of food security. Rev Eur Stud 11:63
4. Koul B, Yakoob M, Shah MP (2022) Agricultural waste management strategies for environmental sustainability. Environ Res 206:112285
5. Reno J (2015) Waste and waste management. Annu Rev Anthropol 44(1):557–572
6. Alston JM, Pardey PG (2014) Agriculture in the global economy. J Econ Perspect 28(1): 121–146
7. Obi FO, Ugwuishiwu BO, Nwakaire JN (2016) Agricultural waste concept, generation, utilization and management. Niger J Technol 35(4):957–964
8. Abd-Elsalam KA, Periakaruppan R, Rajeshkumar S (eds) (2021) Agri-waste and microbes for production of sustainable nanomaterials. Elsevier
9. Ahmed M, Ahmad S, Qadir G, Hayat R, Shaheen FA, Raza MA (2019) Innovative processes and technologies for nutrient recovery from wastes: a comprehensive review. Sustain For 11(18):4938
10. Singh AD, Bakshi P, Kumar P, Kour J, Dhiman S, Ibrahim M et al (2022) Effects of agricultural wastes on environment and its control measures. Agric Kitchen Waste:219–239
11. Gillingham K, Stock JH (2018) The cost of reducing greenhouse gas emissions. J Econ Perspect 32(4):53–72
12. Velasco-Muñoz JF, Mendoza JMF, Aznar-Sánchez JA, Gallego-Schmid A (2021) Circular economy implementation in the agricultural sector: definition, strategies and indicators. Resour Conserv Recycl 170:105618
13. Zhou Y, Xiao R, Klammsteiner T, Kong X, Yan B, Mihai FC et al (2022) Recent trends and advances in composting and vermicomposting technologies: a review. Bioresour Technol 360:127591
14. Chen Y, Cheng JJ, Creamer KS (2008) Inhibition of anaerobic digestion process: a review. Bioresour Technol 99(10):4044–4064
15. Oni BA, Oziegbe O, Olawole OO (2019) Significance of biochar application to the environment and economy. Ann Agric Sci 64(2): 222–236
16. Vishwakarma R, Dhaka V, Ariyadasa TU, Malik A (2022) Exploring algal technologies for a circular bio-based economy in rural sector. J Clean Prod 354:131653
17. Liu Y, Deng YY, Zhang Q, Liu H (2021) Overview of recent developments of resource recovery from wastewater via electrochemistry-based technologies. Sci Total Environ 757: 143901
18. Sun D, Hale L, Kar G, Soolanayakanahally R, Adl S (2018) Phosphorus recovery and reuse by pyrolysis: applications for agriculture and environment. Chemosphere 194:682–691
19. Orner KD, Camacho-Céspedes F, Cunningham JA, Mihelcic JR (2020) Assessment of nutrient fluxes and recovery for a small-scale agricultural waste management system. J Environ Manag 267:110626
20. Chiumenti A, da Borso F, Limina S, Piaia B (2024) Single-stage and multi-stage liquid/solid separation of digestate in full scale biogas plants. In: 2024 ASABE annual international meeting. American Society of Agricultural and Biological Engineers, p 1
21. Borah M, Nath PD, Bhamra GK, Gayithri M, Bhuvaneswari V, Pradeep RM (2025) Microbial-based strategies for remediation of agricultural wastes. In: Development in waste water treatment research and processes. Elsevier, pp 297–321

22. Bhagat R, Walia SS, Sharma K, Singh R, Singh G, Hossain A (2024) The integrated farming system is an environmentally friendly and cost-effective approach to the sustainability of agri-food systems in the modern era of the changing climate: a comprehensive review. Food Energy Security 13(1):e534
23. Dutta L, Bharali S, Barman P, Singh A (2024) An IoT-enabled smart pH monitoring and dispensing system for precision agriculture application. Agric Res 13(2):309–318
24. Kang SWS, Cunningham RP, Miller CB, Brown LA, Cultraro CM, Harned A et al (2024) A spatial map of hepatic mitochondria uncovers functional heterogeneity shaped by nutrient-sensing signaling. Nat Commun 15(1):1799
25. Barenklau KE (2001) Agricultural safety. CRC Press
26. Miner JR, Hazen TE (1977) Transportation and application of organic wastes to land. Soils Manage Org Wastes Waste Waters:377–425
27. Gupta C, Prakash D, Gupta S, Nazareno MA (2019) Role of vermicomposting in agricultural waste management. In: Sustainable green technologies for environmental management. Singapore, Springer Singapore, pp 283–295
28. Alengebawy A, Ran Y, Osman AI, Jin K, Samer M, Ai P (2024) Anaerobic digestion of agricultural waste for biogas production and sustainable bioenergy recovery: a review. Environ Chem Lett:1–28
29. Enaime G, Lübken M (2021) Agricultural waste-based biochar for agronomic applications. Appl Sci 11(19):8914
30. Kataki S, West H, Clarke M, Baruah DC (2016) Phosphorus recovery as struvite from farm, municipal and industrial waste: feedstock suitability, methods and pre-treatments. Waste Manag 49:437–454
31. Kashima H (2020) Nutrient recovery from wasted biomass using microbial electrochemical technologies. Recycle Based Org Agric City:69–90
32. Nguyen LN, Aditya L, Vu HP, Johir AH, Bennar L, Ralph P et al (2022) Nutrient removal by algae-based wastewater treatment. Curr Pollut Rep 8(4):369–383
33. Singh R (2024) (Bio) electrochemical system: a systematic approach from agricultural waste to sewage wastewater treatment with nutrients and hydrogen recovery. J Clean Prod 457: 142387
34. Soni RP, Katoch M, Ladohia R (2014) Integrated farming systems-a review. IOSR J Agric Vet Sci 7(10):36–42
35. Mehta CM, Khunjar WO, Nguyen V, Tait S, Batstone DJ (2015) Technologies to recover nutrients from waste streams: a critical review. Crit Rev Environ Sci Technol 45(4):385–427
36. Awogbemi O, Von Kallon DV (2022) Pretreatment techniques for agricultural waste. Case Studies Chem Environ Eng 6:100229

Chapter 8

Cold Plasma Technology for Sustainable Food Waste Decontamination

Subhajit Das, Ankita Gon, Ayan Chatterjee, and Mahenderan Appukutty

Abstract

Food waste contamination poses noteworthy environmental and public health challenges due to microbial decay and poison production. Cold plasma technology, a developing non-thermal decontamination technique, offers a maintainable solution by successfully incapacitating pathogens while preserving food quality. This technology produces reactive species, such as ozone, nitrogen oxides, and hydroxyl radicals, which disturb microbial cell membranes and genetic material, plummeting spoilage and spreading shelf life. Unlike conservative thermal or chemical treatments, cold plasma operates at ambient temperatures, minimizing nutrient loss and chemical remains. Moreover, it aligns with sustainability goals by tumbling reliance on chemical preservatives and lowering energy consumption. Current studies highlight its effectiveness against bacteria, fungi, and viruses in various food matrices, including fruits, vegetables, dairy, and meat products. However, challenges remain concerning optimization, scalability, and regulatory approval for extensive applications. Further research is needed to improve treatment parameters, confirm consumer safety, and assess long-term conservational influence. Cold plasma technology signifies a promising, eco-friendly approach to mitigating food waste infection and sustainably enhancing food security. Its addition to food processing systems could transfigure waste management strategies and subsidize a more resilient global food supply chain.

Key words Food waste contamination, Microbial decay, Toxin production, Reactive species, Ozone, Nitrogen oxides, Hydroxyl radicals

1 Introduction

Food waste is a global challenge with noteworthy environmental, economic, and social consequences [1]. The cumulative demand for food production to meet the needs of a growing population is convoyed by enormous losses and waste at several stages of the food supply chain [2]. Addressing food waste is not only a matter of resource competence but also of food security and sustainability [3]. A significant portion of food waste results from microbial infection, which hastens spoilage and renders food unfit for consumption. Traditional methods for distillation, such as thermal

Tanmay Sarkar (ed.), *Food Waste Valorization: Green Techniques in Sustainable Management*, Methods and Protocols in Food Science, https://doi.org/10.1007/978-1-0716-4646-5_8,

© The Author(s), under exclusive license to Springer Science+Business Media, LLC, part of Springer Nature 2025

treatments, chemical stabilizers, and irradiation, have boundaries, including nutrient disrepair, the formation of damaging residues, and consumer skepticism [4]. Therefore, the need for advanced and sustainable sanitization techniques is pressing. Cold plasma technology is a promising non-thermal practice for microbial inactivation and food preservation. Plasma, often mentioned as the fourth state of matter, contains ionized gases casing reactive species such as ions, electrons, free radicals, and UV photons. Unlike conventional thermal plasma, cold plasma activates at near-ambient temperatures, making it appropriate for heat-sensitive food products. Cold plasma has established considerable potential in spreading shelf life, plummeting microbial load, and maintaining the nutritious and sensory qualities of food products [5]. Given its eco-friendly nature and unimportant use of chemicals, it aligns with the values of supportable food dispensation and waste lessening [6].

The request for cold plasma in food waste management alternates around its ability to efficiently cleanse food materials before they are castoff. Food waste, particularly unpreserved items such as fruits, vegetables, dairy products, and meat, often becomes inappropriate for consumption due to microbial growth [7]. By applying cold plasma treatment at various stages of the food supply chain, from manufacture to storage and post-consumer waste handling, food spoilage can be significantly reduced, thereby minimalizing overall waste [8]. Moreover, cold plasma technology has been explored for its potential to treat organic waste, making it safer for further dispensation into animal feed, compost, or bioenergy production. The sustainability aspect of cold plasma technology is another vital factor in its increasing interest for food waste distillation [9]. Unlike outmoded decontamination methods that may produce chemical by-products or subsidize to environmental pollution, cold plasma primarily relies on electricity and atmospheric gases such as air, oxygen, or nitrogen [10]. This makes it a clean and energy-efficient approach to microbial decontamination. Furthermore, the potential integration of renewable energy sources with cold plasma systems could enhance its sustainability and reduce operative costs, making it a viable option for large-scale food waste management [11].

Despite its auspicious advantages, several challenges need to be addressed for the extensive adoption of cold plasma technology in food waste purification. One of the key deliberations is the scalability of plasma systems for manufacturing applications [12]. While laboratory-scale studies have established significant efficacy, interpreting these findings into commercial-scale processes necessitates further research and technological advancements. Additionally, controlling approval and purchaser acceptance play a decisive role in the enactment of novel food processing technologies. Certifying that cold plasma-treated food products and waste materials comply with food safety principles and public health guidelines is indispensable for its successful incorporation into the food industry [1].

2 Materials

Cold plasma technology is a developing, eco-friendly solution for food waste decontamination. This non-thermal procedure creates reactive species that effectively incapacitate pathogens, degrade toxins, and spread food waste shelf life without harmful residues [10]. It operates at low temperatures, preserving nutritional veracity while reducing microbial contamination. The technology supports sustainable waste management by minimizing spoilage and enhancing food safety. Its submission in composting and animal feed dispensation further contributes to circular economy practices.

Materials Required

1. Cold plasma generator.
2. Gas supply (e.g., air, oxygen, nitrogen).
3. Power source.
4. Food waste samples.
5. Containment chamber.
6. Microbial culture media.
7. Analytical instruments.

3 Method

3.1 Cold Plasma: Fundamentals and Mechanisms

Cold plasma, also known as non-thermal plasma, is incompletely ionized gas that activates at near-room temperature [13]. Unlike thermal plasmas, where electrons, ions, and neutral species exist at thermal symmetry, cold plasma maintains animated electrons while keeping the overall gas temperature low. This exceptional characteristic makes it valuable for submissions in medicine, materials science, and environmental remediation [14]. Cold plasma is created through various methods, counting dielectric barrier discharge (DBD), atmospheric pressure plasma jets (APPJs), and radio frequency (RF) discharges. These techniques create a plasma state by presenting energy into a gas, leading to ionization, innervation, and detachment of molecules. The reactive species fashioned—such as ozone, hydroxyl radicals, and reactive nitrogen species—play a crucial role in its apparatuses and applications.

3.1.1 Mechanisms of Plasma-Induced Decontamination

Cold plasma achieves refinement primarily through the action of reactive oxygen and nitrogen species. These include ozone, hydroxyl radicals, singlet oxygen, nitric oxide, and peroxynitrite, all of which are extremely reactive and capable of destructive microbial structures [15]. The oxidative stress persuaded by these species disrupts bacterial membranes, proteins, and nucleic acids, leading to cell death. The lipid bilayers of microorganisms undergo peroxidation, which increases membrane permeability, causing leakage of cellular contents and eventual cell lysis [16]. This is predominantly effective against Gram-negative bacteria due to their outer membrane structure, while Gram-positive bacteria, though more resilient, are still susceptible to lengthy exposure.

The interaction of plasma with microbial DNA leads to oxidative amendments, resulting in strand breaks and base modifications that interfere with replication and transcription. In addition, proteins are exaggerated through oxidation of key amino acids, leading to enzyme inactivation and structural deformation. This multi-targeted tactic makes plasma an effective antimicrobial agent that diminishes the risk of confrontation development [17].

3.1.2 Cell Membrane Disruption

Plasma-induced oxidative stress can damage the structural truthfulness of microbial membranes. The lipid bilayer, which serves as a protective barrier, undergoes lipid peroxidation, resulting in increased permeability and leakage of intracellular contents [18]. In Gram-negative bacteria, the outer membrane rich in lipopolysaccharides (LPS) is predominantly vulnerable to oxidative damage, leading to cell lysis. Gram-positive bacteria, though possessing a thicker peptidoglycan layer, also experience membrane destabilization after prolonged plasma exposure [19].

3.1.3 DNA and Protein Damage

Beyond membrane disruption, cold plasma creates radicals that penetrate bacteriological cells, targeting intracellular machineries. Nucleic acids undergo oxidative amendments, leading to DNA strand breaks, base modifications, and cross-linking, finally meddlesome with replication and transcription [20]. Proteins, indispensable for cellular function, are also damaged through oxidation of amino acids, leading to enzyme deactivation and structural distortion [21].

3.1.4 Disruption of Biofilms

Microbial biofilms, which are dense societies of bacteria enclosed in an extracellular polymeric matrix, present a challenge in conservative sterilization [22]. Cold plasma effectually disrupts these biofilms by breaking down the polymeric structure and sensitizing the rooted cells to reactive species. This makes plasma an appreciated tool in medical and industrial settings where biofilm-associated infections and contamination are concerns [23].

3.1.5 Surface Decontamination

Cold plasma has been lengthily used for neutering surfaces in medical apparatuses, food packaging, and industrial equipment. The interaction of plasma-generated species with surfaces leads to the subtraction of organic residues, bacterial cells, and biofilms without the need for exacting chemical disinfectants.

3.2 Interaction of Cold Plasma with Organic Matter

Cold plasma interacts with organic matter through oxidation, operational modification, and chemical conversion [24]. When plasma comes into contact with biological molecules, it persuades changes in proteins, lipids, carbohydrates, and nucleic acids. Proteins undergo oxidation, leading to modifications in enzymatic activity and structural constancy. Lipid peroxidation disrupts membrane veracity, affecting cell viability. Carbohydrates experience cleavage of glycosidic bonds, affecting structural polysaccharides. Nucleic acids suffer from oxidative modifications that can prime to genetic instability, a phenomenon that has potential applications in cancer therapy and antimicrobial conducts [25].

3.2.1 Effects on Biomolecules

Plasma-induced oxidation impressions proteins, lipids, carbohydrates, and nucleic acids. Proteins undergo conformational changes due to amino acid oxidation, which can modify enzymatic activity or trigger dilapidation. Lipid peroxidation disrupts membrane constructions, leading to loss of veracity. Carbohydrates experience oxidative cleavage of glycosidic bonds, affecting polysaccharide stability [26]. Nucleic acids are also vulnerable to damage, resulting in genetic variations that may be connected for specific remedial submissions.

3.2.2 Surface Modification

Cold plasma treatment modifies superficial belongings by familiarizing functional groups that improve hydrophilicity or hydrophobicity [27]. This consequence is exploited in material sciences, where plasma treatment is employed to improve adhesion belongings, remove impurities, and modify polymers. In biomedical applications, plasma-modified exteriors promote cell accessory and tissue renaissance.

3.2.3 Food Preservation and Safety

In the food commerce, cold plasma is used to augment food safety by deactivating decay microorganisms and infective bacteria. Plasma treatment encompasses mantelpiece life by plummeting bacteriological load while preserving nutritional and sensory belongings. Since plasma operates without undue heat, it minimizes the risk of thermal dilapidation, making it an ideal procedure for preserving consumable food products [27].

3.2.4 Environmental Applications

Cold plasma is a commanding tool for degrading environmental contaminants, including unpredictable organic compounds (VOCs), pesticides, and pharmaceutical remains. The responsive species produced during plasma treatment facilitate the cessation of complex organic contaminants into humbler, less destructive compounds. This makes plasma-based technologies appreciated for air and water decontamination [28].

3.2.5 Biomedical Applications

In medicine, cold plasma is discovered for wound healing, cancer therapy, and antimicrobial treatments [29]. Plasma acquaintance can excite cellular responses, endorse tissue renaissance, and selectively induce programmed cell death in cancer cells. These properties make it an emergent technology in non-invasive therapeutic involvements [30].

3.3 Microbial Decontamination of Food Waste Using Cold Plasma

Food waste is a huge global issue that contributes to resource loss and environmental degradation. Nonetheless, the existence of microbiological pollutants in food waste restricts its ability to be recycled into animal feed, biofertilizers, and biofuels and presents health hazards [29]. The nutritional value of food waste is frequently diminished by conventional decontamination techniques like heat treatment and chemical sanitization, which can also leave hazardous residues [13]. An innovative non-thermal technique called cold plasma presents a viable way to remove microorganisms from food waste while maintaining the integrity of its constituent parts [31]. Plasma applications in the food industry are still in their infancy. CP has proven to be effective in treating biofilms and decontaminating foods such as fruits, vegetables, poultry, and meats. Because CP acts quickly on both vegetative cells and spores, Feichtinger et al. (2003) showed that it could be a substitute technique for food surface decontamination. The ability of plasma to flow around the borders of surface fissures may indicate that it is unaffected by surface irregularities, guaranteeing consistent treatment; but, on the other hand, product surface roughness reduced the antibacterial efficiency of CP treatment [32].

By using ionized gases at low temperatures, this technique produces reactive species that can inactivate viruses, fungi, and bacteria (Fig. 1).

3.3.1 Effectiveness Against Bacteria, Viruses, and Fungi

Cell lysis results from the disruption of bacterial cell membranes caused by reactive oxygen and nitrogen species (RONS) in cold plasma [33]. Research has indicated notable decrease in harmful bacteria, including Salmonella species, *Escherichia coli*, and *Listeria monocytogenes*, in food waste matrices. Despite having no cellular structure, viruses can be damaged by cold plasma because their genetic material and protein capsids are oxidatively damaged. Research shows that hepatitis A virus and noroviruses, which are frequently present in tainted food waste, are successfully rendered inactive by cold plasma. Mycotoxin generation and spoiling are caused by fungal contamination, especially from molds and yeasts. Aspergillus and Penicillium species are successfully reduced by cold plasma, which breaks down fungal cell walls and prevents spore germination [34].

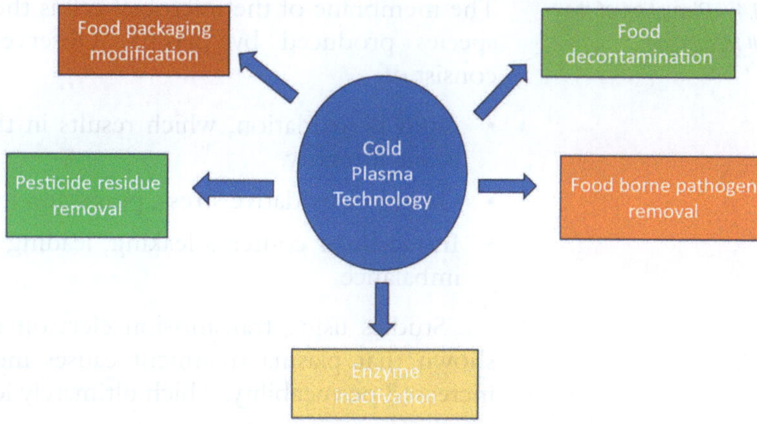

Fig. 1 Applications of cold plasma technology in food safety and processing

3.3.2 Impact on Spoilage Organisms and Pathogens

Yeasts, molds, and lactic acid bacteria are examples of spoilage microorganisms that cause food to deteriorate and acquire undesirable flavors. Research has shown that cold plasma efficiently lowers the number of spoilage organisms in a variety of food matrices [35]. A number of variables, including microbial susceptibility, gas composition, and exposure duration, affect the inactivation efficiency. Cold plasma treatment is a desirable alternative for waste reduction since it prolongs shelf life and preserves food quality [36]. Foodborne pathogens that are harmful to human health include Salmonella species, *Staphylococcus aureus*, *Escherichia coli*, and *Listeria monocytogenes* [37]. Through the disruption of cell membranes and the inhibition of metabolic activity, cold plasma has demonstrated potent bactericidal effects against these pathogens. According to research, cold plasma can significantly lower microbe populations by logarithmic units, improving food safety and lowering the possibility of contamination in food waste streams [38].

3.3.3 Plasma-Induced Structural Changes in Microbes

The fourth state of matter, plasma, is made up of a partly ionized gas that contains ultraviolet (UV) light, electrons, ions, and reactive neutral species. Cold plasma is frequently employed in microbial decontamination because it may function at temperatures close to room temperature [39].

The following are the main reactive species produced in plasma:

Ozone (O_3), singlet oxygen (O_2*), and hydroxyl radicals ($OH^{\cdot}$) are examples of reactive oxygen species (ROS) and reactive nitrogen species (RNS), such as peroxynitrite ($ONOO^-$) and nitric oxide (NO), electric fields, UV rays, and charged particles. Together, these substances cause oxidative stress and structural cellular component breakdown, which aids in microbial inactivation [40].

Microbial Cell Structure Alterations as a Result of Plasma Exposure:

3.3.4 Disruption of the Cell Membrane

The membrane of the microbial cell is the initial target of reactive species produced by plasma. Observed structural alterations consist of:

- Lipid peroxidation, which results in the degradation of membrane integrity.
- Owing to oxidative stress, pores develop.
- Intracellular contents leaking, leading to cell lysis and osmotic imbalance.

Studies using transmission electron microscopy (TEM) have shown that plasma treatment causes membrane weakening and increased permeability, which ultimately leads to cell death.

3.3.5 Enzyme Inactivation and Denaturation of Proteins

Proteins undergo oxidative alteration as a result of plasma exposure, which interferes with microbial metabolism. Among the modifications are:

- Amino acid carbonylation.
- Protein chains being cross-linked, which makes enzymes inactive.
- Breakdown of structural proteins necessary for the survival of cells.

These consequences hinder the operation of microbial cells, making it impossible for them to recuperate or proliferate after exposure [41].

3.3.6 DNA and Genetic Material Damage

The genetic material of microorganisms is also harmed by plasma, which leads to irreversible cell death. Important consequences consist of:

- Breaks in DNA strands brought induced by ROS and UV.
- Changes to the base that result in mutation or loss of genetic integrity.
- Prevention of microbial recovery through inhibition of DNA repair processes.

Mechanisms of Plasma-Mediated Microbial Inactivation.

The process of plasma-induced microbial inactivation involves several steps:

(i) Initial contact: The outer cell structures are impacted by reactive species produced by plasma.

(ii) Membrane permeabilization: Cellular contents leak out, and selective permeability is lost as a result of oxidative damage.

(iii) Intracellular damage: RNS and ROS enter the cell and harm nucleic acids and proteins.

(iv) Apoptotic like death: Necrosis or programmed cell death is brought on by severe cellular injury.

(v) Complete inactivation: Microbial mortality results from cumulative structural alterations that inhibit regeneration.

3.4 Physicochemical Effects of Cold Plasma on Food Waste

For the treatment of heat-sensitive materials, like food waste, cold plasma works at or close to room temperature, unlike thermal plasmas, which need high temperatures [42]. In order to increase degradation efficiency and resource recovery, cold plasma interacts with organic matter to cause a number of physicochemical changes that might affect microbial inactivation, nutritional content, and structural integrity [43].

The modification of the surface characteristics of food waste is one of the main physicochemical impacts of cold plasma. The oxidative changes brought about by cold plasma exposure also help break down complex biomolecules like proteins, lipids, and carbohydrates into simpler forms that are more susceptible to microbial or enzymatic destruction [32]. Additionally, it has been demonstrated that cold plasma changes pH levels, encourages the selective oxidation of nutrients, and produces reactive oxygen and nitrogen species, all of which have an impact on the biochemical makeup of food waste. These modifications have the potential to increase biodegradability, lower pathogenic load, and transform waste into valuable products like biofuels or fertilizers. Knowing how cold plasma affects food waste physiochemically is essential to maximizing its potential to cut waste and advance the ideas of the circular economy [44].

3.5 Changes in Nutrient Composition and Food Structure

With major ramifications for sustainability, food security, and economic stability, food waste is a critical issue in contemporary food systems. A partly ionized gas known cold plasma contains reactive species that may disinfect food surfaces without causing thermal deterioration. More research is necessary to determine how it affects food structure and nutrient composition in order to guarantee that it is effective in lowering food waste while maintaining food quality [45].

3.5.1 Effects of Cold Plasma on Nutrient Composition

(i) Proteins: It has been demonstrated that cold plasma treatment causes structural unfolding, oxidation, and aggregation in proteins, which can improve digestibility and functional characteristics. However, overexposure can cause protein degradation and the depletion of vital amino acids [46].

(ii) Lipids: Lipids can be oxidized by reactive oxygen and nitrogen species (RONS) produced in cold plasma, which may result in lipid peroxidation. Unsaturated fatty acids are more susceptible to this impact, which lowers the nutritional value and produces unpleasant flavors. These effects can be lessened by employing techniques including utilizing antioxidant protection and maximizing treatment time [47].

(iii) Carbohydrates: The solubility and digestibility of carbohydrates, especially polysaccharides, might be affected by changes in their structures caused by cold plasma. Food products' textural qualities may be impacted by changes in the fiber and starch granule composition. But these modifications can also improve the functional qualities and potential of prebiotics [48].

(iv) Vitamins: Cold plasma-induced oxidative processes make water-soluble vitamins (including vitamin C and B-complex vitamins) more prone to deterioration. Though more slowly, fat-soluble vitamins (such as A, D, E, and K) can also experience oxidative alterations. While attaining microbial inactivation, CP parameter optimization can reduce nutrition loss [48].

3.5.2 Effects of Cold Plasma on Food Structure

Food matrices are affected by cold plasma, which alters their texture and microstructure. Food type and plasma treatment conditions have an impact on these outcomes [49].

(i) Microstructural changes: Food surfaces may become more porous as a result of CP, which could improve moisture retention or hasten dehydration (Fig. 2).

(ii) Textural modifications: Changes in the interactions between proteins and carbohydrates can change the texture, elasticity, and hardness of food.

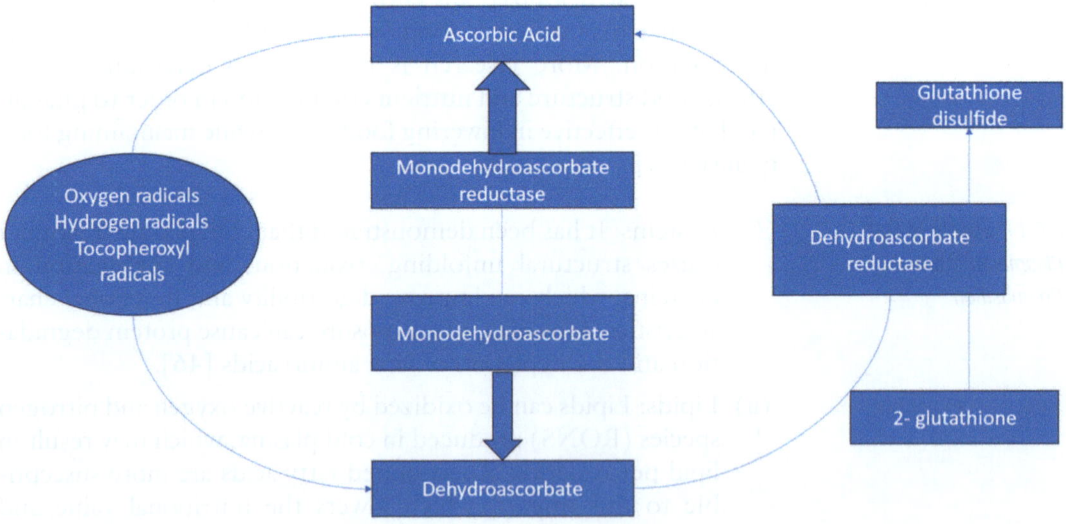

Fig. 2 Ascorbate recycling pathway and its role in antioxidant defense

(iii) Color and appearance: Cold plasma can oxidize pigments, changing the color of food and perhaps affecting consumer acceptance.

By altering the structural characteristics and nutrient composition, cold plasma technology has the potential to improve food preservation and decrease waste. However, to reconcile minimum nutritional loss with microbial inactivation, careful optimization is required. To improve processing conditions and comprehend long-term effects on food safety and quality, more study is needed [50].

3.6 Oxidative Stress and By-product Formation

Numerous RONS, such as ozone (O_3), hydroxyl radicals ($OH^{\cdot}$), hydrogen peroxide (H_2O_2), and nitric oxide (NO), are produced by cold plasma. By interacting with biomolecules, these species cause oxidative stress in organic matter and microbial cells [51].

3.6.1 The Oxidative Mechanisms Include

(i) Lipid peroxidation: Degradation of lipid membranes, which results in microbial death and cell permeability.

(ii) Protein oxidation: Interference with structural integrity and enzymatic function.

(iii) DNA damage: Genetic instability is caused by nucleotide alterations and strand breakage.

3.6.2 A Variety of By-Products Are Created When Cold Plasma and Food Waste Interact

(i) Short chain organic acids: Complex carbohydrate breakdown produces substances like formic acid and acetic acid.

(ii) Volatile organic compounds: When food ingredients oxidize, alcohols, ketones, and aldehydes are released.

(iii) Nitrogenous compounds: Nitrogen interactions that result in the formation of nitrites and nitrates.

(iv) Carbonyl compounds- Aldehydes like formaldehyde and malondialdehyde are produced when proteins and carbohydrates break down.

3.7 Effect on Sensory and Quality Attributes

The way that CP treatment affects sensory and quality features determines how effective it is at reducing food waste. These characteristics—which include texture, color, flavor, and nutritional qualities—determine consumer approval [52].

3.7.1 Effect of Cold Plasma on Sensory Attributes

(i) Texture: Food texture can be altered by cold plasma because of oxidation reactions and moisture loss. It can sometimes make fruits and vegetables firmer, but too much exposure might cause unwanted texture changes.

(ii) Color: The oxidation of pigments caused by reactive oxygen and nitrogen species (RONS) produced during CP treatments might change the appearance of food [53]. For instance, it has been demonstrated that CP keeps some fresh produce's color stable, although it can cause sensitive goods to brown [54].

(iii) Flavor and aroma: CP has the ability to affect the volatile substances that give food its flavor and smell. While some research indicates that oxidation of proteins and lipids causes off-flavors in fatty foods, other studies reveal only slight changes [55].

(iv) Mouthfeel: Variations in the food matrix qualities can lead to changes in mouthfeel. Juiciness perception in fruits and meat items can be impacted by plasma-induced dehydration [56, 57].

(v) Overall acceptability: The balance between microbial inactivation and quality retention determines consumer acceptance. Optimized CP treatment enhances food safety without significantly compromising sensory attributes [57].

3.7.2 Effect on Quality Attributes

(i) Microbial load reduction: By efficiently lowering harmful and spoiling bacteria, CP increases food shelf life and decreases food waste.

(ii) Nutritional composition: Treatment with CP may change the stability of nutrients, especially oxidation-sensitive vitamins (like vitamin C) [58]. Nevertheless, in contrast to traditional thermal treatments, the non-thermal character of CP helps maintain heat-sensitive nutrients [59].

(iii) Oxidative stability: When it comes to fatty foods, lipid oxidation can cause rancidity, which can be lessened by adjusting plasma parameters [60].

(iv) Moisture content: Exposure to plasma can affect food's water retention, which over time might affect texture and weight loss.

3.8 Technological Aspects and Equipment for Cold Plasma Treatment

Cold plasma application in food waste management necessitates specific equipment and a deep comprehension of its technical elements, such as material interactions, operating parameters, and plasma generating techniques [61]. This section examines the main technological elements and tools used in cold plasma treatment, emphasizing different reactor designs, plasma generation methods, and their effects on the processing of food waste [62]. We also look at important elements that are necessary for the effective use of cold plasma in waste management systems, like energy efficiency, scalability, and safety concerns. Researchers and industry experts can successfully improve food waste reduction and the valorization efforts by optimizing cold plasma applications by comprehending these technological factors [63].

3.8.1 Types of Plasma Generators Used for Food Waste Treatment

Food waste is treated using a variety of plasma generator types, each with distinct features and working principles. Among the most often used are:

(i) Dielectric barrier discharge (DBD) plasma generator: Between two electrodes divided by a dielectric barrier, DBD plasma is produced. It produces reactive species like ozone, hydroxyl radicals, and nitrogen oxides by generating non-thermal plasma at atmospheric pressure, which aids in the decomposition of organic matter and the inactivation of microorganisms [64].

(ii) Atmospheric pressure plasma jet (APPJ): APPJ is made up of an ionized gas stream that is released from a nozzle at a high speed [65, 66]. It produces reactive oxygen and nitrogen species (RONS) while operating at atmospheric pressure. APPJ is perfect for handling liquid and semi-solid food waste because it efficiently breaks down organic molecules and improves biodegradability.

(iii) Gliding arc discharge (GAD) plasma generator: An electric arc that travels along diverging electrodes is how GAD creates plasma. It generates plasma with a high density and potent oxidative capabilities. This kind works especially well for increasing the production of methane during anaerobic digestion processes and breaks down complex organic compounds in food waste [66].

(iv) Microwave plasma generator: In order to create a high-energy plasma field, microwave radiation is used to ionize a gas. Because of its great energy efficiency and ability to produce a variety of reactive species, this kind is very effective at breaking down organic matter [67].

3.8.2 Optimization of Processing Conditions

A number of variables, such as the gas type, treatment duration, input power, and plasma source, affect how effective cold plasma is.

(i) Gas composition: The reactivity of plasma species is influenced by the gas used to generate the plasma, such as air, oxygen, nitrogen, or argon. Nitrogen-based plasmas improve biochemical changes for food waste repurposing, whereas oxygen-rich plasmas work well for microbial decrease [68].

(ii) Treatment duration: Long-term exposure to cold plasma may cause excessive nutrient oxidation but can also increase microbial inactivation. It is crucial to determine the threshold treatment duration that strikes a balance between nutrient retention and microbiological safety.

(iii) Voltage and power input: More reactive species are produced by higher voltages, which increases therapy effectiveness. On the other hand, overpower can degrade the components of food waste and ruin their surfaces. For treatment to be effective, voltage must be optimized within a modest range [69].

(iv) Pressure and environmental conditions: Both atmospheric and low-pressure settings are suitable for cold plasma treatment. For large-scale applications, atmospheric plasma is more feasible, but low-pressure plasma improves homogeneity. The ambient temperature and humidity levels also have an impact on plasma reactivity [70].

3.8.3 Scalability and Industrial Applications

There are various obstacles in the way of cold plasma's transition from lab-scale research to industrial-scale uses:

(i) Energy efficiency: In order to make cold plasma systems cost-effective for large-scale operations, power consumption must be optimized.

(ii) Plasma generation and distribution: For constant treatment efficacy, homogenous plasma exposure over vast amounts of food waste is essential.

(iii) Material compatibility: The materials of the reactor and electrodes must be able to endure extended treatment with reactive plasma species.

(iv) Economic viability: Large-scale adoption may be hampered by high startup and operating expenses unless the financial gains outweigh the drawbacks.

(v) Process standardization: Standardized procedures for various kinds of food waste should be established to guarantee regulatory compliance and reproducibility.

Notwithstanding these obstacles, cold plasma has a lot of industrial promise for managing food waste:

(i) Pathogen reduction: Efficient in lowering the microbial load, decontaminating food waste, and minimizing health hazards in applications such as animal feed and trash disposal.

(ii) Reduction of food spoilage: Applications for perishable food waste can help with food redistribution efforts by extending usage prior to decomposition.

(iii) Valorization of food waste: Food waste can be more easily transformed into valuable products like charcoal or bioactive chemicals for use in agriculture with the help of cold plasma.

(iv) Enhancement of anaerobic digestion: Cold plasma pretreatment can improve the production of biogas by dissolving complicated organic molecules.

A ground-breaking method of managing food waste with significant industrial potential is cold plasma technology [71, 72]. Even if scalability is still a problem, continued research and technical developments are opening the door for its widespread use. To fully utilize cold plasma in environmentally friendly waste management, it will be essential to address technological, financial, and regulatory obstacles [72].

3.9 Cold Plasma in Enhancing Biodegradability and Waste Valorization

When it comes to food waste, cold plasma treatment is a viable method to boost the recovery of valuable by-products like biofuels and biofertilizers, speed up decomposition, and increase digestibility in anaerobic digestion systems.

3.9.1 Influence on Composting and Biogas Production

Although cold plasma's use in organic waste management is still relatively new, it has been extensively studied in the fields of food sterilization and wastewater treatment [73].

The Effects of Cold Plasma Treatment on Composting Include

(i) By changing the physicochemical characteristics of organic matter, reactive species produced by plasma can promote the growth of microorganisms.

(ii) Cold plasma pre-treatments have the potential to increase nutrient availability and reduce the composting time by dissolving complicated compounds.

(iii) During composting, cold plasma's antibacterial qualities reduce odor emissions and the pathogen burden.

Cold Plasma Influences the Production of Biogas in Anaerobic Digestion by

(i) Increasing biodegradability: By breaking down lignocellulosic structures, plasma treatment increases the availability of fermentable sugars to microorganisms.

(ii) Improving methane yield: Research indicates that because of enhanced hydrolysis and acidogenesis phases, plasma-pretreated substrates have a greater methane content.

(iii) Balancing microbial consortia: Microbial populations are influenced by cold plasma, which may favor methanogenic archaea that are essential for the production of biogas [74].

3.10 Plasma Treatment for Improved Fermentation Processes

3.10.1 There Are Several Ways in Which Applying Cold Plasma to Food Waste Affects Fermentation Efficiency

(i) Microbial inactivation: A more regulated and predictable fermentation environment is ensured by cold plasma, which produces reactive oxygen and nitrogen species (RONS) that efficiently interfere with microbial cell membranes, DNA, and metabolic pathways, resulting in selective microbial inactivation. Infectious and spoilage microorganisms in food debris can compete with desirable fermentative microbes, lowering process yields [75].

(ii) Substrate modification: Fermentable sugars become more bioavailable when food waste undergoes structural and metabolic changes brought on by cold plasma. More effective fermentation is made possible by the breakdown of complex polysaccharides into simpler sugars, which improves microbial accessibility [76].

(iii) Enhancement of microbial metabolism: It has been demonstrated that exposing fermentative microorganisms to cold plasma-pretreated substrates increases their metabolic activity.

This is probably because the nutrients are more readily available and the spoiling germs are less of a competition. Research shows that plasma therapy can increase the yields of methane, ethanol, and lactic acid in the corresponding fermentation processes [77].

3.10.2 Potential for Nutrient Recovery and Circular Economy

Essential nutrients like potassium, phosphate, and nitrogen can be extracted from food waste with the use of cold plasma. Its potential is demonstrated by the following applications:

(i) Biofertilizer production: Organic waste that has been cold plasma treated can be used as a nutrient-rich fertilizer to improve crop productivity and soil health [78].

(ii) Protein and amino acid extraction: Proteins and amino acids can be more easily extracted for use in feed for animals and human consumption through plasma-induced hydrolysis.

(iii) Conversion to bioactive compounds: The extraction of bioactive substances, including antioxidants and polyphenols, for use in pharmaceutical and nutraceutical applications can be improved by cold plasma [79].

Cold plasma technology converts trash into useful resources, which is consistent with the ideas of the circular economy. Important facets of its integration consist of:

(i) Reducing waste disposal impact: Environmental pollution and greenhouse gas emissions are reduced when food waste is diverted from landfills.

(ii) Enhancing resource efficiency: Recovering nutrients encourages sustainable agriculture and lessens reliance on artificial fertilizers.

(iii) Economic and industrial scalability: In the food processing industry, cold plasma can be used to maximize resource use and produce products with additional value.

To improve the use of cold plasma in nutrient recovery and circular economy models, future research should concentrate on pilot-scale studies, techno-economic feasibility, and lifecycle assessments [80].

In line with the ideas of the circular economy, cold plasma technology provides a fresh and environmentally friendly method of recovering nutrients from food waste. Cold plasma has the ability to improve resource productivity in food systems, lessen waste, and decrease environmental effects by turning waste into useful resources. To fully realize its potential for large-scale applications, more study and technical developments are essential [81].

4 Notes

Cold plasma technology has appeared as an auspicious non-thermal technique for decontaminating food waste, plummeting microbial load, and encompassing the serviceability of discarded food materials. However, despite its advantages, several challenges and boundaries must be addressed to ensure its large-scale submission in food waste management. One of the primary challenges is the inconsistency in food waste composition. Food waste contains of diverse organic materials with different moistness content, surface belongings, and nutrient composition, which can affect the competence of cold plasma treatment. Certain components may shield microorganisms from plasma acquaintance, leading to uneven decontamination. Moreover, disparities in food waste types require tailored plasma parameters, making standardization difficult [82, 83].

Another restraint is the probable impact on food quality. While cold plasma is a non-thermal process, protracted exposure may lead to objectionable changes in texture, color, or flavor due to corrosion reactions. These adjustments can affect the probability of reusing treated food waste in animal feed, bioenergy production, or upcycling processes. Scalability and energy competence are also critical anxieties. Current plasma systems require meticulous surroundings and specialized apparatus, making large-scale application challenging. The energy ingesting of plasma generation must be optimized to ensure economic probability and sustainability. Moreover, the protection of plasma-treated food waste for secondary applications, such as composting or livestock feed, requires further examination [83].

Future progressions in cold plasma technology should focus on educating efficiency through enhanced plasma generation methods, participating with other maintainable decontamination techniques, and confirming minimal quality worsening. Research on the environmental impact, cost-effectiveness, and controlling compliance will be essential for commercial espousal. With continued novelty, cold plasma has the potential to play a significant role in sustainable food waste decontamination and reserve recovery [73].

References

1. Langyan S, Yadava P, Khan FN, Dar ZA, Singh R, Kumar A (2022) Sustaining protein nutrition through plant-based foods. Front Nutr 8:772573
2. Bleakley S, Hayes M (2017) Algal proteins: extraction, application, and challenges concerning production. Food Secur 6(5):33
3. Henchion M, Hayes M, Mullen AM, Fenelon M, Tiwari B (2017) Future protein supply and demand: strategies and factors influencing a sustainable equilibrium. Food Secur 6(7):53
4. Hidayat A, Widayanti VT, Septifani R, Firdausiyah N, Haryati N, Natasya A et al (2023) Peningkatan Kualitas Dan Produktivitas Jamur Tiram Coklat Di Unit Usaha Istana Jamur Balap (Ijb), Kecamatan Kalipare, Kabupaten Malang. In: Seminar Nasional Penelitian

Dan Pengabdian Teknologi Hasil Pertanian, vol 3, No. 1

5. Cho JH, Kim I (2011) Fish meal–nutritive value. J Animal Physiol Animal Nutr 95(6): 685–692
6. Vignaud J, Loiseau C, Hérault J, Mayer C, Côme M, Martin I, Ulmann L (2023) Microalgae produce antioxidant molecules with potential preventive effects on mitochondrial functions and skeletal muscular oxidative stress. Antioxidants 12(5):1050
7. Bauer L, Ranglová K, Masojídek J, Drosg B, Meixner K (2021) Digestate as sustainable nutrient source for microalgae—challenges and prospects. Appl Sci 11(3):1056
8. Deamici KM, de Morais MG, Santos LO, Muylaert K, Gardarin C, Costa JAV, Laroche C (2021) Static magnetic fields effects on polysaccharides production by different microalgae strains. Appl Sci 11(11):5299
9. Suárez Quintana WH, García-Rico RO, García-Martínez JB, Urbina-Suarez NA, López-Barrera GL, Barajas-Solano AF, Zuorro A (2022) Enhancement of metabolite production in high-altitude microalgal strains by optimized C/N/P ratio. Appl Sci 12(13):6779
10. Martins RA, Salgado EM, Gonçalves AL, Esteves AF, Pires JC (2024) Microalgae-based remediation of real textile wastewater: assessing pollutant removal and biomass valorisation. Bioengineering 11(1)
11. Hoyos BS, Hernandez-Tenorio F, Miranda AM, Villanueva-Mejía DF, Sáez AA (2023) Systematic analysis of genes related to selenium bioaccumulation in microalgae: a review. Biology 12(5):703
12. Dolganyuk V, Belova D, Babich O, Prosekov A, Ivanova S, Katserov D et al (2020) Microalgae: a promising source of valuable bioproducts. Biomol Ther 10(8):1153
13. Gomiero T (2016) Soil degradation, land scarcity and food security: reviewing a complex challenge. Sustain For 8(3):281
14. Li C, Mirosa M, Bremer P (2020) Review of online food delivery platforms and their impacts on sustainability. Sustain For 12(14): 5528
15. Sánchez-Teba EM, Gemar G, Soler IP (2021) From quantifying to managing food loss in the agri-food industry supply chain. Food Secur 10(9):2163
16. Keegan E, Breadsell JK (2021) Food waste and social practices in Australian households. Sustain For 13(6):3377
17. Luhar I, Luhar S, Abdullah MMAB (2022) Challenges and impacts of COVID-19 pandemic on global waste management systems: a review. J Comp Sci 6(9):271
18. Kataya G, Cornu D, Bechelany M, Hijazi A, Issa M (2023) Biomass waste conversion technologies and its application for sustainable environmental development—a review. Agronomy 13(11):2833
19. Guo S, Guo H (2024) The driving factors of food waste in Chinese urban households: a qualitative study based on grounded theory. Sustain For 16(14):6091
20. Romero-Perdomo F, González-Curbelo MÁ (2023) Integrating multi-criteria techniques in life-cycle tools for the circular bioeconomy transition of agri-food waste biomass: a systematic review. Sustain For 15(6):5026
21. Mumtaz S, Chu AM, Attiq S, Shah HJ, Wong WK (2022) Habit—does it matter? Bringing habit and emotion into the development of consumer's food waste reduction behavior with the lens of the theory of interpersonal behavior. Int J Environ Res Public Health 19(10):6312
22. Fassio F, Chirilli C (2023) The circular economy and the food system: a review of principal measuring tools. Sustain For 15(13):10179
23. Calicioglu O, Flammini A, Bracco S, Bellù L, Sims R (2019) The future challenges of food and agriculture: an integrated analysis of trends and solutions. Sustain For 11(1):222
24. Atstaja D, Cudecka-Purina N, Koval V, Kuzmina J, Butkevics J, Hrinchenko H (2024) Waste-to-energy in the circular economy transition and development of resource-efficient business models. Energies 17(16): 4188
25. Ikpe V, Shamsuddoha M (2024) Functional model of supply chain waste reduction and control strategies for retailers—the USA retail industry. Logistics 8(1):22
26. Lins M, Puppin Zandonadi R, Raposo A, Ginani VC (2021) Food waste on foodservice: an overview through the perspective of sustainable dimensions. Food Secur 10(6):1175
27. Sobaih AEE (2023) Saudi zero food waste certification: a novel approach for food waste management in Saudi Arabia. Agronomy 13(6):1654
28. Lee HH, Huang PY (2023) Food waste and environmental sustainability of the hotel industry in Taiwan. Sustain For 15(21):15459
29. Bennedetti LV, de Almeida Sinisgalli PA, Ferreira ML, Lemes de Oliveira F (2023) Challenges to promote sustainability in urban agriculture models: a review. Int J Environ Res Public Health 20(3):2110

30. Audet R, Brisebois É (2019) The social production of food waste at the retail-consumption interface. Sustain For 11(14): 3834
31. Todorova Y, Benova E, Marinova P, Yotinov I, Bogdanov T, Topalova Y (2022) Non-thermal atmospheric plasma for microbial decontamination and removal of hazardous chemicals: an overview in the circular economy context with data for test applications of microwave plasma torch. Processes 10(3):554
32. Heydari M, Carbone K, Gervasi F, Parandi E, Rouhi M, Rostami O et al (2023) Cold plasma-assisted extraction of phytochemicals: a review. Food Secur 12(17):3181
33. Katsigiannis AS, Bayliss DL, Walsh JL (2022) Cold plasma for the disinfection of industrial food-contact surfaces: an overview of current status and opportunities. Compr Rev Food Sci Food Saf 21(2):1086–1124
34. Li P, Zhang H, Tian C, Zou H (2024) Experimental investigation of bacterial inactivation of beef using indirect cold plasma in cold chain and at room temperature. Food Secur 13(17): 2846
35. Domonkos M, Tichá P, Trejbal J, Demo P (2021) Applications of cold atmospheric pressure plasma technology in medicine, agriculture and food industry. Appl Sci 11(11):4809
36. Perera KY, Prendeville J, Jaiswal AK, Jaiswal S (2022) Cold plasma technology in food packaging. Coatings 12(12):1896
37. Karthik C, Mavelil-Sam R, Thomas S, Thomas V (2024) Cold plasma technology based eco-friendly food packaging biomaterials. Polymers 16(2):230
38. Abbas HM, Altamim EA, Salama M, Fouad MT, Zahran HA (2024) Cold plasma technology: a sustainable approach to milk preservation by reducing pathogens and enhancing oxidative stability. Sustain For 16(20):8754
39. Bagheri H, Abbaszadeh S (2020) Effect of cold plasma on quality retention of fresh-cut produce. J Food Qual 2020(1):8866369
40. Jiang X, Lee S, Mok C, Lee J (2017) Sustainable methods for decontamination of microcystin in water using cold plasma and UV with reusable TiO2 nanoparticle coating. Int J Environ Res Public Health 14(5):480
41. Lim SWY, Ow SY, Sutarlie L, Lee YY, Suwardi A, Tan CKI et al (2024) Bioaerosol inactivation by a cold plasma ionizer coupled with an electrostatic precipitator. Microorganisms 12(9):1923
42. Hoque M, McDonagh C, Tiwari BK, Kerry JP, Pathania S (2022) Effect of cold plasma treatment on the packaging properties of biopolymer-based films: a review. Appl Sci 12(3):1346
43. Jiang H, Lin Q, Shi W, Yu X, Wang S (2022) Food preservation by cold plasma from dielectric barrier discharges in agri-food industries. Front Nutr 9:1015980
44. Senguler B, Kirkin C, Donmez H, Unal S (2024) Treatment of onion skin waste using dielectric barrier discharge cold plasma processing. Food Sci Nutr 12(10):8387–8393
45. Roobab U, Chacha JS, Abida A, Rashid S, Muhammad Madni G, Lorenzo JM et al (2022) Emerging trends for nonthermal decontamination of raw and processed meat: ozonation, high-hydrostatic pressure and cold plasma. Food Secur 11(15):2173
46. Yepez X, Illera AE, Baykara H, Keener K (2022) Recent advances and potential applications of atmospheric pressure cold plasma technology for sustainable food processing. Food Secur 11(13):1833
47. Özdemir E, Başaran P, Kartal S, Akan T (2023) Cold plasma application to fresh green leafy vegetables: impact on microbiology and product quality. Compr Rev Food Sci Food Saf 22(6):4484–4515
48. Khumsupan D, Lin SP, Hsieh CW, Santoso SP, Chou YJ, Hsieh KC et al (2023) Current and potential applications of atmospheric cold plasma in the food industry. Molecules 28(13):4903
49. Mravlje J, Regvar M, Vogel-Mikuš K (2021) Development of cold plasma technologies for surface decontamination of seed fungal pathogens: present status and perspectives. J Fungi 7(8):650
50. Pankaj SK, Wan Z, Keener KM (2018) Effects of cold plasma on food quality: a review. Food Secur 7(1):4
51. Akhavan-Mahdavi S, Mirzazadeh M, Alam Z, Solaimanimehr S (2023) The effect of chitosan coating combined with cold plasma on the quality and safety of pistachio during storage. Food Sci Nutr 11(7):4296–4307
52. Siddique SS, Hardy GSJ, Bayliss KL (2018) Cold plasma: a potential new method to manage postharvest diseases caused by fungal plant pathogens. Plant Pathol 67(5):1011–1021
53. Ahmadi M, Singer D, Potlitz F, Nasri Z, von Woedtke T, Link A et al (2023) Cold physical plasma-mediated fenretinide prodrug activation confers additive cytotoxicity in epithelial cells. Antioxidants 12(6):1271
54. Singh A, Mehta S, Yadav S, Nagar G, Ghosh R, Roy A et al (2022) How to cope with the challenges of environmental stresses in the era

of global climate change: an update on ROS staves off in plants. Int J Mol Sci 23(4):1995
55. Risa Vaka M, Sone I, García Álvarez R, Walsh JL, Prabhu L, Sivertsvik M, Noriega Fernández E (2019) Towards the next-generation disinfectant: composition, storability and preservation potential of plasma activated water on baby spinach leaves. Food Secur 8(12):692
56. Gola F, Gaiaschi L, Roda E, De Luca F, Ferulli F, Vicini R et al (2023) Voghera sweet pepper: a potential ally against oxidative stress and aging. Int J Mol Sci 24(4):3782
57. Punthi F, Yudhistira B, Gavahian M, Chang CK, Husnayain N, Hou CY et al (2023) Optimization of plasma activated water extraction of Pleurotus ostreatus polysaccharides on its physiochemical and biological activity using response surface methodology. Food Secur 12(23):4347
58. Sindhu RK, Goyal A, Algın Yapar E, Cavalu S (2021) Bioactive compounds and nanodelivery perspectives for treatment of cardiovascular diseases. Appl Sci 11(22):11031
59. Omran B, Baek KH (2021) Nanoantioxidants: pioneer types, advantages, limitations, and future insights. Molecules 26(22):7031
60. Ali A, Riaz S, Sameen A, Naumovski N, Iqbal MW, Rehman A et al (2022) The disposition of bioactive compounds from fruit waste, their extraction, and analysis using novel technologies: a review. Processes 10(10):2014
61. Senadheera TR, Hossain A, Shahidi F (2023) Marine bioactives and their application in the food industry: a review. Appl Sci 13(21):12088
62. Cavallini C, Vitiello G, Adinolfi B, Silvestri B, Armanetti P, Manini P et al (2020) Melanin and melanin-like hybrid materials in regenerative medicine. Nanomaterials 10(8):1518
63. Saini RV, Vaid P, Saini NK, Siwal SS, Gupta VK, Thakur VK, Saini AK (2021) Recent advancements in the technologies detecting food spoiling agents. J Funct Biomater 12(4):67
64. Cavallini C, Vitiello G, Adinolfi B, Silvestri B, Armanetti P, Manini P et al (2020) Melanin and melanin-like hybrid materials in regenerative medicine. Nanomaterials (Basel) 10(8):1518
65. Alhaithloul HAS (2019) Impact of combined heat and drought stress on the potential growth responses of the desert grass Artemisia sieberi alba: relation to biochemical and molecular adaptation. Plan Theory 8(10):416
66. Khatun M, Sarkar S, Era FM, Islam AM, Anwar MP, Fahad S et al (2021) Drought stress in grain legumes: effects, tolerance mechanisms and management. Agronomy 11(12):2374
67. Melo-Sabogal DV, Contreras-Medina LM (2024) Elicitors and biostimulants to mitigate water stress in vegetables. Horticulturae 10(8):837
68. Sharma S, Sharma RK, Gaur K, Cátala Torres JF, Loza-Rosas SA, Torres A et al (2019) Fueling a hot debate on the application of TiO2 nanoparticles in sunscreen. Materials 12(14):2317
69. Surya Krishna S, Harish Chandar SR, Ravi M, Valarmathi R, Lakshmi K, Prathima PT et al (2023) Transgene-free genome editing for biotic and abiotic stress resistance in sugarcane: prospects and challenges. Agronomy 13(4):1000
70. Georgescu AM, Corbu VM, Csutak O (2024) Molecular basis of yeasts antimicrobial activity—developing innovative strategies for biomedicine and biocontrol. Curr Issues Mol Biol 46(5):4721–4750
71. Rahman M, Hasan MS, Islam R, Rana R, Sayem ASM, Sad MAA et al (2022) Plasma-activated water for food safety and quality: a review of recent developments. Int J Environ Res Public Health 19(11):6630
72. Gangemi S, Petrarca C, Tonacci A, Di Gioacchino M, Musolino C, Allegra A (2022) Cold atmospheric plasma targeting hematological malignancies: potentials and problems of clinical translation. Antioxidants 11(8):1592
73. Piazza S, Fumagalli M, Khalilpour S, Martinelli G, Magnavacca A, Dell'Agli M, Sangiovanni E (2020) A review of the potential benefits of plants producing berries in skin disorders. Antioxidants 9(6):542
74. Mildaziene V, Ivankov A, Sera B, Baniulis D (2022) Biochemical and physiological plant processes affected by seed treatment with non-thermal plasma. Plan Theory 11(7):856
75. Zeraat Pisheh F, Falah F, Sanaei F, Vasiee A, Zanganeh H, Tabatabaee Yazdi F, Ibrahim SA (2023) The effect of plasma-activated water combined with rosemary extract (Rosmarinus officinalis L.) on the physicochemical properties of Frankfurter sausage during storage. Food Secur 12(21):4022
76. Pinilla-González V, Rojas-Solé C, Gómez-Hevia F, González-Fernández T, Cereceda-Cornejo A, Chichiarelli S et al (2024) Tapping into nature's arsenal: harnessing the potential of natural antioxidants for human health and disease prevention. Food Secur 13(13):1999
77. Clifford T, Howatson G, West DJ, Stevenson EJ (2015) The potential benefits of red beetroot supplementation in health and disease. Nutrients 7(4):2801–2822

78. Vieira ACA, Ferreira FDS, Araújo JMDD, Dutra LMG, Batista KS, Cordeiro AMTDM, Aquino JDS (2024) Exploring the potential hepatoprotective properties of cactus (Cactaceae) in liver health and disease management: a brief review. Liver 4(2):287–313
79. Martins S, Pereira S, Dinis LT, Brito C (2024) Enhancing olive cultivation resilience: sustainable long-term and short-term adaptation strategies to alleviate climate change impacts. Horticulturae 10(10):1066
80. Kashyap P, Kumar S, Riar CS, Jindal N, Baniwal P, Guiné RP et al (2022) Recent advances in Drumstick (Moringa oleifera) leaves bioactive compounds: composition, health benefits, bioaccessibility, and dietary applications. Antioxidants 11(2):402
81. Trotta T, Porro C, Cianciulli A, Panaro MA (2022) Beneficial effects of spirulina consumption on brain health. Nutrients 14(3):676
82. Bass AV, Falkenberg LJ, Thibodeau B (2024) Seagrasses under stress: independent negative effects of elevated temperature and light reduction at multiple levels of organization. Limnol Oceanogr
83. Kumar R, Sinha R, Sharma PK, Ivy N, Kumar P, Kant N et al (2021) Bioaccumulation of fluoride in plants and its microbially assisted remediation: a review of biological processes and technological performance. Processes 9(12):2154

Chapter 9

Green Valorization Protocols for Producing Bioplastics from Food Waste

Haider Ijaz, Amina BiBi, Farooq Anwar, Aman Ullah, Usama Anwar, Mujahid Farid, Muhammad Nadeem Zafar, and Muhammad Zubair

Abstract

Food waste valorization is a technique for bioplastics production which is evolving tremendously day by day. This method utilizes food waste and assists in reducing food waste and its detrimental consequences. This chapter comprises valorization protocols like the collection of food waste from different resources, pretreatment by various physical, chemical, or biological ways, fermentation by microbial enzymes, polymerization of fermented materials, purification or characterization of polymerized bioplastics, and blending and processing. Bioplastics like P(3HB), P(3HB-Co-3 HV), and PLAs are under substantial consideration to upgrade their characteristics to improve biocompatibility and lower their environmental effects. Advanced technologies like the exploitation of GMOs, bioaugmentation, artificial intelligence, nanotechnology, among others, are integrated into valorization by the researchers to enhance the production capacity, widen their applications, reduce their average formulation cost, boost the life assessment cycle, and many more. The biopolymers synthesized their number of different services in medicine, packaging, automotive coatings, households, etc., which makes the food waste valorization an ideal field of research and suggests the need for extensive production potential.

Key words Valorization, Polylactic acid, Polyglycolic acid, Fatty acids, Adenosine triphosphate

1 Introduction

Food waste is consumable food that is discarded by society during the processes of its production, transportation, processing, or consumption. According to the United Nations Food and Agriculture Organization (FAO) [1] "When food that is safe and healthy for humans to eat is disposed of, it is called food waste." Unfortunately, about one-third of food is wasted annually, which becomes 40% or 2.5 billion tons of the global consumption of food every year. If we can collect this food surplus together, it will become the size of India and Canada combined, and it will be the third largest gas emitter. At the same time, of this huge amount of food being left, almost 780 million people are facing chronic hunger globally per

Tanmay Sarkar (ed.), *Food Waste Valorization: Green Techniques in Sustainable Management*, Methods and Protocols in Food Science, https://doi.org/10.1007/978-1-0716-4646-5_9,
© The Author(s), under exclusive license to Springer Science+Business Media, LLC, part of Springer Nature 2025

annum. In the year 2021, the annual worldwide food consumption rate is around 2.5 billion tons, and, in that year, almost 1.3 billion tons of food wasted, which is 1/3 of the original usage. Evidence closer to home showcased that Pakistan has a hunger rate of 88 out of 107 and produces 36 million tons of food waste annually from different sources like households, restaurants, offices, and other functions like weddings and parties [2]. This food waste consists of 60–70% organic waste, and the water content is 75–85 wt.%. The pH of this food waste is observed between 4.8 and 6.9.

1.1 Sources, Composition, and Challenges of Food Waste

The following flow sheet indicates the types and ingredients of collected food waste. The compositional analysis of these types of wastes is necessary to utilize them as a source of monomers to generate high-potential bioplastics. The composition of these ingredients is as follows: the cereals provide carbohydrates, vitamin B, and minerals like iron and magnesium. Flour provides mainly carbohydrates, moderate proteins, and a small number of vitamins and minerals. Oats provide carbohydrates and dietary fiber and are high in vitamins and minerals like Mn, P, Mg, and Zn. Rice provides carbohydrates and minor amounts of proteins, fats, vitamins, and minerals. Pasta provides carbohydrates and little protein. Dairy products provide proteins, fats, vitamins, and minerals. Eggs provide vitamin A, folates, pantothenic acid (vitamin B5), phosphorus, selenium, calcium, and zinc. Figure 1 gives a detailed overview for extraction initials of these organic monomers from the FW.

For the extraction of inorganic monomers, specific sources could be considered. Table 1 depicts the specific types of sources for the extraction of inorganic compounds from the FW.

The food waste that is being collected from multiple resources could consist of different compositional constituents, i.e., organic and inorganic ones. These organic and inorganic waste components can be used to formulate bioplastics using several different techniques and procedures. The organic waste mostly comprises volatile solids, carbohydrates, proteins, cellulose, lipids, starch, etc., while the inorganic part is composed of ash, calcium, potassium, magnesium, phosphorus, sodium, etc. Their respective concentrations generally found in FW collected are given in Table 2.

Nowadays food waste is being consumed to create bioplastics as it offers great potential and biodegradable plastic with an enhanced lifetime. Food valorization is an ideal strategy which helps us to reduce and reuse food waste and gives us high-demand value-added products that can improve the economic condition of society. The manufacturing of bioplastics is increasing tremendously day by day. In the year 2018, the annual bioplastics fabrication was 2.01 million tons and expanded to 2.23 million tons by 2022 and is expected to rise to 4.7 million tons between now and 2027 by EUBP in the 17th conference in Dec 2022 [5]. With the aid of advanced

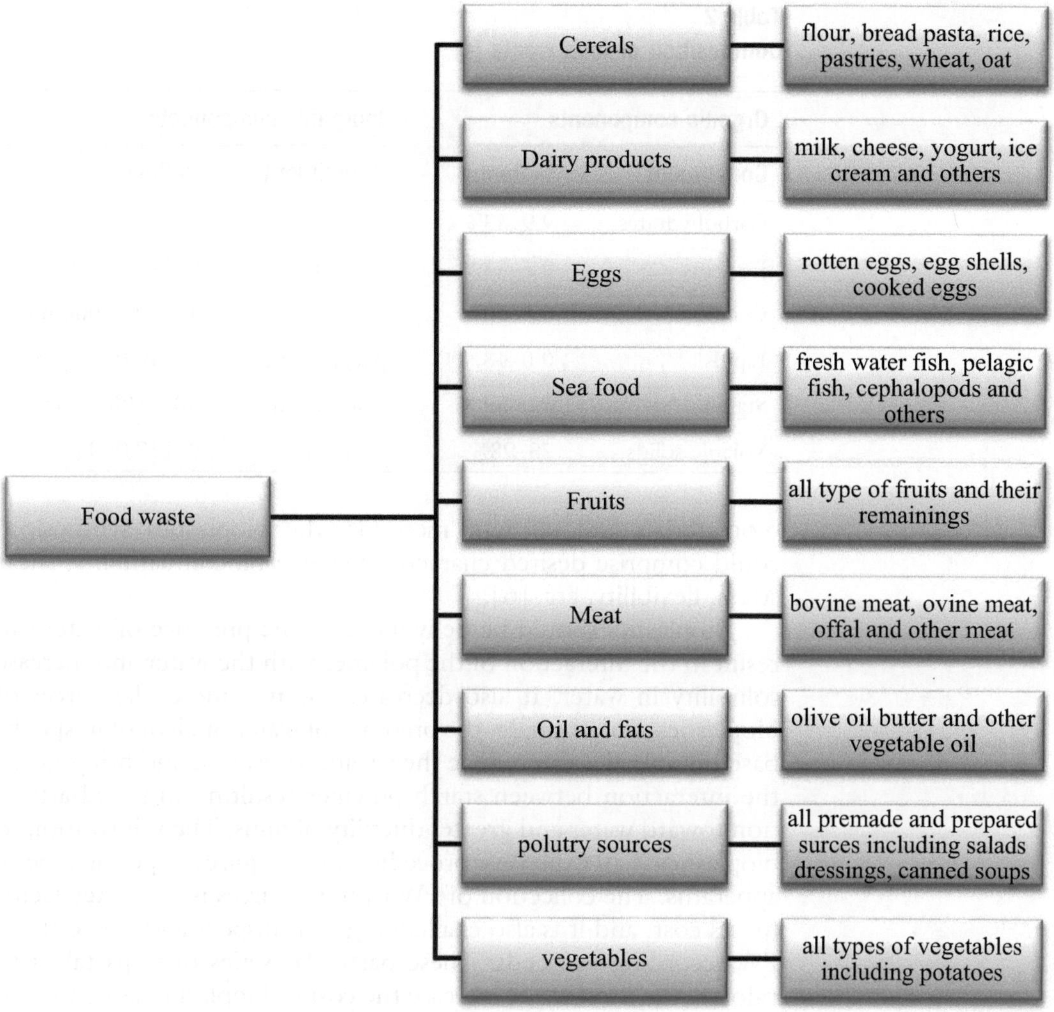

Fig. 1 Food waste resources flow sheet

Table 1
Sources of inorganic wastes [3]

Metals	Sources
Ca	Mollusks, shrimps, eggshells, cuttlebones
K	White beans, sweet potatoes, beets, parsnips, spinach, Swiss chard
Si	Rice husk, rice straws, bamboo leaves, corn cob, wheat straw, sugar cane bagasse
Mg	Mostly vegetables like tomatoes, eggplant, garlic, pepper, cucumber
P	Rye, potatoes, molasses, and fruit-vegetable waste
Na	NaCl salt
Heavy metals	Like Ld, Cd, Ni, Zn, Ar, Al, Hg.

Table 2
Composition of food waste [4]

Organic components		Inorganic components	
Constituent	**% Range**	**Constituent**	**% Range**
Carbohydrates	9.9–83%	Ash	0.9–39.7%
Proteins	0.7–53.9%	Calcium	0–24,400 ppm
Cellulose	0.5–60%	Potassium	400–910,000 ppm
Lipids	0.0–48%	Magnesium	31–40,000 ppm
Starch	2.0–59.8%	Phosphorus	300–1000 ppm
Volatile solids	73–98%	Sodium	0–147,000 ppm

biomaterials and efficient methods, the bioplastics synthesized could comprise desired characteristics like biocompatibility, shear stress, flexibility, etc. [6].

Bioplastics should be dehydrated as the presence of water can result in the interaction of the polymer with the water and increase solubility in water. It also decreases the intermolecular forces of bioplastics. For example, the presence of water or alcohol in starch-based bioplastics can reduce their water resistance and bring down the interaction between starch polymer resulting in raised attraction toward water and greater ductility of films. The valorization to bioplastics is an expensive procedure as it requires a specific type of apparatus. The collection of FW from resources is also a key factor for its cost, and it is also challenging to transport and sort out the FW according to needs. These particular series of steps taken to valorize the food waste increase the cost of bioplastics as compared to conventional plastics. The transport of FW during its valorization can also cause pollution if not managed and transported according to the prescribed and sophisticated precautions. Some bioplastics require proper disposal, which is carried out using advance machinery under strict industrial conditions. For instance, the degradation of PLA under high temperatures and specific microbes.

1.2 Bioplastics and Their Classification

Bioplastics are polymers derived from naturally occurring or renewable materials that can be biodegradable or non-biodegradable. PHAs are one of the most studied bioplastics families and are highly biodegradable linear polyesters produced intracellular by microorganisms as carbon and energy storage (bio-based) with physicochemical characteristics similar to conventional plastics. They are also biodegradable and biocompatible, so they are the next-generation eco-friendly materials [7].The properties of PHAs related to mechanical, thermal, and physical effectiveness rely on

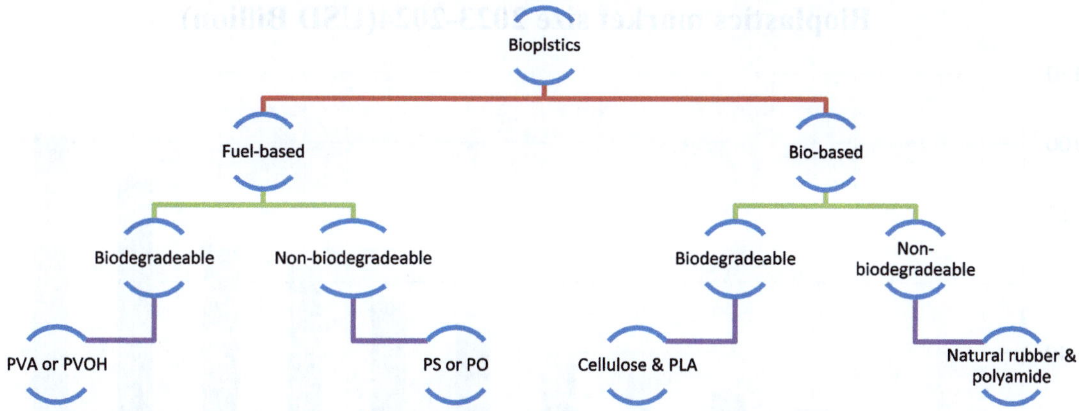

Fig. 2 Tree chart representing a classification of bioplastics

the composition of monomers, microstructure, and molecular weight distribution. The most abundant PHA, polyhydroxybutyrate, is a brittle and stiff polymer, limiting their application The incorporation of other monomers, such as hydroxy valerate (HV), produces a copolymer with improved mechanical properties (more flexible and tougher). So, the change in the PHA microstructure can be seen as a tool to tuning the polymer properties, allowing the application of PHAs in a wide range of industries.

Bioplastics are biodegradable, renewable, or biocompatible plastics which can be categorized based on their fabrication techniques into two main groups, i.e., fuel-based and bio-based as shown in Fig. 2 [8]. Fuel-based bioplastics are a group of bioplastics that require fuel in the form of petroleum or natural gas during their formulation, e.g., PVOH or PVA synthesized by the polymerization of vinyl acetate, polyethene manufactured by additional polymerization of ethylene, and polystyrene from styrene and polyolefin. In contrast, bio-based bioplastics are the bioplastics that are extracted or produced from natural resources like plants, animals, or microbes, e.g., cellulose, PLA, PA, natural rubber.

1.3 Growth of Bioplastics

Nowadays, bioplastics are in high demand because of their higher biodegradability rates and lesser gas emissions. In 2022, 1.8 million tons and in 2023, 2.18 million tons of bioplastics are commercialized, and this valve is supposed to rise to 7.43 million metric tons till 2028 [21]. Studies show that in 2024, bioplastics have 60% production capacity vs a utilizing rate of 58% (1.44 million tons production vs 2.47 million tons production capacity). According to a research report, the global market size of bioplastics was 15.57 billion USD in 2024 and expected at a CAGR of 19.5% from 2025 to 2030 [22]. In the year 2023, the market of bioplastics was 17.13 billion dollars and is expected to approach 104.82 billion USD [23]. Figure 3 depicts the growth and expected rise in bioplastics production from 2024 to 2034 in USD. Despite being

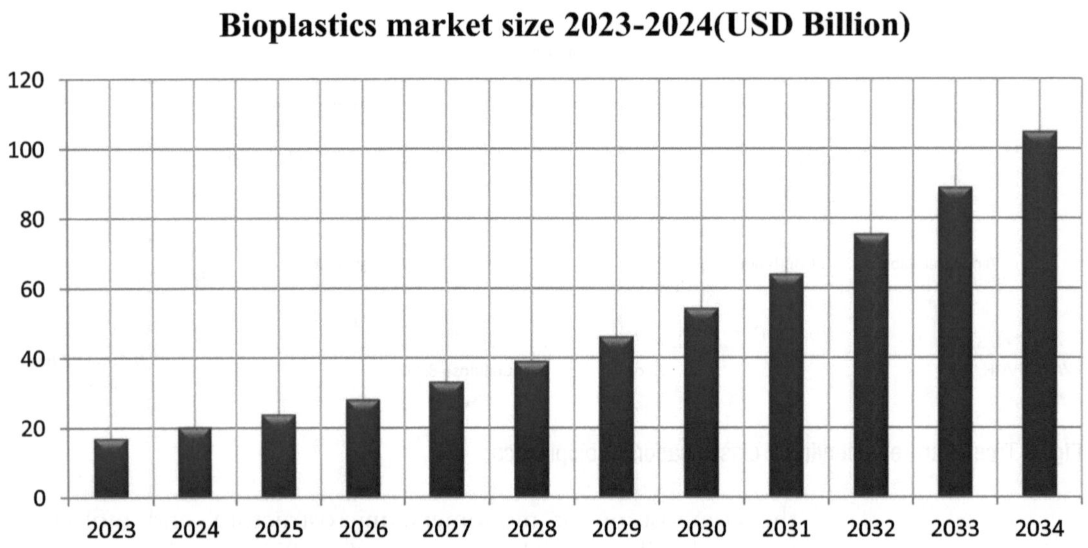

Fig. 3 Market growths of bioplastics

environmentally friendly and highly biodegradable, bioplastics have several different limitations that do not allow us to formulate and commercialize their production easily. We can minimize these effects by proper handling and adopting a preferable methodology, but we cannot wipe them out [24].

2 Materials

Food waste collection may vary depending on the specific situation, waste management policies, community efforts, and infrastructure available in a particular area. Standard methods of collecting food waste include curbside collection. Food waste is collected together with regular municipal waste (trash) and recyclable materials. Residents separate food waste into designated containers and place them on the side of the road for collection by a waste disposal company. In addition, some municipalities have designated disposal sites where individuals can throw away food waste. Certain facilities, such as schools, businesses, and large food operations, may have on-site composting systems installed. They collect food waste generated on their premises and compost it on-site using compost bins or special composting equipment. Additionally, in urban areas with a high concentration of food-related businesses, commercial services exist that specialize in the collection and processing of food waste. These services provide special containers to collect food waste and transport the waste to an appropriate treatment facility for further processing.

The food refuse can be created during the whole process of its production, transportation, processing, preservation, and utilization in multiple stages, i.e., [7]. During the process of harvesting, sowing, and preliminary handling, the food can be affected by multiple factors like improper preservation, environmental conditions, microbial activities, pests, etc., causing the squandering of food. Inappropriate handling and storage: Unprofessional food safety measures like improper storing and unacceptable handling can give rise to notable food loss. These measures can lead to the waste of a variety of consumable food, increase the hunger rate, and disturb the resources. Conditioning and encasing: During the process of preparing and packaging due to its trimming to meet with standards, spoilage, etc. Circulation and marketing: FW, at this stage, is mostly caused by surplus stock, misadministration, and high-quality standards, which lead to the refusal of edible food. At this stage, the food is wasted and directly discarded into landfills, which can negatively affect the environment. Consumer level: At this stage, the food is generally lost by its over-purchasing, inappropriate storage, cooking in excessive amounts, and unnecessary processing by the homes, restaurants, office canteens, etc. The food gathered can be processed and dissociated into groups based on the constitutional composition [8]. The materials that are needed to synthesize the bioplastics are listed below:

- FW constituent for monomer extraction (starch, glucose, acetic acid, etc.).
- Chemicals like HCl or NaOH are used for chemical pretreatment.
- Bacterial cultures and enzymes for digestion and polymerization.
- Autoclave, HAO, or water bath.
- Modeler.

3 Methods

Food waste recycling and valorization have been widely accepted as an emerging sustainable research subject all over the world during recent decades. Several research studies have indicated that some advanced waste food recycling and valorization methods were practiced in some countries only, while very few countries in the world were applying such methods. As a result, food waste which cannot see hidden resources is still untapped, especially at the industrial level [8]. The objective of this method includes [9]. The production of environment-friendly bioplastics, which are relatively easier to degrade in the environment as compared to conventional plastics, and the transformation of food scraps into bioplastics reduce the carbon footprints in the environment and the emission of GHGs. Through food valorization, permanent environmental litter can be eradicated. This method allows us to synthesize high-quality bioplastics of desired characteristics. We can add value to the materials and contribute to upgrading the circular economy. The products obtained via this procedure have a variety of different applications in industries like cosmetics, food, biomedical products, packaging, paints, fertilizers, etc.

3.1 Valorization Technologies

Valorization technologies include the direct extraction of bioplastics from biomass, production via natural or genetically modified organisms, and synthesis from bio-based monomers, petrochemicals, copolymers, and biopolymers as showcased in Fig. 5.

3.1.1 Extraction Directly from Biomass

Bioplastics can be extricated directly from biomass, including corn cop, wheat, oats, rice residues, sugarcane bagasse, palm oil, rapeseed, etc. This process comprises the conversion of naturally occurring polymers like polysaccharides (in potatoes, bread, cereals, rice, grains, pasta, corn, peas, etc.) into bioplastics, e.g., lignocellulose and starch. To improve the properties of bioplastics obtained by this strategy, several types of additives could be added, i.e., plasticizers like glycerol and sorbitol to improve flexibility, natural fiber like flex or hemp for enhancing strength, flame retardants to increase fire resistance, etc. The following are examples for the extraction of bioplastics from biomass:

- Starch is the second most abundant polysaccharide on earth present in wheat, corn, potatoes, rice, rye, etc. It is the most used biopolymer for the generation of bioplastics. Thermoplastic starch is the biodegradable plastic obtained from starch which is used as an alternative for polystyrene. It is composed of a chain of d-glucose units in two forms: amylose and amylopectin. Amylose consists of a straight chain of D-glucose units, while amylopectin comprises a branch chain of units. This branched chain is formed by 1,6-glycosidic linkage, while straight chain linkages are 1,4-glycosidic.
- Cellulose-based bioplastics are also in demand nowadays. To synthesize bioplastics from cellulose, one can consider sources from food waste like wood, seed fibers, grass, algae, fungi, etc. The bioplastics obtained from cellulose can have specific properties like high Young's modulus (approximately 114GPA), high crystallinity, and high polymerization rates. Examples of such polymers are cellophane, cellulose acetates, and cellulose ethers.

3.1.2 Bioplastics from GMOs

Microorganisms like bacteria, which lack prominent nuclei, have the potential to generate polyhydroxyalkanoates (PHAs) as carbon storage. The PHA-producing bacteria can be classified into two classes based on their nutrient requirement, i.e., nutrient dependent and nutrient independent. Nutrient-dependent PHA-producing bacteria are those which require specific types of nutrients like O_2, N_2, PO_4^{3-}.while nutrient bacteria are those microbes which work without the limitation of nutrient requirements. Overall, 250 types of PHA-producing bacteria are recognized in the market, but due to specifications, only a limited number of these can be used industrially, i.e., *Cupriavidus necator, Pseudomonas oleovorans, Bacillus megaterium,* etc. For instance, recombinant *E. coli* is being used for the synthesis of 3HB from soybean oil.

PHAs like P(3HB-Co-3 HV) and P(3HB) can be prepared by polymerization of fermented VFAs with the aid of bacterial cultures. These conversions are carried out in special type of reactors, batch-fed reactors which are common for industrial fermentation. The steps involved in this process are: firstly, the required culture of microbes is developed and grown until a pre-decided mass is reached. This bacterial cell mass concentration is to be utilized for the valorization of VFAs from food waste to the PHAs. Secondly, culture is treated with nutrient-restricted growth in the presence of a carbon source. During this step, the microbial cells will not grow or reproduce but will accumulate PHAs, and PHA produced can be isolated and purified.

3.1.3 Bioplastics Synthesized from Bio-Based Monomers

Bioplastics can be manufactured from bio-based monomers like glucose, amino acids, fructose, and lactic acid. These bioplastics have specific characteristics and are cost effective as the initial raw material is cheap and easy to extract.

PGAs and PLAs are the examples of such bioplastics. Polylactic acids are biodegradable aliphatic polyesters formulated by polycondensation of LA on an industrial scale. It can also be created by the ring-opening polymerization of lactide. The conventional method for the synthesis of PLAs involves the extraction of carbohydrates from sources like rice, wheat, and cereals as elaborated in Fig. 4. Then, after these extracted carbs are converted to dextrose units, followed by the fermentation of these units to LA and finally LA is polycondensated to PLA. These PLAs can be processed in textile, medical, plastic, and packaging industries by melt processing technique. Polyglycolic acid is also biodegradable aliphatic polyester formulated by the ring-opening polymerization of glycolic acid monomers. This polymerization is carried out in the presence of a metal salt catalyst at low concentration. PGAs are mainly being used as medical applications in sutures like polypropylene, catgut, and vicryl and in food packaging.

3.1.4 From Biopolymers and Copolymers

Bioplastics can also be manufactured from biopolymers and used as biocomposites or copolymers to improve the material properties. For example, the use of food gelatin and starch from potatoes with suitable plasticizers, i.e., sorbitol and glycerol (Fig. 5).

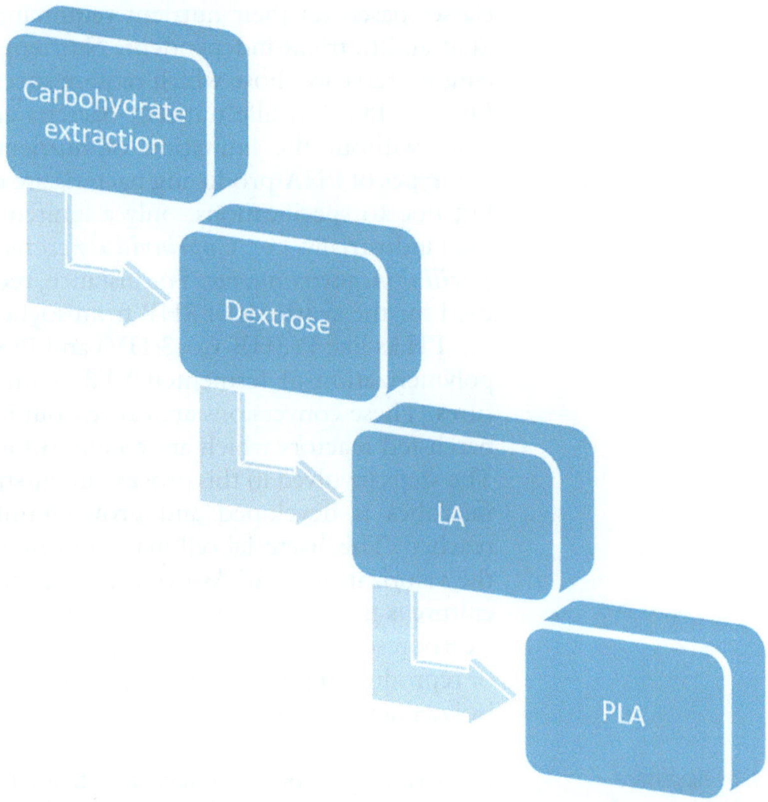

Fig. 4 Synthesis of PLAs from monomeric carbohydrates

3.2 Route for Food Waste Valorization

Food waste valorization requires a successive series of methods and processes to be applied for the conversion of the FW into biocompatible bioplastics of desired characteristics, as illustrated in Fig. 7. These procedures can involve various types of physical, chemical, biochemical, and biological activities involving disintegration, extraction, and catabolism. The occurrence of these events in a specific sequence illustrates a methodology for food waste valorization. Mainly, these processes can be categorized as collection and separation of FW, pretreatment methods, fermentation, polymerization, characterization, blending, and processing as described in Fig. 7.

3.2.1 Pretreatment of FW

The pretreatment of the collected food is then carried out to liberate monomeric units like VFAs, butyric acid, acetic acid, ethanol, and lactic acid. The pretreatment of FW can be done in multiple ways either by treating physically, chemically, or biologically. Explanation for these treatments is given below:

For the physical pretreatment: the FW is prepared to apply further processing physically via enhancing the surface area, reducing the particle size, adjusting the structure, or altering internal

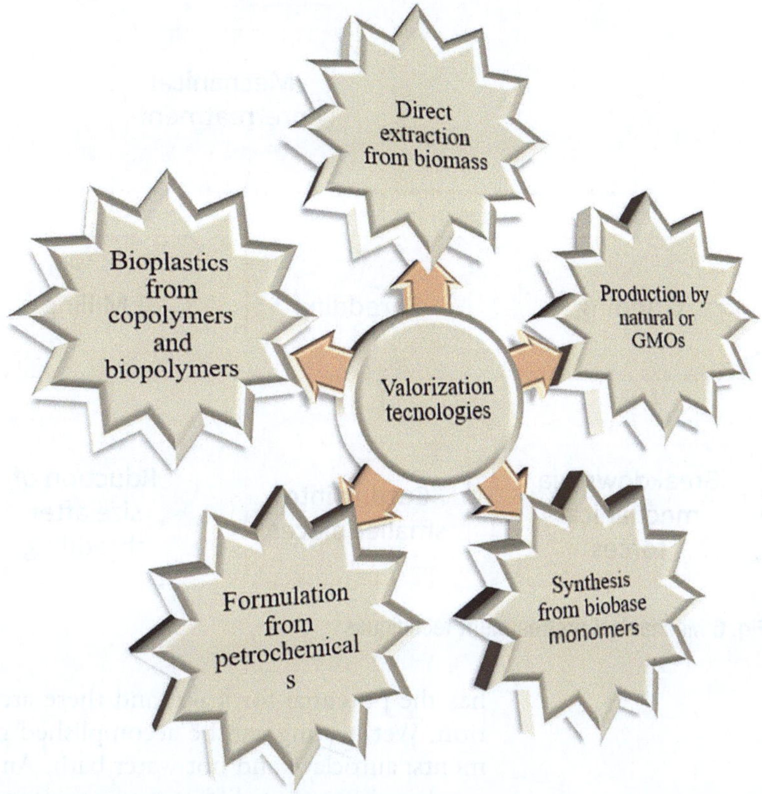

Fig. 5 Technologies for the valorization of food waste

constituents without the involvement of external chemical or biological agents. Physical pretreatment methods can be classified further into mechanical, thermal, and ultrasonic techniques. Figure 6 gives an overview of mechanical practices constituting the reduction of the size of the FW particles via grinding, shredding, milling, and screening. This technique has several different advantages for being simple, cost effective, and easily applicable. While ultrasonic pretreatment is a laboratorial scale procedure in which high frequency sound waves (20–100 KHz) are used. These sound waves are responsible for the breakdown of food waste into their monomers.

Thermal pretreatment comprises the heating of the partially processed FW to remove moisture and for sterilization. Via thermal pretreatment, the softening, solubilization, extraction, and hydrothermal breakdown can be executed. This temperature-aided pretreatment encompasses mainly two kinds of approaches, i.e., wet heating and dry heating [10].

Wet heating, as the name suggests, is a way to treat thermally in which steam works for the breakdown of FW. It is a technique which is not easy to operate demands high installation costs and

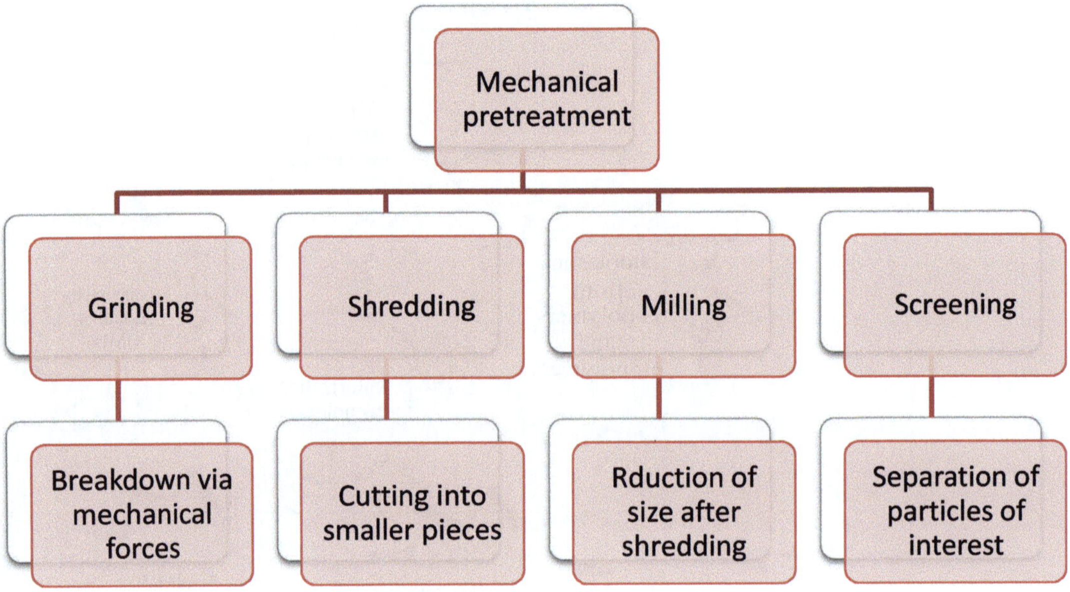

Fig. 6 Mechanical pretreatment techniques

has the potential for leaks, and there are chances for scale formation. Wet heating can be accomplished using two types of instruments: autoclave and hot water bath. An autoclave is a device that can be operated at different temperatures and uses steam for the hydrothermal breakdown of the FW. By altering the pressure and time for the material to stay in the autoclave, a variety of FW can be processed. The hot water bath method does not require very complicated instruments; it only involves the immersion of FW in heated water for softening and solubilization. From this solubilized FW, the units for the fermentation are extracted.

In contrast, the dry method for physical pretreatment calls for the extrication of monomer units at low humidity levels by heated air or a special type of waves. The heating systems for HWB can refer to radiators or under-floor heating systems. Dry heating can be provided by two instruments, i.e., hot air oven and microwave. Hot air ovens as the name indicates use hot air for the extraction procedure. It is based on the circulation of heated air through the FW. HAO can be operated at different temperatures like 80 °C, 90 °C, 100 °C, and 120 °C. The process involves the preheating of the oven, followed by the sample loading and then setting the temperature. The FW can also be treated by using electromagnetic microwaves to lower the moisture content and enhance the structural breakdown. It can also be operated at multiple temperatures, we need to optimize the best temperature based on the sample to be loaded.

Chemical pretreatment is based on the addition of alkali, acid, or oxidizing agents (e.g., hydrogen peroxide) before further processing. It is typically performed by the breakdown of complex

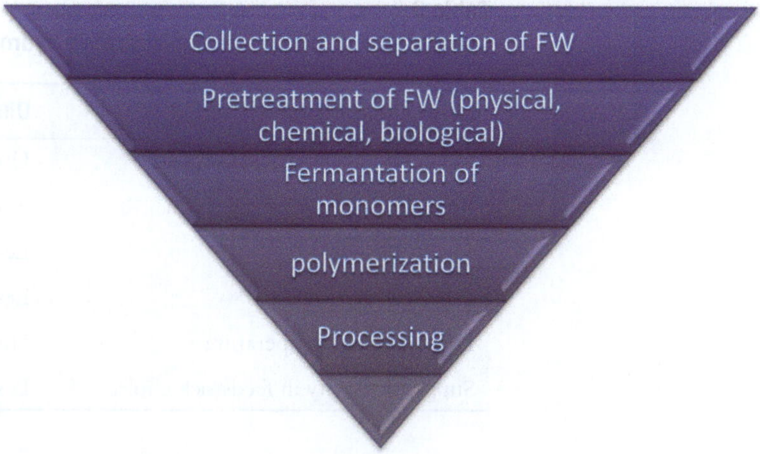

Fig. 7 Pyramid for the route of valorization

molecules to their simpler units, separating the contamination, or by making the surface more active toward the following processes. The chemical pretreatment can be performed by three hydrolysis methods as suggested earlier, i.e., acid hydrolysis, basic hydrolysis, and the addition of oxidizing agents (Fig. 7).

Acid hydrolysis is a technique for chemical pretreatment in which the FW is treated with an acidic solution for its catabolism. Acid hydrolysis can be accomplished in two courses of action, either via the operation of dilute acid or by concentrated/strong acid. First, after spraying the diluted acid on the raw FW, the mixture is heated to 160–220 °C for a few minutes. This process of hydrolysis releases the monomers from the complex molecules of FW. For instance, the hydrolysis of hemicellulose yields the simpler units (monomeric sugars and soluble oligomers).

The acid hydrolysis of this type can be carried out in two different conditions on the basis of temperature. At high temperature (above 160 °C) and low sample loading (5–10 wt %), the process is carried out in a continuous flow manner. At low temperature (equal to 160 °C or less) and high sample loading (10–40 wt %), the process is carried out in a batch manner.

The dilute acid hydrolysis is a method for good recovery performance but also retains quite a few drawbacks for only being suitable for FW consisting of low lignin contents, the acid can cause corrosiveness if not neutralized and its neutralization can make solid contaminants. On the other hand, strong acid hydrolysis uses concentrated acid despite diluted one for the breakdown of complex substances. The acid sprayed on the FW can be HCl H_2SO_4, or any other concentrated acid. Table 3 provides a concise comparison of these hydrolysis techniques.

Basic hydrolysis is a technique for the hydrolysis of FW that utilizes a highly basic solution which is subjected to the raw waste at high pressure and temperature. At this raised temperature and

Table 3
Comparison of dilute and concentrated hydrolysis of food waste

Strong acid hydrolysis of FW	Dilute acid hydrolysis of FW
Uses concentrated acid	Uses diluted enzyme
Require enzymes for digestion	No enzymes are needed
More efficient	Less efficient comparatively
Produce high yield	Less yield
Demands low temperature	High temperature requirement
Supply flexibility in feedstock choice	Less flexibility

pressure, the peptide bonds cleave causing the proteins to denature, fats are hydrolyzed, and carbohydrates are solubilized. To supply high pressure and temperature (180 pa and 150 °C, respectively) SRM (single residue method) is used. Multiple alkalis can be used for basic hydrolysis, but usually $Ca(OH)_2$ and NaOH are used [11]. This strategy for chemical pretreatment is significant for being eco-friendly as it emits fewer gases, nutrients, or pathogens in the environment and its cost of operation is also low. Despite being environment friendly, this technique acquires noteworthy disadvantages, i.e., production of caustic hydrolysate containing high COD and BOD, formation of salts which remain incorporated in products, and slow reaction process. The addition of air or O_2 can improve the process.

Biological pretreatment deploys certain types of microbes for the processing and simplification or catabolism of FW. The microbes like bacteria, fungi, yeast, or mixture of these can be employed to the FW, e.g., the disintegration of cellulose or lignin by the microbial enzymes like LME (lignin modifying enzymes). This pretreatment method consumes less energy, requires low chemicals, and isolates the culture of microbes after the decomposition, while it also requires a long reaction time and provides limited control over the process [12]. We mostly use fungal treatments in this pretreatment activity, i.e., *Pleurotus ostreatus, Ceriporiopsis subvermispora* (white rot fungi), and posing enzymes for the breakdown of lignin. Similarly, *Lactobacillus buchneri* increases the rate for the bioconversion and larval mass of soybean curd residue by 38% and 39%, respectively.

3.2.2 Fermentation of Hydrolyzed OC

Fermentation is a process of synthesizing alcohols, acids, CO_2, and H_2 from the hydrolyzed OC. This procedure involves the breakdown of OC in the presence of microbes like bacteria, yeast, or fungi in anaerobic conditions. Humans have been using fermentation since the Neolithic period, which was around 10,000 years BC, and in 1857, Louis Pasteur introduced fermentation and proposed that microbes are responsible for lactic acid corrosion. In the fermentation of OC extracted from the FW, the microscopic organisms are used to generate acetates, propionates, butyrates, lactates, alcohols, H_2, CO_2, and other solvents [13]. Various types of fermentation are being used for this process based on the starting material.

Mixed VFAs Fermentation

Using mixed inoculation during acidogenesis, VFAs from C_2–C_6 can be generated which can be used to produce PHAs. During the fermentation, different kinds of acidogenic bacteria are active and generate VFAs, lactic acid, and alcohol through various metabolic pathways. In this fermentation technique, three or more relatively high-yield products are obtained, especially butyric acid and acetic acid [14].

$$C_6H_{15}O_6 + H_2O \rightarrow 2CH_3COOH + CO_2 + 4H_2 \quad (1)$$

$$1/2C_6H_{12}O \rightarrow 2CH_3CH_2COOH + CH_3COOH + CO_2 + H_2O \quad (2)$$

$$C_6H_{15}O_6 \rightarrow CH_3CH_2CH_2COOH + 2CO_2 + 2H_2 \quad (3)$$

Butyric Acid Fermentation

In the fermentation, when the main products are butyric acid and acetic acid, it can be termed as butyric acid fermentation. During this process, the glucose is metabolized to pyruvate through the EMP pathway, and 2 moles of ATP and NADPH are released. Then the generated pyruvate molecules are converted to butyric acid via butyrate-producing *clostridia*. The pH required for fermentation ranges 4.7–5 and generally occurs at 5.2 and a potential of −300 to −250 mv.

$$\text{Glucose} \rightarrow \text{pyruvate} + \text{ATP} + \text{NADPH}$$

$$\text{Pyruvate} \rightarrow \text{butyric acid}$$

$$C_6H_{15}O_6 \rightarrow CH_3CH_2CH_2COOH + 2CO_2 + 2H_2 \quad (4)$$

$$4C_6H_{15}O_6 \rightarrow 3CH_3CH_2CH_2COOH + 2CH_3COOH + 8CO_2 + 8H_2 \quad (5)$$

Acetic Acid Fermentation

Globally, 14 million tons of acetic acid per annum are being produced and can only be manufactured from pyruvate through the acetyl-CoA pathway. In the above Eqs. (1), (2), and (5), acetic acid can be produced as a major product from the syntrophic oxidation of ethanol or LCFAs (longer chain fatty acids) like propionate and butyrate. Actinogenic bacteria of the family *Acetobacter aceti* and class *alphaproteobacteria* are used for this fermentation. For the complex FW raw material, the purification of the product is needed.

$$CH_3CH_2CH_2COOH + 2H_2O \rightarrow CH_3COOH + 2H_2 \quad (6)$$

$$CH_3CH_2COOH + 2H_2O \rightarrow CH_3COOH + CO_2 + 3H_2 \quad (7)$$

Ethanol Fermentation

Ethanol fermentation is a form of fermentation in which the hydrolyzed glucose is transformed to ethanol with the aid of yeast or *Zymomonas mobilis* bacteria. This process is carried out in different steps;

- **Conversion of glucose to pyruvate/glycolysis:** In this step, the glucose is broken down to the molecules of pyruvate by the help of different enzymes.

$$\mathbf{C_6H_{12}0_6 \rightarrow pyruvate}$$

- **Reduction of pyruvate to acetyl-CoA:** In this step, the generated pyruvate molecules are reduced to acetyl-CoA by the elimination of hydrogen and carbon dioxide gas. This step is carried out by the enzyme dehydrogenase from the electron carrier protein ferredoxin.

$$\mathbf{Pyruvate \rightarrow acetyl - CoA + H_2 + CO_2}$$

- **Transformation of acetyl-CoA to acetaldehyde:** During this step, the acetyl-CoA is converted to acetaldehyde by the same enzyme.
- **Formulation of alcohol:** In this final step, the acetaldehyde is converted to ethanol by an enzyme dehydrogenase.

To pursue this fermentation, 4 NADHs are utilized and 2 NADHs are regenerated at the end. This reaction, when it occurs in the presence of water, can be coupled with acetic acid. The overall reaction for this procedure can be given as:

$$C_6H_{12}O_6 \rightarrow 2CH_3CH_2OH + 2CO_2$$

$$C_6H_{12}O_6 + H_2O \rightarrow CH_3CH_2OH + CH_3COOH + 2H_2 + 2CO_2$$

Lactic Acid Fermentation

Lactic acid fermentation is preferred as it requires low cast, low temperature, low energy utilization, and low pH. Due to the better efficiency and enhanced significance, the technique is globally being used for the fermentation of lactic acid from glucose. This fermentation involves the first step similar to ethanol fermentation, i.e., the glycolysis to pyruvate, the next step involves the dehydrogenation of pyruvate to lactate via an enzyme lactate dehydrogenase using a molecule of NADH and NAD^+ is released. Lactic acid fermentation is executed at the temperature range of 20–45 °C, at pH 5–7 and in the presence of bacteria of phylum *firmicutes* and *lactobacillus* [15].

$$C_6H_{12}O_6 \rightarrow 2CH_3CHOHCOOH$$

$$C_6H_{12}O_6 \rightarrow CH_3CHOHCOOH + CH_3CH_2OH + CO_2$$

$$C_6H_{12}O_6 \rightarrow CH_3CHOHCOOH + \S\ CH_3CH_2OH$$

The scheme of conversion can be represented by the below equation:

$$C_6H_{12}O_6 \rightarrow C_3H_4O_3 \rightarrow C_3H_6O_3$$

3.2.3 Polymerization

Polymerization is the phenomenon by which smaller molecules called monomers attach each other via some sort of chemical bonding, and as a result, molecules of high molecular weight named polymers are formed. After the fermentation of the glucose or the hydrolyzed products obtained by the pretreatment of FW, they can be polymerized using different techniques to their biopolymers. These biopolymers can then be isolated and treated for multiple properties according to the requirements and can be termed as bioplastics. Based on obtained fermented products from food waste, a variety of polymerization methods can be opted, here we will only discuss two of them, i.e., polymerization to PHAs and polymerization to PLAs.

Polymerization of VFAs to PHAs

The polymerization of VFAs (like AA, BA, PA, and AA) can be done using pure bacterial cultures like *Ralstonia eutropha, Pseudomonas sp. ST2.* To understand the mechanism of polymerization, let's consider the two polymerization procedures for acetic acid and butyric acid [14].

1. Acetic acid can be polymerized to two different biopolymers, P(3HB-Co-3 HV) and P(3HB), through a series of steps by the various types of bacterial cultures. In the first step, the acetic acid is converted to acetyl-CoA via any of the bacterial cultures like *Cupriavidus necator (Ralstonia eutropha), Bacillus megaterium, Methylocystis hirsuta, and Alcaligenes eutrophus.* In the next step, acetyl-CoA is transformed to acetoacetyl-CoA by the *Beta-Ketothiolase* (PhaA). In the third step, acetoacetyl-CoA is changed into (R)-3-Hydroxybutyryl-CoA using cultures of *Escherichia coli, C. necator* or *Bacillus cereus.* The last step is dependent on the specific type of bacteria responsible for the type of product to be synthesized. There are two products that can be formed, i.e., P(3HB-Co-3 HV) and P(3HB) as illustrated in Fig. 8. *Alcaligenes eutrophus* NCIMB 11599, *Ralstonia eutropha* ATCC 17699, or *Methylocystis hirsuta* can be used for the first product and *C. necator, M. hirsuta, Bacillus megaterium, Bacillus sp.* INT 005, or *Pseudomonas putida* for P(3HB).
2. Polymerization of butyric acid yields the same products as for the acetic acid as shown in Fig. 9, the only difference is in the first two steps, where butyric acid is converted to butyryl-CoA and then to the acetoacetyl-CoA by the action of cultures, *C. necator* (ATCC 17699, KCTC 2658), *M. hirsuta, A. eutrophus* NCIMB 111599, *Bacillus thuringiensis, Haloferax mediterranei,* and *E. coli.*

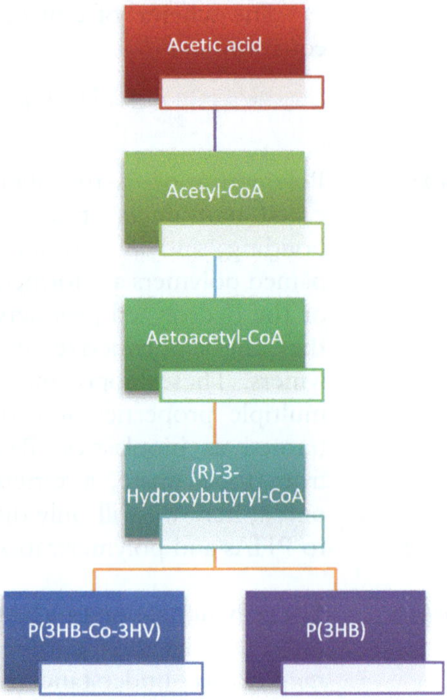

Fig. 8 Flow chart for polymerization of acetic acid

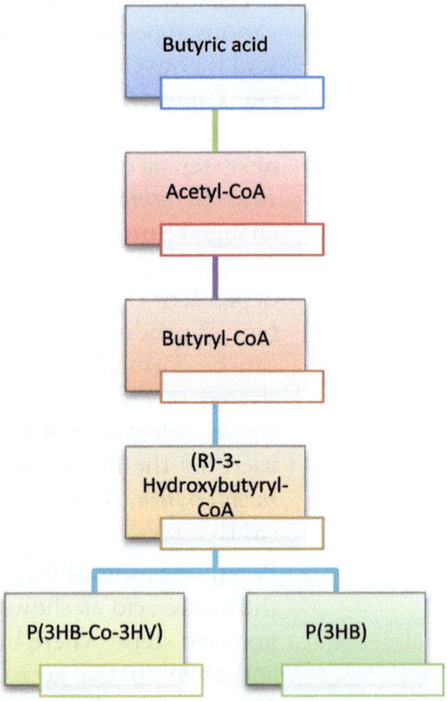

Fig. 9 Flow chart for polymerization of butyric acid

Polymerization of Lactic Acid to PLAs

The polymerization of lactic acid to PLA involves three different types, i.e., direct polymerization, enzymatic polymerization, and ring-opening polymerization [16]. The ring-opening polymerization is the most used method for the polymerization of lactic acid as it is more efficient.

Direct polymerization is a method for the polymerization of lactic acid, which is mainly based on melt polymerization. Melt polymerization is a type of polymerization which is carried out in a melt or agitator which helps in heating and providing the pressure to the lactic acid. The process starts at room temperature and atmospheric pressure, then gradually the pressure starts to decrease and the temperature increases until a pressure of 30 torr and temperature of 180 °C is obtained. These conditions are maintained for 4 h in the direct polymerization or melt polymerization of D and L-lactic acid into PLAs. The following schematic equation elaborates the conversion of lactic acid into PLAs.

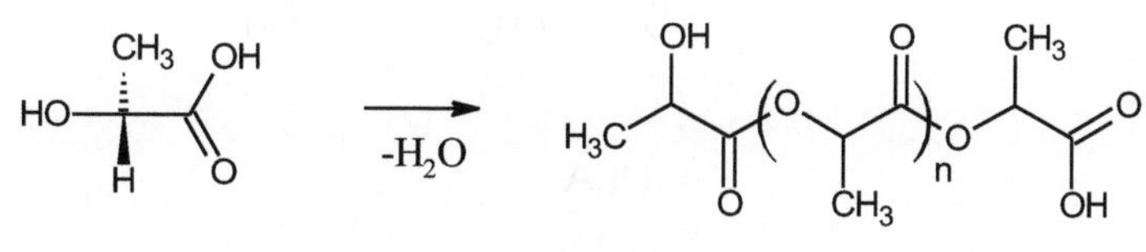

Enzymatic polymerization is a way of eco-friendly polymerization in which we can get high-property PLAs by using enzymes. Mostly the lipase enzyme is used for the polymerization of lactic acid to PLAs. The main function of the lipase is to join the molecules of lactic acid together to form a long chain of lactic acid monomers called polylactic acid. The process for the enzymatic polymerization of lactic acid into PLAs can be explained by the following method [17].

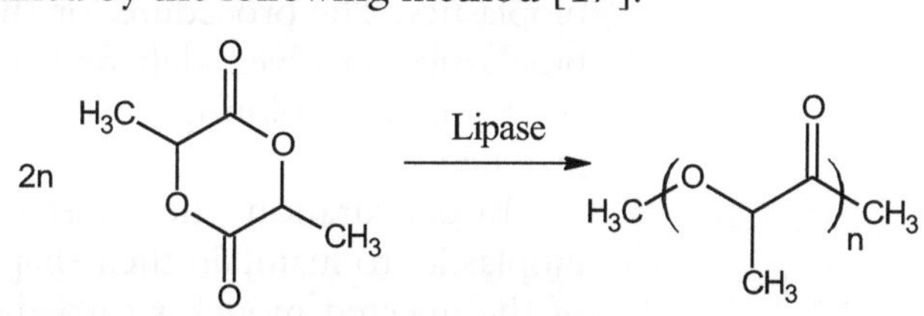

Ring-opening polymerization is a method for the industrial synthesis of PLAs from lactic acid as it is a more efficient technique to formulate bioplastics. This procedure involves several different steps comprising dehydration, lactide formation or thermal cracking and ring opening. In the first step, the lactic acid molecules undergo dehydration or condensation, forming the long-chain polymers named oligomers. During this step, the two molecules of lactic acid combine together, releasing a water molecule. This step can also be called the condensation step [18].

In the next step, the oligomers are formed. Oligomers are converted to lactide rings. Due to the breakdown of long-chain oligomers to low molecular weight lactides, this step is also called thermal cracking. The last step of this polymerization is the ring opening. In this step, the lactides undergo ring opening and form the long-chain PLAs.

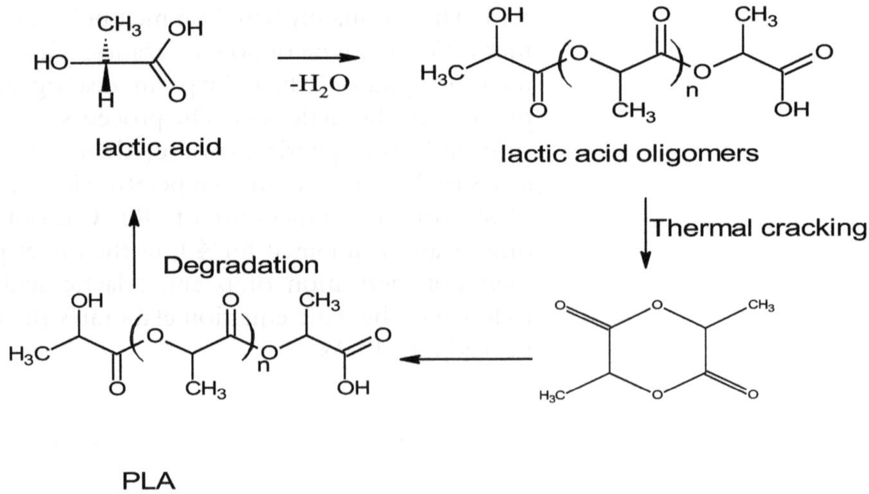

4 Notes

The bioplastics synthesized by the previous processes mostly have different types of contamination and different moisture levels, so to maintain the moisture level and remove contamination, there is a need for the processing and blending of the manufactured bioplastics. The procedure for the processing and blending of the bioplastics involves dehydration, injection molding, foaming, cross-linking, coloring, extrusion blow molding, or flat film production.

In dehydration, the unusual moisture is eliminated from the bioplastics to maintain their shape and rigidity. Then, the foaming of the injected model is carried out to reduce the weight of the bioplastics. After forming, cross-linking is inserted in the bioplastics that help to get specific properties in the bioplastics like thermosetting, elasticity, brittleness, etc. After cross-linking, bioplastics are colored to increase their appeal and to increase sales. In the next step, injection molding of the bioplastics made is carried out into some spaces to mold them and to obtain specific shapes. We can shape the bioplastics in different frameworks based on product feasibility and product requirements.

Bioplastics are synthetic, semisynthetic, or natural polymers which have been formulated extensively over the last 20 years due to their biocompatibility, higher biodegradability, and composite characteristics. In recent times, the valorization of FW to bioplastics has become one of the most active research fields. Bioplastics have applications in every field includes medical, packaging, appliance materials, spray materials, spray products, agriculture products, automation products, chemical media and solvents, etc. In the production of bioplastics, the interlinkage of biotechnology processes is a key strategy aimed at maximizing the use of food waste and increasing the potential revenue of the entire bioprocessing chain. Given that the mass generation of FW is inevitable, the environmental burdens arising from waste disposal (e.g., water contamination and GHG emissions) should be mitigated. Therefore, this review demonstrated the potential of FW as a raw material for bioplastics production to address important environmental problems. Hence, physical, thermo-chemical, and biological methodologies required for the preparation of bioplastics raw materials from FW are reported. Moreover, the production of PHA based on unadulterated/mixed culture and fermentation technologies was emphasized. Modifications of different FW compositions (e.g., cellulose, starch, chitin, and caprolactone) for PHA-derived products were also considered. Most importantly, governments, regulations, companies, public opinions, and consumers should work together to mitigate current environmental burdens arising from FW disposal.

Biorefineries use microbial consortia in bioconversions which may have several multiple traits and lead to an efficient industrial operation. Mainly, the objective for the genetically engineered microbes is to increase the production yield but it could also aim to improve substrate influx or to integrate biochemical pathways of choice. Recently, bioaugmentation is also being used for environmental bioremediation and FW valorization. Bioaugmentation is the introduction of autochthonous, allochthonous or GMOs to the FW to increase biotransformation quantity, quality, and rate. It is significant as it promotes digestibility and overall efficiency. Nanotechnology may also be used to motivate bioconversion in refineries. Cu nanoparticles and enzymes have been used for the conversion of fructose to levulinic acid in aqueous media and circumventing the effects of immobilization of enzymes in fructose [19].

Artificial intelligence is also being integrated into the valorization process for the utilization of multitask neural networks to formulate sustainable bioplastics. Researchers have organized various AI models to foresee the characteristics of bioplastics before their manufacturing, making it easy to obtain bioplastics of the desired choice. AI helps in optimizing the production rate and environment impact assessment of bioplastics. This integration of AI with FW valorization ensures that this technique uses FW from the resources to generate bioplastics and aligns with environmental safety goals [20]. Valorizing FW into bioplastics is an emerging technology and is playing a crucial role in reducing the pollution content of ecosystems and increasing environmental sustainability. Several different companies and organizations are valorizing the OFW for the generation of biocompatible bioplastics.

References

1. Fao F (2018) Food and agriculture organization of the United Nations. Rome, p 403. http://faostat.fao.org
2. Canali M et al (2016) Food waste drivers in Europe, from identification to possible interventions. Sustain For 9(1):37
3. Li X et al (2024) Enhanced dewaterability of food waste digestate by biochar/potassium ferrate treatments: performance and mechanisms. J Environ Manag 354:120268
4. Socas-Rodríguez B et al (2021) Food by-products and food wastes: are they safe enough for their valorization? Trends Food Sci Technol 114:133–147
5. Domínguez-Solera E et al (2025) The biodegradability of plastic products for agricultural application: European regulations and technical criteria. Clean Technol 7(1):11
6. Silva SB et al (2025) A comprehensive review on valorization of chestnut processing wastes into bio-based composites and bioplastics. Polym Compos
7. Wang L, Cui Y-W (2024) Simultaneous treatment of epichlorohydrin wastewater and polyhydroxyalkanoate recovery by halophilic aerobic granular sludge highly enriched by Halomonas sp. Bioresour Technol 391: 129951
8. Tsang YF et al (2019) Production of bioplastic through food waste valorization. Environ Int 127:625–644
9. Bruno D et al (2025) Valorization of organic waste through black soldier fly: on the way of a real circular bioeconomy process. Waste Manag 191:123–134
10. Saha B, Khwairakpam M, Kalamdhad AS (2021) Thermal pre-treatment–a prerequisite for the reduction of hydrolysis stage during anaerobic digestion of Ageratum conyzoides. Mater Sci Energy Technol 4:34–45
11. Arias JZ et al (2018) Ambient alkaline hydrolysis and anaerobic digestion as a mortality management strategy for whole poultry carcasses. Waste Manag 81:71–77
12. Peguero DA et al (2022) A review of pretreatment methods to improve agri-food waste bioconversion by black soldier fly larvae. Front Sustain Food Syst 5:745894
13. Wang Q et al (2020) Oriented fermentation of food waste towards high-value products: a review. Energies 13(21):5638
14. Szacherska K et al (2021) Volatile fatty acids as carbon sources for polyhydroxyalkanoates production. Polymers 13(3):321
15. Pivsa-Art S et al (2013) Synthesis of poly (D-lactic acid) using a 2-steps direct polycondensation process. Energy Procedia 34:604–609
16. Montané X et al (2020) Synthesis and synthetic mechanism of Polylactic acid. Phys Sci Rev 5(12):20190102
17. Chuensangjun C et al (2012) Lipase-catalysed polymerization of lactic acid and the properties of the polymer. Adv Mater Res 506:154–157
18. Horváth T, Marossy K, Szabó TJ (2022) Ring-opening polymerization and plasticization of poly (L-lactic) acid by adding of glycerol-dioleate. J Therm Anal Calorim 147(3): 2221–2227
19. Giwa AS et al (2024) Recent advances in food waste transformations into essential bioplastic materials. Molecules 29(16):3838
20. Golden, J.S., et al. (2024) An economic impact analysis of the US biobased products industry: 2023 update

Chapter 10

Green Valorization of Dairy Waste: Protocols for Sustainable Protein Recovery

Sammra Maqsood, Waseem Khalid, Muhammad Tayyab Arshad, Douglas S. Kalman, Mondher Mzoughi, and Kübra Sağlam

Abstract

The organic load of dairy waste, mainly by-products like whey, buttermilk, and casein sludge, is a major environmental matter. The present investigation into sustainable waste management and resource reclamation options becomes pertinent in light of rising concern over untreated dairy effluents. Green valorization of dairy leftover and protein recovery through environmentally friendly approaches is the subject matter of this chapter. It also covers the necessary steps in the protein extraction process, including advanced methods such as enzymatic hydrolysis and ultrafiltration, as well as pretreatment, which comprises filtration and pH adjustment, among others. Other green valorization strategies are explored with an exceptional focus on its compensations over more conventional methods such as efficiency and sustainability, among others. This section further examines recovered proteins' process of purification methods, such as chromatography and dialysis. In addition to that, nutrient and functional assessment is made as well as reviewing environmental benefits coupled with the sound economic feasibility involved in scaling such procedures for application at industrial sites. It reports challenges and hurdles in presenting cases of valorization of dairy industry effluent waste to date. Last but not least, the potential for future development and the prospect of emerging technologies that can enhance the sustainability and efficiency of dairy waste valorization are discussed. This chapter then delves into innovative and maintainable approaches to protein recovery in the food and nutraceutical industries, thus providing a comprehensive overview of the topic of green valorization of dairy waste.

Key words Dairy, Waste, Green valorization, Chromatography, Protein extraction

1 Introduction

Whey, buttermilk, and curd water are part of the wastes generated by dairy industries, although they significantly add to the overall food produced on earth. Out of every hundred volumes of milk, 85–90% comprises whey itself during cheese manufacture, while another 50% of milk nutrients, such as lactose, proteins, and minerals, occur in the by-product [1]. The United States and the European Union are the main producers of the millions of tons of

Tanmay Sarkar (ed.), *Food Waste Valorization: Green Techniques in Sustainable Management*, Methods and Protocols in Food Science, https://doi.org/10.1007/978-1-0716-4646-5_10,
© The Author(s), under exclusive license to Springer Science+Business Media, LLC, part of Springer Nature 2025

whey annually produced globally [2]. Correspondingly, large quantities of curd water and buttermilk are produced during the production of paneer and butter, respectively [3]. Though such by-products rich in nutrients are detrimental and become waste if not properly disposed of, thereby affecting environmental disputes [4]. This cumulative ultimatum for dairy products, due to expansion of population and per capita consumption, is exacerbating the growing problem of dairy waste. India, the largest milk producer in the world, is confronted with the difficulty of handling the by-products of its various dairy production activities [5]. The EU, whose focus is on the manufacture of cheese and butter, as along with Americas and Oceania, constitutes significant percentage [7]. This large dairy waste production thus requires efficient systems to mitigate adverse effects while unlocking its prospective in becoming a valued reserve. Dairy waste is created from several steps of processing such as cheese, yogurt, and butter production [7]. Although buttermilk is a by-product of butter production, whey is the liquid portion unglued from curd in cheese production [6]. Other sources comprise milk standardization and cleaning waste streams in dairy factories and curd water from paneer and chhana production [8]. The conformation of each by-product varies; buttermilk is rich in proteins and phospholipids, while whey is rich in lactose and protein [9]. At a distance from these primary sources, the waste stream is further affected by effluents from spills, equipment cleaning, and discards [10]. Optimization of resource recovery and valorization approaches requires an understanding of the makeup and volume of these trashes. Untreated dairy effluents are thus responsible for serious environmental problems due to their high chemical oxygen demand (COD) and biochemical oxygen demand (BOD). These measures indicate the biodegradable content, which harms aquatic environments upon discharge into water bodies because it leads to the depletion of oxygen in such bodies [11, 15]. Trainings reveal that untreated dairy wastewater often surpasses the permissible rate of flow that results in eutrophication and water quality deterioration [12]. Furthermore, upon decomposition of biodegradable matters in effluent, methane, a potent greenhouse gas is radiated, thus exacerbating climatic alteration [13]. Since the high salinity of dairy waste leads to salinization and reduced agricultural productivity, inappropriate disposal of the material also harms the soil [14]. Secondly, untreated wastes pose health dangers to the people through microbiological contamination and smell-causing components, which require immediate application of sustainable waste management methods [16]. Many small- and medium-scale dairy enterprises continue to face the challenge of complying with legislation intended to prevent industrial effluent from being discharged. The two major hindrances are a privation of substructure and the exclusive cost of sophisticated treatment approaches

[17]. Logistical experiments such as trash assortment and transportation make administration determinations much more problematic in areas with dispersed dairy farms [18]. Waste treatment and valuation adopting green technologies can be considered as a major part of the solutions to these problems. Regulation structures must, therefore, be improved for both encouragements of maintainable practices in the dairy business and compliance [19]. Cooperation among the researchers, the policymakers, and the industry players must strive for long-term environmental sustainability [20]. Green valorization is a means of converting milk waste into merchandise that have high value and serve to reduce some adverse environmental impact in the income generation process. Unlike traditional wastewater management techniques focusing on disposal and mere treatment [21], green valorization focuses recovery and reuse, mainly of nutrients, including proteins as well as other bioactive materials. This is rendering to the notions of the circular economy, where trash is distorted into resources in a manner that decreases environmental impacts [22]. High-value chemicals may be improved from dairy by-products using green methods like membrane separation and enzymatic hydrolysis [23]. These approaches consume less energy and resources and contribute to solving difficulties of trash disposal. The solution must be accomplished for the dairy industry to developed sustainably [25]. The funding for investigation and technology development is essential to assist in a transition to green valorization [24]. Rendering to advanced techniques such as biorefinery systems [26], it has been established that comprehensive recovery of proteins, lactose, and minerals is possible in dairy wastes. These recovered resources are additional applied in the food, feed, and pharmaceutical industries, creating new revenue sources [27]. Teamwork between the public and private segments will be key in overcoming some of the obstacles like extraordinary startup costs and scalability glitches. The transformation to a sustainable dairy business will be accelerated by policies inspiring resource recovery and creativity in valorizing waste [28]. Future advances may revolutionize completely the processes used in waste administration, ornamental the environment and bottom line. Dairy waste proteins are of important nutritional value that is corresponding to their native sources. For instance, whey proteins are perfect for functional foods and dietary supplements because they are rich in essential amino acids [29]. Additionally, these proteins have bioactive possessions that make them of interest for health-related applications, including immune-stimulating and antioxidative activities [30]. Indication of their adaptability and commercial potential is the augmented usage of recovered proteins in medical meals, sports nutrition, and baby formula [31]. Solicitations in animal feed also have nutritional reimbursements; protein-rich by-products improve livestock nutrition and decrease dependence on traditional feed sources. This dual role in animal and human nutrition underscores the standing

of protein recovery in solving the problems related with global food security [32]. Interest in the commercialization of proteins from dairy waste has grown because of their economic value. The demand in food, beverage, and nutraceutical industries should drive the market growth of whey worldwide [33]. Successful examples from valorization effort productivity are recognized through the utilization of whey in protein powder and functional beverage [34]. However, procedures used in recovering are optimized with costs lowered, thereby increasing yield; economies of scale and improvements through technology in the processes of extraction and purification optimize the returns from an investment. Industrials might transform wastage administration from a cost area into a money maker by combining recovery of protein from present dairy procedures [35].

1.1 Objectives of the Chapter

The main objective of this chapter will be the main focus on protein salvage to help mitigate environmental and economic issues, thereby providing an inclusive review of sustainable processes for the valorization of dairy waste. Highlighting the reputation of green technology in resource recovery, it is attentive in probing into the scope, origin, and environmental belongings of dairy waste. The chapter demonstrates useful techniques to extract, purify, and use proteins originating from dairy wastes in an exertion to highlight nutritional value and commercial worth. The current chapter goals at providing a wholesome view on sustainability in dairy waste management and supporting circular budget objectives based on case studies, sustainability assessments, and economic analysis.

2 Materials

2.1 Types of Dairy Waste Used

Dairy waste, which includes the significant by-products whey, buttermilk, skim milk, and casein sludge from diverse phases of dairy production, comprises rich sources of proteins, lactose, and minerals. Whey is the liquid portion detached during the manufacturing of cheese and is a great resource for value adding [42, 43]. These milk solids and distinctive phospholipids in buttermilk hold great industrial probable as they are by-products of butter manufacture [43, 44]. Of course, skim milk and casein sludge are further plentiful and are rich sources of nutrients as they result from cream separation and casein precipitation, correspondingly [45, 46]. The volumes of dairy wastes around the biosphere leave much to be said for effective administration and value adding [42]. These waste streams differ in composition from produce compositions and dispensation methods. For instance, whey can further be categorized into sweet and acid whey based on pH and method. Likewise, the buttermilk differs in composition based on whether it is consequent from the manufacture of sweet cream or

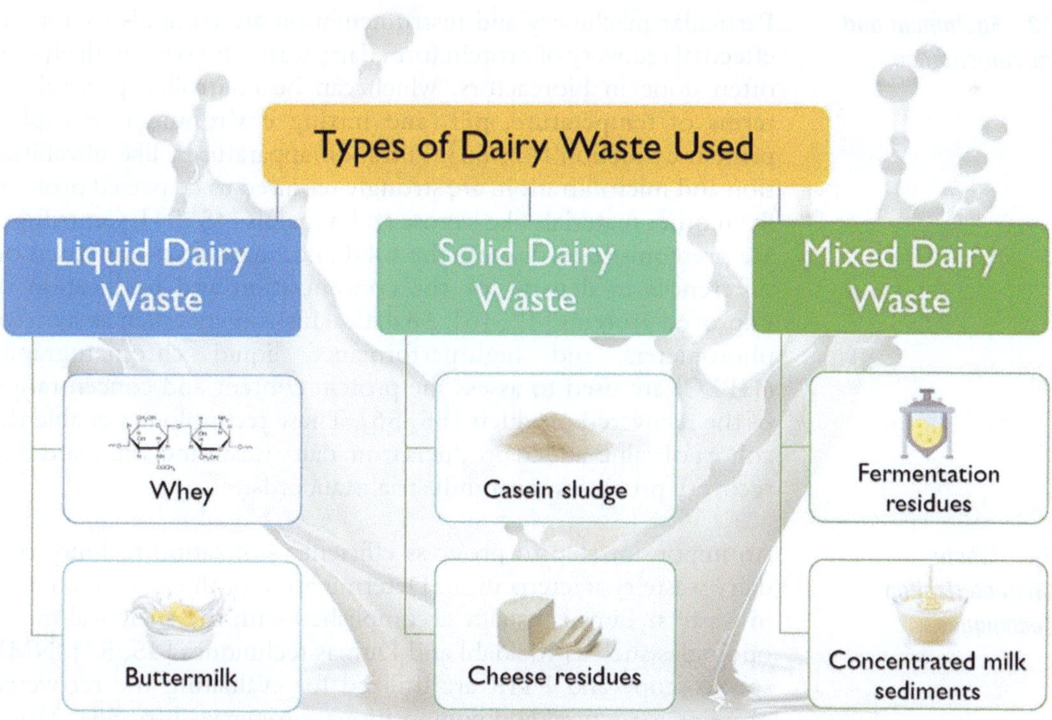

Fig. 1 Types of dairy waste used

cultured butter [44, 47]. Because of this compositional variability, exact recovery processes adapted to the unique physiognomies of each by-product are desirable [43, 45]. Figure 1 describes the types of dairy waste used in dairy production.

2.2 Enzymes and Chemicals

Chemicals and enzymes often used in protein recovery from dairy waste comprise proteases, acids, and bases. The proteolytic enzymes are characteristically used to break down proteins into bioactive peptides for many solicitations, like trypsin, pepsin, and microbial proteases [48, 49]. Based on green chemistry principles, these enzymes have been exposed to improve the digestibility and functionality of proteins with least consumption of energy [46, 50]. To certify ideal conditions for enzymatic activity and precipitation, acids and bases like sodium hydroxide or hydrochloric acid are used to alter pH during protein extraction [49, 51]. The choice of chemicals and enzymes depends on the intended use and characteristics of the recovered proteins. For instance, enzymes with broad pH stability are preferred for industrial-scale applications because they can acclimate to the assorted pH levels of various waste streams [48, 50]. Moreover, the enclosure of bio-based reagents or green catalysts is becoming increasingly significant in order to augment sustainability and decrease chemical waste [46, 52].

2.3 Equipment and Instrumentation

Particular machinery and instrumentation are compulsory for the effectual recovery of protein from dairy waste. Enzyme hydrolysis is often done in bioreactors, which can be controlled precisely in terms of temperature, pH, and mixing environment to exploit protein extraction [42, 53]. Filtration apparatuses, like ultrafiltration and microfiltration, are strongly required to dispersed proteins from other materials like lactose and minerals [45, 54]. Centrifuges are prerequisite to discrete the solid and liquid fractions based on differences in density for the concentration and purification of improved proteins [42, 46]. Analytical instruments such as spectrophotometers and high-performance liquid chromatography (HPLC) are used to assess the protein content and concentration of the recovered products [55, 56]. These technologies enable the cohort of value-added products from dairy waste and safeguard that recovery processes meet industrial standards.

2.4 Waste Characterization Techniques

An important step to progress efficient valorization techniques is dairy waste characterization. Determination of the protein content in waste streams is usually accomplished with the analytical methodologies such as Kjeldahl and Dumas techniques [45, 57]. NMR spectroscopy and FTIR are utilized for evaluating the recovered proteins' structural and compositional properties [46, 58]. Moreover, measurement of zeta potential and particle size analysis will elucidate constancy and functional characteristics of protein solutions [59, 60]. Waste characterization is thereby simplified to assistance in the commercialization of dairy waste-based products by accurately making sure that these are acquiescent with regulations and are augmented in terms of retrieval techniques [43, 55]. Figure 2 and Table 1 illustrate about the materials and methods used in processing dairy waste.

3 Methods

3.1 Pretreatment of Dairy Waste

Pretreatment is the first step to be taken in refining the recovery of proteins from dairy waste. The process involves filtration, sedimentation, and pH modification to formulate the waste for further processing [57, 63]. Sedimentation will make it at ease to separate the larger particles; filtration will eliminate solid contaminants, and pH modification will make the medium ideal for chemical or enzymatic reactions. These methods are vital in enhancing the efficiency of subsequent extraction methods [57, 58, 60, 63]. For instance, Xing et al. exemplified that treatment of dairy manure with acidity improves the efficiency of subsequent biohydrogen invention from dairy manure [65].

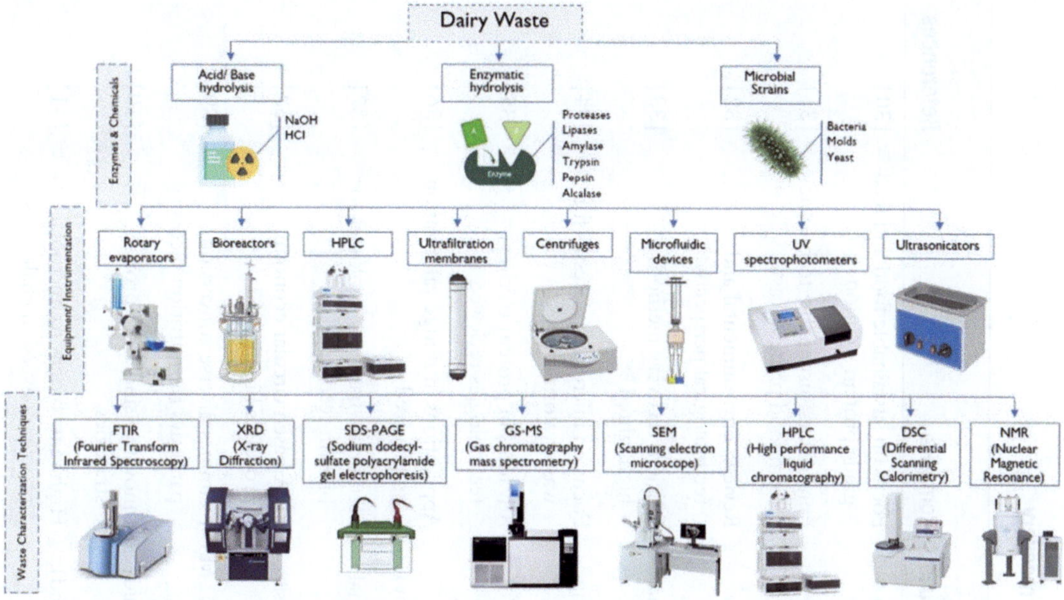

Fig. 2 Materials and methods used in processing dairy waste

3.2 Protein Extraction Processes

Proteins are isolated from dairy waste through different means, and the most notable are enzymatic hydrolysis and ultrafiltration [68]. The enzymatic hydrolysis is a method through which proteins are hydrolyzed with certain proteases to form peptides and amino acids that are easier to discrete and sanitize [71]. On the other hand, ultrafiltration separates and concentrates proteins through the use of membrane technology and sizes. This method can effectively separate whey proteins from dairy by-products [67, 68, 70]. Barry et al.'s research focuses on the amalgamation of ultrafiltration and enzymatic hydrolysis for the pilot scale developed of dairy products [71, 78].

3.3 Green Valorization Approaches

Green valorization emphasizes on environmentally friendly and sustainable dairy leftover processing approaches. Many of these techniques incorporate green chemistry principles such as avoiding or diminishing the use of hazardous chemicals and energy intensive procedures [79]. Low-energy ultrasonication along with enzymatic hydrolysis and additional processes with green solvents are two promising methods that improve the functional properties of the recovered proteins while reducing their adverse environmental impact [75, 76, 78]. For example, Amado et al. used enzymatic hydrolysis coupled with ultrafiltration to improve bioactive peptides with antihypertensive and antioxidant activities from wastewaters of seafood processing, and this method may also be adapted for dairy waste [74].

Table 1
Overview of materials, methods, and technologies for dairy waste valorization and protein recovery

Dairy waste used	Enzymes and chemicals	Equipment/ instrumentation	Waste characterization techniques	Remarks	References
Whey, buttermilk	Proteases, lipases	Bioreactors, centrifuges	Protein content, fat composition analysis	Focus on sustainable food and feed-grade protein solutions	[30]
Skim milk, casein sludge	Acid/base hydrolysis	Filtration units, UF systems	Rheological and thermal property analysis	Emphasizes sustainable technologies from food protein waste	[31]
Whey protein concentrates	Papain, alcalase	Chromatography tools	SDS-PAGE, FTIR spectroscopy	Reviews environmental and technological perspectives	[32]
Whey	Trypsin, pepsin	Ultrafiltration membranes	Kjeldahl method, calorimetric analysis	Efficient recovery methods highlighted	[33]
Buttermilk, casein waste	Acid treatments	Rotary evaporators, centrifuges	FTIR, DSC (Differential Scanning Calorimetry)	Focus on microbial processes for waste valorization	[34]
Dairy effluents	Alkaline proteases	Crossflow membranes	Chromatography, protein concentration assays	Advances in protein-recovery technologies reviewed	[35]
Sludge	Enzymatic hydrolysis agents	Pressurized reactors	XRD (X-ray diffraction), SEM (Scanning Electron Microscopy)	Applications in sludge valorization emphasized	[36]
Dairy wastewater	Amylase, lipase	Nanofiltration systems	Particle size analysis, zeta potential measurement	Membrane-based technologies for protein recovery	[37]
Cheese whey, effluents	PHA precursors, acids	Anaerobic digesters, bioreactors	TGA (Thermogravimetric Analysis), EDS	PHA-based circular economy processes highlighted	[38]
Various dairy residues	Proteases, chemical hydrolyzers	Fermenters, filtration units	Chromatographic separation, GC-MS	Examines diverse conversion pathways for nutrition recovery	[39]
Ricotta whey	Haloferax strains, NaCl	Fermentation tanks	HPLC (High-Performance Liquid Chromatography)	Innovative single-cell protein strategies	[40]
Dairy manure	Acid pretreatment	Anaerobic reactors	COD (Chemical Oxygen Demand), microbial enumeration	Highlights biogas to microbial protein conversion challenges	[41]

Dairy by-products	Enzyme cocktails	Dryers, spray concentrators	Thermal analysis, compositional assays	Integration in multi-sectoral biorefinery pathways	[42]
Whey, buttermilk	Alkaline proteases, papain	UV spectrophotometers	Rheological properties, functional peptide analysis	Trends in whey and buttermilk valorization discussed	[43]
Second cheese whey	Acidic and enzymatic treatments	Reverse osmosis, dialysis	LC-MS (Liquid Chromatography-Mass Spectrometry)	Valorization of specific whey fractions	[44]
All dairy wastes	Alcalase, bromelain	Ultrasonicators, agitators	Functional and therapeutic property characterization	Comprehensive review on applications of dairy wastes	[45]
Milk sludge	Microbial strains	Continuous reactors	Protein solubility and digestibility studies	Advances in agricultural waste technologies	[46]
Milk processing waste	Ionic liquids, proteolytic enzymes	Microfluidic devices	Elemental analysis, NMR (Nuclear Magnetic Resonance)	Potential integration with agri-food residues	[47]
Whey, casein residues	Plant-based proteases	Fractionation tools	Hydrolysate characterization, peptide quantification	Biocatalysis methods for biomass valorization	[48]
Protein concentrates	Papain, bromelain	Enzyme reactors	Free amino acid profiling, isoelectric focusing	New applications for plant proteases in dairy waste valorization	[49]
Biomass pulp residues	Cellulases, proteases	Screw press, separators	Fiber composition, enzyme efficiency tests	Recovery of green biomass protein highlighted	[50]

3.4 Purification of Recovered Proteins

Protein purification is a must to ensure that the recovered proteins are functional and nutritionally sound [80, 81]. Communal methods comprise precipitation, dialysis, and chromatography. Using chromatography techniques such as affinity chromatography or ion-exchange chromatography, it is possible to separate proteins rendering to their charge or binding characteristics with great specificity and efficiency [84, 89]. After dialysis eliminates small solutes, it leaves behind pure proteins, though proteins can easily be disconnected from other substances using precipitation produced by salts or solvents [85, 90]. These procedures are essential in the preparation of high purity protein isolates that can be realistic in medical and food associated uses [82].

3.5 Assessment of Protein Quality

The final step is to evaluate the nutritional and functional eminence of improved proteins. In nutritional analysis, digestibility and composition of amino acids are measured. Functional analysis trials solubility, emulsification, and gelation properties. Protein quality is dignified with standard parameters, such as the Digestible Indispensable Amino Acid Score (DIAAS) and the Protein Digestibility-Corrected Amino Acid Score (PDCAAS) [92, 94, 97]. These studies by Hertzler et al. reflect how significant such assessments are in confirming that the improved proteins meet industrial and nutritional supplies [77, 98]. Meade et al. have discussed how processing impacts protein function and quality of nutrition, bringing out the essential for effective approaches [90–92]. Figure 3 shows about the Digestible Indispensable Amino Acid Score (DIAAS) and the Protein Digestibility-Corrected Amino Acid Score (PDCAAS), while Table 2 illustrates about green valorization approaches for dairy waste and protein recovery: methods, techniques, and assessment.

3.6 Scale-Up Considerations

There are a number of experiments that need to be addressed when transporting dairy waste valorization from laboratory scale investigation to industrial applications, such as technological, financial, environmental, and regulatory matters [87, 88]. Technically, adaptation of laboratory processes to handle advanced volumes and different feedstocks is compulsory to ensure robust and efficient operations at a larger scale [93, 94]. Scale-up would be contingent on the reactor design, process optimization, and sophisticated monitoring systems for operational steadiness and product quality [104, 108, 109]. Automation would help recover the process efficiency further by plummeting human error and exploiting the use of resources [110]. Another key aspect of scale-up is financial viability [89]. Commercial feasibility of valorized products, such as improved proteins and bio-based resources, requires cost analysis of equipment, raw material handling, and energy supplies [84, 86]. Economic value produces revenue based on meeting customers' requirements and expectations, and comprehensive

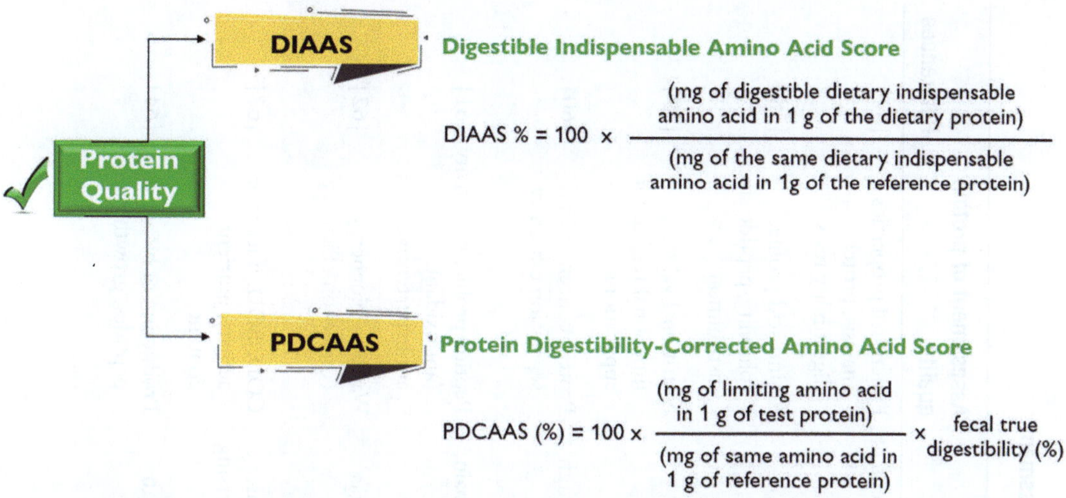

Fig. 3 Digestible Indispensable Amino Acid Score (DIAAS) and the Protein Digestibility-Corrected Amino Acid Score (PDCAAS)

cost analysis identifies efficiencies that can be developed to improve economic viability [113, 114, 117]. The environment must be distributed in a sustainable way. Energy efficient technology and renewable energy sources must be exploited because expansion of industrial operations might lead to higher energy consumption and waste cohort [100–102]. According to the principles of the circular budget, careful management of subordinate waste streams must also be made to stop mutilation to the environment [104, 116, 120]. One of the most major hurdles is related to regulatory compliance, particularly for products to be used for food and feed applications, which mandate more stringent safety and quality standards [112]. Zero-waste processes are also achievable with environmental and waste disposal regulations [108, 113, 114]. Integration of new valorization technologies with prevailing dairy processing infrastructure would thus decrease the cost and diminish the need for new amenities. Potential ways to improve sustainability and performance may include modular approaches to retrofitting the existing infrastructure and synergies with biorefinery actions such as the fabrication of biogas [105, 115, 118]. With a potential of being costly and high-tech in scalability, risk management is a prerequisite [106]. Risk reduction can also be facilitated through contingency planning as well as pilot-scale studies with the resolution of any problem ahead of full-scale deployment [107, 111, 117]. Cutting-edge strategies like digital twin technology which delivers valuable evidence prior to actual installation and permits virtual simulation of industrial processes can further augment scaling efforts [113, 119]. The circular bargain models recycle resources and energy within the system, further refining the sustainability of prolonged activities [105, 117]. To put it briefly,

Table 2
Green valorization approaches for dairy waste and protein recovery: methods, techniques, and assessment

Method	Description	Techniques/processes	Green valorization approaches	Assessment of protein quality	References
Pretreatment of dairy waste	Methods to prepare dairy waste for further processing	Filtration, sedimentation, pH adjustment	Sustainable waste management techniques	Functional properties analysis, protein digestibility tests	[57]
Protein extraction	Processes to isolate proteins from dairy by-products	Enzymatic hydrolysis, ultrafiltration, acid/alkaline treatment	Enzyme-assisted extraction, non-toxic solvents	Amino acid profile, solubility, peptide size distribution	[58]
Green valorization	Sustainable approaches for waste management and protein recovery	Bioreactors, fermenters, membrane filtration systems	Use of renewable energy, biological agents like enzymes or microbes	Nutritional value, functional testing for applications	[59]
Membrane-based separation	Advanced filtration techniques to separate proteins	Ultrafiltration, reverse osmosis	Zero-waste concept through membrane filtration	Protein concentration, separation efficiency	[60]
Enzymatic hydrolysis	Use of enzymes to break down complex proteins into smaller peptides	Proteases, pepsin, trypsin	Green enzymes (e.g., papain, bromelain) used for hydrolysis	Peptide profile, bioactivity (antioxidant, antihypertensive)	[61]
Anaerobic digestion	Biological treatment method for dairy wastewater with microbial activity	Anaerobic digesters	Methanogenesis for biogas production	Waste-to-bioenergy conversion efficiency	[62]
Wastewater treatment	Treatment processes for managing wastewater from dairy production	Chemical, biological, or physical methods	Water recycling, wastewater reuse for further processing	COD, BOD, microbial activity, nutrient content	[63]
Fermentation techniques	Microbial fermentation to produce high-value compounds from waste	Lactic acid bacteria, fungi	Circular economy through microbial biomass conversion	Production of bioactive peptides, growth rates	[64]

Protein fractionation	Isolation of specific protein fractions for high-value applications	Chromatography, ultrafiltration	Valorization of whey protein concentrates	Protein fraction purity, functional assays	[65]
Solvent extraction	Use of solvents to recover proteins or peptides from waste	Supercritical CO_2 extraction, organic solvents	Use of food-safe and non-toxic solvents	Recovery efficiency, peptide size and purity	[66]
Membrane bioreactors	Integrated systems for waste treatment and protein recovery	Crossflow filtration, hybrid membrane systems	Enhanced treatment efficiency with minimal energy usage	Membrane fouling resistance, recovery yield	[67]
Thermal treatment	Using heat to extract or purify proteins	Drying, spray drying, evaporation	Use of renewable heat sources for drying	Protein structure, denaturation degree	[68]
Ultrafiltration coupled hydrolysis	Combined techniques for peptide extraction from protein-rich waste	Thermolysin enzyme, ultrafiltration	Green processing to extract bioactive peptides	Molecular weight distribution, peptide bioactivity	[69]
Ion-exchange chromato-graphy	Separation of proteins based on charge differences	Ion-exchange resins	Non-toxic, eco-friendly ion-exchange materials	Elution profile, purity of protein fractions	[70]
Precipitation	Protein separation via solubility changes	Ammonium sulfate precipitation, acid precipitation	Low-energy methods using non-hazardous agents	Protein yield, purity of the precipitate	[71]
Functional peptide synthesis	Synthesis of bioactive peptides from dairy proteins	Enzymatic hydrolysis, fermentation	Green peptide synthesis for health benefits	Antioxidant, antihypertensive, or antimicrobial activity	[72]
Peptide profiling	Characterizing peptides produced during hydrolysis	LC-MS, HPLC, peptide mapping	Eco-friendly methods for high-throughput profiling	Peptide length, hydrophobicity, bioactivity	[73]
Functional protein applications	Utilizing proteins for food and nutraceutical applications	Hydrolysis, fermentation	Use of protein by-products for food ingredients	Solubility, emulsifying, and foaming properties	[74]

(continued)

Table 2 (continued)

Method	Description	Techniques/processes	Green valorization approaches	Assessment of protein quality	References
Biorefinery integration	Integrating multiple processes for efficient dairy waste management	Enzymatic treatments, fermentation, anaerobic digestion	Full waste-to-value conversion	Scale-up feasibility, cost-effectiveness	[75]
Biohydrogen production	Conversion of organic waste into hydrogen via microbial processes	Acid pretreatment, anaerobic fermentation	Biohydrogen as an alternative energy source	Hydrogen yield, conversion efficiency	[76]

captivating these influences into consideration forms the backbone for successful industrial solicitation of dairy waste valorization [115, 116]. Scale-up enterprises can move closer to a sustainable and economically feasible solution with the help of technical innovation, economic analysis, environmental stewardship, and regulatory obedience.

4 Notes

1. **Note 1: Sustainability Aspects**

   Dairy leftover green valorization technologies have important impacts on the environment by plummeting ecological footprints and promoting a circular economy [95, 96]. With the processes like enzymatic hydrolysis and membrane filtering that decrease waste disposal glitches and the emission of methane from traditional treatments, improving proteins and bioactive materials is less challenging [42, 44, 53]. To further improve sustainability, new biorefinery designs integrate energy efficient methods such as anaerobic digestion and renewable energy practice to renovate dairy waste into high-value products and biofuels [55, 62, 104]. An example of how to decrease resource waste and environmental degradation in the manufacture of valuable merchandises is the utilization of whey and buttermilk, which were before considered low value by-products [43, 48, 75]. Whey and buttermilk can be valorized using various green technologies for the sustainable production of proteins, bioactive peptides, biochemicals, biodegradable packaging materials, and others [4].

   Several green technologies can be used in the management and process of dairy wastes toward protein recovery. They are characterized by uniformity and standard operation and considered cost-effective and environmentally friendly. These include membrane technologies (microfiltration, ultrafiltration, nanofiltration, reverse osmosis and diafiltration), ultrasound-assisted extraction, microwave-assisted extraction, and enzymatic hydrolysis. Membrane technologies are generally used in the fractionation of different components of dairy by-products and in the recovery of proteins. Ultrasound-assisted extraction can be used to modify the structure of the extracted proteins, homogenize their particle sizes, increase their solubility, and enhance their thermal stability. Microwave-assisted extraction can enhance the structural and functional properties of the extracted proteins. Enzymatic hydrolysis of the extracted proteins with various enzymes can enhance their functional properties and produce bioactive compounds such as peptides with antioxidant, antibacterial, antihypertensive, and immunomodulatory activities

[127, 137]. Besides, enzymatic hydrolysis can improve the digestibility, solubility, and the emulsifying characteristics of proteins obtained from dairy by-products [140]. In addition, the use of drying techniques such as spray dying is another green approach for the green valorization of dairy wastes for protein production. Spray drying can be used to produce whey protein concentrates and whey protein isolates in powdered form. The drying process can increase the shelf life of the product, enhance its quality and stability, facilitate the transportation process, and decrease the consumption of energy, thereby making the protein recovery from dairy wastes more sustainable [140].

In addition to their use in protein recovery, dairy by-products such as whey can be valorized in other ways through various green technologies in order to arrive to a sustainable use of natural resources, protect the environment, and favorize the principles of circular economy [127]. Similar to proteins, lactose can also be recovered from dairy by-products such as whey and applied in various ways into food and pharmaceutical products. The recovery of nutrients other than proteins such as lactose from dairy by-products makes their valorization for protein recovery more sustainable and economically viable [119]. Whey is rich in various nutrients such as lactose, proteins, and peptides. Lactose in particular can act as a source of carbon that can promote microbial growth. Therefore, whey can be used to produce various value-added products through microbial fermentation using lactic acid bacteria or yeasts. Obtained products can be bioactive compounds that can be used in functional foods or high-value chemicals [127, 140]. For instance, precision fermentation, a novel biotechnological process based on programming microbes to generate specific organic components, can be used to produce bioactive food ingredients with high purity from dairy by-products [125]. Besides, due to its richness in nutrients, dairy wastewater can be used for microalgae production. This can protect the environment and at the same time generate value-added products like microalgae which can be used for various purposes [135, 136].

Cheese whey can be used for the production of biodegradable plastic components like polymers, polylactide, and polyhydroxyalkanoates which can be utilized in creating sustainable packaging materials that are safe for the human health and for the environment [44]. This can be done through anaerobic digestion of dairy by-products like whey. Additionally, whey protein concentrates and isolates can be used in edible films production [140]. Dairy by-products can be converted into biohydrogen via dark and photo-fermentation processes. Dark fermentation is realized under anaerobic conditions using anaerobic or facultatively anaerobic bacteria. Photo-fermentation involves the use of photoheterotrophic bacterial strains that are able to produce biohydrogen and carbon dioxide from organic acids under light presence. The transformation of dairy wastes into biohydrogen is a green way of energy production that goes with the approach of circular economy [127]. Dairy wastes can also be used to produce other biofuels such as biodiesel and biomethane via fermentation or anaerobic digestion processes [34]. As a conclusion, a viable approach that integrate various valorization ways of dairy wastes in the dairy plants can make the recovery of proteins from dairy bio-products more profitable and sustainable [104].

2. **Note 2: Economic Feasibility**

Recovered proteins require a market value and cost-effective management for economic viability in valorizing dairy waste. Although early investment in advanced equipment is high, protocols such as enzymatic hydrolysis and ultrafiltration have been proved to produce a high yield of high eminence proteins [47, 71, 73]. It would cover production costs with market demand, mainly coming from the food and nutraceutical industries for functional proteins and bioactive peptides [46, 52, 67]. Furthermore, value added to existing dairy processing technologies by incorporating valorization technology provides cost savings related to operations and transportation [104]. Success cases, through intelligent investments in innovation, illustrate long-term profitability [42, 43, 53].

Some studies investigated the economic feasibility of the valorization of dairy wastes for protein and nutrients production. However, more researches are needed to confirm the economic feasibility and viability of this approach and to make it more profitable and sustainable. Macedo et al. [132] integrated membrane technologies to valorize goat and sheep cheese whey with the purpose of enhancing the eco-efficiency of small/medium-size cheese dairy plants. The study evaluated the efficacy of using ultrafiltration/diafiltration for protein recovery. A cost-benefit analysis (CBA) and a sensitivity analysis (SA) were realized to assess the profitability of using membrane technology for the purpose of producing liquid protein concentrate, powdered protein concentrates, and whey cheese containing probiotics. Results showed the possibility of obtaining protein concentrates with 64% and 80% of crude protein from goat and sheep cheese whey, respectively. In addition, the results of CBA and SA confirmed the economic viability of such system in small to medium cheese production units. Kim et al. [128] used a redox-mediated electrochemical desalination method in order to valorize dairy wastes in protein recovery. The study focused on the removal of salt as well as the recovery of whey proteins using a redox-mediated electrodialysis system. The techno-economic analysis that was performed confirmed the economic viability of this system with 51–73% lower energy use and 51–62% lower operating costs comparing to normal desalination methods. Aydiner et al. [120] investigated the techno-economic viability of innovative membrane for the recovery of water from whey wastewater and the transformation of concentrated whey into powder via spray drying technique. For whey concentration, two combinations of membrane systems were tested which are forward osmosis/membrane distillation and membrane distillation/reverse osmosis. Both combinations gave concentrates with total solids of 21% and 25.8%, respectively. Both systems were found more efficient and profitable than ultrafiltration/reverse osmosis system with an estimated return of 12–13 million $, an annual net profit of around 800,000 $ and payback period under 1 year. The economic feasibility for the green recovery of proteins and other nutrients was also studied by Raut et al. [138] and BolicoPletsch et al. [121].

Sar et al. [34] reported that techno-economic analyses showed that integrating microbial processes in dairy plants for the bioconversion of dairy by-products into valuable bio-products is feasible and can be efficient. For biogas production in particular, this approach can give positive net present value (NPV) and payback period less than 10 years to the plant. In the study of Meneses and Flores [133], the use of a system that combines ultrafiltration and reverse osmoses for the recovery of water from cheese whey for reuse in equipment cleaning and separation of lactose and proteins for spray drying aims was found economically feasible and profitable. A cost analysis was realized for production levels of cheese which have an annual whey generation of 1, 20, and 225 million liters, respectively. Results showed that these units generated revenues of $0.18, $3.05, and $33.4 million per year from this strategy.

3. **Note 3: Case Studies**

Several successful industrial implementations validate the feasibility and compensations of dairy waste valorization [98, 99]. For instance, membrane filtration for whey protein retrieval has become extensively used, manufacturing high-value products for food and nutritional supplement solicitations [43, 47, 73]. A comparable environmentally promising route for biodegradable plastics is the microbial fermentation process rummage-sale to yield polyhydroxyalkanoates (PHA) from whey [56, 62]. Valorization approaches, which also showcase scalability and market potential, were established in pilot experiments on the production of phospholipid improved dairy ingredients [71, 104]. In these examples, cutting-edge approaches are modified to the regional requirements and resources accessible in support of sustainable dairy industry growth, both financially and environmentally [97].

Different technologies were used for the recovery of proteins from dairy wastes in various research studies. Uald Lamkaddam et al. [141] used progressive freeze concentration for the recovery of proteins and lactose from cheese whey. The process was optimized and 20.52% (w/w) of total solids was concentrated at optimum conditions, with 13.93% (w/w) of protein content was obtained in the concentrated part and 6.24% (w/w) of lactose in the ice part. Wu et al. [142] used xanthan gum fractionation to recover casein from dairy wastewater. Xantan was used to stabilize casein through electrostatic interaction. The recovery rate of casein was around 86.51%. In the study of Das et al. [124], membrane technology was used to recover proteins and lactose from cheese whey with high purity and yield. Proteins were separated in the retentate via ultrafiltration in a diafiltration mode while lactose which was obtained in the permeate phase was concentrated through nanofiltration. Both products were transformed into powder using the freeze-drying technique and their quality was verified using FTIR analysis. Results showed that with this method, fouling could be reduced, and membranes can be re-used many times. In addition, proteins and lactose can be recovered with a yield of 80% and 90%, respectively. Chollangi and Hossain [123] also used ultrafiltration to recover proteins and lactose from whey. Lech et al. [130] used an approach that integrates membrane technology and chromatographic techniques for the separation of whey proteins and their fractionation. After this process of high purity whey protein recovery was tested at the laboratory scale, it was also elaborated at an industrial scale.

In other studies, other by-products were recovered along with proteins from dairy wastes. Kim et al. [128] used a redox-mediated electrochemical desalination method in order to valorize dairy wastes in protein recovery. The redox-mediated electrodialysis system uses reversible redox reaction for desalination and concentration of salt for reutilization purposes as well as for the recovery of proteins with high purity. The system was able to remove 99% of salt and to recover 98% of whey proteins with stable performance over time. In the study of Meneses and Flores [133], a system combining ultrafiltration and reverse osmoses was used to recover good quality water from cheese whey to use it for cleaning purposes and to separate protein and lactose for valorization through spray drying. The system succeeded to recover 47% of water with acceptable physicochemical and microbiological properties which can be used in equipment cleaning. The same approach was used in the study of Sarghini et al. [139] where membrane technology with reverse osmosis was used for separation of water intended to be used for cleaning purposes from dairy wastes and the concentration of proteins, which, after enzymatic treatment with transglutaminase, can be used as a fat replacer resistant to elevated temperatures.

Some studies also worked on the valorization of dairy wastes for the production of microalgae, the production of nutrients using microbial fermentation, or the generation of bioenergy. Hemalatha and Venkata Mohan [126] used the phytoremediation ability of Duckweed to treat dairy wastewater and its use as nutritional biomass. The process succeeded to remove organic acids, nitrates, and phosphates with high efficiency. The generated biomass was pretreated with mild acid, and the hydrolysate was used as a substrate for *Bacillus subtilis* to produce microbial proteins. In the study of Luo et al. [131], a membrane treatment process of two stages was used to recycle dairy wastewater. Ultrafiltration was employed to concentrate proteins and lipids which were cultivated with microalgae for biofuel biodiesel generation. In the second phase, nanofiltration was used to concentrate lactose in the retentate that can be utilized in biogas production via anaerobic digestion and to separate reusable water in the permeate. This two-stage membrane process was more efficient and reduced fooling compared to single-stage nanofiltration method. Kusmayadi et al. [129] used a similar approach that integrates anaerobic digestion and microalgae using to treat dairy wastewater and produce biochemicals.

4. **Note 4: Challenges and Limitations**

There is great potential for valorizing dairy waste, but large-scale protein recovery poses significant logistical and technical challenges. Among the major hurdles remain the high initial investment and technological complexity of leading-edge technologies, such as enzymatic hydrolysis and ultrafiltration [47, 72, 86]. The variability in waste volumes and configuration among dairy firms complicates attaining process consistency [44, 58, 66]. Environmental damage is avoided by extra approaches and planning to man secondary leftover streams like used membranes or concentrated brines [63, 104]. Market entry would take longer and be more luxurious due to regulatory barricades requiring strict adherence to quality and safety necessities, particularly in food grade items [57, 76, 81].

Some techniques which are considered green methods for protein recovery from dairy wastes have some limitations and possible problems that should be solved to preserve their efficiency. The performance of membrane techniques can be limited by the problem of fouling which is generally caused by concentration polarization, cake formation, or blocking of pores. Protein aggregation can increase fouling occurrence. The problem of fouling can be solved by pretreatment of dairy by-products before valorization. However, since the most commonly used method for this purpose is the chemical pretreatment, there is a need to use green pretreatment methods which are friendly to the environment such as ultrasonic pretreatment and turbulence promoters. The same thing applies for membrane cleaning [33]. In addition, the activity of some enzymes may not be high enough to give high hydrolysis rate that is required for the successful and efficient recovery of proteins from dairy wastes. Therefore, the production and use of enzymes with higher activity can solve this problem [134]. Another issue is that the valorization of dairy wastes for other purposes along with protein production can be challenging and complicated. For example, the production of biopolymers from dairy wastes can be more complicated than the production of conventional polymers. It requires high collaboration between various stakeholders and different sectors to ensure successful and sustainable valorization of these wastes [38].

The rapid growth of the human population in the world is accompanied with a fast expansion of the dairy sector and therefore an increase in the global amounts of dairy waste products generated by dairy factories. This huge amount of dairy wastes and by-products makes their management quite difficult and challenging. Another challenge is the variability of these amounts from a season to another because of the seasonal nature of milk. Thus, more efforts are required and novel methods needs to be implemented for the successful management of dairy wastes and the recovery of proteins from them [108]. The industrialization of food waste valorization, including dairy wastes faces difficulties related with quantifying the amounts of wastes which makes the establishment of a suitable logistic system and a good economic sustainability assessment more challenging. Food wastes such as dairy wastes are prone to fast deterioration and the absence of a suitable logistic system and advanced technologies can limit the possibility of their valorization and make this process less feasible and unsafe. In addition, it is generally assumed that the valorization of wastes has a positive effect on the environment and economy. However, this is highly dependent on the techniques used during the process. Therefore, the sustainability of the methods used should be assessed and verified [122].

5. **Note 5: Future Prospects**

New methods and emerging technologies promise to break the current barriers and unlock previously unexplored opportunities for the value-added processing of dairy waste. Machine learning models and digital twin simulations offer predictive insights for process optimization and risk reduction during scale-up [54, 79, 85]. Innovations in green solvent extraction methods, such as ultrasound-assisted processes, could be the way toward higher sustainability and efficiency, including novel approaches involving ionic liquids [46, 75, 78]. Stakeholder teamwork in the advance of circular bioeconomy models will assist a balance among environmental obligation and profitability [53, 77, 104]. Additional investigation should be directed toward cumulative the utilization of improved proteins in bioplastics, cosmetics, and pharmaceuticals to additional expand revenue streams [48, 50, 82]. This whole method for addressing environmental, economic, technical, and monitoring trials of dairy leftover valorization highlights the probable for maintainable industrial transformation and reserve effectual performs.

In response to new regulations and laws concerning waste management and protection of the environment as well as the awareness of consumers of health problems that can results from incorrect waste management and their interest in more healthy and functional foods, the food industry is moving toward more sustainable strategies in managing wastes and their valorization in the production of healthy ingredients. Dairy by-products with their richness in nutrients have a great potential in this context. Their valorization can generate various food ingredients that can be used in the development of high quality functional and healthy food products that respond the consumer demands. However, more studies are needed to investigate the effect of the addition of such ingredients on the physicochemical and nutritional properties as well as the consumer acceptance of the developed products [32]. The increase of using emergent technologies like 3D printing represents a promising approach for successful incorporation of proteins derived from dairy wastes in functional foods. The 3D printing technique can preserve the original physicochemical and sensory properties of foods enriched with ingredients like dairy waste proteins [125].

Although there was a rapid growth in the use of novel technologies for the recovery of proteins and other nutrients from dairy wastes, more research is needed to make this process more intensive. This can be done through the development of more sophisticated membranes that have large surface area and the use of systematic and programmed processes such as in the enzymatic production of bioactive peptides. In addition, the production of transgenic enzymes with higher activity can increase the hydrolysis rate of proteins while more research on the immmobilization of enzymes for the production of bioactive peptides can increase their performance [134]. Furthermore, in order to improve the efficiency of protein and nutrient recovery from dairy wastes, dairy manufacturers can collaborate together and make a network for establishing a central factory that is specialized in the collection of dairy by-products from different dairy products producers to valorize them in the production of high-value compounds that are intended for food industry and pharmaceutical industry uses. With this way, we save time and efforts, reduce treatment costs, and make the valorization of dairy by-products more feasible and economically viable and profitable [3].

References

1. Patra F, Duary RK (2020) Waste from dairy processing industries and its sustainable utilization. In: Sustainable food waste management: Concepts and Innovations. Springer, pp 127–154
2. Garg S, Rumjit NP, Thomas P, Sikander, Lai CW, George PJ (2020) Green technologies for the treatment and utilisation of dairy product wastes. In: Sustainable green chemical processes and their allied applications. Springer Nature, pp 311–338
3. Durham RJ, Hourigan JA (2007) Waste management and co-product recovery in dairy processing. In: Handbook of waste management and co-product recovery in food processing. Woodhead Publishing, pp 332–387
4. Hameed A, Anwar MJ, Perveen S, Amir M, Naeem I, Imran M et al (2023) Functional, industrial and therapeutic applications of dairy waste materials. Int J Food Prop 26(1): 1470–1496
5. Mehrotra R, Trivedi A, Mazumdar SK (2016) Study on characterisation of Indian dairy wastewater. Int J Eng Appl Sci Technol 1(11):77–88
6. Poonia A (2020) Emerging opportunities for effective valorization of dairy by-products. In: Sustainable food waste management: concepts and innovations. Springer, pp 267–287
7. Sirmacekic E, Atilgan A, Rolbiecki R, Jagosz B, Rolbiecki S, Gokdogan O et al (2022) Possibilities of using whey wastes in agriculture: case of Turkey. Energies 15(24): 9636
8. Jakopović KL, Jurina IB, Božanić R (2025) Trends in utilization of whey and buttermilk—valuable by-products of the dairy industry. In: Nutraceutics from agri-food by-products. Wiley, pp 245–283
9. Rad SJ, Lewis MJ (2014) Water utilisation, energy utilisation and waste water management in the dairy industry: a review. Int J Dairy Technol 67(1):1–20
10. Oliveira D, Fox P, O'Mahony JA (2019) Byproducts from dairy processing. In: Byproducts from agriculture and fisheries: adding value for food, feed, pharma, and fuels. Wiley, pp 57–106
11. Raghunath BV, Punnagaiarasi A, Rajarajan G, Irshad A, Elango A, Mahesh Kumar G (2016) Impact of dairy effluent on environment—a review. In: Integrated waste management in India: status and future prospects for environmental sustainability, pp 239–249
12. Ahmad T, Aadil RM, Ahmed H, urRahman U, Soares BC, Souza SL et al (2019) Treatment and utilization of dairy industrial waste: a review. Trends Food Sci Technol 88:361–372
13. Shete BS, Shinkar NP (2013) Dairy industry wastewater sources, characteristics & its effects on environment. Int J Curr Eng Technol 3(5):1611–1615
14. Tikariha A, Sahu O (2014) Study of characteristics and treatments of dairy industry waste water. J Appl Environ Microbiol 2(1):16–22
15. Ritambhara, Zainab, Vijayaraghavalu S, Prasad HK, Kumar M (2019) Treatment and recycling of wastewater from dairy industry. In: Advances in biological treatment of industrial waste water and their recycling for a sustainable future. Springer, pp 91–115
16. Ilyas M, Ahmad W, Khan H, Yousaf S, Yasir M, Khan A (2019) Environmental and health impacts of industrial wastewater effluents in Pakistan: a review. Rev Environ Health 34(2):171–186
17. Roufou S, Griffin S, Katsini L, Polańska M, Van Impe JF, Valdramidis VP (2021) The

(potential) impact of seasonality and climate change on the physicochemical and microbial properties of dairy waste and its management. Trends Food Sci Technol 116:1–10

18. Dineshkumar M, Ghosh SK, Dey PK, Dutta A (2025) Sustainable practices and performances of dairy effluent treatment in India. J Solid Waste Technol Manag 51(1):86–282
19. Kolev Slavov A (2017) General characteristics and treatment possibilities of dairy wastewater—a review. Food Technol Biotechnol 55(1):14–28
20. Bansal A, Kumar A (2022) Exploring challenges in effective wastewater treatment for dairy industries. In: Sustainability of water resources: impacts and management. Springer International Publishing, Cham, pp 237–247
21. He J, Xia S, Li W, Deng J, Lin Q, Zhang L (2023) Resource recovery and valorization of food wastewater for sustainable development: an overview of current approaches. J Environ Manag 347:119118
22. Usmani Z, Sharma M, Gaffey J, Sharma M, Dewhurst RJ, Moreau B et al (2022) Valorization of dairy waste and by-products through microbial bioprocesses. Bioresour Technol 346:126444
23. Lanjekar PR, Panwar NL, Patel MR, Divyangkumar N (2024) Exploring sustainable energy: an overview of biochemical and thermochemical conversion of dairy and food waste. Environ Pollut Manag 1:152
24. Adesra A, Srivastava VK, Varjani S (2021) Valorization of dairy wastes: integrative approaches for value added products. Indian J Microbiol 61(3):270–278
25. Sharma V, Singh A, Grenier M, Singh V, Thakur M (2023) Waste valorization in food industries: a review of sustainable approaches. In: Sustainable food systems (Volume II) SFS: novel sustainable green technologies, circular strategies, food safety & diversity. Springer, pp 161–183
26. Awasthi MK, Paul A, Kumar V, Sar T, Kumar D, Sarsaiya S et al (2022) Recent trends and developments on integrated biochemical conversion process for valorization of dairy waste to value added bioproducts: a review. Bioresour Technol 344:126193
27. Usmani Z, Sharma M, Awasthi AK, Sharma GD, Cysneiros D, Nayak SC et al (2021) Minimizing hazardous impact of food waste in a circular economy—advances in resource recovery through green strategies. J Hazard Mater 416:126154
28. Ribeiro TB, Voss GB, Coelho MC, Pintado ME (2022) Food waste and by-product valorization as an integrated approach with zero waste: future challenges. In: Future foods. Academic, pp 569–596
29. Chandra R, Castillo-Zacarias C, Delgado P, Parra-Saldívar R (2018) A biorefinery approach for dairy wastewater treatment and product recovery towards establishing a biorefinery complexity index. J Clean Prod 183: 1184–1196
30. Piercy E, Verstraete W, Ellis PR, Banks M, Rockström J, Smith P et al (2023) A sustainable waste-to-protein system to maximise waste resource utilisation for developing food-and feed-grade protein solutions. Green Chem 25(3):808–832
31. Peydayesh M, Bagnani M, Soon WL, Mezzenga R (2022) Turning food protein waste into sustainable technologies. Chem Rev 123(5):2112–2154
32. El-Aidie SA, Khalifa GS (2024) Innovative applications of whey protein for sustainable dairy industry: environmental and technological perspectives—a comprehensive review. Compr Rev Food Sci Food Saf 23(2):e13319
33. Ganju S, Gogate PR (2017) A review on approaches for efficient recovery of whey proteins from dairy industry effluents. J Food Eng 215:84–96
34. Sar T, Harirchi S, Ramezani M, Bulkan G, Akbas MY, Pandey A, Taherzadeh MJ (2022) Potential utilization of dairy industries by-products and wastes through microbial processes: a critical review. Sci Total Environ 810:152253
35. Yusree FIFM, Peter AP, Mohd Nor MZ, Show PL, Mokhtar MN (2021) Latest advances in protein-recovery technologies from agricultural waste. Foods 10(11):2748
36. Xiao K, Zhou Y (2020) Protein recovery from sludge: a review. J Clean Prod 249:119373
37. Shahid K, Srivastava V, Sillanpää M (2021) Protein recovery as a resource from waste specifically via membrane technology—from waste to wonder. Environ Sci Pollut Res 28: 10262–10282
38. Yadav B, Chavan S, Atmakuri A, Tyagi RD, Drogui P (2020) A review on recovery of proteins from industrial wastewaters with special emphasis on PHA production process: sustainable circular bioeconomy process development. Bioresour Technol 317: 124006
39. Javourez U, O'Donohue M, Hamelin L (2021) Waste-to-nutrition: a review of current and emerging conversion pathways. Biotechnol Adv 53:107857

40. Sekoai PT, Roets-Dlamini Y, O'Brien F, Ramchuran S, Chunilall V (2024) Valorization of food waste into single-cell protein: an innovative technological strategy for sustainable protein production. Microorganisms 12(1):166
41. Shahzad HMA, Almomani F, Shahzad A, Mahmoud KA, Rasool K (2024) Challenges and opportunities in biogas conversion to microbial protein: a pathway for sustainable resource recovery from organic waste. Process Saf Environ Prot 185:644
42. Deshwal GK, Alam T, Panjagari NR, Bhardwaj A (2021) Utilization of cereal crop residues, cereal milling, sugarcane and dairy processing by-products for sustainable packaging solutions. J Polym Environ 29:2046–2061
43. Mediboyina MK, O'Neill S, Holden NM, Murphy F (2024) Prospective life cycle assessment of an integrated biorefinery for production of lactic acid from dairy side streams. Sustain Prod Consum 50:376–390
44. Pires AF, Marnotes NG, Rubio OD, Garcia AC, Pereira CD (2021) Dairy by-products: a review on the valorization of whey and second cheese whey. Foods 10(5):1067
45. Mozejko-Ciesielska J, Marciniak P, Moraczewski K, Rytlewski P, Czaplicki S, Zadernowska A (2022) Cheese whey mother liquor as dairy waste with potential value for polyhydroxyalkanoate production by extremophilic Paracoccus homiensis. Sustain Mater Technol 33:e00449
46. Moatsou G (2024) Emerging technologies for improving properties, shelf life, and analysis of dairy products. Foods 13(7):1078
47. del Mar Contreras M, Lama-Muñoz A, Gutiérrez-Pérez JM, Espínola F, Moya M, Castro E (2019) Protein extraction from agri-food residues for integration in biorefinery: potential techniques and current status. Bioresour Technol 280:459–477
48. Avramidou K (2020) Biocatalysis for biomass valorization: protein hydrolysates and sugar esters from agri-food wastes. Università degli Studi di Milano
49. David Troncoso F, Alberto Sánchez D, Luján Ferreira M (2022) Production of plant proteases and new biotechnological applications: an updated review. ChemistryOpen 11(3): e202200017
50. Dotsenko G, Lange L (2017) Enzyme enhanced protein recovery from green biomass pulp. Waste Biomass Valor 8:1257–1264
51. Dugmore TI, Clark JH, Bustamante J, Houghton JA, Matharu AS (2018) Valorisation of biowastes for the production of green materials using chemical methods. In: Chemistry and chemical technologies in waste valorization, pp 73–121
52. Venugopal V, Sasidharan A, Rustad T (2023) Green chemistry to valorize seafood side streams: an ecofriendly roadmap toward sustainability. J Agric Food Chem 71(46): 17494–17509
53. Bhuvaneshwari S, Majeed F, Jose E, Mohan A (2022) Different treatment methodologies and reactors employed for dairy effluent treatment—a review. J Water Process Eng 46:102622
54. Hansen EB, Jacobsen C, Lund O, Marcatili P, Moreno PJG (2017) PROVIDE a project aiming at protein valorization through informatics, hydrolysis, and separation. In: DTU Sustain 2017. Technical University of Denmark
55. Giulianetti de Almeida MP, Mockaitis G, Weissbrodt DG (2023) Got whey? Sustainability endpoints for the dairy industry through resource biorecovery. Fermentation 9(10):897
56. Raho S, Carofiglio VE, Montemurro M, Miceli V, Centrone D, Stufano P et al (2020) Production of the polyhydroxyalkanoate PHBV from ricotta cheese exhausted whey by Haloferax mediterranei fermentation. Foods 9(10):1459
57. Reig M, Vecino X, Cortina JL (2021) Use of membrane technologies in dairy industry: an overview. Foods 10(11):2768
58. Britz TJ, van Schalkwyk C, Hung YT (2004) Treatment of dairy processing wastewaters. In: Handbook of industrial and hazardous wastes treatment. CRC Press, pp 673–705
59. Bella K, Rao PV (2021) Anaerobic digestion of dairy wastewater: effect of different parameters and co-digestion options—a review. Biomass Convers Biorefinery 13:1–26
60. Sarkar B, Chakrabarti PP, Vijaykumar A, Kale V (2006) Wastewater treatment in dairy industries—possibility of reuse. Desalination 195(1–3):141–152
61. Vialkova EI, Sidorenko OV, Glushchenko ES (2019) Qualitative composition and local pretreatment of dairy wastewaters. In: IOP conference series: materials science and engineering, vol 687(6). IOP Publishing, p 066049
62. Kaur N (2021) Different treatment techniques of dairy wastewater. Groundw Sustain Dev 14:100640

63. Poništ J, Dubšíková V, Schwarz M, Samešová D (2021) Methods of processing whey waste from dairies. A review. Environ Protect Eng 47(4):69
64. Selmer-Olsen E, Ratnaweera HC, Pehrson R (1996) A novel treatment process for dairy wastewater with chitosan produced from shrimp-shell waste. Water Sci Technol 34(11):33–40
65. Xing Y, Li Z, Fan Y, Hou H (2010) Biohydrogen production from dairy manures with acidification pretreatment by anaerobic fermentation. Environ Sci Pollut Res 17:392–399
66. Mohammad SS, da Silva Ferreira MV, Jacintho Barbosa MI, Barbosa Junior JL (2023) Characteristics of enzymatic hydrolysis of protein from different food sources and potential separation techniques. Curr Nutr Food Sci 19(6):590–601
67. Trusek-Holownia A, Lech M, Noworyta A (2016) Protein enzymatic hydrolysis integrated with ultrafiltration: thermolysin application in obtaining peptides. Chem Eng J 305:61–68
68. Martínez-Maqueda D, Hernández-Ledesma B, Amigo L, Miralles B, Gómez-Ruiz JÁ (2013) Extraction/fractionation techniques for proteins and peptides and protein digestion. In: Proteomics in foods: principles and applications. Springer Nature, pp 21–50
69. Wang H, Wang J, Lv Z, Liu Y, Lu F (2013) Preparing oligopeptides from broken rice protein by ultrafiltration-coupled enzymatic hydrolysis. Eur Food Res Technol 236:419–424
70. Eckert E, Han J, Swallow K, Tian Z, Jarpa-Parra M, Chen L (2019) Effects of enzymatic hydrolysis and ultrafiltration on physicochemical and functional properties of faba bean protein. Cereal Chem 96(4):725–741
71. Barry KM, Dinan TG, Kelly PM (2017) Pilot scale production of a phospholipid-enriched dairy ingredient by means of an optimised integrated process employing enzymatic hydrolysis, ultrafiltration and super-critical fluid extraction. Innov Food Sci Emerg Technol 41:301–306
72. Argüello MA, Álvarez S, Riera FA, Álvarez R (2003) Enzymatic cleaning of inorganic ultrafiltration membranes used for whey protein fractionation. J Membr Sci 216(1–2):121–134
73. Saxena A, Tripathi BP, Kumar M, Shahi VK (2009) Membrane-based techniques for the separation and purification of proteins: an overview. Adv Colloid Interf Sci 145(1–2):1–22
74. Amado IR, Vázquez JA, González MP, Murado MA (2013) Production of antihypertensive and antioxidant activities by enzymatic hydrolysis of protein concentrates recovered by ultrafiltration from cuttlefish processing wastewaters. Biochem Eng J 76:43–54
75. More PR, Jambrak AR, Arya SS (2022) Green, environment-friendly and sustainable techniques for extraction of food bioactive compounds and waste valorization. Trends Food Sci Technol 128:296–315
76. de SouzaMesquita LM, Contieri LS, e Silva FA, Bagini RH, Bragagnolo FS, Strieder MM et al (2024) Path2Green: introducing 12 green extraction principles and a novel metric for assessing sustainability in biomass valorization. Green Chem 26(19):10087–10106
77. Langsdorf A, Volkmar M, Holtmann D, Ulber R (2021) Material utilization of green waste: a review on potential valorization methods. Bioresour Bioprocess 8(1):1–26
78. Al Khawli F, Pateiro M, Domínguez R, Lorenzo JM, Gullón P, Kousoulaki K et al (2019) Innovative green technologies of intensification for valorization of seafood and their by-products. Mar Drugs 17(12):689
79. Rana AK (2022) Green approaches in the valorization of plant wastes: recent insights and future directions. Curr Opin Green Sustain Chem 38:100696
80. Ferreira-Santos P, Zanuso E, Genisheva Z, Rocha CM, Teixeira JA (2020) Green and sustainable valorization of bioactive phenolic compounds from pinus by-products. Molecules 25(12):2931
81. Arancon RAD, Lin CSK, Chan KM, Kwan TH, Luque R (2013) Advances on waste valorization: new horizons for a more sustainable society. Energy Sci Eng 1(2):53–71
82. Roselló-Soto E, Koubaa M, Moubarik A, Lopes RP, Saraiva JA, Boussetta N et al (2015) Emerging opportunities for the effective valorization of wastes and by-products generated during olive oil production process: non-conventional methods for the recovery of high-added value compounds. Trends Food Sci Technol 45(2):296–310
83. Romani A, Pinelli P, Ieri F, Bernini R (2016) Sustainability, innovation, and green chemistry in the production and valorization of phenolic extracts from Olea europaea L. Sustainability 8(10):1002
84. Goldring JD (2019) Concentrating proteins by salt, polyethylene glycol, solvent, SDS

precipitation, three-phase partitioning, dialysis, centrifugation, ultrafiltration, lyophilization, affinity chromatography, immunoprecipitation or increased temperature for protein isolation, drug interaction, and proteomic and peptidomic evaluation. In: Electrophoretic separation of proteins: methods and protocols. Wiley, pp 41–59
85. Liu S, Li Z, Yu B, Wang S, Shen Y, Cong H (2020) Recent advances on protein separation and purification methods. Adv Colloid Interf Sci 284:102254
86. Ersson B, Rydén L, Janson JC (2011) Introduction to protein purification. In: Protein purification: principles, high resolution methods, and applications. Wiley, pp 1–22
87. Smith DM (2017) Protein separation and characterization procedures. Food Anal:431–453
88. Novák P, Havlíček V (2016) Protein extraction and precipitation. In: Proteomic profiling and analytical chemistry. Elsevier, pp 51–62
89. Kaufmann M (1997) Unstable proteins: how to subject them to chromatographic separations for purification procedures. J Chromatogr B Biomed Sci Appl 699(1–2):347–369
90. Wingfield PT (2015) Overview of the purification of recombinant proteins. Curr Protoc Protein Sci 80(1):6–1
91. Evans DR, Romero JK, Westoby M (2009) Concentration of proteins and removal of solutes. Methods Enzymol 463:97–120
92. Meade SJ, Reid EA, Gerrard JA (2005) The impact of processing on the nutritional quality of food proteins. J AOAC Int 88(3):904–922
93. Tome D (2012) Criteria and markers for protein quality assessment—a review. Br J Nutr 108(S2):S222–S229
94. Adhikari S, Schop M, de Boer IJ, Huppertz T (2022) Protein quality in perspective: a review of protein quality metrics and their applications. Nutrients 14(5):947
95. Bos C, Gaudichon C, Tomé D (2000) Nutritional and physiological criteria in the assessment of milk protein quality for humans. J Am Coll Nutr 19(Suppl 2):191S–205S
96. Gigliotti JC, Jaczynski J, Tou JC (2008) Determination of the nutritional value, protein quality and safety of krill protein concentrate isolated using an isoelectric solubilization/precipitation technique. Food Chem 111(1):209–214
97. Millward DJ, Layman DK, Tomé D, Schaafsma G (2008) Protein quality assessment: impact of expanding understanding of protein and amino acid needs for optimal health. Am J Clin Nutr 87(5):1576S–1581S
98. Hertzler SR, Lieblein-Boff JC, Weiler M, Allgeier C (2020) Plant proteins: assessing their nutritional quality and effects on health and physical function. Nutrients 12(12):3704
99. Friedman M (1996) Nutritional value of proteins from different food sources. A review. J Agric Food Chem 44(1):6–29
100. Gehring CK, Gigliotti JC, Moritz JS, Tou JC, Jaczynski J (2011) Functional and nutritional characteristics of proteins and lipids recovered by isoelectric processing of fish by-products and low-value fish: a review. Food Chem 124(2):422–431
101. Womeni HM, Tiencheu B, Linder M, Nabayo C, Martial E, Tenyang N et al (2012) Nutritional value and effect of cooking, drying and storage process on some functional properties of Rhynchophorus phoenicis. Int J Life Sci Pharma Res 2:203
102. McAuliffe GA, Takahashi T, Beal T, Huppertz T, Leroy F, Buttriss J et al (2023) Protein quality as a complementary functional unit in life cycle assessment (LCA). Int J Life Cycle Assess 28(2):146–155
103. Fleddermann M, Fechner A, Rößler A, Bähr M, Pastor A, Liebert F, Jahreis G (2013) Nutritional evaluation of rapeseed protein compared to soy protein for quality, plasma amino acids, and nitrogen balance—a randomized cross-over intervention study in humans. Clin Nutr 32(4):519–526
104. Asunis F, De Gioannis G, Dessì P, Isipato M, Lens PN, Muntoni A et al (2020) The dairy biorefinery: integrating treatment processes for cheese whey valorisation. J Environ Manag 276:111240
105. Costa CF, Amorim CL, Duque AF, Reis MA, Castro PM (2022) Valorization of wastewater from food industry: moving to a circular bioeconomy. Rev Environ Sci Biotechnol 21(1):269–295
106. Meena RAA, Ghosh A, Sathishkumar P, Jayabalan R (2020) Scaling up of food waste valorization market outlooks: key concerns. In: Food waste to valuable resources. Academic, pp 401–416
107. Koutinas AA, Vlysidis A, Pleissner D, Kopsahelis N, Garcia IL, Kookos IK et al (2014) Valorization of industrial waste and by-product streams via fermentation for the production of chemicals and biopolymers. Chem Soc Rev 43(8):2587–2627
108. Stasinakis AS, Charalambous P, Vyrides I (2022) Dairy wastewater management

in EU: produced amounts, existing legislation, applied treatment processes and future challenges. J Environ Manag 303:114152

109. Haris NIN, Salleh SF, Nolasco-Hipolito C, Awang Husaini AAS, Wondi MH, Harun NH, Abdul Raof N (2024) A review on lactic acid production via microbial fermentation from biowaste: insights on scaling up to industrial process. Biofuels Bioprod Biorefinery 19:437
110. Kasmi M (2018) Biological processes as promoting way for both treatment and valorization of dairy industry effluents. Waste Biomass Valor 9(2):195–209
111. Asunis F (2020) Application of integrated biochemical processes for the valorisation of sheep dairy bioresidues in the framework of waste biorefinery concept
112. Duque AF, Campo R, Val del Rio A, Amorim CL (2021) Wastewater valorization: practice around the world at pilot-and full-scale. Int J Environ Res Public Health 18(18):9466
113. Nayak A, Bhushan B (2019) An overview of the recent trends on the waste valorization techniques for food wastes. J Environ Manag 233:352–370
114. Buchanan D, Martindale W, Romeih E, Hebishy E (2023) Recent advances in whey processing and valorisation: technological and environmental perspectives. Int J Dairy Technol 76(2):291–312
115. Kuo CC, Chen D, Jiménez-Flores R, Wick M, Campanella O (2024) Valorization of byproducts from meat and dairy industries through fermentation to produce peptides. Sustain Food Technol 2(5):1469–1475
116. Jha S, Nanda S, Acharya B, Dalai AK (2022) A review of thermochemical conversion of waste biomass to biofuels. Energies 15(17):6352
117. Capanoglu E, Nemli E, Tomas-Barberan F (2022) Novel approaches in the valorization of agricultural wastes and their applications. J Agric Food Chem 70(23):6787–6804
118. Siegrist A, Green A, Gold M, Mathys A (2023) Recent findings on environmental sustainability and conversion efficiency of waste-to-protein pathways. Curr Opin Green Sustain Chem 41:100833
119. Addai FP, Lin F, Wang T, Kosiba AA, Sheng P, Yu F et al (2020) Technical integrative approaches to cheese whey valorization towards sustainable environment. Food Funct 11(10):8407–8423
120. Aydiner C, Sen U, Topcu S, Ekinci D, Altinay AD, Koseoglu-Imer DY, Keskinler B (2014) Techno-economic viability of innovative membrane systems in water and mass recovery from dairy wastewater. J Membr Sci 458:66–75
121. BolicoPletsch G, Setti D, de Lima JD, Adamczuk Oliveira G, Pimentel TC (2023) A multi-criteria model for the techno-economic evaluation of fat removal technologies from dairy waste. SAGE Open 13(2): 1–15
122. Caldeira C, Vlysidis A, Fiore G, De Laurentiis V, Vignali G, Sala S (2020) Sustainability of food waste biorefinery: a review on valorisation pathways, techno-economic constraints, and environmental assessment. Bioresour Technol 312:123575
123. Chollangi A, Hossain MM (2007) Separation of proteins and lactose from dairy wastewater. Chem Eng Process Process Intensif 46(5): 398–404
124. Das B, Sarkar S, Sarkar A, Bhattacharjee S, Bhattacharjee C (2016) Recovery of whey proteins and lactose from dairy waste: a step towards green waste management. Process Saf Environ Prot 101:27–33
125. Kheto A, Adhikary U, Dhua S, Sarkar A, Kumar Y, Vashishth R, Shrestha BB, Saxena DC. A review on advancements in emerging processing of whey protein: Enhancing functional and nutritional properties for functional food applications. Food Safety and Health. 2025 Jan;3(1):23-45.
126. Hemalatha M, Mohan SV (2022) Duckweed biorefinery—potential to remediate dairy wastewater in integration with microbial protein production. Bioresour Technol 346: 126499
127. Iñiguez-Moreno M, Poonia A (2023) Green technologies for treatment and utilization of whey towards sustainable exploitation. Whey Valor:63–81
128. Kim N, Jeon J, Elbert J, Kim C, Su X (2022) Redox-mediated electrochemical desalination for waste valorization in dairy production. Chem Eng J 428:131082
129. Kusmayadi A, Lu PH, Huang CY, Leong YK, Yen HW, Chang JS (2022) Integrating anaerobic digestion and microalgae cultivation for dairy wastewater treatment and potential biochemicals production from the harvested microalgal biomass. Chemosphere 291: 133057
130. Lech M, Niesobska A, Trusek-Holownia A (2016) Dairy wastewater utilization: separation of whey proteins in membrane and chromatographic processes. Desalin Water Treat 57(48–49):23326–23334
131. Luo J, Ding L, Qi B, Jaffrin MY, Wan Y (2011) A two-stage ultrafiltration and

nanofiltration process for recycling dairy wastewater. Bioresour Technol 102(16): 7437–7442

132. Macedo A, Bilau J, Cambóias E, Duarte E (2021) Integration of membrane processes for by-product valorization to improve the eco-efficiency of small/medium size cheese dairy plants. Foods 10(8):1740
133. Meneses YE, Flores RA (2016) Feasibility, safety, and economic implications of whey-recovered water in cleaning-in-place systems: a case study on water conservation for the dairy industry. J Dairy Sci 99(5):3396–3407
134. Nath A, Mondal S, Kanjilal T, Chakraborty S, Curcio S, Bhattacharjee C (2015) Synthesis and functionality of proteinacious nutraceuticals from casein whey—a clean and safe route of valorization of dairy waste. Chem Eng Res Des 97:192–207
135. Ozcelik D, Suwal S, Ray C, Tiwari BK, Jensen PE, Poojary MM (2024) Valorization of dairy side-streams for the cultivation of microalgae for value added food products. Trends Food Sci Technol 146:104386
136. Pandey A, Srivastava S, Kumar S (2019) Phyco-remediation of dairy effluents and biomass valorization: a sustainable approach. In: Application of microalgae in wastewater treatment. Springer Nature, pp 195–213
137. Papaioannou EH, Mazzei R, Bazzarelli F, Piacentini E, Giannakopoulos V, Roberts MR, Giorno L (2022) Agri-food industry waste as resource of chemicals: the role of membrane technology in their sustainable recycling. Sustainability (Switzerland) 14(3): 1–21
138. Raut SS, Yadav RD, Dhamole PB, Agrawal SG (2024) Techno-economic assessment of whey proteins production from dairy waste using sodium carboxymethyl cellulose of different degree of substitution. Biomass Convers Biorefinery 15:5811–5825
139. Sarghini F, Sorrentino A, Di Pierro P (2013) An integrated mechanical-enzymatic reverse osmosis treatment of dairy industry wastewater and milk protein recovery as a fat replacer: a closed loop approach. J Agric Eng 44:283–286
140. Soumati B, Atmani M, Benabderrahmane A, Benjelloun M (2023) Whey valorization—innovative strategies for sustainable development and value-added product creation. J Ecol Eng 24(10):86–104
141. UaldLamkaddam I, Vega E, Colón J, Ponsá S, Llenas L, Mora M (2023) Progressive freeze concentration of cheese whey for protein and lactose recovery. Int Dairy J 139:105572
142. Wu Z, Yin H, Liu W, Huang D, Hu N, Yang C, Zhao X (2020) Xanthan gum assisted foam fractionation for the recovery of casein from the dairy wastewater. Prep Biochem Biotechnol 50(1):37–46

Chapter 11

Microbial Fermentation Protocols for Eco-Friendly Feedstock Conversion

Selva Kumar T, Saranyadevi Subburaj, Balakrishnan Navina, and Aminu Abdullahi

Abstract

Microbial fermentation is an important technology for the conversion of renewable feedstocks into bio-based products, with a sustainable alternative to traditional chemical methods. The current study defines optimized microbial fermentation protocols to achieve maximum conversion efficiency of green feedstocks such as lignocellulosic biomass, crop residues, and industrial waste streams. The protocols integrate state-of-the-art microbial strain screening, genetic engineering, and process design to achieve maximum productivity with minimum energy requirements and environmental impact. Emphasis is laid on the development of fermentation strategies with the least chemical supplements and maximum resource yield. Key developments include adaptive fermentation control systems, custom-designed nutrient formulations, and scalable bioprocess designs that provide consistent product yields with variable feedstock conditions. Comparative studies show the protocols provide enhanced conversion rates and product titers compared to conventional methods with substantial greenhouse gas savings and waste reduction. The work also highlights the importance of the use of microbial consortia and engineered symbiotic interactions in facilitating efficient utilization of substrates and by-product reduction. Case studies demonstrate successful implementation in biofuel, biodegradable plastics, and value-added bi-chemicals. The study is helping to develop microbial fermentation as a pillar technology for the bioeconomy to facilitate circular and sustainable production platforms. The protocols described here provide a complete strategy to researchers and industry players who wish to utilize microbial fermentation to convert environmentally friendly feedstocks, paving the way to greener industrial processes.

Key words Fermentation, Eco-friendly conversion, Microbial fermentation, Feedstocks preparation and applications

1 Introduction

Feedstocks are raw materials employed in industrial processes for fuel, chemicals, and other chemical products production. Sustainable and renewable feedstocks are obtained from biological sources with an inherent capability of self-replenishment, thereby offering alternative options to conventional fuel-based products [1, 2]. Such

Tanmay Sarkar (ed.), *Food Waste Valorization: Green Techniques in Sustainable Management*, Methods and Protocols in Food Science, https://doi.org/10.1007/978-1-0716-4646-5_11,
© The Author(s), under exclusive license to Springer Science+Business Media, LLC, part of Springer Nature 2025

renewable feedstocks involve biomass resources such as crops, agricultural residues, forest residues, and organic waste materials. Biorefinery activities allow such material to be converted into value-added products such as chemicals, biofuels, and biopolymers [3]. Crude palm oil, rapeseed oil, waste cooking oil, non-edible animal fat, algae oil, and lignocellulosic biomass are some of the common feedstocks for such an organization [2, 4]. Interestingly, some of these renewable resources such as lignin and industrial waste are not food items and therefore do not compete with agriculture [1, 5]. There are extremophilic bacteria as well with the potential to be used in the future in biodegradable polymer production from renewable feedstocks such as polyhydroxyalkanoates (PHAs) further promoting sustainable bioplastic production [6]. Sustainable and renewable feedstocks are required for a transition to a non-fossil fuel-based economy. Such systems have reduced carbon emissions and reduced greenhouse gas amounts [2]. Certain challenges entail finite supply and the technological requirements of feedstocks to be processed more efficiently [2, 7]. Higher consumption of green products, particularly those produced with renewable raw materials, is due to heightened environmental awareness. Their consumption is encouraged by the government since they are believed to foster sustainable development [8]. Social norms, environmental concern, and perceived effectiveness shape attitudes toward green tourism destinations and products [9]. Quite unexpectedly, consumers may have positive attitudes toward green products but may not necessarily be willing to pay more for them. For instance, young Indian consumers who were highly educated had a positive attitude toward green holidays, yet there was a weak link between this positive attitude and being willing to pay more for such a holiday [9]. In summary, although stockpiling of sustainable feedstocks is not discussed as such, the growing market for green products indicates a rise in demand for these products. Their effect on consumer choice is multifactorial, and environmental concerns, social norms, and economics are among the factors included. Further studies emphasizing green feedstocks and their storage are needed to provide more detailed information on this topic.

1.1 Microbial Fermentation Overview

Microbial fermentation has also proven to be a feasible pathway for the conversion of bio-based feedstocks into value-added products, thereby mitigating dwindling fossil fuel reserves and the issue of pollution [10]. Scientists are investigating the production of renewable materials like coffee waste, renewable plant material, and algae toward the synthesis of biofuels and biochemicals [11]. These materials are economically and eco-friendly due to their availability, low-cost, and environment friendly nature than traditional materials. The substrate and its treatment largely determine fermentation and product yield. Instances could be quoted like hydrogen peroxide and acetic acid treatment of coffee waste with certain enzymes can effectively achieve a conversion rate of 78.9% to lactic acid [12]. Likewise, residual microalgae can be used for yeast oil production with a 90–100% glucose conversion through an environmentally friendly enzyme-based fermentation process [13]. It gives the process an eco-friendly approach using microbial fermentation with the biodegradables leading to a sustainable process against using coal or any fossil for energy. Challenges such as high costs, low productivity, and the need for better preparation methods must be overcome to make these processes more economically viable. Contributions to ecosystem-relevant production will be made through improvements in material preparation, fermentation strategies, and organism modifications to addressing these challenges.

1.2 Types of Microorganisms

Humongous diversified microbes are engaged into crucial roles in the process of microbial feedstock conversion. They include bacteria, fungi, and yeasts. Each type of microbe employs different technique and hence the final product which it yields could differ [14]. Most of these microbes find applications in bioremediation, production of biofuels, cosmetics, fine chemicals, food supplements, pharmaceuticals, etc. Many microbes have adapted to thrive in extreme environments such as hot springs and deep sea vents. These microbes have different metabolic attributes. Therefore, classifying microbes according to their metabolic capabilities could perfectly fit while considering microbial feedstock conversion [15]. With respect to this, they can be categorized into seven different types such as Fermentative Microorganisms, Cellulolytic Microorganisms, Methanogenic Archaea, Oleaginous Microorganisms, Acidogenic Bacteria, Ligninolytic Microorganisms, and Genetically Engineered Microorganisms. Bacteria are the most prominent ones that are widely employed in this microbial feedstock conversion. In particular, *Lactobacillus* sp. and *Acetobacter* sp. are dominant in solid-state fermentation processes, such as vinegar production [16]. During Baijiu fermentation, a myriad of microorganisms participates in spontaneous fermentations, whereupon they bring about unique flavoring [17]. There are several yeasts and fungi used in fermentation, resulting in the generation of many foods and drinks [18, 19]. It could be noted that microbes for application in fermentation can also be classified on the basis of their function during fermentation. These are employed as biosurfactants or bio-emulsifiers with vastly important functions in biomedicine and industry [20]. They are selected for the generation of targeted enzymes or metabolites that endow the targeted characteristics to fermented products [14, 19]. The choice of selection depends on the fermentation or conversion method, the appearance of the final product, and the environment. Advances in the diversity and operation of microorganisms provide us with the opportunity to utilize these microbes for novel applications in food production and other purposes [15, 19].

1.3 Fermentation Types

A wide range of fermentation methods can be employed to convert microbes into useful products, and each comes with its own advantages and applications. Solid-state fermentation is an exciting alternative for biorefineries based on lignocellulosic biomass. It has notably achieved good results in converting corn stover into carboxylates during as well when combined with processes such as extrusion pretreatment, microbial consortia, and hydrolytic enzymes [21]. This approach works particularly well for recycling agricultural waste and other solid biomass feedstock. Syngas fermentation is another big player synthetically. In this case, it uses the principle of synthesizing gas, or syngas, which you get from steam of different types of feedstocks. This approach is well-studied in the laboratory and has transitioned to commercial application for the generation of liquid fuels and platform chemicals, but it is not without, high yield, diffusion, and substrate inhibition bottlenecks [22]. Separate hydrolysis and fermentation, simultaneous saccharification and fermentation, and consolidated bioprocessing and mixed culture fermentation are some of the integrated processes considered to improve the efficiency of biomass conversion. These methods all seek to reduce the time and money spent in the processing step from lignocellulosic biomass to ethanol and products therein [23]. So, in a way, the kind of feedstock and production needed are going to determine what kind of fermentation is to be done. For lignocellulosic biomass, solid solid-state fermentation does the job well; for non-solid biomass syngas fermentation provides versatility in feedstock types that can be accommodated. Developing these fermentation technologies further is a significant step in producing biofuels and biochemicals from a broader array of origins in a more sustainable manner.

2 Feedstock Selection for Eco-Friendly Conversion

Green conversion feedstock choice relies on the consideration of numerous factors for sustainability and environmental benefits: The choice of feedstock has a significant influence on the overall efficiency and sustainability of biofuel production. Forest and agricultural residues, energy crops, and municipal solid waste lignocellulosic biomass have been explored as a cheap and sustainable feedstock for the production of biofuels and bioplastics like polyhydroxyalkanoates (PHAs) [24]. These non-food feedstocks neither compete with food crops nor consume waste materials. Algae is another third-generation feedstock with great potential since it has high productivity and can be cultivated on non-arable land [25, 26]. Surprisingly, the selection process sometimes becomes very complicated and context-dependent. For instance, while high lipid content is the prime concern in the case of microalgae-based biodiesel production, cost, policy considerations, and geography are some of the other factors. Fuzzy Multi-criteria Decision-making algorithms have been employed in the evaluation of different microalgal strains, and *Chlorella vulgaris* was among the best performers on technological and economic grounds [25]. In general, green conversion feedstocks are best if they are renewable, plentiful, and not competitive with agricultural production. They must also be convertible to efficient processes and yield high-quality products. Lignocellulosic biomass, algae, and waste streams are promising options. But the decision has to consider a number of parameters, including technological feasibility, economic viability, environmental impact, and social concerns to ensure truly sustainable and green production of biofuels [27, 28].

2.1 Types of Feedstocks

Microbial fermentation can potentially use a wide range of feedstocks to be fermented to value-added products. The feedstocks fall into the following broad categories: Lignocellulosic biomass, agricultural and forestry residues, and herbaceous energy crops represent one of the dominant types of microbial fermentation feedstocks [4]. They are rich in complex carbohydrates that require pretreatment to make them available to enzymes to be converted to fermentable sugars. Raw materials with starch and sugar are also commonly utilized for the bioethanol production via fermentation [4]. Most importantly, waste streams from diverse sources, primarily lignocellulosic biomass, have been declared as prospective feedstocks for the microbial fermentation-based production of polyhydroxyalkanoates (PHAs) [29]. Not only is it an inexpensive method for PHA production but also addresses the issue of waste treatment and sustainability. In general, microbial fermentation feedstocks range from traditional sugar- and starch-based substrates to more advanced lignocellulosic biomass and waste streams. Feedstock selection is based on availability, cost, and final product. Utilization of alternative feedstocks, i.e., waste streams, can enhance the economic viability and sustainability of microbial fermentation processes [29, 30].

3 Materials

In bioprocessing processes, materials involved in microbial fermentation, feedstock pretreatment, and process optimization are essential for the breakdown of cellulose structures and depolymerization of lignin. Chemicals like acids, alkalis, and organic solvents are involved in feedstock pretreatment to break down complex lignocellulosic biomass into fermentable sugars. Cellulase and other enzymes further hydrolyze cellulose, increasing the amount of sugar released for microbial conversion [31]. Thermal pretreatment becomes more convenient with heating equipment while grinding and milling equipment increases the surface area for chemical and enzymatic reactions by breaking down particle size. Microbial strains like *Saccharomyces cerevisiae*, *Escherichia coli*, and *Clostridium* are essential for microbial fermentation because they ferment carbohydrates into valuable chemicals like organic acids and ethanol. While pH buffers provide the optimal fermentation environment, culture media, carbon sources, and nitrogen sources provide optimal microbial growth and metabolism. For improving microbial productivity, fermenters provide a controlled environment with temperature, agitation, and pH control systems. Chromatography systems, enzymatic test kits, and analytical reagents are used in process optimization for monitoring and analyzing fermentation products. Temperature, dissolved oxygen, and pH sensors all provide real-time data to adjust [32]. Determining the optimal settings and optimizing process efficiency depends heavily on statistical design tools and data-collecting technologies. All these supplies and equipment function together to provide a well-integrated system for efficient and sustainable bioprocessing processes [33].

3.1 Summary of Materials

1. **Feedstock specimens**: Biomass like agricultural waste, lignocellulosic substances, algae, food waste, or industrial by-products.
2. **Chemicals for chemical pretreatment**: Dilute acids (e.g., sulfuric acid, hydrochloric acid), Alkalis (e.g., sodium hydroxide, ammonia), Organic solvents (e.g., ethanol, acetone), Ionic liquids, Enzymes (e.g., cellulase, hemicellulase, ligninase).
3. **Water and buffers** for pH control.
4. **Heating equipment**: Autoclave, steam explosion chamber.
5. **Grinding/milling tools**: Mechanical grinder or ball mill.
6. **Reactors**: Pretreatment reactors for chemical/thermal treatments.
7. **Analytical kits**: To measure lignin, cellulose, and hemicellulose content.
8. Safety equipment.
9. **Personal protective equipment (PPE)**: Gloves, goggles, and lab coats, fume hood for chemical-based treatments.
10. **Microbial cultures**: Bacteria (e.g., *Escherichia coli*, *Clostridium*), yeast (e.g., *Saccharomyces cerevisiae*), fungi (e.g., *Aspergillus* species).
11. **Culture media**: LB broth, YPD medium, or custom media for specific microorganisms.

12. **Sources**: Pretreated feedstock, glucose or xylose, yeast extract, peptone, ammonium salts, trace elements and vitamins, magnesium, iron, zinc, essential growth factors, pH buffers, phosphate buffers, sodium bicarbonate.
13. **Equipment**: Fermenters/bioreactors, shake flasks and incubators, pH meters, dissolved oxygen probes, sterilization equipment (autoclave), gas cylinders (for anaerobic conditions), statistical design software: Design of experiments (DOE) software (e.g., Minitab, JMP), Bioreactors with control systems (for pH, temperature, agitation), Data acquisition systems for process monitoring, Sensors (pH, temperature, dissolved oxygen), Chromatography systems (GC, HPLC), Spectrophotometer for biomass and product analysis, Response surface methodology (RSM).

4 Methods

4.1 Feedstock Pretreatment Methods

The creation of biofuel, waste management, and material processing all depend on feedstock pretreatment techniques. Through the preparation of the raw material (feedstock) for conversion or processing, these techniques aim to increase the finished product's efficiency, yield, and quality. An explanation of several feedstock pretreatment techniques is provided below. This involves mechanical, thermal, or physical operations to alter the feedstock's physical properties. In one form or another, pretreatment is required for all raw feedstocks to produce hydrogen efficiently [34]. The lignocellulosic materials' cellulose content is not easily used to produce hydrogen. Because of its inflexible crystal structure, it is pretreated physically, chemically, or biologically. In a similar vein, organic waste and wastewater derived from industrial and food waste are not utilized directly. There are numerous feedstock pretreatment techniques available to address these issues (Fig. 1). A greater percentage of easily accessible feedstocks for the microbial fermentation process that produces hydrogen is the outcome of these pretreatment process sequences. Depending on the feedstock, a mix of pretreatment techniques is occasionally used to get better results than single pretreatment techniques [35].

4.1.1 Physical Methods

To process the feedstocks without the use of chemicals or enzymes, physical pretreatment involves applying physical pressures. Physical techniques for the pretreatment of feedstocks include thermal, microwave, freeze and thaw, and ionizing radiation. Size reduction is the process of cutting, grinding, and milling to enhance conversion efficiency and surface area. Briquetting or pelletizing to enhance the handling and storage of feedstock. Thermal pretreatment includes heating procedures like drying, torrefaction, and

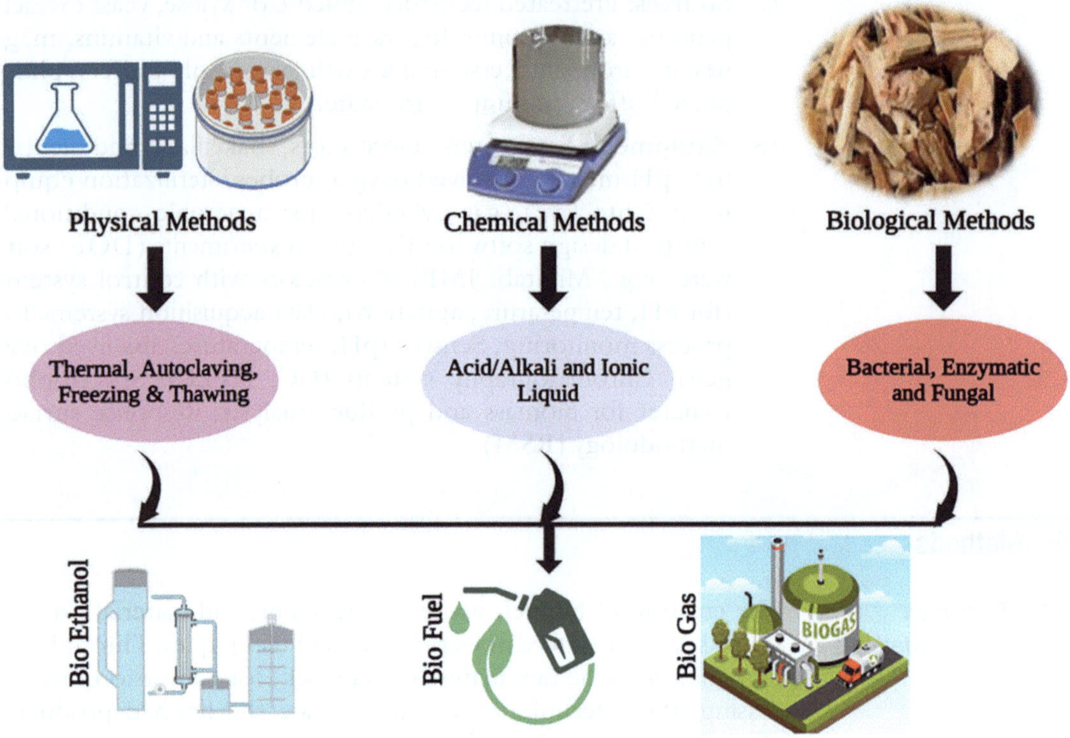

Fig. 1 Different feedstock conversion pretreatment methods

steam explosion. It is mainly used in the manufacturing of composite materials, the processing of biomass, and the creation of biofuel [36].

Thermal treatment: The most popular technique for hydrolyzing or solubilizing organic compound biomass to produce hydrogen is thermal treatment. By liquefying the biomass, thermal pretreatment breaks down the complex substrate into a simpler one and facilitates anaerobic digestion. The breakdown of the chemical connections between the cell wall and the membrane causes thermal denaturation at high temperatures [37]. In addition to increasing hydrogen generation, the greater temperature also deactivates harmful microorganisms that do not produce hydrogen. Few substrates are thermally responsive. Hence, the temperature used to putrefy biomass varies depending on the substrate. The most common temperature is 121 °C for 20–30 min; however, the temperature range for waste-activated sludge is 100–175 °C, with a duration of roughly 15–60 min for the creation of hydrogen. The temperature of cellulose-based compounds ranges widely, from 50 to 220 °C, and their duration ranges from 20 min to 24 h [38]. For cellulose-based feedstocks, thermal pretreatment is always done in conjunction with other techniques. When lignocellulose feedstocks are thermally treated over 110 °C, they emit harmful substances such as furfural, carbonic acid, and phenolic chemicals in addition to chemical hydrolysis. The harmful substances can be eliminated via adsorption with activated carbon or precipitation with calcium chloride [39].

Autoclaving: For many years, autoclaving has been the standard sterilizing method used in labs. This method uses a steam sterilization process at 121 °C and 15 psi pressure to burst microalgal cells or lyse microbial cells [40]. Depending on the necessity to use less energy, autoclaving is done at various temperatures and periods. The autoclaving process also removes undesirable microorganisms from the feedstocks and preserves the nutrients in the waste without the need for additional chemicals. Autoclaving is being used for heat treatment of food and municipal wastes. This process converts food waste into a fiber-like substance that can be used as organic fertilizer or to improve hydrogen generation. Numerous investigations have shown that more hydrogen is produced when microalgae biomass from *Scenedesmus obliquus* is sterilized before fermentation. Methanol is an inhibiting substance.

Found in pure glycerol, it can be significantly eliminated by autoclaving at 121 °C for 15 min before the fermentation of hydrogen [40].

Freezing and thawing: Waste-activated sludge and microalgal biomass are treated by freezing and thawing. Large ice crystals from the freeze-drying process increase cell enlargement, causing cell breakage. When the cell thaws at normal temperature, the internal components are released via the broken cell. The research found that thawing and freezing wastewater sludge might increase the hydrogen output by 40%, but there was no discernible improvement in the solubilization of the sludge [41]. Heavy operation and high energy consumption limit its usage in industrial applications, and there is very little evidence in the literature supporting hydrogen synthesis using microalgae in laboratory-scale trials.

4.1.2 Chemical Methods

Because of their low energy requirements and ease of scaling up, chemical pretreatments are typically chosen. Cell disruption, pigment extraction from algae, and the hydrolysis of lignocellulosic components from food waste all include the use of a variety of organic and inorganic solvents. However, this session is limited to acid and alkaline therapy because of its ease of use and lack of need for contemporary equipment in contrast to the addition of an oxidizing agent [42].

Acid/alkali pretreatment: By combining the liquefaction and saccharification processes into a single stage, acid/alkaline pretreatment uses cell wall disintegration and feedstock carbohydrate hydrolysis simultaneously. While sodium or potassium hydroxide and aqueous ammonia are frequently used in alkali pretreatment, sulfuric acid, hydrochloric acid, hydrogen peroxide, and phosphoric acid are used in acid pretreatment. While the base saponifies the ester bonds of cell wall lipids, acid hydrolysis splits the polymer molecule, causing the cellular matrix to disintegrate and the cellular crystallinity to be destroyed [43]. Previous studies reported that treating the activated sludge with 0.5% (w/v) HCl produced the most hydrogen in 6 h [44]. Claimed that, in comparison to the control, the amount of hydrogen produced rose by 3 and 2.6 times when food waste was treated at pH 2 and pH 12, respectively. According to a small number of research, waste-activated sludge treated with an alkaline solution at pH 12 produces 82.4% hydrogen. Through a fermentation process, the lignocellulose components break down into cellulose and hemicellulose for the microorganism's immediate use. At 40 °C, the strong acid dissolves the cellulose and lignin in rice straw and progressively increases the production of hydrogen, but the lignin does not degrade in diluted acid (2% concentration). Eighty percent of cellulose is efficiently and affordably solubilized by the acid treatment. Alkali treatment weakens and changes the crystalline and unstructured structure of cellulose, increasing the lignocellulose material's thermodynamic stability [45]. By expanding the surface area of cellulose and hemicellulose and removing lignin, alkali treatment in conjunction with enzyme or ultrasound may improve the anaerobic digestion of sugarcane and the production of biohydrogen. When used alone, acid and alkali treatment require a lengthy residence period and is typically performed at room temperature. Acid/alkali treatment is typically used in conjunction with thermal treatment for ease of use and high efficiency. Before fermentation, the pretreatment feedstocks should go through a neutralizing procedure to alter the pH, which is safe for the environment and/or the economy. In addition to acid/alkali treatment, feedstocks are pretreated with oxidizing agents. Through the ozonolysis process, the feedstock is dissolved using the potent oxidizing power of ozone gas. By reducing the inhibitor in the feed, ozone applied to wheat straw increases hydrogen generation by 158%. Simple sugars can be produced when the glycosidic linkages in carbohydrates are broken by the nascent oxygen produced by H_2O_2. One of ozonolysis's disadvantages is that it needs a lot of ozone and is a very costly procedure [46].

Ionic liquid pretreatment: The ability of ionic liquid pretreatment to break down cellulose under normal circumstances and produce the highest number of sugars and carbs has drawn attention in recent studies. Additionally, it tends to stay liquid over a wide range of temperatures. It has excellent thermal and chemical stability and is also known as a green solvent. The ionic liquid lowers processing costs because it is regarded as a nonvolatile, nonflammable, and recyclable substance [47]. LCBs have been pretreated using a variety of ionic liquids (ILs), including those based on imidazolium ($[(C_3N_2)Xn]^+$), pyridinium ($[(C_5N)Xn]^+$), pyrrolidinium ($[(C_4N)Xn]^+$), ammonium ($[NX_4]^+$), phosphonium ($PX_4]^+$), and sulfonium ($[SX_3]^+$). Since acidic ILs have a remarkable effect on lignin depolymerization by breaking ether linkages and upsetting the lignin structure, they have greater potential for LCB pretreatment. Using pyridinium chloride, an acidic IL, it has broken down the biomass's resistant structure by establishing strong hydrogen bonds and making room for solvent penetration, dissolving a variety of polysaccharides, including the cellulose component. Pure cellulose would dissolve in ILs due to the formation of electron donor–electron acceptor complexes between the ionic liquid and the hydroxyl groups that are naturally present in cellulose. These interactions during pretreatment may cause disruptions to its intra- and inter-H-bond network, increasing enzymatic digestibility [47]. Even for the extremely complicated structure of biomass, cellulase is thought to be deactivated by an ionic liquid concentration of greater than 10% (v/v). When employed at concentrations up to 30% (v/v), the aqueous ionic liquid, also known as 1-ethyl-3-methylimidazolium methyl phosphonate ($[C2mim][MeO(H)PO_2]$), is thought to be more compatible with enzymatic action and exhibits synergistic alacrity of cellulase diversification utilizing aqueous ionic liquids [48].

4.1.3 Biological Methods

Biological pretreatment, which does not require high temperatures or chemicals, is frequently used to treat cellulose-based feedstocks and algal biomass. However, the physical or chemical preparation is typically followed by the biological pretreatment. The pretreatment step of the biological technique necessitates a lengthy retention period, which ultimately leads to a time-consuming procedure, and the microorganisms and enzymes easily use carbohydrates [36].

Enzymatic pretreatment: Enzymes are referred to as biocatalysts because of their mode of activity. After undergoing various physical or chemical pretreatments, the feedstocks are transformed into acceptable raw materials, and the enzymatic pretreatment process is necessary as a stable method for producing hydrogen. Laccases, xylanases, and cellulose enzymes typically break down the lignin, hemicellulose, and cellulose components found in food waste and other lignocellulose-based feedstocks. Biomass swelling results from the chemical extraction of lignin because the toxic derivative of lignin in the feedstock prevents the hydrolysis of the feedstock. The lignin's structure is changed by the biomass swelling, which makes it possible for the cellulolytic enzyme to break it down. These enzymes break down the feedstock into compounds like glucose, xylose, arabinose, or mannose that have five or six carbons [49]. These five and six carbons could be used by the microbe to make biohydrogen through fermentation. Either a pure type of enzyme or bacteria that release enzymes are employed for the pretreatment procedure, which involves fermentation. Nonetheless, maintaining the right conditions for waste hydrolysis is crucial to the enzymatic pretreatment procedure. Enzyme treatment improves biodegradability, while the polyphenols, melanoids, and heavy metals in the molasses wastewater may hinder microbes. A study found that utilizing cellulose enzyme to hydrolyze rice straw yields 54.18 g/L of sugar and 2.58LH2/L of hydrogen [50].

Bacterial pretreatment: Because of its ease of genetic modification and environmental tolerance, biological pretreatment combined with bacterial treatment demonstrated greater lignin breakdown than fungal pretreatment. Numerous bacteria were utilized for bacterial pretreatment, including *Rhodococcus* sp., *Ceriporiopsis* sp., *Pandoraea* sp., *Galactomyces* sp., *Mycobacterium* sp., and *Sphingobium* sp. SYK-6 demonstrates effective lignin and polycyclic aromatic hydrocarbon breakdown before cellulose. As the carbon supply needed for the process, glucose is present in the medium in which the bacterial strain *Mycobacterium smegmatis* L2-K2 develops [51]. Cellulase, xylanase, amylase, and several other lignolytic enzymes necessary for the subsequent process of bioethanol synthesis have been produced by these bacteria after they have consumed the cell wall of biomass [52]. Pretreating biomass with *Bacillus firmus*, a bacterium known to produce xylanolytic enzymes, can boost the yield of glucan recovery by up to 74% and increase the removal of xylan by up to 30%. This, in turn, enhances the availability of cellulose for its corresponding enzymes [53]. *Galactomyces* sp. generates β-glucosidase, xylanase, and cellulase up to 15 mg/g of glucan. This has raised the output of glucose from 100 g of biomass to 74.1% and the yield of reducing sugars to 81%. The hydrolysis and fermentation processes' successive yields have grown due to the combination of biological species. Manganese peroxidase, a laccase produced by the bacterial strain *Pandoraea* sp., breaks down the lignocellulosic biomass's resistant structure and extracts lignin and hemicellulose from it. The process of lignin oxidation involves a broad range of small molecules that function as diffusible mediators and directly interact with lignin to generate both phenyl and phenoxyl radicals on the biomass. A sequence of bond scission events is subsequently triggered by these radicals. Bacteria and fungi may have distinct reactions to lignin; in other words, small-molecule oxidants may not be able to fully regulate the bacterial system [54]. *Streptomyces griseorubens* is another bacterial strain utilized for biological pretreatment; its degradative properties rely on several variables, including temperature, pH, residence time, and enzyme loading. An increase in the amount of accessible cellulose up to 64.77% and a corresponding decrease in the contents of hemicellulose and lignin up to 86.13% and 93.31%, respectively, were found to have a greater effect on the delignification rate of rice straw biomass as the incubation time increased [55].

Fungal pretreatment: Biological pretreatment with a variety of microorganisms, including fungi that cause red, white, and brown rot. While brown fungi aim to break down cellulose and hemicellulose with minimal lignin content degradation, they can degrade lignin. Similarly, cellulose, hemicellulose, and lignin have all been broken down at the same rate by white rot fungus [56]. Laccase, polyphenol oxidase, and manganese peroxidase enzyme production enable the white rot fungus to break down lignin with high efficiency. Fungal growth and metabolism, as measured by biomass particle size, incubation duration, and moisture content, are the primary determinants of fungal pretreatment efficacy [57]. For the biomass to engage in lignolytic processes, the moisture content must be appropriate. In terms of pretreatment effectiveness, strain *P. ostreatus* has a 33.4% lignin removal capacity, whereas strain *T. reesei* has a lesser removal capacity of 23.6% after 30 days of incubation. *Ceriporiopsis subvermispora*, *Phellinus pini*, *Phlebia* spp., *Pleurotus* spp., and *Trametes hirsute* are among the fungi that are utilized to break down lignin. Up to 70% more glucan is recovered by *Myrothecium roridum* [58]. *Irpex lacteus*, another white rot fungus belonging to the class Basidiomycetes, exhibits the highest xylan digestibility of approximately 82.1%, the largest lignin reduction of 41.5% in the case of maize spoilage, and the highest hydrolysis result following a 15-day fungal pretreatment [59]. Another fungus, *Trichoderma harzianum*, which is cultivated in potato dextrose media and utilized as a nitrogen source in biomass pretreatment, can produce cellulase, xylanase, and glucosidase for the hydrolysis process with the highest sugar yield and improve fermentation in 96–120 h of reaction time [59]. According to reports, the peroxidase enzyme secreted by *Phanerochaete chrysosporium* exhibits the highest lignolytic activity. After soaking at 100 °C for 1 h, the microbial pretreatment, which was aided by 4% NaOH or 1% H_2SO_4, produced the highest cellulose content of 39.4%. Therefore, it has been shown that many fungi can improve the digestibility of polysaccharides, but only a few species can break down large amounts of lignin while maintaining cellulose and some hemicellulose [60]. A similar fungal pretreatment employs the filamentous fungus *Fusarium oxysporum*, which produces a range of enzymes and has a great capacity to tolerate zinc metals. Because zinc oxide nanoparticles are utilized as reducing, capping, and stabilizing agents during fungal pretreatment, they are therefore used to increase the production of bioethanol. Cellulose recovery can reach 48.8% during the combined procedure, whereas lignin degradation ranges from 25% to 16.7% [61].

4.1.4 Mechanical Pretreatment

Increasing the surface area and decreasing crystallinity are the ultimate objectives of the mechanical pretreatment method, which is accomplished by a variety of procedures like milling, grinding, and chipping. For the efficient use of a substrate, the mechanical pretreatment approach is always supported by physical, chemical, or biological techniques.

Ultrasonication: In the ultrasonication process, ultrasonic waves are introduced for the cell rupture and particle solubilization in the frequency ranges of 10–20 kHz. The electrical energy is transformed into physical energy, resulting in the production of the bubble and cavitation in the waste due to the generation of pressure. Massive energy is released during cavitation, and strong vibration causes shear pressures and a high localized temperature to occur, shattering the cell walls. At higher frequencies, the ultrasonic waves cause damage to the biological components by producing free radicals [62]. In the food waste industry, ultrasonication is typically used to dissolve activated sludge and to lyse bacterial and algal cells for the extraction of proteins and lipids using an ice bath. By breaking the hydrogen bonds in lignocellulose feedstock, ultrasonography breaks down lignocellulosic structures and shortens the fermentation time needed to produce hydrogen. As stated in their research, they could increase the hydrogen yield by 60% and the COD solubilization in waste-activated sludge by 5% by employing ultrasonication [63]. A study used a pretreatment of ultrasonication in distillery wastewater and applied RSM methodology, leading to a 1.4-fold increase in hydrogen generation [64]. In the dark fermentation process, *Scenedesmus obliquus* YSW15 yields 2.5 mL/L of hydrogen following a 15-min pretreatment with ultrasonication (40 kHz, 2.2 kW). In a similar vein, numerous studies have shown that strong ultrasonic wave frequencies can disturb algae cells and increase hydrogen production by 25%. Aside from process variables, the size, shape, age, and density of the cell suspension all have a significant impact on how well ultrasonic waves break cells.

4.2 Microbial Fermentation Methods

A bioprocess called microbial fermentation uses microorganisms to transform organic material into useful products like organic acids, biofuels, bioplastics, and other biochemicals. The kind of feedstock, the intended product, and the process parameters all influence the choice of microbial strain and fermentation technique. The key objective is to transform feedstock's complex polysaccharides into fermentable sugars, which can then be converted into the desired biochemicals. Optimize the growth of microorganisms and metabolic pathways to achieve maximum product yield. Prevent by-product formation and improve process efficiency [65].

Feedstock preparation: Microbial treatment of feedstocks involves the biological transformation of raw materials using microorganisms to improve their fermentability and suitability for other bioprocesses. To achieve maximum utilization of feedstock for biofuel, bioplastic, and biochemical production, microbes have a central role to play in the degradation of complex biomass fractions, solubilization of nutrient sources, and detoxification of inhibitors. The main objective in feedstock preparation is the conversion of complex lignocellulosic biomass to fermentable sugars. Improve nutritional availability of feedstock to the microorganisms for growth. Toxic inhibitors that form as a result of pretreatment should be eliminated or inactivated. To improve enzymatic access, increase the feedstock's surface area and porosity [66].

Microbial culture preparation: The selection of microbes for feedstock preparation is based on their capacity to break down biomass components and generate desirable metabolites. Effective at breaking down cellulose and hemicellulose is *Bacillus subtilis*. The thermophilic cellulolytic bacteria that can break down lignocellulose is *Clostridium thermocellum*. *Pseudomonas putida* can break down lignin's aromatic components. Genetically modified strains of *Escherichia coli* that break down lignin and hemicellulose. *Trichoderma reesei* fungus produces a lot of cellulase to break down cellulose. The enzymes cellulase, xylanase, and pectinase are produced by *Aspergillus niger*. The white rot fungus *Phanerochaete chrysosporium* effectively breaks down lignin [67]. Although it can be modified for lignocellulosic feedstocks, *Saccharomyces cerevisiae* is frequently employed for sugar fermentation. Hexose and pentose sugars are fermented by *Pichia stipitis* from lignocellulosic feedstock.

In bioprocessing applications such as the manufacture of enzymes, organic acid, and biofuel, the efficient conversion of feedstocks depends on the establishment of microbial cultures. The best possible microbial growth, metabolic activity, and feedstock breakdown efficiency are guaranteed by properly prepared cultures. The culture media provide the necessary nutrients for microbial growth and activity. The process involves choosing the right microbial strain followed by creating the sterile growing media that are ideal for the chosen microbe. Subsequently, the microbial strain is added into the medium, followed by the incubation at the ideal temperature, pH, and oxygen concentrations [68].

Fermentation process: Numerous feedstock types can be transformed into important bio-products like enzymes, organic acids, biofuels, and bioplastics via microbial fermentation techniques. There are two main methods of fermentation: batch and fed batch. Each has unique characteristics, benefits, and drawbacks. The goal product, microbial strain, and feedstock type are taken into consideration when choosing these procedures [65].

Batch fermentation: Substrates are added at the beginning, and the process runs to completion. Feedstock, microbial inoculum, and nutrients are loaded into a bioreactor at the start of the batch fermentation process. Except for necessary process control modifications like pH or temperature, the fermentation continues without any additional additives. The first step is to get rid of impurities and sterilize the feedstock and bioreactor. Microbial culture is added to the sterile medium during inoculation. Subsequently, under carefully monitored circumstances, microorganisms are used to break down the material. Further, consistent monitoring of temperature, pH, microbial development, and product creation. At last, when the nutrients run out, or inhibitory by-products build up, fermentation comes to an end. This was followed by gathering the fermentation broth and processing it further to extract the desired product [69].

Continuous fermentation: Involves the continuous addition of substrates and removal of products, allowing for higher efficiency. An equivalent volume of the fermentation broth, which contains microbial biomass, metabolic by-products, and wasted nutrients, is simultaneously removed from the bioreactor while fresh feedstock is continually supplied into it. This process is known as continuous microbial fermentation. By keeping the bioreactor in a constant condition, this technique enables continuous production for long periods. The process includes feedstock preparation, where raw materials are pretreated (if necessary) to increase bioavailability for microbial fermentation [70]. Further, for inoculum development, a healthy and active microbial culture is prepared and introduced into the bioreactor. Subsequently, the continuous addition of sterilized feedstock and nutrients was performed. Moreover, maintenance of constant conditions for optimal microbial activity, such as temperature, pH, and dissolved oxygen. Then, simultaneous removal of the fermentation broth at a rate proportional to feed addition. Lastly, downstream processing was used to separate the microbial cells and recover the target bio-product [71].

4.3 Process Optimization Strategies

Microbial conversion of feedstock is required to produce biofuels, biochemicals, and other commodities from renewable feedstocks. Numerous process optimization technologies are employed to optimize productivity, yield, and cost. These range from systems biology technologies to bioreactor design, fermentation control, strain improvement, substrate optimization, product recovery, and computational strategies. For maximum financial and environmental return from microbial conversion processes, these technologies should be comprehensively understood and applied [72].

Strain improvement: Strain improvement is a fundamental component of microbial feedstock conversion. Microbial strains are enhanced by selective breeding and genetic modification to optimize their performance in industry. Through the accurate control of metabolic pathways, genetic engineering technologies such as CRISPR-Cas9, homologous recombination, and TALENs maximize substrate consumption efficiency and product titers. For instance, lignocellulosic sugars are fermented to bioethanol by engineered *Escherichia coli* strains with minimal by-product formation. Another successful technique is adaptive laboratory evolution (ALE), in which stress-resistant variants are generated by subjecting microbial cultures to selection pressures such as high substrate concentrations, inhibitors, or extremely high pH. For maximum microbial efficiency, metabolic engineering aims to redirect metabolic fluxes, inhibit competing pathways, and overexpress key enzymes [73].

Substrate optimization: The selection and optimization of feedstock are critical for efficient microbial conversion. To enhance the bioavailability of complex substrates, pretreatment methods are often necessary. Physical pretreatments, such as milling and thermal treatments, increase the biomass surface area. Chemical pretreatments, including acid or alkaline processes, facilitate the breakdown of lignin and hemicellulose, while biological pretreatments employ lignin-degrading fungi to enhance substrate accessibility. Nutrient supplementation plays a vital role in maintaining the optimal carbon-to-nitrogen ratio and providing the essential micronutrients for microbial growth. Additionally, the strategy of feedstock mixing, which combines various biomass sources, helps to balance nutritional content and reduce costs. This approach ensures that production can be sustained throughout the year, accommodating feedstocks that may vary in availability with the seasons [66].

Process optimization: Effective management of the fermentation process is essential for achieving high yield and productivity. Different fermentation methods, such as batch, fed-batch, and continuous cultures, offer distinct advantages. While batch fermentation is simple and widely used, it faces limitations due to product inhibition and substrate depletion. Fed-batch fermentation addresses these challenges by allowing for the gradual addition of substrates, thereby extending the production duration. Continuous fermentation maintains steady-state conditions, enhancing productivity but requiring sophisticated control systems. Key parameters such as temperature, dissolved oxygen levels, and pH must be meticulously regulated. Temperature influences growth rates and product formation, while pH impacts microbial metabolism and enzyme function. In aerobic processes, controlling dissolved oxygen is crucial, and optimizing agitation and aeration ensures uniform conditions and effective oxygen transfer [74].

Process surveillance and management: Maintaining optimal fermentation conditions necessitates advanced monitoring and control systems. Real-time sensors for temperature, dissolved oxygen, pH, and metabolite concentrations provide critical data for dynamic process management. Automated control systems adjust operational parameters in response to fluctuations, ensuring stable conditions. The increasing application of data analytics and machine learning to analyze process data, identify trends, and predict potential issues is noteworthy. Machine learning algorithms facilitate predictive maintenance and the optimization of fermentation parameters, thereby reducing downtime and enhancing process efficiency [75].

Design and scale-up of bioreactors: The design of bioreactors plays a crucial role in the conversion of microbial feedstock. Well-designed bioreactors ensure effective waste removal, nutrient distribution, and the transfer of mass and heat. Stirred-tank bioreactors are commonly employed due to their flexibility and scalability, while airlift and packed-bed reactors can be beneficial in specific scenarios. Microfluidic systems are gaining popularity for their ability to rapidly screen fermentation conditions on a microscale, thus shortening development timelines. Scale-up strategies must address challenges such as shear stress, heat dissipation, and limitations in oxygen transfer [69]. The successful transfer of laboratory-scale procedures to industrial production is ensured by appropriate reactor design and scaling procedures.

Systems biology approaches: Techniques derived from systems biology provide an in-depth insight into cellular responses and microbial metabolic processes. Metabolic flux analysis (MFA) enables the measurement of metabolite flow through various pathways, allowing for the identification of bottlenecks that require enhancement. Comprehensive insights into gene expression, protein levels, and metabolite concentrations under various situations are offered by omics technologies, such as transcriptomics, proteomics, and metabolomics. By simulating metabolic networks, genome-scale metabolic models (GEMs) help with process optimization and logical strain engineering. These methods speed up the creation of high-performance microbial strains and facilitate data-driven decision-making [76].

Product recovery and downstream processing: For microbial feedstock conversion to be economically viable, downstream processing and product recovery must be done effectively. Pervaporation and gas stripping are examples of in situ product removal methods that keep microbial production high and avoid product inhibition. Membrane separation techniques are effective for both concentrating products and retaining cells. Integrated bioprocessing merges production and recovery processes, resulting in reduced processing times and costs. For example, simultaneous saccharification and fermentation (SSF) combines enzymatic hydrolysis and fermentation into a single step, enhancing overall process efficiency [77]. Further streamlining is achieved through consolidated bioprocessing (CBP), which employs a single microorganism to conduct both fermentation and hydrolysis concurrently.

Integration and intensification of processes: The integration and intensification of processes lead to reductions in production time, resource usage, and operational complexity. Notable examples of such integrated approaches include consolidated bioprocessing (CBP) and simultaneous saccharification and fermentation (SSF). SSF minimizes process duration and enzyme requirements by merging enzymatic hydrolysis with fermentation. In contrast, CBP utilizes microorganisms capable of both hydrolysis and fermentation, eliminating the need for external enzymes. Additionally, coupled fermentation and gas stripping techniques enhance microbial productivity by removing inhibitory volatile compounds during fermentation. These integrated strategies significantly improve the overall efficiency of microbial feedstock conversion [78].

Computational and statistical optimization: The optimization of processes heavily relies on computational and statistical techniques. The Design of Experiments (DoE) provides a structured approach to identify critical process variables and their interactions. The Response Surface Methodology (RSM) facilitates the development of mathematical models for the concurrent optimization of multiple factors. Advanced computational techniques, such as genetic algorithms (GAs) and artificial neural networks (ANNs), are employed to model complex, nonlinear bioprocesses and identify optimal operating conditions [79]. These methodologies accelerate the optimization process and reduce the experimental workload.

4.4 Applications of Eco-Friendly Feedstock Conversion

The conversion of eco-friendly feedstocks has diverse applications across multiple industries, significantly contributing to environmental sustainability by decreasing dependence on fossil fuels and reducing waste generation. This process involves transforming sustainable, biodegradable, and renewable raw materials into commercially viable products while minimizing environmental harm. As concerns regarding pollution, resource depletion, and climate change intensify, businesses are increasingly adopting green technologies that utilize bio-based feedstocks. Key applications include the production of biofuels, bioplastics, and animal feed, which play a crucial role in fostering sustainable development and reducing reliance on non-renewable resources [55]. A detailed examination of the use of eco-friendly feedstock conversion in these sectors follows and also exemplified in Fig. 2.

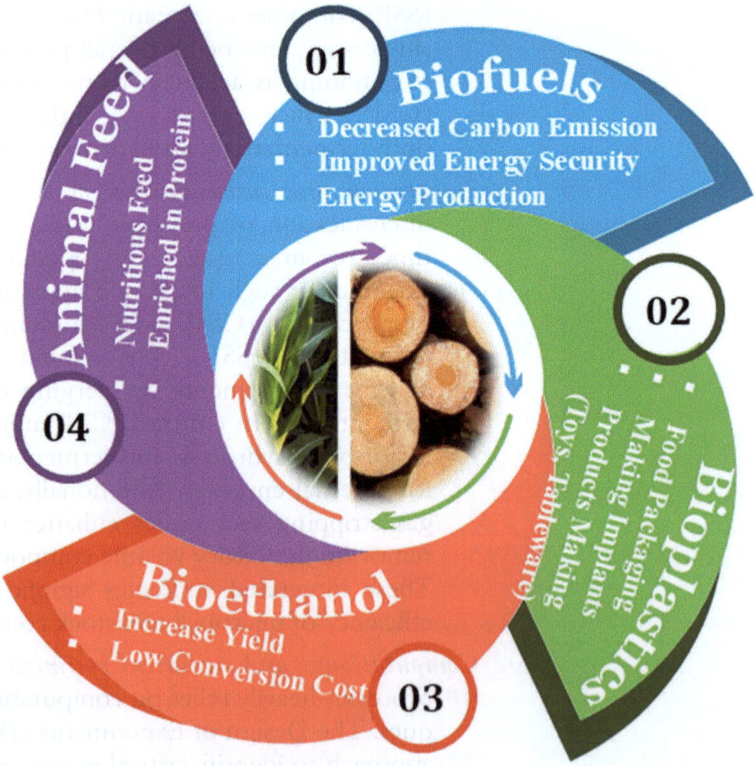

Fig. 2 Applications of eco-friendly feedstock conversion

Biofuels: Renewable biomass sources, such as agricultural by-products and algae, are processed into bioethanol, biodiesel, and biogas, offering cleaner alternatives to traditional fuels. Green feedstock conversion is the technique that in the chemical industry contributes to producing biodegradable plastics, bio-based chemicals, and solvents by converting organic matter in the effort of pollution minimization and decreasing carbon footprint. For pharmaceutical use, it applies a plant-based raw material during the synthesis process for APIs. Furthermore, eco-friendly conversion technologies are pivotal in the food business for manufacturing natural additives and useful ingredients. Waste valorization through processes like anaerobic digestion also permits the creation of renewable energy and useful by-products, such as biofertilizers, contributing to circular economy projects [62]. All things considered, implementing environmentally friendly feedstock conversion promotes sustainable growth, improves resource efficiency, and lessens environmental effects in a variety of industries.

The types of biofuels include bioethanol and biodiesel. Bioethanol is often mixed with gasoline to lower carbon emissions; bioethanol is made by fermenting the sugars and starches present in crops like corn and sugarcane. Biodiesel is made from leftover cooking oil, animal fats, and vegetable oils; biodiesel requires little modification to run diesel engines. Biogas is a flexible fuel that can be used for heating, producing electricity, and even powering vehicles. It is produced by the anaerobic digestion of organic waste. The advanced biofuels are made from non-food biomass sources like algae, municipal waste, and agricultural residues [62].

Applications of biofuels: The applications of biofuels are employed in transportation by combining biofuels with traditional fuels, carbon emissions are decreased and energy security is improved. In a power generation point, it acts as a cleaner substitute for coal and natural gas, biofuels are utilized in power plants to produce energy. A variety of industrial operations employ bio-based fuels to generate steam and heat.

Benefits to the environment and the economy include a smaller carbon footprint since biofuels produce fewer pollutants than fossil fuels. Producing biofuel from organic waste and agricultural residues lowers the amount of waste that ends up in landfills. By offering a market for agricultural waste and energy-specific crops, the biofuel sector boosts rural economies and generates jobs.

Bioplastics: Bioplastics are made from sustainable feedstocks like cellulose, sugarcane, and maize starch. Bioplastics provide a more environmentally friendly way to address the problem of plastic pollution than conventional plastics, which are petroleum-based and non-biodegradable. The types of bioplastics are starch-based bioplastics, which are biodegradable, and packaging-appropriate bioplastics that are derived from tapioca, corn, or potato starch. Polylactic acid (PLA) is made from fermented plant sugars, PLA finds extensive application in medical implants, throwaway dinnerware, and food packaging. Polyhydroxyalkanoates (PHA) are made by microbiologically fermenting carbohydrates and lipids; PHA is biodegradable and can be used for a variety of purposes, such as agricultural films and packaging. Cellulose-based bioplastics are made from plant fibers and are utilized in fibers, films, and coatings [80]. The use of bioplastics includes packaging due to their minimal environmental effect and biodegradability; bioplastics are being utilized more and more for food packaging, bottles, and containers. Bioplastic-based mulch sheets and biodegradable films lessen soil pollution and streamline waste disposal. Bioplastics are utilized to make implants, drug delivery systems, and biodegradable sutures. Bioplastics are being used in the production of products, including toys, tableware, and phone cases. When compared to conventional plastics, bioplastics have a less carbon footprint. A lot of bioplastics break down organically, which eases the strain on landfills. Bioplastics lessen reliance on limited fossil fuels by using renewable feedstocks and agricultural leftovers.

Animal feed: The creation of sustainable animal feed also heavily relies on the conversion of environmentally friendly feedstock. The livestock sector may lessen its impact on the environment while preserving the productivity and health of its animals by using food waste, agricultural residues, and other renewable resources. Nutritious animal feed can be made from crop wastes such as maize stalks, rice husks, and wheat bran. With the right processing, food scraps and surplus from the food sector can be safely turned into animal feed. Feeds made from microalgae and insects are becoming more and more popular because of their high protein content and minimal environmental impact. Glycerol and distillers dried grains with solubles (DDGS), by-products of the production of bioethanol and biodiesel, can be fed to animals. The use of balanced meals for pigs, poultry, and cattle is created using environmentally acceptable feedstocks [81]. Fish and shrimp can be raised using algae and insect-based diets, which lessens the need for conventional fishmeal. Formulations for pet food use sustainable components such as plant-based proteins and food waste derivatives. Methane emissions and landfill volumes can be decreased by turning food waste and agricultural waste into animal feed. The cattle business can lessen its need for traditional feed materials like soy and maize by using by-products and alternative feed sources. Sustainable animal production methods are facilitated by environmentally friendly feed options.

5 Notes

In the feedstock conversion, common techniques include chemical (acid/alkali treatments), biological (using fungi/enzymes), physical (milling, steam explosion), and physicochemical approaches. Various approaches for microbial fermentation are essential for turning biomass into biochemicals and biofuels. Batch, fed-batch, continuous, solid-state (SSF), and submerged fermentation (SmF) are among the methods; the choice of technique is determined by the needs of the substrate and the desired end product.

In terms of process optimization, integrated bioprocessing, microbial strain engineering, enzyme optimization, process parameter management, and techno-economic analysis to increase economic viability and efficiency. In terms of applications in biofuels (ethanol, biodiesel), bioplastics, biochemicals (lactic and succinic acid), animal feed, biofertilizers, and waste management solutions are examples of applications of eco-friendly feedstock conversion that support a sustainable bioeconomy and circular resource use.

An innovative solution to some of the most important environmental issues of our day is provided by eco-friendly feedstock conversion. The use of biofuels, bioplastics, and animal feed supports the shift to a circular economy in addition to reducing waste and maximizing resource efficiency. Even if there are still obstacles to overcome, many sustainable solutions can reach their full potential with sustained innovation, funding, and governmental support. Eco-friendly feedstock conversion will only become more important as businesses and consumers place a higher priority on environmental stewardship.

References

1. Chatterjee A, Delorenzo DM, Carr R, Moon TS (2019) Bioconversion of renewable feedstocks by Rhodococcus opacus. Curr Opin Biotechnol 64:10–16. https://doi.org/10.1016/j.copbio.2019.08.013
2. Kikuchi Y, Torizaki N, Tähkämö L, Enström A, Kuusisto S (2022) Life cycle greenhouse gas emissions of biomass- and waste-derived hydrocarbons considering uncertainties in available feedstocks. Process Saf Environ Prot 166:693–703. https://doi.org/10.1016/j.psep.2022.08.054
3. Truong CC, Verma VK, Mishra P, Suh Y-W, Mishra DK (2022) Chapter 24—Catalytic transformation of biomass-based feedstocks in green solvents. In: Biomass, biofuels, biochemicals. Elsevier, pp 673–720. https://doi.org/10.1016/b978-0-12-824419-7.00004-2
4. Bušić A, Novak S, Morzak G, Belskaya H, Komes D, Kundas S, Marđetko N, Ivančić Šantek M, Šantek B (2018) Bioethanol production from renewable raw materials and its separation and purification: a review. Food Technol Biotechnol 56(3):289–311. https://doi.org/10.17113/ftb.56.03.18.5546
5. Zhang B, Wu X-Y, Wang F, Jiang Y, Li Z (2020) Recent progress on chemical production from non-food renewable feedstocks using Corynebacterium glutamicum. Front Bioeng Biotechnol 8(115):606047. https://doi.org/10.3389/fbioe.2020.606047
6. Możejko-Ciesielska J, Ray S, Sankhyan S (2023) Recent challenges and trends of polyhydroxyalkanoate production by extremophilic bacteria using renewable feedstocks. Polymers 15(22):4385. https://doi.org/10.3390/polym15224385
7. Usman M, Cross JS, Cheng S, Boonyubol S (2023) Evaluating green solvents for bio-oil extraction: advancements, challenges, and

future perspectives. Energies 16(15):5852. https://doi.org/10.3390/en16155852

8. Chen S, Qiu H, Xiao H, He W, Mou J, Siponen M (2020) Consumption behavior of eco-friendly products and applications of ICT innovation. J Clean Prod 287:125436. https://doi.org/10.1016/j.jclepro.2020.125436
9. Nowacki M, Chawla Y, Kowalczyk-Anioł J (2021) What drives the eco-friendly tourist destination choice? The Indian perspective. Energies 14(19):6237. https://doi.org/10.3390/en14196237
10. Kazmi A, Sultana T, Ali A, Nijabat A, Li G, Hou H (2025) Innovations in bioethanol production: a comprehensive review of feedstock generations and technology advances. Energ Strat Rev 57:101634
11. Madavi TB, Chauhan S, Madathil V, Sankaranarayanan M, Navina B, Velmurugan NK et al (2025) Microbial methanotrophy: methane capture to biomanufacturing of platform chemicals and fuels. Next Energy 8:100251. https://doi.org/10.1016/j.nxener.2025.100251
12. Sayoud S, Derbal K, Panico A, Pontoni L, Fabbricino M, Pirozzi F, Benalia A (2025) The effect of hydrogen peroxide on biogas and methane produced from batch mesophilic anaerobic digestion of spent coffee grounds. Fermentation 11(2):60. https://doi.org/10.3390/fermentation11020060
13. Konwar G, Thakur V, Baghmare P, Verma JS, Geed SR, Namdeo A (2025) Biochemical conversion and cultivation strategies for microalgae biomass: challenges and future prospects in biofuel production. In: Microalgal biofuels. Woodhead Publishing, pp 165–184. https://doi.org/10.1016/B978-0-443-24110-9.00008-6
14. Begum PS, Rajagopal S, Razak MA (2020) Chapter 11—Emerging trends in microbial fermentation technologies. In: Recent developments in applied microbiology and biochemistry. Elsevier, pp 113–119. https://doi.org/10.1016/b978-0-12-821406-0.00011-4
15. Malik JA, Rahman MM, Rais N (2022) Chapter 1—Microbial biotechnology: an introduction. In: Microbes and microbial biotechnology for green remediation. Elsevier, pp 3–11. https://doi.org/10.1016/b978-0-323-90452-0.00035-9
16. Han D, Wang Y, Zhu Y, Wang K, Liu P, Guo Z, Chen K, Yu Y, Rong C, Dai S, Yu Z, Yang Y (2024) Metagenomics profiling of the microbial community and functional differences in solid-state fermentation vinegar starter (seed Pei) from different Chinese regions. Front Microbiol 15:1389737. https://doi.org/10.3389/fmicb.2024.1389737
17. Kang J, Huang X, Li R, Zhang Y, Chen X-X, Han B-Z (2024) Deciphering the core microbes and their interactions in spontaneous Baijiu fermentation: a comprehensive review. Food Res Int 188:114497. https://doi.org/10.1016/j.foodres.2024.114497
18. Flachs A, Orkin JD (2019) Fermentation and the ethnobiology of microbial entanglement. Ethnobiol Lett 10(1):35–39. https://doi.org/10.14237/ebl.10.1.2019.1481
19. Ramírez Rojas AA, Schindler D, Swidah R (2022) Microbes of traditional fermentation processes as synthetic biology chassis to tackle future food challenges. Front Bioeng Biotechnol 10(2):982975. https://doi.org/10.3389/fbioe.2022.982975
20. Adetunji AI, Olaniran AO (2020) Production and potential biotechnological applications of microbial surfactants: an overview. Saudi J Biol Sci 28(1):669–679. https://doi.org/10.1016/j.sjbs.2020.10.058
21. Marone A, Prompsy P, Joseph-Aimé M, Barakat A, Trably E, Carrère H, Bernet N, Fayoud N, Guillon F, Escudié R (2018) Enhancement of corn stover conversion to carboxylates by extrusion and biotic triggers in solid-state fermentation. Appl Microbiol Biotechnol 103(1):489–503. https://doi.org/10.1007/s00253-018-9463-x
22. Yasin M, Cha M, Chang IS, Atiyeh HK, Munasinghe P, Khanal SK (2019) Chapter 13—Syngas fermentation into biofuels and biochemicals. In: Biofuels: alternative feedstocks and conversion processes for the production of liquid and gaseous biofuels. Elsevier, pp 301–327. https://doi.org/10.1016/b978-0-12-816856-1.00013-0
23. Chen H, Wang L (2017) Chapter 7—Microbial fermentation strategies for biomass conversion. In: Technologies for biochemical conversion of biomass. Elsevier, pp 165–196. https://doi.org/10.1016/b978-0-12-802417-1.00007-7
24. Al-Battashi HS, Nguyen QD, Sivakumar N, Al-Bahry S, Gupta VK, Annamalai N, Tripathi BN (2019) Lignocellulosic biomass (LCB): a potential alternative biorefinery feedstock for polyhydroxyalkanoates production. Rev Environ Sci Biotechnol 18(1):183–205. https://doi.org/10.1007/s11157-018-09488-4
25. Kokkinos K, Karayannis V, Moustakas K (2021) Optimizing microalgal biomass feedstock selection for nanocatalytic conversion into biofuel clean energy, using fuzzy multicriteria decision making processes. Front

Energy Res 8:622210. https://doi.org/10.3389/fenrg.2020.622210

26. Zhang J, Zhang Z, Yang J, Zhang H, Zhang Y (2021) Research status and future development of biomass liquid fuels. Bioresources 16(2):4523–4543. https://doi.org/10.15376/biores.16.2.zhang
27. Alptekin FM, Celiktas MS (2022) Review on catalytic biomass gasification for hydrogen production as a sustainable energy form and social, technological, economic, environmental, and political analysis of catalysts. ACS Omega 7(29):24918–24941. https://doi.org/10.1021/acsomega.2c01538
28. Jha S, Acharya B, Dalai AK, Nanda S (2022) A review of thermochemical conversion of waste biomass to biofuels. Energies 15(17):6352. https://doi.org/10.3390/en15176352
29. Li M, Wilkins MR (2020) Recent advances in polyhydroxyalkanoate production: feedstocks, strains and process developments. Int J Biol Macromol 156:691–703. https://doi.org/10.1016/j.ijbiomac.2020.04.082
30. Hakeem IG, Sharma T, Surapaneni A, Joshi JB, Shah K, Ball AS, Sharma A, Sharma A (2023) Techno-economic analysis of biochemical conversion of biomass to biofuels and platform chemicals. Biofuels Bioprod Biorefin 17(3): 718–750. https://doi.org/10.1002/bbb.2463
31. Ojo AO (2023) An overview of lignocellulose and its biotechnological importance in high-value product production. Fermentation 9(11):990
32. Voidarou C, Antoniadou M, Rozos G, Tzora A, Skoufos I, Varzakas T, Lagiou A, Bezirtzoglou E (2020) Fermentative foods: microbiology, biochemistry, potential human health benefits and public health issues. Food Secur 10(1):69
33. Dar AA, Yadav P, Aswani A, Tashi Wangmo A (2024) Optimizing processes and products: the role of DOE. Insight Statistics 7(1):644
34. Ramos A, Monteiro E, Rouboa A (2022) Biomass pre-treatment techniques for the production of biofuels using thermal conversion methods—a review. Energy Convers Manag 270:116271
35. Mujtaba M, Fraceto LF, Fazeli M, Mukherjee S, Savassa SM, de Medeiros GA, Santo Pereira AD, Mancini SD, Lipponen J, Vilaplana F (2023) Lignocellulosic biomass from agricultural waste to the circular economy: a review with focus on biofuels, biocomposites and bioplastics. J Clean Prod 402: 136815
36. Abolore RS, Jaiswal S, Jaiswal AK (2024) Green and sustainable pretreatment methods for cellulose extraction from lignocellulosic biomass and its applications: a review. Carbohydr Polym Technol Applic 7:100396
37. Scapini T, Dos Santos MS, Bonatto C, Wancura JH, Mulinari J, Camargo AF, Klanovicz N, Zabot GL, Tres MV, Fongaro G, Treichel H (2021) Hydrothermal pretreatment of lignocellulosic biomass for hemicellulose recovery. Bioresour Technol 342:126033
38. Asabuwa Ngwabebhoh F, Saha N, Saha T, Saha P (2022) Bio-innovation of new-generation nonwoven natural fibrous materials for the footwear industry: current state-of-the-art and sustainability panorama. J Nat Fibers 19(13): 4897–4907
39. Martín C, Dixit P, Momayez F, Jönsson LJ (2022) Hydrothermal pretreatment of lignocellulosic feedstocks to facilitate biochemical conversion. Front Bioeng Biotechnol 10: 846592
40. Swenson VA, Stacy AD, Gaylor MO, Ushijima B, Philmus B, Cozy LM, Videau NM, Videau P (2018) Assessment and verification of commercially available pressure cookers for laboratory sterilization. PLoS One 13(12): e0208769
41. Aljabri H, Cherif M, Siddiqui SA, Bounnit T, Saadaoui I (2023) Evidence of the drying technique's impact on the biomass quality of Tetraselmis subcordiformis (Chlorophyceae). Biotechnol Biofuels Bioprod 16(1):85
42. Bora A, Rajan AS, Kumar P, Muthusamy G, Alagarsamy A (2024) Microalgae to bioenergy production: recent advances, influencing parameters, utilization of wastewater—a critical review. Sci Total Environ 946:174230
43. Roy R, Rahman MS, Raynie DE (2020) Recent advances of greener pretreatment technologies of lignocellulose. Curr Res Green Sustain Chem 3:100035
44. El-Kebeer AA, Mahmoud UF, Ismail S, Jalal AA, Kowal P, Al-Hazmi HE, Hassan GK (2024) Maximizing bio-hydrogen and energy yields obtained in a self-fermented anaerobic bioreactor by screening of different sewage sludge pretreatment methods. Processes 12(1):118
45. Babcock-Jackson L, Konovalova T, Krogman JP, Bird R, Díaz LL (2023) Sustainable fertilizers: publication landscape on wastes as nutrient sources, wastewater treatment processes for nutrient recovery, biorefineries, and green ammonia synthesis. J Agric Food Chem 71(22):8265–8296

46. Karthikeyan PK, Bandulasena HC, Radu T (2024) A comparative analysis of pre-treatment technologies for enhanced biogas production from anaerobic digestion of lignocellulosic waste. Ind Crop Prod 215: 118591
47. Taokaew S, Kriangkrai W (2022) Recent progress in processing cellulose using ionic liquids as solvents. Polysaccharides 3(4):671–691
48. Baruah J, Nath BK, Sharma R, Kumar S, Deka RC, Baruah DC, Kalita E (2018) Recent trends in the pretreatment of lignocellulosic biomass for value-added products. Front Energy Res 6: 141
49. Almajanni Y, Amiri H, Taheri-Kafrani A (2025) Efficient and cost-effective biodegradation of phenolic compounds in lignocellulosic biomasses using laccase immobilized onto magnetically recoverable nanocellulose-functionalized iron-oxide nanoparticles. Ind Crop Prod 223:120256
50. Nirmala N, Praveen G, AmitKumar S, SundarRajan P, Baskaran A, Priyadharsini P, SanjayKumar S, Dawn S, Pavithra KG, Arun J, Pugazhendhi A (2023) A review on biological biohydrogen production: outlook on genetic strain enhancements, reactor model and techno-economics analysis. Sci Total Environ 896:165143
51. Rath S, Pradhan D, Du H, Mohapatra S, Thatoi H (2024) Transforming lignin into value-added products: perspectives on lignin chemistry, lignin-based biocomposites, and pathways for augmenting ligninolytic enzyme production. Adv Compos Hybrid Mater 7(1):27
52. Benatti AL, Polizeli MD (2023) Lignocellulolytic biocatalysts: the main players involved in multiple biotechnological processes for biomass valorization. Microorganisms 11(1):162
53. Teeravivattanakit T, Baramee S, Ketbot P, Waeonukul R, Pason P, Tachaapaikoon C, Ratanakhanokchai K, Phitsuwan P (2022) Digestibility of Bacillus firmus K-1 pretreated rice straw by different commercial cellulase cocktails. Prep Biochem Biotechnol 52(5): 508–513
54. Grgas D, Rukavina M, Bešlo D, Štefanac T, Crnek V, Šikić T, Habuda-Stanić M, Landeka Dragičević T (2023) The bacterial degradation of lignin—a review. Water 15(7):1272
55. Shukla A, Kumar D, Girdhar M, Kumar A, Goyal A, Malik T, Mohan A (2023) Strategies of pretreatment of feedstocks for optimized bioethanol production: distinct and integrated approaches. Biotechnol Biofuels Bioprod 16(1):44
56. Qi J, Li F, Jia L, Zhang X, Deng S, Luo B, Zhou Y, Fan M, Xia Y (2023) Fungal selectivity and biodegradation effects by white and brown rot fungi for wood biomass pretreatment. Polymers 15(8):1957
57. Benavides V, Ciudad G, Pinto-Ibieta F, Robledo T, Rubilar O, Serrano A (2024) Enhancing laccase and manganese peroxidase activity in white-rot fungi: the role of copper, manganese, and lignocellulosic substrates. Agronomy 14(11):2562
58. Ferdeş M, Dincă MN, Moiceanu G, Zăbavă BŞ, Paraschiv G (2020) Microorganisms and enzymes used in the biological pretreatment of the substrate to enhance biogas production: a review. Sustainability 12(17):7205
59. Mushlihah S, Husain DR, Langford A, Tassakka AC (2020) Fungal pretreatment as a sustainable and low cost option for bioethanol production from marine algae. J Clean Prod 265:121763
60. Zhang Q, Wei Y, Han H, Weng C (2018) Enhancing bioethanol production from water hyacinth by new combined pretreatment methods. Bioresour Technol 251:358–363
61. Gupta K, Chundawat TS (2020) Zinc oxide nanoparticles synthesized using Fusarium oxysporum to enhance bioethanol production from rice-straw. Biomass Bioenergy 143: 105840
62. Lari Z, Ahmadzadeh H, Hosseini M (2019) Cell wall disruption: a critical upstream process for biofuel production. In: Advances in feedstock conversion technologies for alternative fuels and bioproducts. Woodhead Publishing, pp 21–35
63. Kotay SM, Das D (2009) Novel dark fermentation involving bioaugmentation with constructed bacterial consortium for enhanced biohydrogen production from pretreated sewage sludge. Int J Hydrog Energy 34(17): 7489–7496
64. Gadhe A, Sonawane SS, Varma MN (2014) Ultrasonic pretreatment for an enhancement of biohydrogen production from complex food waste. Int J Hydrog Energy 39(15): 7721–7729
65. Thomas AP, Kasa VP, Dubey BK, Sen R, Sarmah AK (2023) Synthesis and commercialization of bioplastics: organic waste as a sustainable feedstock. Sci Total Environ 904: 167243
66. Khunnonkwao P, Thitiprasert S, Jaiaue P, Khumrangsee K, Cheirsilp B, Thongchul N (2024) The outlooks and key challenges in renewable biomass feedstock utilization for

value-added platform chemical via bioprocesses. Heliyon 10(10):e30830

67. Jahangeer M, Rehman MU, Nelofer R, Nadeem M, Munir B, Smułek W, Jesionowski T, Qamar SA (2024) Biotransformation of lignocellulosic biomass to value-added bioproducts: insights into bio-saccharification strategies and potential concerns. Topics Catal:1–22. https://doi.org/10.1007/s11244-024-01941-9
68. Tsegaye KN, Alemnew M, Berhane N (2024) Saccharomyces cerevisiae for lignocellulosic ethanol production: a look at key attributes and genome shuffling. Front Bioeng Biotechnol 12:1466644
69. Palladino F, Marcelino PR, Schlogl AE, José ÁH, Rodrigues RD, Fabrino DL, Santos IJ, Rosa CA (2024) Bioreactors: applications and innovations for a sustainable and healthy future—a critical review. Appl Sci 14(20):9346
70. Olszewska-Widdrat A, Alexandri M, López-Gómez JP, Schneider R, Venus J (2020) Batch and continuous lactic acid fermentation based on a multi-substrate approach. Microorganisms 8(7):1084
71. Couvert O, Divanac'h ML, Lochardet A, Thuault D, Huchet V (2019) Modelling the effect of oxygen concentration on bacterial growth rates. Food Microbiol 77:21–25
72. Kamalesh R, Shaji A, Saravanan A, Vickram AS, Yaashikaa PR (2024) Advances in engineered microbes for sustainable biofuel production: current research and future outlook on lignocellulose utilization. Ind Crop Prod 222: 119988
73. Jeong BR, Jang J, Jin E (2023) Genome engineering via gene editing technologies in microalgae. Bioresour Technol 373:128701
74. Bolmanis E, Dubencovs K, Suleiko A, Vanags J (2023) Model predictive control—a stand out among competitors for fed-batch fermentation improvement. Fermentation 9(3):206
75. Reyes SJ, Durocher Y, Pham PL, Henry O (2022) Modern sensor tools and techniques for monitoring, controlling, and improving cell culture processes. Processes 10(2):189
76. de Falco B, Giannino F, Carteni F, Mazzoleni S, Kim DH (2022) Metabolic flux analysis: a comprehensive review on sample preparation, analytical techniques, data analysis, computational modelling, and main application areas. RSC Adv 12(39):25528–25548
77. Straathof AJ, Janković T, Kiss AA (2025) Distillation for in situ recovery of volatile fermentation products. Trends Biotechnol. https://doi.org/10.1016/j.tibtech.2024.12.009
78. Susmozas A, Martín-Sampedro R, Ibarra D, Eugenio ME, Iglesias R, Manzanares P, Moreno AD (2020) Process strategies for the transition of 1G to advanced bioethanol production. Processes 8(10):1310
79. Yolmeh M, Jafari SM (2017) Applications of response surface methodology in the food industry processes. Food Bioprocess Technol 10(3):413–433
80. Zhao X, Wang Y, Chen X, Yu X, Li W, Zhang S, Meng X, Zhao ZM, Dong T, Anderson A, Aiyedun A (2023) Sustainable bioplastics derived from renewable natural resources for food packaging. Matter 6(1):97–127
81. Nath PC, Ojha A, Debnath S, Sharma M, Nayak PK, Sridhar K, Inbaraj BS (2023) Valorization of food waste as animal feed: a step towards sustainable food waste management and circular bioeconomy. Animals 13(8):1366

Chapter 12

Biorefinery Protocols for Green Biofuel Production from Food Waste

Aminu Abdullahi, Saranyadevi Subburaj, and Tarangini Korumilli

Abstract

The escalating global energy demand, coupled with environmental and regulatory principles governing the use of fossil fuels have intensified the need to uncover sustainable and clean energy alternatives. Food waste, which forms significant portion of global waste, has emerged as promising biomass for generating eco-friendly bioenergy through advanced biorefinery technologies. This chapter offers a comprehensive evaluation of the diverse biorefinery techniques employed to convert food waste into valuable bioenergy products, including biodiesel, bioethanol, and biogas. It delves into significant biochemical and thermo-chemical conversion technologies, including anaerobic digestion, fermentation, transesterification, and pyrolysis with emphasis on their essential mechanisms and efficiencies in bioenergy production from food waste (FW). Important physical and biochemical parameters such as temperature, pH, and substrate concentration affecting the process, biofuel yields, and quality are also discussed. It also, further investigates the ecological and fiscal ramifications of incorporating biorefinery technologies within contemporary waste management systems, emphasizing their significance in alleviating greenhouse gas emissions and promoting a circular bioeconomy. Through a comprehensive examination, this chapter seeks to promote the advancement of sustainable biofuel generation using food waste.

Key words Biorefinery, Bioenergy, Food waste, Biofuels, Bioeconomy

1 Introduction

Food waste presents not only an environmental and economic challenge but also engenders ethical dilemmas, as the food that is wasted could alternatively be utilized to combat global hunger. Food waste (FW) refers to materials that have become spoiled, contaminated, discarded, or lost, thus making them unfit for any type of human consumption [1]. As reported by the United Nations Food and Agriculture Organization (UNFAO), an

The original version of the chapter has been revised. A correction to this chapter can be found at https://doi.org/10.1007/978-1-0716-4646-5_17

Tanmay Sarkar (ed.), *Food Waste Valorization: Green Techniques in Sustainable Management*, Methods and Protocols in Food Science, https://doi.org/10.1007/978-1-0716-4646-5_12,
© The Author(s), under exclusive license to Springer Science+Business Media, LLC, part of Springer Nature 2025, Corrected Publication 2025

estimated 1.3 billion tons of food waste are produced, which constitutes nearly one-third of the total food generated worldwide for human consumption. According to the FAO (2020), approximately 14% of all food produced globally spoils the supply chain before reaching to market [2]. This significant level of waste not only imposes a substantial economic strain but also exacerbates critical environmental challenges, including increased greenhouse gas emissions, diminishing water resources, and suboptimal land usage.

Climate change and energy security challenges due to over dependence on fossil base fuel have demanded urgent need to explore sustainable alternatives. FW is a rich source of organic materials, such as carbohydrates, proteins, lipids, and cellulose, which can be converted into various forms of biofuels, such as biogas, bioethanol, biohydrogen, and biodiesel [3]. In contrast to traditional biofuel sources such as grains and agricultural waste, which are potential human and animal feed resources and necessitate considerable land, water, and energy, for cultivation, the utilization of food waste as biomass for bioenergy production offers a more sustainable and economically viable alternative [4]. Instead of allowing food waste to undergo anaerobic decomposition in landfills, which results in the release of methane, a formidable greenhouse gas, it can be transformed into bioenergy through various technologies such as anaerobic digestion, fermentation, and transesterification [5]. This technique not only reduces environmental pollution, but also reconstitutes waste into economic benefits.

Food waste is a heterogeneous and dynamic material with varying compositions and properties, which greatly impact its suitability for bioenergy production. A thorough understanding of these characteristics is essential to design efficient management approaches and treatment methods. The subsequent sections discuss the composition and attributes of food waste, emphasizing its variability and consequences for its utilization.

The feasibility of utilizing food waste as a substrate for biofuel production has been validated by numerous research studies [6, 7]. For instance, the anaerobic digestion techniques employed for food waste can generate biogas containing like methane [8], which is suitable for electricity generation or serves as a renewable clean alternative natural gas. Similarly, the fermentation of carbohydrate-rich food waste can lead to the production of bioethanol [9], and waste rich in lipids are utilized as biomass for biodiesel production through transesterification [10].

These techniques not only produce bioenergy but also result in some essential by-products, such as glycerol and organic acids, which can be used as fertilizers or employed in numerous industrial applications to augment the overall sustainability and circular economy.

While food waste shows promise for biofuel production, several obstacles hinder its widespread adoption. The diverse composition of food waste makes processing complex, efficient systems for collection and sorting are necessary, and both financial and technological hurdles impede the scaling up of production processes [11].

Furthermore, regulatory structures and societal awareness are critical in advancing the integration of food waste conversion to biofuel technologies. It's also imperative for research institutes and energy industries to focus on innovations and refining conversion technologies for enhance efficiency and economic viability.

This chapter examines the current advancements in technologies designed to transform food waste into bioenergy. It delves into the environmental and economic advantages of these processes while also addressing the challenges needed to realize their optimum potential.

1.1 Composition of Food Waste (FW)

Food waste typically contains a mix of carbohydrates, proteins, fats, and fibers [12]. While animal-based food waste is rich in protein and lipids but low in starch, plant-based waste can vary significantly, being either high in lignin or carbohydrates [13]. Carbohydrate-rich food residues are suitable for biogas generation, while lignin-rich residues are ideal for bioethanol generation [14]. The cellulose present predominantly in plant food waste can be hydrolyzed to release fermentable sugars for bioethanol generation. Moreover, proteins and lipid from animal waste are potential substrate to biofuel production, particularly in anaerobic digestion processes [15]. Depending on the type of FW, the total solid (TS) and volatile solid (VS) levels detected ranges from 18.1% to 30.9% and 17.1% to 26.35%, respectively, which implies that water accounts for an estimated 70–80% of the food waste composition [16]. Food waste are often characterized with moisture content which is critical for energy generation through anaerobic digestion and can be as high as 87% [2]. Table 1 describes some essential composition of plant and animal-based FW.

1.2 Properties of Food Waste (FW)

1. Degradability: The degradation properties of food waste are crucial for effective waste management and environmental sustainability. Food waste is primarily composed of organic materials, making it highly susceptible to biodegradation through microbial actions [13]. In addition, its high-moisture content makes it prone to degradation and can be efficiently processed through various treatment technologies such as composting and anaerobic digestion. Understanding these properties is essential for optimizing waste treatment and bioconversion and volarization of food waste to essential products. Plant-based food waste generally degrades faster than animal waste, which has implications for composting and anaerobic digestion processes [17].
2. Functional Properties: Parts of fruits and vegetables often considered waste possess functional properties, including antioxidant and anti-inflammatory effects, enhancing their value beyond mere disposal. Numerous studies have highlighted the functional properties of food waste, challenging the

Table 1
Properties of plant versus animal-based food waste

Properties	Plant-based food waste	Animal-based food waste
Major components	High in carbohydrates, fiber, and water	High in proteins, fats, and moisture
Organic matter	Rich in cellulose, hemicellulose, lignin, and pectin	Contains proteins, lipids, and collagen
Moisture	Generally high (varies by fruit, vegetable, or grain)	Typically high, especially in meat and dairy
Protein	Low to moderate (higher in legumes, grains, and nuts)	High (meat, dairy, and seafood waste)
Lipid content	Low to moderate (depends on nuts, seeds, and avocados)	High (especially in meat, dairy, and fish waste)
Carbohydrate	High (sugars, starches, and fiber)	Low (except in dairy waste like lactose)
Fiber	High (cellulose, lignin, pectin)	Negligible
Ash	Moderate (depends on mineral composition)	Rich (bone waste contributes to ash content)
Decomposition rate	Rapid (especially fruits and vegetables)	Slower due to protein and fat composition
Environmental impact	Lower methane production, but can cause leachate issues	Higher methane and ammonia emissions
Nutrient recovery potential	High for composting and bioenergy	High for production of biogas and fertilizer

perception of its insignificance as mere waste. For instance, [18] investigated the nutritional and functional properties of household fruits and vegetables, revealing a range of beneficial attributes. These include anticancer, antimicrobial, antiviral, and anti-inflammatory effects, as well as potential benefits in preventing neurological disorders, obesity, diabetes, and aging. Such findings underscore the significant health advantages of food waste beyond its basic nutritional value.

3. Chemical Properties: Food waste exhibits diverse chemical properties that can be harnessed for various applications, including environmental remediation, soil enhancement, and energy production. The characterization of food waste reveals its potential as a resource rather than a liability, with specific properties that can be utilized in innovative ways. Plant waste has lower ash but higher volatile matter, while animal waste contains more nitrogen, making it a valuable nutrient source for soil enhancement. Using FTIR analysis, the study identifies key functional groups including carboxylic acids, amines, and hydroxyl groups that enhance the reactivity and adsorption capabilities of these waste materials.

1.3 Trends in Food Waste Generation and Management

Food waste (FW) has become a global concern for both developed and developing countries. The generation and management of food waste represent a pressing global challenge. In terms of ecological impact, food waste contributes to approximately 8% of the total greenhouse gas emissions produced globally [2]. However, sustainable practices are increasingly emphasized, focusing on waste reduction, advanced treatment methods, and valorization techniques to transform waste into valuable resources [19]. According to (FAO) 2019, assessment of food waste approximately 14% of global food are unused or wasted [20]. Current trends toward reduction in food waste focus on integrating behavioral, cultural, and technological perspectives into comprehensive, interdisciplinary solutions.

1. Household contributions: Households are significant contributors to food waste, influenced by sociodemographic factors and consumer behaviors. The amount of food wasted by individuals is substantial, with estimates indicating approximately 74 kg per person annually [2]. Household food waste generally occurs in three distinct stages:
   (a) Logistic: During the procurement phase, the process of over procurement, due to factors such as bulk purchasing, promotional incentives, or inadequate meal preparation, frequently results in an accumulation of food that cannot be utilized prior to deterioration. Unplanned purchases and insufficient examination of food expiration timelines further intensify this predicament.
   (b) Storage: Inadequate storage techniques, including insufficient refrigeration, inappropriate packaging, or disorganized arrangement of food products, can significantly expedite the process of spoilage. Furthermore, a deficiency in understanding the ideal storage conditions pertinent to various food categories also plays a substantial role in generating avoidable waste.
   (c) Consumption or serving: At the phase of consumption, the prevalence of excessively large portions, particularized dietary preferences, and an aversion to the utilization of leftover food culminate in substantial quantities of consumable food being wasted. Moreover, cultural perceptions regarding food esthetics, including the rejection of produce deemed imperfect or "ugly," exacerbate the issue.
2. Food waste from industry: The food service industry is the third largest contributor to food waste, with projections indicating that waste could double by 2050 if no action is taken [21]. The industrial facet of the food supply chain comprises processing, manufacturing, packaging, and distribution, each of which reveals distinct possibilities for waste production.

(a) Processing food waste: During the course of food processing, various by-products including peels, seeds, and trimmings are frequently deemed as waste and discarded. The presence of inefficient machinery, inadvertent human errors, and rigorous quality control protocols can additionally contribute to considerable levels of waste generation. For instance, food product that is irregularly shaped or improperly sized may be dismissed despite its edibility.

(b) Food waste from packaging: The process of packaging is integral to maintaining the integrity of food quality; however, the prevalence of excessive or non-recyclable packaging is a significant contributor to environmental waste. Furthermore, any damage incurred during the packaging operations may result in food being deemed unsellable, consequently necessitating its disposal.

(c) Food waste due to logistics: Deficiencies in transportation and storage, including inadequate temperature regulation or delays, can lead to deterioration of goods. Excessive inventory levels and suboptimal inventory management significantly intensify waste during this phase.

1.4 Significance of Biorefinery

The biorefinery strategy constitutes an essential element of the circular bioeconomy, emphasizing the necessity of resource sustainability and the retrieval of agro-industrial by-products. This strategy not only lessens environmental degradation but also strengthens energy security and fosters sustainable development [22].

Significance of biorefineries includes:

1. Environmental sustainability: The implementation of biorefineries plays a crucial role in combatting greenhouse gas emissions and facilitates the utilization of renewable resources.
2. Waste management: By harnessing organic food waste, transforming it into biofuels, biochemicals, and various other economically advantageous products. Biorefinery significantly help in reducing landfill dependency and mitigating environmental degradation.
3. Economic opportunities: They generate products that possess additional value, thereby diversifying sources of revenue and augmenting economic values.

2 Materials

The FW biomass for the production of biofuels can be sourced from local eateries, residential households, marketplaces, and food processing industries. This food waste comprises carbohydrates (inclusive of starch and cellulose), lipids, and proteins. Microbial species, such as *Saccharomyces cerevisiae* (bioethanol generation) and *Clostridium* spp. (use in the production of butanol) are involved in the fermentation and can be obtained from commercial suppliers or isolated within controlled laboratory settings. Commercial enzymes like amylase, cellulase, and pectinase serve essential functions in hydrolysis. Additionally, chemical agents such as sulfuric acid (H_2SO_3), notable organic acids, and alkalis (like sodium hydroxide) are key for the pretreatment and hydrolysis of FW biomass. In thermochemical conversion processes, catalysts such as zeolites, metal oxides (ZnO, Al_2O_3), and activated carbon might be required.

2.1 Summary of Materials

1. **Feedstock**: Leftovers, fruit and vegetable waste.
2. **Chemicals**: Sulfuric acid (H_2SO_4), alkali, and organic acids for pretreatment to break down hemicellulose to enable enzymatic hydrolysis.
3. **Enzymes**: Cellulose, amylase, pectinase, and lipase to break starch into simple sugars for fermentation.
4. **Microbial culture**: Bacteria (e.g., *Clostridium* spp. and *Zymomonas mobilis*), fungi (e.g., *Saccharomyces cerevisiae*) as fermenters of sugars into bioethanol or alcohol fuels.
5. **Microbial culture media**: YPD (Yeast Extract Peptone Dextrose), LB broth, and selective media like reinforced clostridial medium (RCM).
6. **Nutrient sources**: Pretreated biomass/feedstock, glucose, volatile fatty acids, essential nutrients e.g., Yeast extract, sodium (Na^+), nickel (Ni^{2+}), and trace metal elements (e.g., Zinc (Zn^{2+}), copper (Cu^{2+})).
7. **Catalyst**: Zeolites, alumina, novel catalyst (e.g., platinum (Pt) and palladium (Pd)).
8. **Water and buffers**: Phosphate buffers, carbonate buffers, acetate buffers, and tris buffers for pH adjustment.
9. **Reactors**: Examples include batch reactors, continuous stirred-tank reactors, continuous-flow digesters, and fixed-bed and fluidized-bed reactors.
10. **Analytical kits**: Proximate kit analysis for cellulose, hemicellulose, and reducing sugars test kits.
11. **Biosafety materials**: Safety goggles/face shields, closed-toe shoes, hand gloves, biohazard spill kits.
12. **Equipment**: Homogenizer, autoclave, thermal, hydrolyzer, pH meter, thermometer, centrifuge, automated bioreactor (with controlled temperature, pH, and agitation), distillation column, membrane filter, gasifier, pyrolysis reactor, moisture analyzer.

3 Method

3.1 Feedstock Pretreatment

Pretreatment constitutes an essential phase within biorefinery protocols as it significantly improves the availability of biomass constituents for hydrolysis and ensuring effectiveness of subsequent processes such as hydrolysis, aerobic digestion, and fermentation.

The most commonly employed methods of biomass pretreatment are: (1) Physical, (2) Chemical, (3) Biological, and (4) Physicochemical techniques.

3.1.1 Physical Pretreatment

Physical pretreatment involves the use of mechanical or thermal techniques to break down FW biomass, making it more susceptible to subsequent processing. The technique entails the processes of shredding, milling, or grinding to decrease particle size, thus increasing the surface area for enzymatic or microbial interactions. The major advantage of physical pretreatment is the non-involvement of chemicals, which helps reduce the environmental impact of pretreatment. However, high energy consumption, difficult to scale up and the inability to completely act on highly lignocellulosic biomass has lowered its wide adoption in biofuel production.

Notable methods include extrusion, steam explosion (SE), and hydrothermal pretreatment.

1. Extrusion: This established techniques employed in the generation of both char and gaseous products. It involves processing of lignocellulosic biomass under at temperatures beyond 300 °C, utilizing a synergistic approach of shearing and mixing to facilitate the removal and reduction of biomass fibers [23].
2. Steam explosion (SE): This is an eco-friendly pretreatment method for lignocellulosic biomass, involving water, high temperature, and high steam pressure. Chemicals are not involved, thereby minimizing environmental impact. In this process, biomass is treated in a reactor for seconds to minutes, followed by rapid pressure release. This causes partial hemicellulose hydrolysis and lignin breakdown, enhancing cellulose accessibility for further processing.
3. Hydrothermal: Biomass can be subjected to hydrothermal pretreatment, which employs high-temperature water or steam to break down hemicellulose and partially decompose lignin, thus making cellulose more accessible for hydrolysis. The treated biomass is then quickly cooled to maintain the structural changes. When temperatures surpass 170 °C during this procedure, acid is formed from acetyl groups in hemicellulose which further acts as a catalyst for the hydrolysis reaction. This acetic acid further weakens the biomass structure by eliminating additional hemicellulose, consequently improving the ability to convert cellulose.

4. Microwave-assisted pretreatment: Microwave irradiation creates heat directly inside food waste, making it a quick and energy-efficient process. It heats the biomass volumetrically, microwave heating interacts with both the surface and the core of the heated material and this disrupts hydrogen bonds to break down complex structures making it more accessible for subsequent processing [24].

3.1.2 Chemical Pretreatment

Chemical pretreatment methods are regarded as cost-effective due to their efficiency in degrading complex substrates. These methods involve the utilization of alkalis, acids, and ionic liquids to break down the structural component of the biomass.

1. Acids pretreatment: The most commonly adopted chemical pretreatment is acid and alkaline pretreatment particularly for lignocellulosic biomass. Furthermore, this approach is particularly significant within biorefinery system, as it enhances the effective separation of the FW cellulose, hemicellulose, and lignin. Acids employed mechanisms of hemicellulose hydrolysis into simpler sugars (such as xylose and arabinose), partial solubilization of lignin and reduced cellulose crystallinity. Sulfuric acid (H_2SO_3) and hydrochloric acid (HCl) are most commonly used because of their effectiveness and cost friendliness. However, issues surrounding neutralization coupled with the formation of inhibitory substances such as furfural and hydroxymethylfurfural, which can suppress subsequent fermentation processes, remains a critical challenge for the adoption of acid pretreatment.
2. Alkaline pretreatment: Alkaline pretreatment primarily modifies the lignin architecture, enhancing the accessibility of cellulose and hemicellulose by saponification of intermolecular ester bonds within lignin and hemicellulose, lignin solubilization, and swelling cellulose fibers. Sodium hydroxide (NaOH), potassium hydroxide (KOH), ammonium hydroxide (NH_4OH), and calcium hydroxide ($Ca(OH)_2$) also known as lime pretreatment are most dominant examples of alkaline pretreatment.
3. Ionic liquids (ILs) pretreatment: This method has demonstrated efficacy across a variety of biomass including agricultural and plant-based FW. ILs have demonstrated the capability to selectively extract lignin from rice straw and fruit waste thereby considerably enhancing saccharification efficiencies. Their capacity for reusability across several cycles positions ILs as a sustainable alternative.

3.1.3 Biological Methods

Biological pretreatment deploys microorganisms or enzymes to effectively break down the complex chemical structures of lignocellulosic biomass, making it more susceptible to hydrolysis for biofuel production [25]. It's an effective technique for pretreatment of plant-based FW, which often characterized with mixture of lignocellulosic materials.

Microorganisms such as fungi and bacteria plays important role in decomposing cellulose, hemicellulose, and lignin. Notably, filamentous fungi, especially those classified as Basidiomycetes, are characterized by their exceptional enzymatic systems that facilitate efficient delignification and biomass transformation. Also, enzymes such as cellulases, hemicellulases, and ligninolytic enzymes are used to break down the biomass into simpler sugars particularly for bioethanol production. Unlike physical and chemical pretreatment methods, biological pretreatment does not require energy or release toxic chemicals, making it an eco-friendly option. However, its time consuming and the needs for high enzyme loadings due to the recalcitrance of lignin can increase the cost of the overall production.

3.2 Biofuel Production from FW

Subsequent to the pretreatment of FW biomass, various conversion technologies can be employed to produce biofuels. The use of FW as a primary resource in biorefineries is currently in the nascent phase of innovation, with various research efforts assessing its financial sustainability on a broad spectrum. Table 2 present examples of biofuels produce from FW using different conversion techniques.

Table 2
Example of reported work on FW conversion to biofuels

Food waste	Biofuel type	Conversion method	References
Fruit and vegetable waste	Biogas (methane)	Anaerobic digestion	[29]
Dairy waste (cheese whey, milk residues)	Bioethanol	Fermentation	[30]
Used cooking oil	Biodiesel	Transesterification	[31]
Kitchen waste slurry	Ethanol	Fermentation	[27]
Cattle slaughterhouse waste	Biogas	Anaerobic digestion	[32]
Coffee grounds	Biodiesel	Transesterification	[33]
Cereal waste	Bioethanol and biogas	Hydrolysis and anaerobic digestion	[34]
Coffee bakery waste	Biodiesel	Transesterification	[35]
Pineapple waste (*Ananas comosus*)	Bioethanol	Fermentation	[36]
Eggshell	Biohydrogen	Steam and air gasification	[37]
Fruits and vegetables, cooked pasta, rice, cheese, and biscuits	Biohythane	Dark fermentation (DF) and anaerobic digestion (AD)	[38]

3.2.1 Biochemical Methods

The process of converting food waste (FW) into biofuels via biochemical approaches is a green-based sustainable solution to tackle waste management. This approach employs biochemical strategies that make use of microorganisms and enzymes to degrade organic substances in food waste, which are subsequently converted into biofuels such as biogas and bioethanol. This section presents commonly used techniques methods for the conversion of FW into significant biofuels.

1. *Anaerobic Digestion*

   Anaerobic digestion (AD) represents a biological process that entails the decomposition of organic substances, including food waste, through the action of microorganisms in an oxygen-free environment [26]. It leads to the production of biogas, predominantly methane gas, which can serve as alternative clean energy. The comprehensive process comprises four essential phases as follows:

   (a) Hydrolysis: This represent initial stage of the overall AD process also refers to as liquefaction. It is a biochemical mechanism that facilitates the degradation of intricate organic substances into soluble constituents. During this phase, the fermentative microbes mediate the conversion of insoluble complex organic materials sourced from food waste into soluble organic compounds, as well as break down complex polymers into simpler monomers. The enzymatic activities of hydrolytic and fermentative organisms serve as catalysts for this process.

   (b) Acidogenesis: This process involves the transformation of basic monomers into volatile fatty acids. The soluble organic by-products resulting from hydrolysis are converted into basic organic compounds, primarily volatile short-chain fatty acids, throughout the fermentation process [27].

   (c) Methanogenesis: Methane is generated through the microbial activities of microbes commonly referred to as methanogens. They transform microbes convert soluble matter into methane [26]. The first step for methane production is by breaking down acetic acid which yields carbon dioxide and methane. The second phase is the reduction of CO_2 with hydrogen. This second phase generates more methane, but it is limited by the quantity of hydrogen in the digester.

3.2.2 Enzymatic Hydrolysis and Fermentation

Fermentation represents a significant biochemical process for the conversion of organic substrates into biofuels, such as bioethanol, biogas, biodiesel, and biohydrogen. Multiple fermentation methods can be utilized, depending on the exact biofuel type, chemical composition of the substrate, and planned production scale.

1. *Anaerobic Fermentation*

   Anaerobic fermentation is an efficient and versatile technology for the production of bioethanol from organic food waste. It transforms carbohydrates into multiple essential products, such as biohydrogen, biogas, and ethanol, through the metabolic actions of microorganisms in an oxygen-free space.

2. *Simultaneous Saccharification and Fermentation (SSF)*

   Simultaneous Saccharification and Fermentation (SSF) is an integrated methodology that involves simultaneous hydrolysis and fermentation in a unified step, thereby optimizing processing duration and enhancing efficiency. For instance, co-culture systems, which contain aerobic fungi (for instance, *Aspergillus niger*) and facultative anaerobic yeasts (e.g., *Saccharomyces cerevisiae*), have proven to significantly enhance ethanol yield from tofu FW. Engineered solid-state fermentation (ESSF) entailing the halting aeration at predetermined time intervals, demonstrate superior productivity compared to conventional solid-state fermentation (CSSF), resulting in elevated ethanol concentrations and enhanced product yields.

3. *Sequential Saccharification and Simultaneous Fermentation (SSSF)*

   Sequential Saccharification and Simultaneous Fermentation (SSSF) represents an advanced form of SSF, demonstrated efficacy in bioconversion of pineapple waste through initial saccharification followed by fermentation, comparatively yielding high ethanol than untreated substrates, thereby highlighting its potential for large-scale bioethanol production.

4. *Repeated-Batch Fermentation*

   Repeated-batch fermentation constitutes a bioprocessing technology employed in the industrial synthesis of valuable bioproducts. In contrast to conventional batch fermentation, which ceases following a solitary cycle, repeated-batch fermentation facilitates multiple fermentation cycles utilizing the same original culture [28]. This approach significantly improves productivity, minimizes idle time, and maximizes resource efficiency. For instance, immobilized yeast cells, such as *Saccharomyces cerevisiae*, have been used in repeated-batch fermentation.

3.2.3 Thermochemical Conversion Approaches

1. *Pyrolysis*

   Pyrolysis, is a thermochemical process of conversion, provides a green solution by changing food waste into bio-oil, biochar, and syngas. Pyrolysis takes place in the absence of oxygen at high temperatures, resulting in the thermal breakdown of organic material in FW [39]. This method not only minimizes waste but also helps produce renewable energy, which makes it a viable alternative to conventional waste management practices. The abundance of organic content and calorific value in FW has made it potential good feedstock to produce bio-oil via pyrolysis. Nevertheless, difficulties in the form of high-water content and composition variability need to be overcome for effective conversion.

The overall steps involved are:

(a) Pretreatment: This may include reduction of particle size, grinding, determination of moisture content, limiting it to between 10%, drying, weighing, etc.

(b) Drying: It is critical since it involves the removal of water that may be a serious interference with the stability, pH value, calorific value, or corrosiveness of the yield of the product.

(c) Thermal processing: Feedstock enters the reaction chamber from the feeder, the elevated temperature causes depolymerization of the macromolecules and phase separation of all three types of products which are phase-separated individually as char, oil, and vapor.

(d) Product recovery: Bio-oil is recovered after condensation of the vaporized fragments and the uncondensed residues left behind leave the system along with the carrier gas, while the solid residues are unloaded periodically from the system so that it can disturb oil yield less.

Important types of pyrolysis are fast pyrolysis, slow pyrolysis, and flash pyrolysis

- *Fast Pyrolysis*

  It is a complicated method that involves the simultaneous and sequential reactions of organic materials in an inert (nonreactive) environment. The thermal breakdown process is conducted at diverse temperatures, varying from 350 to 550 °C to 800 to 1000 °C [40]. The lengthy chains of carbon, hydrogen, and oxygen are produced as condensable vapor, gases, and solid residues. The biomass is placed in the reactor that pyrolyzes it for the ideal residence time, at a specific temperature after it has been dried and decreased to the appropriate size. The vapor generated are transferred through and subsequently collected in the condenser, which serves as a unique quenching device. While the solid char is gathered in a char cyclone, the residual gas fraction is flared. Figure 1 exemplifies the fast pyrolysis process.

- *Slow Pyrolysis*

  Carbonization, another name for slow pyrolysis, is primarily the method of heating the substrate

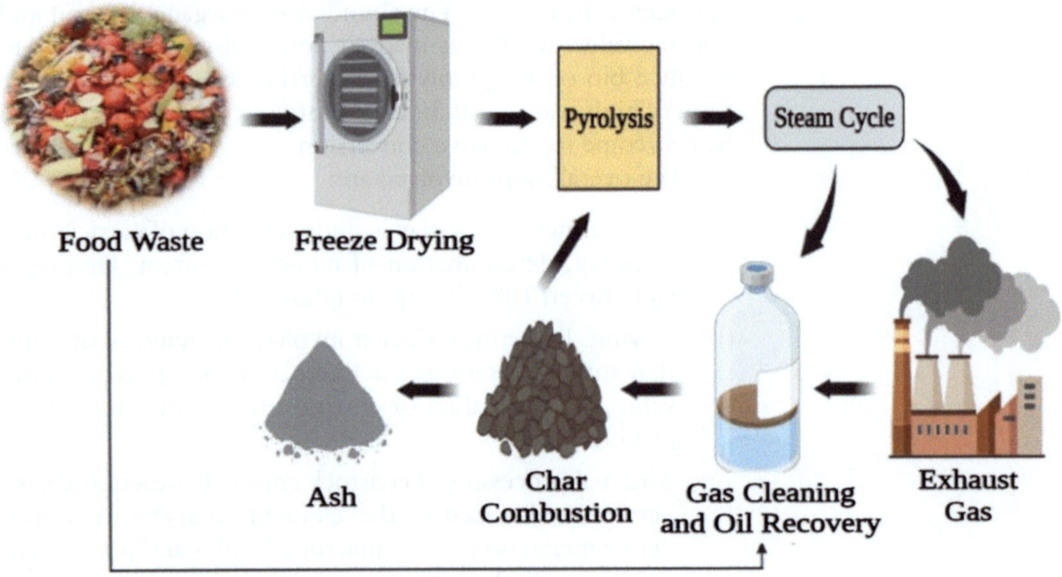

Fig. 1 Schematic diagram of fast pyrolysis process

(FW biomass) in an anoxic (oxygen-free) or oxygen-limited atmosphere. In essence, slow pyrolysis is a conventional method of producing charcoal with the primary goal of creating smokeless fuel. The primary purpose of the slow pyrolysis technique is to produce more char. The primary gaseous components are dust, hydrocarbons, CO, NOx, CO_2, H_2, and CH_4. The feedstock composition and process setup have a substantial effect on the number of products generated by the process [41]. The slow pyrolysis process described schematically in Fig. 2 is performed in a two-chambered setup.

Electronic controllers are programmed to heat the material with a linear increase in temperature (*10 °C/min) can be done up to the maximum (*500 °C in the case of slow pyrolysis) which has to be sustained for a specified time. A steady flow of nitrogen gas purges the gases and vapors produced during reaction time through the second chamber. The gases and vapors constitute the tar fraction of the product. This heavy tar fraction is distilled and condensed; the distillate is decanted further to yield settled tar and aqueous phase. The solid residence time varies from a few hours to a few days.

The benefits of this technique can be run on a variety of feedstock, including biomasses which are difficult to process in low quality. A slow pyrolysis

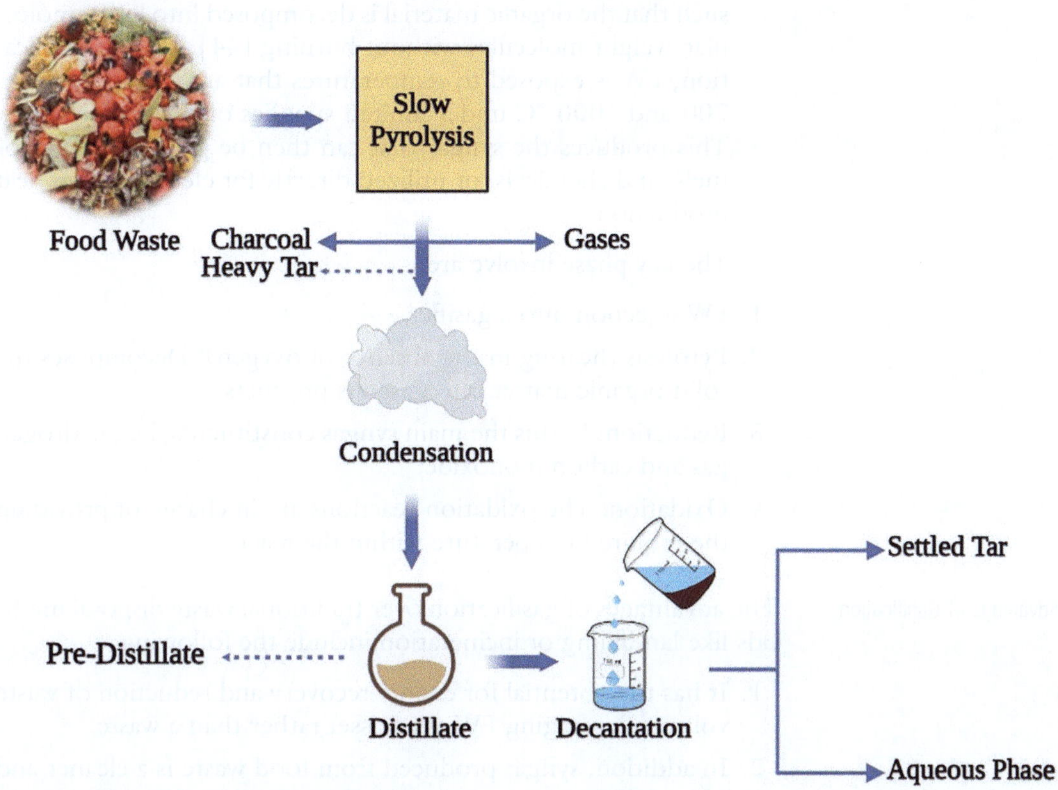

Fig. 2 Slow pyrolysis process flow diagram

reactor can feed feedstock containing approximately 50% moisture. Apart from energy generation, the products can be utilized for other uses in sectors such as agriculture, in the production of fertilizers, and pesticides [42].

- *Flash Pyrolysis*

 Flash pyrolysis is an essential technique for biomass conversion into eco-friendly biofuels [43]. It is frequently marketed as an improvement on rapid pyrolysis with a capable yield of bio-oil. It is distinguished by the reactive flash volatilization method (RFV), which uses thermal breakdown in an anoxic environment at high heating rates between 400 and 1000 °C and a very brief vapor residence time to transform non-volatile solids and liquids into volatile chemicals.

2. *Gasification: Syngas Generation*

 Gasification is a new technology for making valuable syngas from FW, which is a mixture of gases consisting mainly of carbon monoxide, hydrogen, and carbon dioxide. It involves heating of FW under controlled oxygen-limited conditions such that the organic material is decomposed into lower molecular weight molecules without burning [44]. During gasification, FW is exposed to temperatures that are usually between 700 and 1000 °C under limited supplies of oxygen or steam. This produces the syngas that can then be processed for biofuels, and chemicals, or utilized directly for electricity and heat production.

The key phase involve are:

1. FW injection into a gasifier.
2. Pyrolysis (heating in the absence of oxygen): Decomposes the solid organic matter into gaseous products.
3. Reduction: Forms the main syngas constituents, i.e., hydrogen gas and carbon monoxide.
4. Oxidation: The oxidation reactions are in charge of providing the required temperature within the reactor.

Advantage of Gasification

The advantages of gasification over traditional waste disposal methods like landfilling or incineration include the following:

1. It has the potential for energy recovery and reduction of waste volume, leveraging FW as an asset rather than a waste.
2. In addition, syngas produced from food waste is a cleaner and more efficient energy source than fossil fuels and helps in fostering a circular economy.
3. Food waste gasification also helps address the global issue of food waste management, which is environmentally and economically important.
4. It can also be able to help reduce greenhouse gases when combined with carbon capture technologies.

Some key challenges affecting the efficiency of FW gasification are:

1. Variability in feedstocks
2. Optimization of reactor design
3. Economic viability during scale-up

3.2.4 Transesterification for Biodiesel Synthesis

Transesterification is an established chemical reaction that is widely applied in the synthesis of biodiesel from diverse feedstocks, such as FW oils. This is a process where the triglycerides, primary components of fats and oils, react with alcohol (in most cases, methanol or ethanol) with a catalyst to yield biodiesel (methyl or ethyl esters) and glycerol as a by-product. In the process of food wastage, cooking oils discarded can be extracted with oils, animal fat from processed meat, or food residue with oils as promising feedstocks in this process [45].

Some forms of FW (particularly animal based) are known to be rich in oils and fats, and this has made them good substrate of transesterification for generating renewable energy (Biodiesel). The overall process involves blending of FW oils with alcohol, usually methanol, under heat and pressure, in the presence of acid or base catalyst. The triglycerides react with the molecules of alcohol to produce biodiesel and glycerol.

The FW oil use in the production of biodiesel offers some inherent benefits.

1. For one, it manages to handle the issue of food waste disposal as the oils and fats are diverted away from landfill dumping and can be prevented from contributing to the release of methane.
2. Secondly, it prevents the dependence on virgin vegetable oils or animal fats that require higher energy to produce [21].
3. Biodiesel produced from food waste oils via transesterification is a cleaner alternative to fossil fuels as it has better combustion and produces fewer emissions in the form of sulfur, carbon monoxide, and particulates.
4. Biodiesel produced from food waste oils also assists in the circular economy by transforming industrial waste into a valuable renewable energy source.

Associated challenges include:

1. Pretreatment of FW oils for transesterification. Such oils can have large amounts of FFAs, poisoning the catalyst used in the process, and needing further pretreatment processes, such as acid catalysis or esterification, to transform FFAs into biodiesel.
2. Catalyst development.
3. Process optimization.
4. FW oils are also likely to be of blended composition and be contaminated with species that can affect the quality and uniformity of the resulting biodiesel.

3.3 Challenges of Using FW for Biofuel Production

Even with numerous advantages, there are also some challenges for food waste gasification, especially in comparison with other feedstocks like biomass or coal.

1. Feedstock variability

   The biggest constraint in using gasification for food waste is its extremely variable nature. Food waste can consist of a wide range of materials with different moisture content, calorific value, and chemical composition. The variability is hard to handle, as it is not easy to maintain consistent and optimal performance from the gasifier. Pretreatment processes, such as shredding, drying, or sorting, might be necessary to offer a more uniform feedstock.

2. High-moisture level

   Food waste typically has a high-moisture level, which can reduce the effectiveness of the gasification process. Water must be evaporated from the feedstock before efficient gasification, and this consumes additional energy. The moisture level can be reduced by pre-drying food waste before gasification, but this gives an added step and cost to the process.

3. Tar and particulate formation

   Food waste gasification may lead to the production of tar and particulates, which may clog the gasifier and reduce syngas production efficiency. Tar and particulates can also cause downstream equipment operation problems, such as engines or turbines. To avoid these problems, gasifiers need to be designed to accommodate tar and particulates, and gas cleaning systems can be necessary to remove them before utilization of the syngas [46].

4. Economic feasibility

   Gasification can potentially be an economic waste-to-energy solution, but currently, the technology is comparatively pricey compared to alternative waste management approaches such as landfilling or anaerobic digestion. Capital expenditure on building gasifiers, their operation, and maintenance may be substantial, and the technology can be uneconomic for smaller- or decentralized-scale operations. Still, as time goes on and economies of scale improve, gasification is predicted to reduce its cost.

3.4 Optimization Strategies in Biorefinery Processes

Optimization techniques for biorefinery processes have the important responsibility of improving efficiency, economic profitability, and sustainability in biorefinery production. A biorefinery is similar to a petroleum refinery but is one where biomass (agriculture, forest residues, and waste food) is converted into an assortment of beneficial products such as biofuels, biochemicals, bioplastics, and other bio-materials. Optimization of biorefinery processes consists of enhancing some production aspects like resource use, energy use, feedstock flexibility, product yield, and reduction of waste [47]. These measures, in addition to making biorefineries profitable, enable biorefineries to contribute toward sustainable development effectively by minimizing industrial process environmental degradation.

3.4.1 Influence of Key Process Parameters on Efficiency

Optimization strategies in biorefinery operations are meant to maximize the efficiency of several operations, including feedstock conversion, energy recovery, product yield, and waste reduction. A key component of such optimization strategies understands how the most important process parameters influence overall efficiency.

The most important process parameters in important processes that impact the effectiveness of conversion operations, energy, product quality, and environmental sustainability include temperature, pressure, reaction time, catalyst type, substrate concentration, and pH. Optimization of these parameters is central to maximizing cost-effectiveness, sustainability, and operational efficiency in biorefineries.

1. *Temperature*

   Temperature is one of the most important parameters in biorefinery processes, particularly for both thermochemical and biochemical conversions. Temperature influences reaction rate, product yield, and reaction selectivity toward target products. In thermochemical reactions like pyrolysis, gasification, and transesterification, temperature has a direct influence on the decomposition of complex molecules into simpler, more valuable products [48]. For example, organic matter is decomposed at higher temperatures (usually 450–600 °C) into bio-oil, syngas, and biochar. When the temperature is too high, there is a possibility of over-cracking of bio-oil into gases, which reduces the yield of bio-oil.

2. *Pressure*

   Pressure has an important role to play, especially in the case of high-temperature operations such as gasification and hydrothermal liquefaction. During gasification, the pressure within the gasifier regulates the residence time of the gases and their interaction with solid feedstocks. Running at higher pressure has the potential to improve syngas yield and quality, but it consumes more energy and potentially raises operational costs. High pressure (as well as temperature) allows hydrothermal liquefaction for the conversion of wet biomass into bio-oil, particularly high-moisture content feedstocks like algae and food waste. Pressure affects the solubility of gases such as oxygen or carbon dioxide in biochemical reactions, and this may affect fermentation [23]. Pressurized bioreactors are employed in certain biorefineries for the maximization of product yield and microbial growth, e.g., butanol production by acetone-butanol-ethanol (ABE) fermentation, which is enhanced at higher pressures.

3. *Reaction/fermentation time*

   The reaction time or residence time is another significant parameter. For both thermochemical and biochemical reactions, the conversion efficiency is often a function of how long the feedstock is subjected to reaction conditions.

   In microbe fermentation, the length of the fermentation time has a direct impact on the yield of products. For example, ethanol fermentation usually lasts for 48–72 h as yeast ferments sugar to form ethanol. Long fermentation, though, can cause product inhibition whereby the build-up of ethanol or other by-products will reduce microbial growth and efficiency in fermentation [49]. A balance between conversion rates product quality and energy usage in minimizing is achieved in optimizing reaction time. Generally, a shorter reaction time is desirable to shorten the process period and enhance throughput, but attention should be exercised not to leave products unrecovered or unreacted.

4. *Catalyst type and concentration*

   Catalysts are central to biorefinery processes through the acceleration of chemical reactions and enhancing selectivity toward target products. The concentration and type of catalysts applied can have a great impact on the process efficiency.

5. *Acid or base catalysis (transesterification)*

   The catalyst selection (acid or base) is important in biodiesel production from oils. A base catalyst (such as sodium hydroxide) is widely used due to its high efficiency for the conversion of low-free fatty acid content oils to biodiesel. Acid catalysts (such as sulfuric acid) are needed for high-free fatty acid feedstocks to avoid soap formation. Catalyst concentration affects the rate of reaction and yield.

6. *Substrate concentration*

   The substrate (feedstock) concentration can significantly influence the efficiency of biochemical as well as thermochemical processes. For example, the concentration of sugars (or other fermentable substrates) during fermentation influences directly the microbial metabolic rate and overall biofuel yield, e.g., ethanol. However, elevated substrate concentrations may induce osmotic stress or product inhibition, both of which lower the efficiency of microbial fermentation. Equally, in the transesterification of oils to produce biodiesel, oil concentration in the reaction mixture dictates the rate of reaction. Though high oil concentrations can lead to higher biodiesel yields, excessive viscosity can impede the mixing of reactants and therefore necessitate additional energy or longer reaction periods. Thermochemical processes such as gasification, substrate concentration, or biomass loading, influence the efficiency of conversion. Increased biomass concentration can enhance energy recovery but might also cause operational difficulties, e.g., incomplete conversion or tar formation, which needs extra cleaning and purification steps.

7. *pH*

   pH is also a key factor in biochemical as well as in some chemical operations of biorefineries. During biochemical reaction processes like fermentation, pH plays an important role in microbial behavior. Various microbes prefer different optimum pH ranges to grow and generate the product. For example, *Saccharomyces cerevisiae* (yeast) in ethanol fermentation prefer about pH 4–5, while the microorganisms adopted to produce organic acids can show a preference for varied pH [50]. In transesterification reactions, the reaction mixture pH is also an important factor in regulating the process efficiency. Base catalysis needs a weakly alkaline condition, while acid catalysis needs acidic conditions. When the pH is not maintained in the optimal range, the reaction rate may fall or unwanted by-products can be produced, leading to yield reduction and energy requirements.

8. *Mixing and agitation*

   Agitation or mixing is a process parameter that guarantees proper contact among the reactants (e.g., oil, methanol, and catalyst in transesterification or substrate and microbe in fermentation). Poor mixing can result in inefficient reaction rates, incomplete conversion, and low yields of products. In transesterification, mixing keeps methanol and the oil well mixed, increasing the reaction between the triglycerides and methanol. In fermentation, adequate mixing also ensures the even distribution of nutrients and oxygen (if necessary) to microorganisms, encouraging their growth and metabolism.

9. *Water content*

   Water content is a frequently neglected parameter, but it has the potential to influence the efficiency of biorefining operations considerably. For example, in fermentation, excess water content dilutes the substrate, retarding microbial activity and lowering product concentration [51]. During thermochemical processes like gasification or pyrolysis, the moisture in the biomass dictates the energy requirements to evaporate the water and increases the potential for tar generation. The water content of reaction mixtures and feedstocks should be regulated for optimal efficiency for biorefining processes. Drying of feedstocks, including food wastes or algae, before conversion could enhance conversion yields and the efficiency of energy utilization.

3.5 Recent Developments in Process Optimization Techniques

Recent developments in process optimization techniques for biorefinery processes have been driven by the need to maximize the efficiency, sustainability, and economic attractiveness of processing biomass into high-value bio-based products. With biorefineries being viewed as key contributors to the pursuit of a sustainable bioeconomy, optimizing the operation of these systems has become a significant area of research. The goal of such optimization programs is to maximize the yield and quality of biofuels, biochemicals, and other bio-based products with minimized energy consumption, waste production, and environmental impact [22].

Biorefining involves several complex processes, from pretreatment of the feedstock to conversion technologies (e.g., fermentation, pyrolysis, gasification, and transesterification), product recovery, and waste treatment. In the last few years, process optimization has experienced tremendous technological advances with a focus on simplifying each step to achieve greater output, lower cost, and higher sustainability. The following discusses some of the most significant process optimization methods that have been created within the framework of biorefining.

3.5.1 Advanced Process Control (APC) and Automation Applications

Advanced process control (APC) and automated techniques are increasingly being used in biorefineries to optimize the monitoring and regulation of processes. The technologies make use of real-time data, sensors, and mathematical models to optimize various phases of production, such as feedstock processing, conversion, and product purification [52].

1. Real-time Monitoring: The availability of sensors and data analytics software allows operators to monitor key process parameters (temperature, pressure, pH, substrate concentration, etc.) in real-time. Real-time feedback allows for better control and adjustment of the process conditions such that the system operates at maximum efficiency at the optimal parameters for operation. For instance, in fermentation, monitoring of dissolved oxygen and pH online enables these parameters to be regulated such that microbial activity is maintained at optimal levels, resulting in higher product yields.
2. Model Predictive Control (MPC): Model predictive control is increasingly being used in the biorefinery sector. MPC utilizes mathematical models to predict future process behavior and regulate the inputs so that the process is kept at its optimal operating point. For example, in transesterification for biodiesel production, MPC can be utilized to optimize reaction time, methanol-to-oil ratio, and catalyst concentration to produce maximum biodiesel yield with minimal by-product formation [53].
3. Automation in Continuous Processing: As the biorefinery sector continues to shift more toward continuous processing from batch processing, automation has come to the forefront for throughput maximization, reduction of human errors, and enhanced process reliability. Automating fermentation, distillation, and biofuel manufacturing processes streamlines procedures, reduces labor costs, and enhances safety.
4. Artificial Intelligence and Machine Learning: Artificial Intelligence (AI) and Machine Learning (ML) have become effective tools for biorefinery process optimization. AI and ML can be used in a variety of ways to enhance the efficiency, precision, and sustainability of biorefining processes.

(a) Predictive Modeling: Machine learning models are capable of handling vast amounts of historical data and establishing patterns to predict the outcome of different process conditions. For example, machine learning models aided by AI can predict the optimal fermentation condition for the highest ethanol yield or forecast the most efficient catalyst for biodiesel production. By learning from historical performance data, AI systems can also forecast and avoid process failure, hence reducing downtime and increasing system reliability.

(b) Process Optimization Algorithms: AI and ML can be employed to develop optimization algorithms that dynamically modify process parameters in real time as a function of sensor inputs. For instance, deep learning networks can be trained to forecast optimal feedstock composition for gasification, and the optimal operating conditions to achieve maximum syngas yield and tar formation. Such self-optimizing systems can enhance the efficiency of both thermochemical and biochemical processes.

(c) Data-Driven Decision-Making: Advanced machine learning algorithms also make possible more refined data analysis and enable operators to make data-driven decisions. They can bring together several different data streams (e.g., temperature, composition of the feedstock, reaction kinetics) and offer conclusions about process inefficiencies, process bottlenecks, and points for improvement.

5. Process Intensification (PI): Process intensification is the application of innovative technologies and methods that compact, intensify, and make biorefining processes more efficient and cost-effective by optimizing the performance of a single step or the whole system. PI seeks to minimize equipment size, energy requirements, raw material consumption, and environmental footprint while maximizing production rates. Various process intensification strategies are currently being investigated in biorefineries are given below.

(a) Multifunctional Reactors: Multifunctional reactors combine a series of several process stages within a single device. A typical multifunctional reactor in a biodiesel reaction, for instance, may offer the reaction step and the step of separation merged into a combined unit that curtails energy expenditure and enhances overall efficiency. Likewise, bioreactors merging both fermentation and the distillation stage can maximize overall bioethanol production via reduced heat waste and enhanced integration of energy [54].

(b) Supercritical Fluid Processing: Supercritical fluids (most often CO_2 or ethanol) possess special characteristics that enable them to function both as a solvent and reactant in chemical reactions. Supercritical fluid processing is under consideration for a variety of biorefining applications, such as lipid extraction, biomass pretreatment, and production of biofuels. The technology can potentially lower significantly the reliance on harmful solvents and increase reaction rates.

6. High-End Separation Technologies: Innovations in separation technologies have aided process intensification through the enhancement of product purity, decrease in energy utilization, and reduction in waste generation in biorefineries. Conventional methods of separation like distillation and filtration are being supplemented or replaced by more efficient and selective ones.

   (a) Membrane Filtration: Membrane technologies such as ultrafiltration, nanofiltration, and reverse osmosis are increasingly being used to remove valuable products from post-reaction mixtures or fermentation broths. These operations are extremely energy-efficient and allow the selective recovery of biochemicals, biofuels, or other products. As an example, reverse osmosis can be used to separate ethanol from fermentation broths with less energy consumption than traditional distillation operations [55].

   (b) Electrochemical Separation: Electrochemical separation processes are also being envisioned in biorefining applications. These processes utilize an electric field to separate charged species or improve the generation of target products. In the case of producing biodiesel, for example, electrochemical processes can be used to separate glycerol from biodiesel and improve the purity of the final product.

   (c) Solvent Extraction and Ionic Liquids: Solvent extraction, especially with ionic liquids, has been lucrative in increasing the recovery of the product from biomass. Ionic liquids are capable of dissolving wide ranges of organic and inorganic molecules, a property that sees them as favorites in biomass pretreatment and recovery of biofuels or value biochemicals [26]. The use of ionic liquids will reduce the demand for harsh chemicals, increase selectivity, and decrease the ecological footprint of the separation.

7. Integrated Biorefinery Systems: Integrated biorefineries have gained wide popularity in the past few years. Integrated processes incorporate several conversion steps (e.g., thermochemical, biochemical, and physicochemical) in a single plant so that

raw materials and energy can be used more effectively. Optimization of integrated systems aims to make optimal use of biomass resources, minimize waste, and obtain maximum yields of products.

(a) Waste-to-Energy Integration: Biorefineries are increasingly using waste-to-energy technologies which enable the conversion of biomass residues and waste streams (e.g., lignin, glycerol, and organic municipal waste) to value-added products, including biofuels, biogas, and biochemicals. This lowers waste disposal expenses and enhances the overall economics of biorefinery operations.

(b) Coupling of Thermochemical and Biochemical Processes: Thermochemical and biochemical processes combined (such as gasification or pyrolysis with fermentation) have advantages in terms of energy use and resource recovery. For instance, syngas produced from the gasification of biomass can serve as a fermentable substrate to produce biofuels or biochemicals, while waste heat from either process can serve to drive the other, hence a more energy-efficient process.

(c) Circular Economy Strategy: New biorefineries are more often planned using a circular economy strategy in which waste products from one stage are the feedstock for another. For instance, waste streams from biodiesel production (e.g., glycerol and biodiesel waste) can serve as a feedstock for other biorefinery stages, i.e., microbial fermentation for bioplastics or biochemicals, toward a greener operation.

3.6 Environmental and Economic Considerations

Greenhouse gas emission reduction utilizing the production of biofuels is one critical approach in the global struggle against climate change and fossil fuel dependence. Biofuels, generated from FW, provide an alternative to conventional fossil fuels, with the ability to minimize carbon emissions along their life cycle.

3.6.1 Economic Considerations of Biorefinery Operations

Biorefining economic feasibility has strong interrelations with some key cost drivers, whose optimization is necessary to allow biorefineries to compete with traditional, fossil-based processing. These include

1. Cost of feedstocks
2. Costs of technology development
3. Investment
4. Operating costs

3.7 Life Cycle Assessment (LCA) for Biofuel Production from FW

Applying LCA to the production of biofuels from FW allows one to have an overall view of the environmental impacts across these stages. The life cycle stages are examined, namely feedstock collection, conversion, energy recovery, transport, and waste treatment. The LCA can either account for one stage or encompass the whole chain of biofuel production from food waste generation to final biofuel utilization [56]. LCA is typically divided into three broad stages:

(1) Aim and scope definition (2) Inventory analysis (3) Impact assessment Interpretation

1. Aim and Scope Definition: The aim of conducting an LCA on producing biofuels out of waste food is to examine its environmental sustainability relative to traditional fossil fuel production as well as other forms of renewable energy sources. During this phase, system boundaries have to be identified and functional units determined, the latter of which here may include the production of a certain volume of biofuel (e.g., 1 L of bioethanol).

   The scope of the analysis covers the biofuel type that is being produced (e.g., bioethanol, biodiesel, and biogas). The food waste feedstock employed in biofuel production followed by the conversion technology applied (e.g., fermentation, anaerobic digestion).

2. Inventory Analysis: In this stage, information is gathered on the inputs and outputs at every stage in the biofuel manufacturing process. The inventory comprises material and energy inputs (e.g., feedstock, water, chemicals, energy sources), and outputs (e.g., biofuels, by-products, waste, and emissions). The intention is to measure the resource utilization and emissions at every stage of the process, including such factors as:

   (a) Feedstock Acquisition and Transportation: This comprises the gathering, sorting, and transport of food waste to the biofuel production plant. Distance travelled, transport mode (truck, rail), and energy consumption in this step can be the cause of the overall environmental footprint.

   (b) Energy Consumption: The energy consumption processes in such conversions tend to be very energy-intensive, drawing power from electricity, heat, or steam. The type of energy used for the process, renewable or non-renewable, contributes to the environmental burden. Production of bioethanol through fermentation or biodiesel production through transesterification is energy-demanding and could be either fossil-derived or renewable.

(c) Greenhouse Gas (GHG) Emissions: The GHG emissions associated with food waste collection, transportation, and conversion to biofuels are assessed, including carbon dioxide (CO_2), methane (CH_4), and nitrous oxide (N_2O). For example, if food waste is left to decompose in a landfill without energy recovery, it can emit significant amounts of methane, a potent GHG.

(d) Waste and By-products: An integral component of LCA is determining how effectively the biorefinery manages waste and by-products. For instance, in anaerobic digestion for biogas, the digestate (wastewater organic material remaining after biogas production) can be applied as fertilizer, preventing nutrient pollution and lowering synthetic fertilizer demand.

3. Impact Assessment: Here, the information obtained during inventory analysis is translated into quantifiable environmental effects. Several impact categories are often used in LCA of biofuel production are as follows:

(a) Global wearing potential (GWP): The category evaluates how biofuel production contributes to global warming by examining the GHG emissions throughout the whole life cycle. One of the strengths of biofuels produced from food waste is their capability to minimize GHG emissions when pitted against fossil fuels. Recycling food waste that would otherwise rot in landfills prevents methane emissions, which have a much higher global warming potential than CO_2.

(b) Energy use: This impact category measures the energy efficiency of the biofuel production process by comparing the energy inputs used for feedstock acquisition, conversion, and end-use to the energy content of the finished biofuel product. The energy return on investment, or EROI, is the most important gauge of the process's sustainability—more energy yield from an input tends to represent higher efficiency.

(c) Water usage and pollution: This indicator refers to the amount of water that is utilized during food waste processing, which includes washing, fermentation, or anaerobic digestion. The LCA also considers the risk of water pollution via the discharge of wastewater or effluents from processing.

(d) Eutrophication and acidification: Production and disposal of biofuels can contribute to nutrient loading in aquatic ecosystems, resulting in eutrophication, or acidification of terrestrial and aquatic ecosystems. Such effects are generally associated with the emission of nitrogen and phosphorus compounds during biofuel production or fertilizers applied to produce feedstocks.

(e) Environmental impacts and issues: There are various environmental benefits and challenges in producing biofuel from food waste as follows.

(f) Reduction of landfill emissions: Food waste diverted from landfills to biofuel production decreases the emission of methane, which otherwise would be released from the anaerobic decomposition of organic waste in landfills. Methane is an extremely powerful greenhouse gas, and its utilization for energy production diminishes its effect on climate change [57].

(g) Energy recovery: Food waste is a good source of organic material, which can be readily converted into renewable energy in the form of biofuels such as biogas, bioethanol, and biodiesel.

(h) Waste minimization: The use of food waste as a feedstock reduces landfill needs and minimizes the environmental effects associated with waste disposal.

(i) High processing costs: Food waste conversion into biofuels is typically energy-intensive, and in some instances, the technology efficiency of the conversion process may still be in development. Reduction of these costs is one of the main reasons for enhancing economic viability.

(j) Feedstock quality and availability: It is not always suitable food waste for use as a biofuel feedstock, and quality variations within feedstock affect yield and efficiency. Furthermore, the sporadic nature of the supply of food waste could pose difficulty for the continuous mass production of biofuel.

4 Notes

Food waste (FW) constitutes a considerable ecological issue on an international level, thereby demanding immediate and sustainable management strategies. Comprehensive initiatives are being undertaken to tackle FW in manners that are both ecologically and financially feasible. This chapter has examined the characteristics and composition of FW, in addition to the various types of biofuels that may be produced from it. Additionally, various technologies and optimization processes for converting FW into bioenergy have been thoroughly examined. The discussion also highlights the economic and environmental factors that influence the costs associated with refining FW into valuable products. By understanding these aspects, stakeholders can develop more efficient and sustainable strategies to transform FW into renewable energy, thereby mitigating its environmental impact while creating economic opportunities.

References

1. Patel A, Hrůzová K, Rova U et al (2019) Sustainable biorefinery concept for biofuel production through holistic volarization of food waste. Bioresour Technol 294:122247. https://doi.org/10.1016/j.biortech.2019.122247
2. Szymkowiak A, Borusiak B, Pierański B et al (2022) Household food waste: the meaning of product's attributes and food-related lifestyle. Front Environ Sci 10:918485. https://doi.org/10.3389/fenvs.2022.918485
3. Zeng J, Zeng H, Wang Z (2022) Review on technology of making biofuel from food waste. Int J Energy Res 46:10301–10319. https://doi.org/10.1002/er.7868
4. Mahapatra S, Kumar D, Singh B, Sachan PK (2021) Biofuels and their sources of production: a review on cleaner sustainable alternative against conventional fuel, in the framework of the food and energy nexus. Energy Nexus 4: 100036. https://doi.org/10.1016/j.nexus.2021.100036
5. Lee SY, Sankaran R, Chew KW et al (2019) Waste to bioenergy: a review on the recent conversion technologies. BMC Energy 1:4. https://doi.org/10.1186/s42500-019-0004-7
6. Hafid HS, Rahman NAA, Shah UKM et al (2017) Feasibility of using kitchen waste as future substrate for bioethanol production: a review. Renew Sust Energ Rev 74:671–686. https://doi.org/10.1016/j.rser.2017.02.071
7. Dhiman S, Mukherjee G (2021) Present scenario and future scope of food waste to biofuel production. J Food Process Eng 44:e13594. https://doi.org/10.1111/jfpe.13594
8. Leung DYC, Wang J (2016) An overview on biogas generation from anaerobic digestion of food waste. Int J Green Energy 13:119–131. https://doi.org/10.1080/15435075.2014.909355
9. Kazemi Shariat Panahi H, Dehhaghi M, Guillemin GJ et al (2022) Bioethanol production from food wastes rich in carbohydrates. Curr Opin Food Sci 43:71–81. https://doi.org/10.1016/j.cofs.2021.11.001
10. Carmona-Cabello M, Leiva-Candia D, Castro-Cantarero JL et al (2018) Valorization of food waste from restaurants by transesterification of the lipid fraction. Fuel 215:492–498. https://doi.org/10.1016/j.fuel.2017.11.096
11. Pham TPT, Kaushik R, Parshetti GK et al (2015) Food waste-to-energy conversion technologies: current status and future directions. Waste Manag 38:399–408. https://doi.org/10.1016/j.wasman.2014.12.004
12. Kumar Gupta R, Ali EAE, Abd El Gawad F et al (2024) Valorization of fruits and vegetables waste byproducts for development of sustainable food packaging applications. Waste Manag Bull 2:21–40. https://doi.org/10.1016/j.wmb.2024.08.005
13. Moonsamy TA, Rajauria G, Priyadarshini A, Jansen MAK (2024) Food waste: analysis of the complex and variable composition of a promising feedstock for valorisation. Food Bioprod Process 148:31–42. https://doi.org/10.1016/j.fbp.2024.08.012
14. Pereira BS, Castrisana RN, de Freitas C et al (2021) Chemical composition determines the bioenergy potential of food waste from pre- and post-production. J Mater Cycles Waste Manag 23:1365–1373. https://doi.org/10.1007/s10163-021-01215-6
15. Ngabala FJ, Emmanuel JK (2024) Potential substrates for biogas production through anaerobic digestion—an alternative energy source. Heliyon 10:e40632. https://doi.org/10.1016/j.heliyon.2024.e40632
16. Zhang C, Su H, Baeyens J, Tan T (2014) Reviewing the anaerobic digestion of food waste for biogas production. Renew Sust Energ Rev 38:383–392. https://doi.org/10.1016/j.rser.2014.05.038
17. Elgarahy AM, Eloffy MG, Alengebawy A et al (2023) Sustainable management of food waste; pre-treatment strategies, techno-economic assessment, bibliometric analysis, and potential utilizations: a systematic review. Environ Res 225:115558. https://doi.org/10.1016/j.envres.2023.115558
18. Yılmaz ENÖ, Pehlivan DÖÜT (2024) Using household fruit and vegetable waste in recipes to reduce kitchen food waste and their nutritional and functional values. Akademik Gıda:33–44. https://doi.org/10.24323/akademik-gida.1554427
19. Ling Wen Xia F, Supri S, Djamaludin H et al (2024) Turning waste into value: extraction and effective valorization strategies of seafood by-products. Waste Manag Bull 2:84–100. https://doi.org/10.1016/j.wmb.2024.06.008
20. Yukesh Kannah R, Merrylin J, Poornima Devi T et al (2020) Food waste valorization: biofuels and value added product recovery. Bioresour Technol Rep 11:100524. https://doi.org/10.1016/j.biteb.2020.100524

21. Lopez Barrera E, Hertel T (2021) Global food waste across the income spectrum: implications for food prices, production and resource use. Food Policy 98:101874. https://doi.org/10.1016/j.foodpol.2020.101874
22. Ezeorba TPC, Okeke ES, Mayel MH et al (2024) Recent advances in biotechnological valorization of agro-food wastes (AFW): optimizing integrated approaches for sustainable biorefinery and circular bioeconomy. Bioresour Technol Rep 26:101823. https://doi.org/10.1016/j.biteb.2024.101823
23. Chen J, Ma X, Liang M et al (2024) Physical–chemical–biological pretreatment for biomass degradation and industrial applications: a review. Waste 2:451–473. https://doi.org/10.3390/waste2040024
24. Ethaib S (2024) Microwave-assisted pretreatment for lignocellulosic biomass energy conversion path. Bioresour Technol Rep 28: 102006. https://doi.org/10.1016/j.biteb.2024.102006
25. Chaurasia PK, Bharati SL, Singh S et al (2025) Fungal pretreatment methods for organic wastes: advances and challenges in biomass valorization. RSC Sustain 3:1234. https://doi.org/10.1039/D4SU00582A
26. Kuo J, Dow J (2017) Biogas production from anaerobic digestion of food waste and relevant air quality implications. J Air Waste Manage Assoc 67:1000–1011. https://doi.org/10.1080/10962247.2017.1316326
27. Iqbal SA, Rahaman S, Rahman M, Yousuf A (2014) Anaerobic digestion of kitchen waste to produce biogas. Proc Eng 90:657–662. https://doi.org/10.1016/j.proeng.2014.11.787
28. Luongo V, Policastro G, Ghimire A et al (2019) Repeated-batch fermentation of cheese whey for semi-continuous lactic acid production using mixed cultures at uncontrolled pH. Sustainability 11:3330. https://doi.org/10.3390/su11123330
29. Scano EA, Asquer C, Pistis A et al (2014) Biogas from anaerobic digestion of fruit and vegetable wastes: experimental results on pilot-scale and preliminary performance evaluation of a full-scale power plant. Energy Convers Manag 77:22–30. https://doi.org/10.1016/j.enconman.2013.09.004
30. Okamoto K, Nakagawa S, Kanawaku R, Kawamura S (2019) Ethanol production from cheese whey and expired milk by the brown rot fungus Neolentinus lepideus. Fermentation 5:49. https://doi.org/10.3390/fermentation5020049
31. Monika, Banga S, Pathak VV (2023) Biodiesel production from waste cooking oil: a comprehensive review on the application of heterogenous catalysts. Energy Nexus 10: 100209. https://doi.org/10.1016/j.nexus.2023.100209
32. Ware A, Power N (2016) Biogas from cattle slaughterhouse waste: energy recovery towards an energy self-sufficient industry in Ireland. Renew Energy 97:541–549. https://doi.org/10.1016/j.renene.2016.05.068
33. Acosta Suasnabar Eusterio H, Camarena Taxa Lizzette P, Ordonez Galvez Juan J, Benites Alfaro E (2023) Oil extracted from coffee grounds to obtain biodiesel as renewable energy. Chem Eng Trans 105:517–522. https://doi.org/10.3303/CET23105087
34. ElMekawy A, Diels L, De Wever H, Pant D (2013) Valorization of cereal based biorefinery byproducts: reality and expectations. Environ Sci Technol 47:9014–9027. https://doi.org/10.1021/es402395g
35. Karmee SK, Linardi D, Lee J, Lin CSK (2015) Conversion of lipid from food waste to biodiesel. Waste Manag 41:169–173. https://doi.org/10.1016/j.wasman.2015.03.025
36. Aili Hamzah AF, Hamzah MH, Che Man H et al (2021) Recent updates on the conversion of pineapple waste (Ananas comosus) to value-added products, future perspectives and challenges. Agronomy 11:2221. https://doi.org/10.3390/agronomy11112221
37. Valizadeh S, Lam SS, Ko CH et al (2021) Biohydrogen production from catalytic conversion of food waste via steam and air gasification using eggshell- and homo-type Ni/Al_2O_3 catalysts. Bioresour Technol 320:124313. https://doi.org/10.1016/j.biortech.2020.124313
38. Ghimire A, Luongo V, Frunzo L et al (2022) Biohythane production from food waste in a two-stage process: assessing the energy recovery potential. Environ Technol 43:2190–2196. https://doi.org/10.1080/09593330.2020.1869319
39. Shahzad HMA, Asim Z, Khan SJ et al (2024) Thermochemical and biochemical conversion of agricultural waste for bioenergy production: an updated review. Discov Environ 2:134. https://doi.org/10.1007/s44274-024-00171-w
40. Houston R, Oyedeji O, Abdoulmoumine N (2022) Detailed biomass fast pyrolysis kinetics integrated to computational fluid dynamic (CFD) and discrete element modeling framework: predicting product yields at the bench-scale. Chem Eng J 444:136419. https://doi.org/10.1016/j.cej.2022.136419
41. Huang Y-F, Chiueh P-T, Kuan W-H, Lo S-L (2013) Microwave pyrolysis of rice straw: products, mechanism, and kinetics. Bioresour

Technol 142:620–624. https://doi.org/10.1016/j.biortech.2013.05.093

42. Rahimi Z, Anand A, Gautam S (2022) An overview on thermochemical conversion and potential evaluation of biofuels derived from agricultural wastes. Energy Nexus 7:100125. https://doi.org/10.1016/j.nexus.2022.100125
43. Ighalo JO, Iwuchukwu FU, Eyankware OE et al (2022) Flash pyrolysis of biomass: a review of recent advances. Clean Techn Environ Policy 24:2349–2363. https://doi.org/10.1007/s10098-022-02339-5
44. Waste-based intermediate bioenergy carriers: syngas production via coupling slow pyrolysis with gasification under a circular economy model. https://www.mdpi.com/1996-1073/14/21/7366. Accessed 28 Feb 2025
45. Wan Osman WNA, Rosli MH, Mazli WNA, Samsuri S (2024) Comparative review of biodiesel production and purification. Carbon Capture Sci Technol 13:100264. https://doi.org/10.1016/j.ccst.2024.100264
46. Singh Sikarwar V, Zhao M, Clough P et al (2016) An overview of advances in biomass gasification. Energy Environ Sci 9:2939–2977. https://doi.org/10.1039/C6EE00935B
47. Ahmed SF, Kabir M, Mehjabin A et al (2023) Waste biorefinery to produce renewable energy: bioconversion process and circular bioeconomy. Energy Rep 10:3073–3091. https://doi.org/10.1016/j.egyr.2023.09.137
48. Wang Y, Wu JJ (2023) Thermochemical conversion of biomass: potential future prospects. Renew Sust Energ Rev 187:113754. https://doi.org/10.1016/j.rser.2023.113754
49. Kent JA (2003) Industrial fermentation: principles, processes, and products. In: Kent JA (ed) Riegel's handbook of industrial chemistry. Springer US, Boston, MA, pp 963–1045
50. Narendranath NV, Power R (2005) Relationship between pH and medium dissolved solids in terms of growth and metabolism of Lactobacilli and Saccharomyces cerevisiae during ethanol production. Appl Environ Microbiol 71:2239–2243. https://doi.org/10.1128/AEM.71.5.2239-2243.2005
51. Uçkun Kiran E, Trzcinski AP, Ng WJ, Liu Y (2014) Bioconversion of food waste to energy: a review. Fuel 134:389–399. https://doi.org/10.1016/j.fuel.2014.05.074
52. Enhancing sustainable energy through cutting-edge waste biorefinery technologies. https://www.mdpi.com/2624-781X/5/4/59. Accessed 28 Feb 2025
53. Enabling catalysts for biodiesel production via transesterification. https://www.mdpi.com/2073-4344/13/4/740. Accessed 28 Feb 2025
54. Aspects of reaction engineering for biodiesel production. https://www.mdpi.com/2073-4344/14/10/701. Accessed 28 Feb 2025
55. Le NL, Nunes SP (2016) Materials and membrane technologies for water and energy sustainability. Sustain Mater Technol 7:1–28. https://doi.org/10.1016/j.susmat.2016.02.001
56. Liu F, Shafique M, Luo X (2023) Literature review on life cycle assessment of transportation alternative fuels. Environ Technol Innov 32:103343. https://doi.org/10.1016/j.eti.2023.103343
57. Mar KA, Unger C, Walderdorff L, Butler T (2022) Beyond CO_2 equivalence: the impacts of methane on climate, ecosystems, and health. Environ Sci Pol 134:127–136. https://doi.org/10.1016/j.envsci.2022.03.027

Chapter 13

Sustainable Edible Film and Coating Protocols from Food Waste Polymers

Yas Kohansal, Zahra Mansouri Kermanshahi, Mohammad-Taghi Golmakani, and Hadi Hashemi

Abstract

Due to the customer awareness and rising concerns regarding the increased application of plastics which lead to the increased carbon emission and finally global warming, researches have been looking for an alternative to plastics in the food industry as well as other sectors. Edible films/coatings have risen as a reasonable solution to said problem. These packaging are edible, non-toxic, biodegradable, and environmentally friendly. Main materials for the formation of edible packaging include biopolymers, such as polysaccharides, proteins, and lipids. These biopolymers come from various sources including plant, animal, and marine origins. However, considering the amount of food waste produced worldwide, this particular source has gained popularity as a sustainable source to obtain polymers needed for film/coating production. Other than biopolymers, which are the main components of film formation, additives such as antioxidants, antimicrobial agents, plasticizers, and solvents are also important components in this process since each compound can influence different properties of films/coatings including, its gas and water vapor barrier properties, mechanical attributes (tensile strength and elongation percentage), and its appearance (color, transparency, and odor). Basic methods of film formation are categorized into two groups: wet and dry methods. Moreover, coating deposition on different products can be done through various techniques including dipping, spreading, spraying, panning, and fluidized-bed processing. This chapter aims to introduce the essential materials in the production of films/coatings and their individual influence on the final product. Moreover, common methods of film formation and their specific pros and cons are explained.

Key words Biopolymers, Coatings, Edible, Films, Food waste

1 Introduction

Plastic packaging has been a source of concern in the last decade since it contributes to growing non-biodegradable waste in the environment. Plastics, such as polyethylene and polypropylene are cheap and easy to handle; therefore, they have gained a lot of popularity in the food packaging industry. These materials are petroleum originated, which are not sustainable or eco-friendly. In addition, they add up to the carbon emissions profile, thus

Tanmay Sarkar (ed.), *Food Waste Valorization: Green Techniques in Sustainable Management*, Methods and Protocols in Food Science, https://doi.org/10.1007/978-1-0716-4646-5_13,
© The Author(s), under exclusive license to Springer Science+Business Media, LLC, part of Springer Nature 2025

further polluting the environment. However, in today's world with customers being environmentally conscious, researches in food industries have been searching for more sustainable and biodegradable alternatives for food packaging [1–4]. Edible coatings/films emerged as a solution to these environmental concerns. First prototype of edible packaging dates back to 1992, in which waxes were applied on the surface of fruits and vegetables as a protective layer. Edible packaging is used to extend the shelf life of foods and protect them against microbial spoilage and degradation [5]. Biopolymers are the raw materials in the formation of edible coatings/films. They are green, sustainable, renewable, and natural materials, including polysaccharides (cellulose, starch, chitosan, alginate, and carrageenan), proteins (gelatin, collagen, milk proteins, and meat proteins), and lipids (waxes, vegetable oils, and resins). In addition to their advantages, using biopolymers provide us with many benefits. For instance, they can be utilized as carriers for bioactive compounds such as antimicrobial, antioxidant, flavoring, and coloring molecules, therefore producing active and intelligent packaging. However, it is essential to choose a suitable polymer depending on the product and its conditions since each biopolymer exhibits specific characteristics regarding mechanical and barrier properties [6, 7]. Food waste is considered a renewable source of these biopolymers considering the large quantities of food waste produced in food sector. Main food waste sources for obtaining biopolymers include potato waste, mango seed, citrus peel, coffee waste, straw, sugar bagasse, pumpkin seed, banana peel, avocado seed, corn stover, carrot waste, and peanut husk [8]. Edible coatings can be applied on a wide variety of products including meat, poultry, seafood, diary, fruits, and vegetables. Main methods of film formation include wet (solvent casting) and dry method (extrusion), while several methods have been reported for coating applications, including dipping, spreading (brushing), panning, spraying, and fluidized-bed processing. Each method offers its own advantages and disadvantages; thus, it should be chosen based on the specific type of the product and its conditions. This chapter delves deep into the materials needed for the production of edible films/coatings and its common production methods and additives [9].

Cellulose: Cellulose as the most abundant organic compound is derived from a broad range of sources, like leaves or food waste. It is composed of D-glucose units linked by β-1,4 glycoside bonds [10, 11]. Cellulose has a specific crystalline structure and it is biodegradable, renewable, and a stable material in case of producing edible coatings. Despite all these advantages, cellulose application is limited because of its low solubility in water and most organic solvents. Derivatization as a viable solution can be applied to achieve cellulose derivatives with enhanced solubility. Most common derivatives used in film forming include methylcellulose, carboxymethylcellulose, hydroxypropyl methylcellulose, hydroxypropyl, and cellulose acetate [12]. Generally, these derivatives exhibit similar characteristics to the films obtained from other polysaccharides, such as transparency, odorlessness, and tastelessness, yet they offer a weak moisture barrier as a result of their hydrophilic nature [13]. Methylcellulose and hydroxypropyl cellulose have provided reasonable barriers against oxygen, carbon dioxide, and also lipids. However, they exhibited high water vapor permeability, which can be improved through the addition of lipids [14]. Overall, cellulose and its derivatives have permeability to water in both gas and liquid states, which can be overcome by chemical or physical modifications, with the goal of enhanced properties, like higher hydrophobicity and thermal resistance, enhanced barrier traits, and mechanical strength [15].

Starch: Starch as the main source of carbohydrate storage in green plants is composed of amylose (straight chain) and amylopectin (branched chain), in which the former has more influence on its film-forming characteristics. α-D-Glucose is the monomer in starch. Recently, starch has gained a lot of popularity in the production of edible coatings due to its abundance, biodegradability, and reasonable price [16]. Main sources of this biopolymer are various vegetables and their byproducts, such as cereals (maize, cassava, wheat, and potato), sorghum, rice, sweet potato, yams, barley, sago, fruits, tubers, legumes, and pseudo-cereals [17]. The origin of starch, combined with environmental parameters, play a huge role in indicating its structure and properties [18]. Starch granules have the ability to absorb water and swell until they reach a point called critical concentration, in which the swollen granules take over all the volume at 95 °C. Consequently, gelatinization is a must to achieve a consistent film-forming solution [15]. There are several shortcomings regarding the application of starch as a raw material in producing edible coatings. Firstly, this biopolymer has been reported to have undesirable mechanical properties, namely high tensile strength values and low elongation percentages. Furthermore, because of starch hydrophilicity, its coatings exhibit poor water vapor and gas barriers, thus they are not suitable for the packaging of dry foods or O_2-sensitive products [19, 20]. However, several solutions have been experimented to reduce said limitations, such as using different plasticizers (e.g., glycerol or other alternative plasticizers which provide improved thermal stability, low density and toxicity at room temperature), using starch from different sources, chemical modification of starch (to lower its hygroscopic properties), and adding compounds (polyols, polysaccharides, and proteins) that may enhance its mechanical and hydrophobic properties [21].

Alginates: Alginates are natural anionic polysaccharides originated from brown seaweed (*Phaeophyceae*). This polymer has two main constituents; β-D-mannuronic acid (M) and α-L-guluronic acid (G). The ratio of M and G blocks significantly affects the final properties of the produced edible coatings. Alginates exist in various forms such as calcium, sodium, and magnesium salts of alginic acid. It is important to mention that the M to G block ratio could affect some parameters in the final coating/film, for example, higher G values in alginate leads to the formation of coatings with reduced water vapor permeability [22–24]. Apart from having numerous advantages, like being biodegradable, biocompatible, non-toxic, and inexpensive, alginate has the ability to form crosslink and produce gel. Their ability to form gels is significantly correlated with the amount of G blocks. Edible coatings based on alginate are proven to be weak barriers against moisture. However, calcium alginate can produce films with reduced water solubility, while sodium alginate-based coatings display great mechanical strength and provide good water vapor barrier and cohesiveness [22, 25, 26].

Carrageenan: Carrageenan is a linear and marine-originated polysaccharides of D-galactopyranosyl units that can be found in red seaweeds of *Rhodophyceae* family [27]. There are three main types of carrageenans, including lambda, kappa, and iota, which are different in the arrangements of their sulfate group. Among them, the highest gel-forming ability belongs to the kappa-carrageenan (containing 25–30% sulfate concentration) and consequently it is the carrageenan usually utilized in the production of food packaging. The mechanism of carrageenan's film forming is reported to be influenced by gelation during drying [28]. This biopolymer contains sulfate groups in its structure; it is also water soluble and has a broad range of applications in food and pharmaceuticals, as emulsifiers, gelling agents, and stabilizers. Carrageenan-based edible coatings can be used in sausage casings, dry solid foods, oily foods, meat, fish, and poultry to prevent dehydration [27, 29]. This biopolymer owns its great film-forming abilities to its good solubility and specific helical structure. Similar to alginate-based films, their film-forming properties may be improved through the addition of plasticizers or other polymers. They are also reported to be of great use in functional food packaging [30].

Chitosan: Chitosan as another linear and animal origin polysaccharide is mainly composed of β-linked *N*-acetyl-glucosamine and *N*-glucosamine units. Chitosan can be produced via alkaline deacetylation of chitin [31]. Chitin is an insoluble compound, however after its conversion to chitosan, it becomes soluble in acidic solutions [32]. In addition, this polymer possesses antimicrobial and antioxidant attributes. Studies showed that chitosan-based films could effectively inhibit the growth of *Listeria monocytogenes* and *Escherichia coli* in meat products. Among its advantages, we could mention biocompatibility, non-toxicity, bioadhesivity, and biodegradability. Chitosan-based coatings display a selective permeability to oxygen and dioxide carbon. They also exhibit good mechanical properties. These types of coatings are commonly applied to apples, pears, strawberries, cucumbers, bell peppers, peaches, and plums. In contrast, chitosan hydrophilicity leads to poor moisture barrier, which can be improved through addition of plasticizers, lipids, or modifications. Since chitosan does not produce thermoplastic films, heat-sealing is impossible, which limit its application at commercial levels [33, 34]. Furthermore, it was reported that 0.5–1.5% chitosan content leads to the production of coatings applicable in meat products, even in longer periods of storage, like 9 days [35].

Whey Protein: Whey protein is a byproduct of dairy industry and includes 20% of the milk proteins. Important proteins in whey include beta-lactoglobulin, alpha-lactalbumin, bovine serum albumin, and immunoglobulins. These proteins offer antibacterial properties when they are used in coatings, which make them an active material. Among its technological benefits, whey protein's ability to form gels, foams, and emulsions have attracted a lot of attention. Whey protein is mostly found in two forms, including whey protein concentrate and whey protein isolate, in which the later contains more protein concentration. Whey protein-based coatings are advantageous since they are biodegradable, edible, and offer good mechanical and barrier characteristics. Moreover, these kinds of films are colorless, odorless, flexible, and translucent. They are able to lower the rate of microbial spoilage in fruit and vegetable packaging. Whey protein-based films have reported to coat different products, such as kilka fish and salmon, thus causing a reduction in their lipid oxidation rate. In addition, combining essential oils and whey protein have proven to increase its antibacterial properties [36, 37].

Casein: Casein is another milk originated protein which includes 80% of milk's proteins. It can be divided into four fractions, namely s-, α-, β-, and κ-casein. Casein-based films are responsible for the most transparent coatings among protein-based films [38]. They also demonstrate good elasticity and firmness values. Moreover, they are good barriers against oxygen, however similar to other protein-based films, Casein-based films are poor water vapor barriers [39]. Common form of casein has mentioned to be caseinate (calcium caseinate or sodium caseinate), which is made through casein precipitation at the minimum pH of 4.6, after which the pH is returned to 7 with the help of an alkali solution [40]. Interestingly, calcium caseinate-based films are better barriers due to the existence of calcium cation that improves protein–protein interactions. These films also have rigid texture. On the other hand, sodium caseinate-based films are better at mechanical and tensile properties. Shortcomings in casein-based coatings may include shrinkage during the drying process. Addition of plasticizers could enhance these film's brittleness after drying [41].

Gelatin: Gelatin is a yellow-colored, water-soluble protein that can be obtained from the hydrolysis of collagen. It is an animal-based protein and it can be found in pork, fish, and bovine. Based on the production process of gelatin, we have two kinds of gelatin called, type A gelatin, which originates from an acid-treated collagen and the other one is type B gelatin, which is a product of alkali-treated collagen [42]. Gelatin-based films are reported to have reasonable mechanical properties. However, they are not good barriers against oxygen and moisture since they have high hygroscopicity [31]. Finally, the influence of integrating crosslinkers, plasticizers, vegetable oils, and phenolic compounds with gelatin has been deeply studied. It was reported that introducing hydrophilic polymers to gelatin-based films leads to improved degradability values, while cross-linking decreases it [43]. The gelatin-based films were reported to prevent shrinkage in beef patty samples during 30 days of storage, which can be attributed to the existence of hydrophilic amino acids in gelatin that have high capacity to bind water molecules. These kinds of films are also capable of preventing microbial growth, therefore acting as a preservation method [44, 45].

Zein: Zein is a prolamin protein byproduct of corn processing and mainly composed of glutamine (21–26%), leucine (20%), and proline (10%). Zein-based coatings provide low water vapor permeability as opposed to other protein-based films, which can be attributed to its non-polar amino acid, which includes the majority of proteins in its structure. These coatings are also thermally stable, have smooth and glossy appearance, and exhibit barrier properties against O_2, CO_2, and oil or grease [46]. Despite its excellent barrier properties, zein-based coatings have undesirable mechanical attributes since they are fragile and brittle films. Zein-based films can be produced via both wet and solvent process. Type of solvent can impact the final film properties, for instance, zein coatings prepared with acetone proved to be stronger, but offered less flexibility in comparison with ethanol [47, 48].

Soy Protein: Soy protein is a byproduct of soy oil refining process, and it is mainly composed of globular proteins. It can be found in different forms, including soy protein isolate, soy flour, soy protein concentrate, which differ in their protein content and other impurities [10]. Since the existence of components other than protein, such as carbohydrate, could interfere with the production of coatings, the common film producing form is soy protein isolate, which goes through different processes to increase its protein content [49]. In contrast to other protein-originated coatings, soy protein isolate produces uniform, flexible, and clear films. They are also better barriers against gases [50]. Moreover, these coatings exhibit good tensile strength and elongation percentages that can be attributed to the high molecular weight of soy protein. Because of poor thermal resistance of soy proteins, dry formation method cannot be suitable for their production. Soy protein-based film has proven to prevent weight loss in sausages [30, 51].

Wax: Waxes are esters of a long-chain alcohol and fatty acid that poses high molecular weights. They act as a preservative coat on different food products. Because of their high hydrophobicity, waxes are deemed effective in reducing moisture diffusivity in coatings. Waxes can be natural (carnauba wax, candelilla wax, rice bran wax, and beeswax) or synthetic (paraffin wax and petroleum wax). It is worthy to mention that natural originated waxes are a safer option to use in the production of edible films compared to the synthetic ones [52]. Natural waxes offer prolonged shelf life in fruits. A study demonstrated that using Candelilla wax as a coating on freshly cut fruits can improve its firmness and appearance and also help with the weight loss [53].

Vegetable Oils: Vegetable oils occupy various roles in the production of edible coatings. They can act as plasticizers, water-resistant material; they can also reduce gas permeability to keep the product fresh and enhance mechanical properties, such as flexibility and elasticity [54]. Moreover, vegetable oils can be used to decrease the brittleness of films and increase their processability during the casting method. They also contribute to the production of uniform coatings, prevent the migration of volatile compounds from the film to the product, enhance adhesion between the film and material; thus, making sure the coverage of film is desirable. It is important to note that use of vegetable oils in edible films has lost its appeal due to the rancidity issue of oils that has negative impact on sensorial properties of product [9].

2 Materials

Polysaccharides: Polysaccharides are among the common materials used in the production of edible films/coatings and they have achieved worldwide prominence due to their advantages. They exist in profuse quantities in nature and can be originated from various sources, including microbes, plants, animals, and marines [55]. More transparency can be achieved via polysaccharides compared to other biopolymers. Also, they can produce inodorous and tasteless coatings, which is significantly related to the sensory perception of coated foods [14]. Polysaccharides have shown to be remarkable oxygen barriers, as a result of having several hydrogen chains that contribute to their density. Moreover, there are hydrogen binding sites in polysaccharide's structure which is beneficial in case of incorporated functional components, such as flavor, color, antimicrobial, and antioxidant substances [56]. On the other hand, these materials offer weak blockage against water vapor because of their hydrophilic properties. However, this drawback could be resolved by combining a more hydrophobic macromolecule, such as lipids for improved moisture barrier properties [6]. Finally, despite their high water vapor permeability, and considering their low gas permeability, polysaccharides can easily coat the cross sections of fruits and vegetables, and act as a natural packaging [57]. Cellulose, starch, alginates, carrageenan, and chitosan are the main polysaccharides.

Protein: Protein, as another biopolymer that can be recovered from food waste, has great functional and structural properties, which makes it an excellent candidate in the production of edible coatings. Protein-based films have displayed great mechanical attributes as opposed to lipids and polysaccharides. They can also provide a good barrier against gases like O_2 and CO_2; however, they exhibit high permeability to moisture. Their resistant to water diffusion can be improved via introducing hydrophobic compounds such as beeswax and glycerol in the film-forming solution [57, 58]. Moreover, due to their wide-ranging amino acid functional groups, proteins contain several active sites that make chemical modifications possible to achieve enhanced stability in coatings. Another unique quality of proteins is their ability to form intermolecular connections and arrangements in different directions, which results in more rigid structure in the final coating, thus better mechanical properties are achieved. Not only proteins provide foods with a coating, they also add to their nutritional value. Proteins come from different sources with their specific qualities, including animal-based proteins (casein, whey protein, meat proteins, egg albumin, gelatin) and plant-based proteins (wheat, soy, zein). Overall, protein's structure, weight, solubility, and charge have great influence in indicating its film-forming potentials [59]. For the creation of edible films generated from animals, protein sources, such as whey protein, casein, gelatin, and egg albumin have been found to be promising candidates. A number of plant-based sources, including, but not limited to, peanuts, soy, corn, rice, and wheat have been mentioned as suitable materials for use in food polymer applications [49].

Lipid: Lipids are natural, hydrophobic compounds found in plants, animals, or insects. These molecules cannot support their own structure, and they are not capable of polymerization. Lipids contain a broad range of compounds, including animal and vegetable oils, waxes, resins, fatty acids, essential oils, and lecithin. Lipid-based coatings are great barriers against moisture, due to the presence of non-polar and hydrophobic long-chain fatty acids in their structure. However, they are unable to form coatings on their own and are mostly combined with other materials [60]. Lipids provide glossy appearance, reduce moisture loss and overall cost of the final product. These qualities have earned lipids great popularity in recent decades. Lipids are effective coatings in meat products to minimize fat oxidation. Moreover, resulted coatings often have higher viscosity and they are also brittle and may need adjustment [30, 54, 57].

Functional Additives: Active packaging can be achieved through addition of functional compounds including antioxidant and antimicrobials. Functional components offer prolonged shelf life, prevent spoilage and lipid oxidation, and enhance mechanical and barrier properties of edible coatings [61].

Antioxidants: Antioxidants are substances that can extend the shelf life of foods and have gained popularity as an additive in active film packaging. These compounds have free radical scavenging properties, and they can prevent further reactions of free radicals by donating an electron to them. They can also slow down the lipid oxidation process in foods. The most common antioxidant used in edible coatings is phenolic compounds, which includes flavonoids, tannins, and phenolic acids. In addition to antioxidant properties, phenolic compounds often possess anti-inflammatory, antihypertensive, antioxidative, and antiaging benefits [62, 63]. The efficiency of these compounds can be evaluated through in vitro assays like DPPH, FRAP, and ABTS [64]. Various studies have reported the use of α-cinnamaldehyde in chitosan films, betacyanin in PVA-starch based film, and quercetin in packaging films, all lead to an increase in their antioxidant properties [65, 66].

Antimicrobial Agents: Antimicrobial agents are defined as compounds that have the ability to extend food's shelf life via preventing the growth of microorganisms. Integrating antimicrobial compounds with edible film technology has gained a lot of popularity since it solves two problems simultaneously, including reduction of plastic consumption and microbial spoilage of food products [67]. Plant-derived antimicrobial agents have superiority over their synthetic equivalent, and they also receive higher consumer acceptability. Essential oils (such as thymol, carvacrol, p-cymene, and γ-terpinene in Thymus) are plant-derived metabolite that exhibit strong antimicrobial activities against a broad spectrum of microbes including *Staphylococcus aureus*, *Escherichia coli*, *Streptococcus* spp., *Bacillus cereus*, *Enterococcus faecalis*, and *Klebsiella pneumoniae*. There are several parameters influencing their antimicrobial activity, such as extraction method and chemical composition [63]. The effectiveness of essential oils can be attributed to their phenolic groups, which can damage the cell membrane of microorganism and cause leakage of its interior; therefore, eliminate them successfully [68]. Organic acids and their salts are among antimicrobial compounds that can be added to the film-forming solutions. They offer lower costs and reasonable efficacy. Moreover, majority of them are considered to be GRAS. Some examples include lauric, acetic, sorbic, citric, benzoic, or propionic acids. Their ability to reduce pH and consequently altering cell's integrity is responsible for their antimicrobial properties [69]. Whey protein, lactoperoxidase, lactoferrin, lysozyme, and polypeptides are among animal-based antimicrobial agents that can be used in edible coatings. In which lysozyme shows antimicrobial effects against Gram-positive bacteria through diminishing the peptidoglycan layer of cells, lactoperoxidase is mostly effective against enteric bacteria, and lactoferrin through its iron chelating properties against Gram-negative bacteria [70, 71]. Even the biopolymer itself can contribute to the antimicrobial properties of edible packaging. For instance, chitosan offers antagonistic properties against *Bacillus cereus*, *Salmonella typhimurium*, *Listeria monocytogenes*, *Escherichia coli*, *Staphylococcus aureus*, and various mold and yeast. Chitosan's antimicrobial characteristics can be attributed to its positively charged amino fractions interacting with cell membrane and causing its death. Moreover, this biopolymer is capable of metal chelating, thus, making necessary metals unavailable for bacteria and inhibiting its growth. Finally, it is noteworthy to mention that the combined use of antimicrobial can be beneficial to achieve a synergic antimicrobial effect that is effective against more types of microorganisms [50].

Plasticizers: Plasticizers are inevitable part of edible films, especially in case of proteins or polysaccharide-based coatings. The mentioned coatings suffer brittleness mostly due to the excess intermolecular interactions of polymer chains. Plasticizers are used to enhance mechanical and barrier properties in these coatings [72]. Plasticizers are low molecular weight and hydrophilic compounds that when added to the film-forming solution, improve the flexibility and mechanical properties of the resultant films. Plasticizers are hygroscopic materials capable of attracting water, which is a plasticizer itself, and make complexes containing plasticizer and water. Different types of plasticizers utilized in edible packaging include monosaccharide-based plasticizers (glucose and fructose), oligosaccharide-based plasticizers (maltodextrin), polyol-based plasticizers (glycerol and sorbitol), and lipid-based plasticizers (vegetable oils) [73].

3 Methods

It is important to distinguish these common terms used in food packaging: film and coating. The first word refers to a wrapping material around the food and generally it should not contain any flavor and color of its own. They are considered as primary packaging and are acquired through two main processes: wet process (solvent casting) and dry process (extrusion). On the other hand, coatings usually refer to direct deposition of film-forming solution onto the surface of products via various techniques including dipping, spreading, spraying, panning, and fluidized-bed processing [9].

3.1 Film-Forming Methods

Solvent Casting (Wet Formation): Casting method (displayed in Fig. 1) is the most common technique used in laboratories, as it is simple and inexpensive. This method is based on dissolving

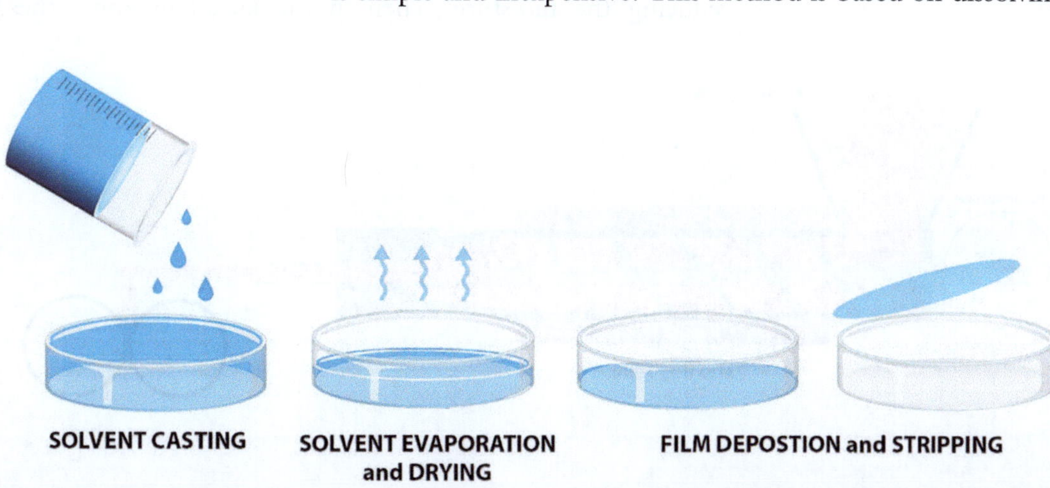

Fig. 1 Different stages of solvent casting process

polymers in a proper solvent and then drying and evaporation of that solvent, so a film can be produced. The process of the solvent casting method is as follows: first stage is dissolving biopolymers in a suitable solvent. Other additives such as antimicrobials, antioxidants, plasticizers, and emulsifiers are also added in this stage with the purpose of producing active or intelligent films or to enhance flexibility and mechanical properties of the obtained films. After that, the film-forming solution is poured into a flat mold and it is left to dry (this step is called casting) [74]. pH adjustment, minimal heat treatment, and homogenization might be required in the first step. Second stage is drying which can be done through various instruments other than ambient air, including ovens, microwaves, vacuum dryers, etc. [9]. Finally, dried films are peeled from the surface of molds and they are ready to be used as edible packaging. The drying conditions, such as drying rate and temperature, have significant effect on the properties of final formed film. For instance, studies have reported that higher temperatures and shorter times produced undesirable mechanical properties in the produced films [75]. However, drying period of 12–24 h has been suggested to achieve films with optimum mechanical properties. Due to this preferred drying time, this method is not suitable for large-scale and industrial applications [76].

Extrusion (Dry Process): Extrusion method is the main procedure applicable at commercial levels. This method benefits from the least possible amount of moisture content and does not need solvents, which is why it is called the dry method. This process involves three main steps (displayed in Fig. 2): feeding zone, kneading zone, and heating zone. The extruded films often display improved structural and physicochemical properties in this process [50]. Different film participants are mixed together in the feeding zone, in which compressed air is used as a way of reducing the moisture. Then in the kneading zone, these

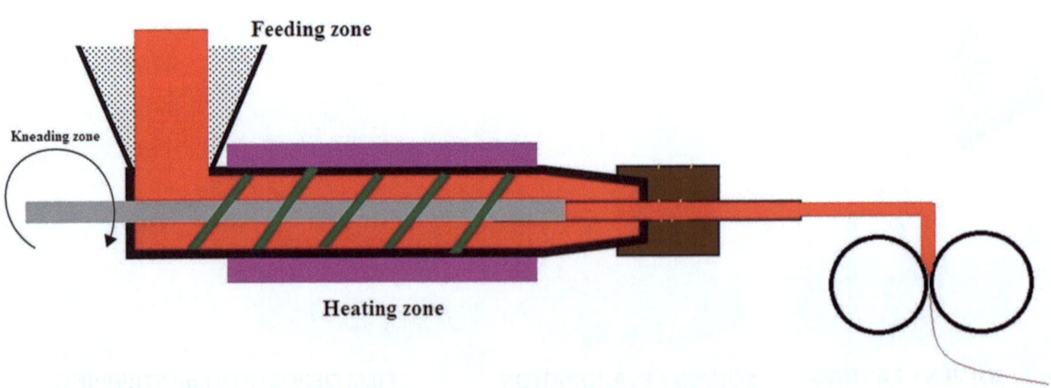

Fig. 2 Extrusion method

components go through the strain and temperature caused by the process, which results in their conversion to a molten state and flexible state, due to the increased density and heating (above the glass transition temperature). Finally, the melted mixture is extruded through a nozzle as a result of applied rotating force. The final films are formed after cooling [77]. Extruded films offer several advantages. They require less drying time compared to the wet process. This process results in a better appearance and mechanical properties in final films. Furthermore, extrusion process needs less energy consumption and it is inexpensive and can be scaled up. Moreover, application of high viscosity polymers is possible via this method, and there is no need to use any solvents. However, disadvantages of extrusion method are mainly attributed to its limitation when choosing a polymer since the used polymer should be resistant to heat and have low moisture content [9]. For instance, chitosan-based films cannot be produced via this method due to chitosan's lack of thermoplastic behaviors. In addition, initial costs of equipment and maintenance expenses are quite high in this process [78].

3.2 Edible Coating Methods

Dipping Method: Dipping is the simplest and the most common laboratory method for coating of fruits and vegetables. It is an inexpensive method, which is mainly composed of three basic stages. First stage is immersion and dwelling, in which the product is plunged in the coating solution for an adequate amount of time, with a controlled speed. In the second stage, deposition, a layer of film-forming solution is formed on the surface of product. Last stage is evaporation, in which the excess solvent will evaporate either at ambient temperature or with the help of a dryer [27, 79]. The dip-coating process is shown in Fig. 3. Different characteristics of coatings obtained through dipping method are influenced by many factors, including time and speed of immersion and withdrawal, number of dipping cycles, film-forming solution conditions (such as

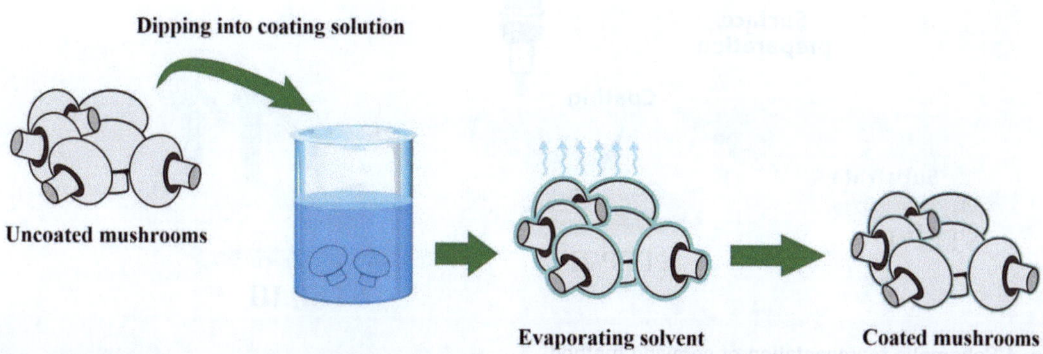

Fig. 3 Schematic process of dipping method

density, viscosity, surface tension), and lastly the drying conditions [80, 81]. For instance, higher immersion speed leads to a thicker coating due to the increased viscous forces in the solution. If the coating solution has higher viscosity, it is better to reduce the number of dipping cycles since viscous solutions tend to agglomerate more; therefore, producing thicker layers of coating. The acquired coating layer may be brittle and breakable when using high viscosity solution, in these cases a more dilute suspension combined with multiple dipping cycle could be the answer [82]. Dipping method might not be suitable for cut surfaces of fruits since its hydrophilic surface is not aligned with the hydrophobic coating solution. Moreover, this method has several drawbacks, including contamination of film-forming solution via dirt and microbes of the fruit surface, large amount of precursor solution needed for coating, formation of undesirable heavy layers of coating that could damage the food or decrease its respiration, and finally detachment of coated layer is possible after the process [83–85].

Spraying Method: This method is applicable to low viscosity solutions, in which the process ensures a homogenous coating of the product. The conversion of liquid to tiny droplets via spraying technology leads to a higher ratio of surface to volume, making it possible for these tiny droplets to cover larger parts of the product [60]. Spraying is made possible via a nozzle or an atomizer. Moreover, multilayer coatings can be achieved through this particular method [86]. The spraying process is demonstrated in Fig. 4. Size of the droplets, flow rate of the film-forming solution, air flow rate, temperature, air quality, and drying time are among important parameters that need to be controlled to ensure a desirable outcome in the coating

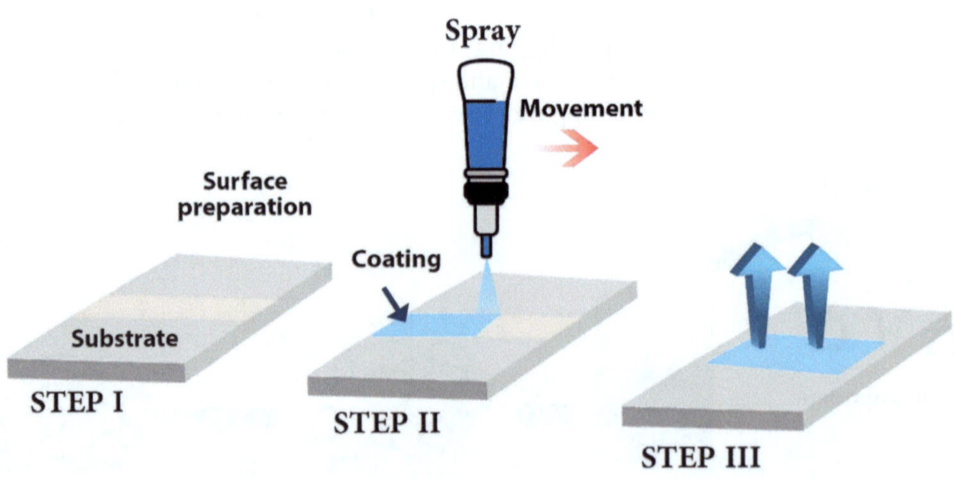

Fig. 4 Schematic representation of spraying method

process [87, 88]. Furthermore, atomization condition holds great importance in the spraying process. Spraying method can be applied in three different settings based on the fluid used to convert the film-forming solution from liquid state to droplets.

1. Air spray atomization: This affordable technique includes using air with high speed to produce fine droplets of the film-forming solution as this high velocity air creates friction with the said solution. Air jet nozzles are applicable in this method [89].
2. Air-assisted airless atomization: The atomization process in this method involves two steps, first a specific nozzle is used to breakdown the film-forming solution into smaller droplets. Then a tiny amount of compressed air is applied to obtain the final and more small droplets. This method could be beneficial when the solution in use has high viscosity and needs higher pressure or temperature [90].
3. Pressure atomization: In contrast with previous methods, this process uses pressure instead of air flow to breakdown the liquid solution into fine droplets. That is why this method is also called airless atomization. Small nozzles utilize high pressures to produce droplets from the film-forming solution. To avoid damaging the film-forming solution the pressure is kept below 3.5 bar [91].

Fluidized-Bed Processing Method: This process (Fig. 5) involves the spraying of film-forming solution into the chamber where fluidized powder exists. Fluidization process happens when a liquid flow moves through fluidized particles (which act as a bubbling fluid at this stage). This method can be divided into three groups, namely top spray (more applicable and efficient in food industry), bottom spray, and rotary fluidized bed [9]. One of the important parameters during this process is the size of the powder, which should exceed 100 μm, since in dimensions less than that either stable fluidization cannot happen correctly, or agglomeration occurs. Heat treatment is part of this method and it can either be done in a batch or a continuous process [92]. Fluidized-bed method solves a very crucial problem, which occurs easily during panning method, called clustering of the coated process. Among its disadvantages, complexity and higher cost of maintenance can be mentioned [84].

Panning Method: Panning method (Fig. 6) in the first place was developed by Greek Arabian society to be used in drug industry. This technique simply involves adding the food and film-forming solution (via a nozzle) in a rotating pan to tumble and mix, so a uniform coated layer is achieved. Afterwards, drying process is applied via forced air circulation either at ambient or higher temperatures. Hard panning, soft panning, and chocolate panning are among this category [93]. In addition to

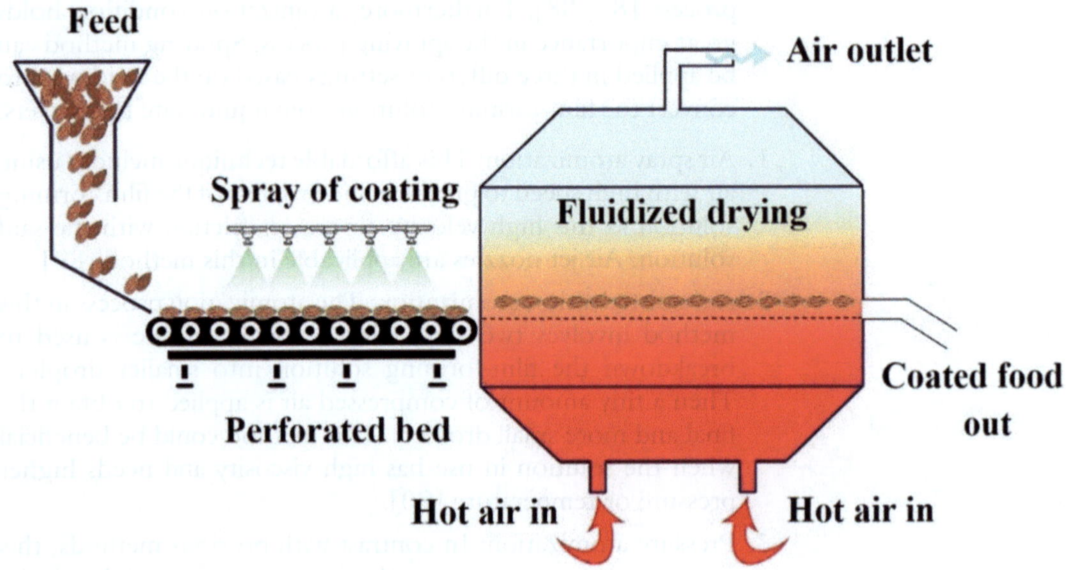

Fig. 5 Demonstration of fluidized-bed processing

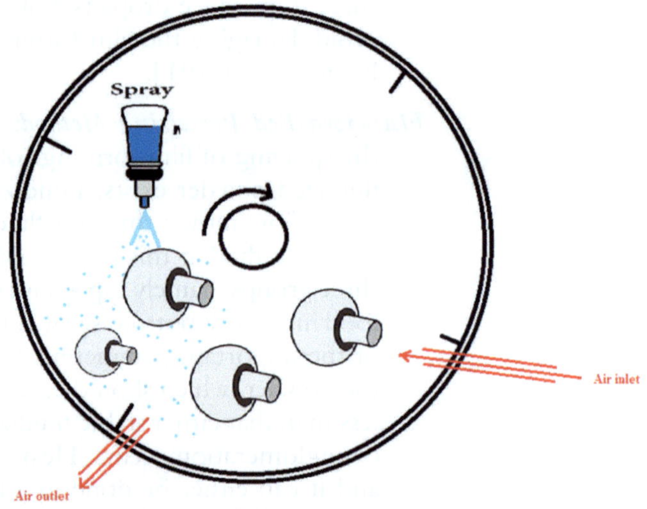

Fig. 6 Panning method

pharmaceuticals, this method is also applicable in confectionery. For instance, coating peanuts and almonds as confectionery cores, with gum Arabic solution which produced a homogenous layer. Through this method high transparency, flexibility, and a glossy appearance is achievable. Moreover, coating extruded foods in various shapes is possible. Also, it is possible to coat a large quantity of foods in a single step in this method. However, extensive process time, possibility of cluster formation, and unreasonable cost are among its drawbacks [9, 94].

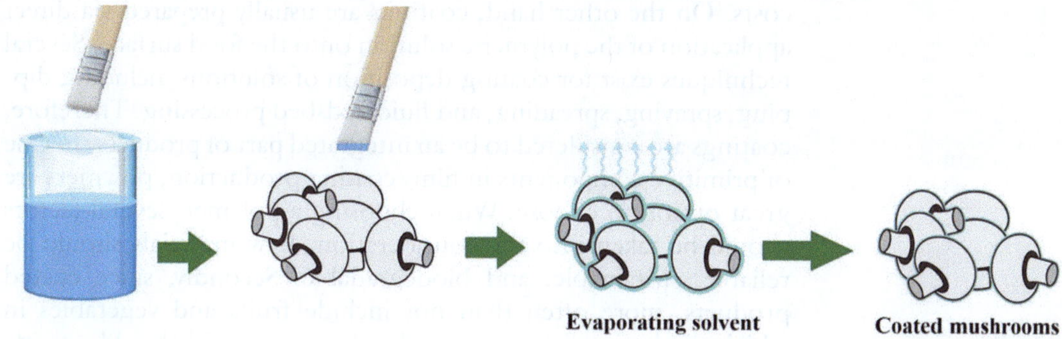

Fig. 7 Spreading method

Spreading Method: This method (Fig. 7) involves spreading or brushing of the film-forming solution on the surface of the food with a controlled speed and then drying of the coated film. This method can be applicable for viscous solution and also as an alternative to casting for producing larger films. A special blade is connected to the brushing or spreading tool, which ensures a controlled film thickness during the process. Drying stage is done via hot air circulation and it takes place on the support. Moreover, this method is applicable to the coatings based on proteins and polysaccharides [95]. There are two parameters concerning this process, including spreading rate and wetting degree which both are estimated by the contact angle analysis. The overall method is under the influence of the geometry of the product, temperature, relative air moisture, viscosity, density, and surface tension of the film-forming solution. Spreading is a harder method to apply films when the film-forming solution has high viscosity since viscosity literally means the resistance of a liquid against flow [96].

4 Notes

Edible films and coatings are described as polymeric layers with very thin thickness which are considered safe and non-toxic for humans. These coatings act as a barrier against gases or moisture to keep the coated food in its optimum condition during its storage time [6]. There is a distinction between films and coatings, the former is prepared separately through wet or dry processing; then it is used to cover food products after its drying and processing. Moreover, films can be produced as single, double, and multiple layers [14]. Consequently, the multiple layer technology in film production offers greater protection regarding its barrier properties against moisture and gases. On the other hand, multiple layer film is not common since it requires additional processing steps and costs. On the other hand, coatings are usually prepared via direct application of the polymeric solution onto the food surface. Several techniques exist for coating deposition of solutions including dipping, spraying, spreading, and fluidized-bed processing. Therefore, coatings are considered to be an integrated part of products. In case of primitive components in film/coating production, polymers are great option to choose. When choosing a polymer, several factors should be taken into the consideration. Raw materials should be reliable, sustainable, and biodegradable. Secondly, since coated products, more often than not include fruits and vegetables in which respiration is important, the chosen material should provide sufficient control over this phenomenon. The initial material should be easy to process, based on the specific method used and its required parameters. For instance, thermoplasticity and low moisture content are essential parameters in dry method of film formation. The additive used in the process should be compatible with the rest of the materials and have the ability to combine and produce homogenous solutions. Finally, the obtained film/coating should extend the shelf life of product, prevent oxidation and microbial spoilage.

Edible coating, like any other innovation, has its own advantages and disadvantages.

Some advantages are mentioned below:

Edible coatings lead to glossy and better appearance in the coated products, especially if the coating is based on lipids.

Edible packaging causes a reduction in respiration rate of coated fruits and vegetables, therefore prevent them from ripening early through controlling the release of ethylene gas.

Decreasing the annual usage of plastic may be among the most important advantages of these biodegradable films which are made out of sustainable biopolymers.

Edible coatings/films are a feasible method to prevent the diffusion of volatile compounds such as aroma molecules from the food.

In the case of coated fruits, edible packaging can moderately enhance the firmness of the fruit's texture.

As one of its main goals, edible packaging of foods contributes to the extension of their shelf life during the storage period.

In contrast to petroleum-based and conventional plastic packaging, edible coatings/films are free of toxins and free radicals.

Since one of the main applications of edible coatings/films is considered to be their barrier properties, moisture loss is reduced in coated foods. Especially when the coatings contain lipid components.

Incorporation of different bioactive compounds such as antioxidant or antimicrobial molecules can enhance the shelf life extension of food by this packaging.

Edible packaging can prevent the coated fruits from minimal surface damages.

Several disadvantages of edible packaging are as follows:

Gas and water vapor interactions may lead to some modifications and defects in the structure of edible film/coating.

Increase in the thickness of coated layer can alter the sensorial perception of coated foods regarding its texture, flavor, and overall appearance.

The used ingredients may trigger an allergic reaction in some individuals.

The end product could sometimes be expensive based on the types of ingredients and production method that was applied [97].

References

1. Campos-Requena VH, Rivas BL, Pérez MA, Garrido-Miranda KA, Pereira ED (2015) Polymer/clay nanocomposite films as active packaging material: modeling of antimicrobial release. Eur Polym J 71:461–475
2. Jaramillo CM, Gutiérrez TJ, Goyanes S, Bernal C, Famá L (2016) Biodegradability and plasticizing effect of yerba mate extract on cassava starch edible films. Carbohydr Polym 151:150–159
3. Bashir K, Jan K, Aggarwal M (2017) Thermo-rheological and functional properties of gamma-irradiated wholewheat flour. Int J Food Sci Technol 52(4):927–935
4. Moghadam M, Salami M, Mohammadian M, Khodadadi M, Emam-Djomeh Z (2020) Development of antioxidant edible films based on mung bean protein enriched with pomegranate peel. Food Hydrocoll 104: 105735
5. Dong C, Wang B, Li F, Zhong Q, Xia X, Kong B (2020) Effects of edible chitosan coating on Harbin red sausage storage stability at room temperature. Meat Sci 159:107919
6. Petkoska AT, Daniloski D, D'Cunha NM, Naumovski N, Broach AT (2021) Edible packaging: sustainable solutions and novel trends in food packaging. Food Res Int 140:109981
7. Cazón P, Velazquez G, Ramírez JA, Vázquez M (2017) Polysaccharide-based films and coatings for food packaging: a review. Food Hydrocoll 68:136–148
8. Gautam K, Vishvakarma R, Sharma P, Singh A, Kumar Gaur V, Varjani S, Kumar Srivastava J (2022) Production of biopolymers from food waste: constrains and perspectives. Bioresour Technol 361:127650
9. Suhag R, Kumar N, Petkoska AT, Upadhyay A (2020) Film formation and deposition methods of edible coating on food products: a review. Food Res Int 136:109582
10. Zhao G, Lyu X, Lee J, Cui X, Chen W-N (2019) Biodegradable and transparent cellulose film prepared eco-friendly from durian rind for packaging application. Food Packag Shelf Life 21:100345
11. Erginkaya Z, Kalkan S, Ünal E (2014) Use of antimicrobial edible films and coatings as packaging materials for food safety. In: Malik A, Erginkaya Z, Ahmad S, Erten H (eds) Food processing: strategies for quality assessment. Springer, New York, pp 261–295
12. Mahajan BVC, Tandon R, Kapoor S, Sidhu MK (2018) Natural coatings for shelf-life enhancement and quality maintenance of fresh fruits and vegetables—a review. J Postharvest Technol 6(1):12–26
13. Asgher M, Qamar SA, Bilal M, Iqbal HM (2020) Bio-based active food packaging materials: sustainable alternative to conventional petrochemical-based packaging materials. Food Res Int 137:109625
14. Mohamed SAA, El-Sakhawy M, El-Sakhawy MA-M (2020) Polysaccharides, protein and

lipid-based natural edible films in food packaging: a review. Carbohydr Polym 238:116178

15. Pelissari FM, Ferreira DC, Louzada LB, dos Santos F, Corrêa AC, Moreira FKV, Mattoso LH (2019) Chapter 10—Starch-based edible films and coatings: an eco-friendly alternative for food packaging. In: Silva Clerici MTP, Schmiele M (eds) Starches for food application. Academic, pp 359–420
16. Jiang T, Duan Q, Zhu J, Liu H, Yu L (2020) Starch-based biodegradable materials: challenges and opportunities. Adv Ind Eng Polym Res 3(1):8–18
17. Tagliapietra BL, Felisberto MHF, Sanches EA, Campelo PH, Clerici MTPS (2021) Non-conventional starch sources, current opinion in food. Science 39:93–102
18. Versino F, Lopez OV, Garcia MA, Zaritzky NE (2016) Starch-based films and food coatings: an overview. Starch-Stärke 68(11–12): 1026–1037
19. Samsudin H, Hani NM (2017) Use of starch in food packaging. In: Starch-based materials in food packaging. Academic
20. Molavi H, Behfar S, Shariati MA, Kaviani M, Atarod S (2015) A review on biodegradable starch based film. J Microbiol Biotechnol Food Sci 4:456–461
21. Rostamabadi H, Demirkesen I, Colussi R, Roy S, Tabassum N, de Oliveira Filho JG, Bist Y, Kumar Y, Nowacka M, Galus S, Falsafi SR (2024) Recent trends in the application of films and coatings based on starch, cellulose, chitin, chitosan, xanthan, gellan, pullulan, Arabic gum, alginate, pectin, and carrageenan in food packaging. Food Front 5(2):350–391
22. Baybaş D, Serdaroğlu G, Semerci B (2021) The composite microbeads of alginate, carrageenan, gelatin, and poly(lactic-co-glycolic acid): synthesis, characterization and Density Functional Theory calculations. Int J Biol Macromol 181:322–338
23. Goh CH, Heng PWS, Chan LW (2012) Alginates as a useful natural polymer for microencapsulation and therapeutic applications. Carbohydr Polym 88(1):1–12
24. Agüero L, Zaldivar-Silva D, Peña L, Dias ML (2017) Alginate microparticles as oral colon drug delivery device: a review. Carbohydr Polym 168:32–43
25. Gheorghita R, Gutt G, Amariei S (2020) The use of edible films based on sodium alginate in meat product packaging: an eco-friendly alternative to conventional plastic materials. Coatings 10(2):166
26. Zhou T, Wang H, Han Q, Song Z, Yu D, Li G, Liu W, Dong C, Ge S, Chen X (2024) Fabrication and characterization of an alginate-based film incorporated with cinnamaldehyde for fruit preservation. Int J Biol Macromol 274:133398
27. Tavassoli-Kafrani E, Shekarchizadeh H, Masoudpour-Behabadi M (2016) Development of edible films and coatings from alginates and carrageenans. Carbohydr Polym 137:360–374
28. Hassan RA, Heng LY, Tan LL (2019) Novel DNA biosensor for direct determination of carrageenan. Sci Rep 9(1):6379
29. Jancikova S, Dordevic D, Jamroz E, Behalova H, Tremlova B (2020) Chemical and physical characteristics of edible films, based on κ- and ι-carrageenans with the addition of lapacho tea extract. Foods (Basel, Switzerland) 9(3):357
30. Song D-H, Hoa VB, Kim HW, Khang SM, Cho S-H, Ham J-S, Seol K-H (2021) Edible films on meat and meat products. Coatings 11(11):1344
31. Maurizzi E, Bigi F, Quartieri A, De Leo R, Volpelli LA, Pulvirenti A (2022) The green era of food packaging: general considerations and new trends. Polymers 14(20):4257
32. Nešić A, Cabrera-Barjas G, Dimitrijević-Branković S, Davidović S, Radovanović N, Delattre C (2020) Prospect of polysaccharide-based materials as advanced food packaging. Molecules 25(1):135
33. Economou V, Tsitsos A, Theodoridis A, Ambrosiadis I, Arsenos G (2022) Effects of chitosan coatings on controlling Listeria monocytogenes and methicillin-resistant Staphylococcus aureus in beef and mutton cuts. Appl Sci 12(22):11345
34. Xie Q, Liu G, Zhang Y, Yu J, Wang Y, Ma X (2023) Active edible films with plant extracts: a updated review of their types, preparations, reinforcing properties, and applications in muscle foods packaging and preservation. Crit Rev Food Sci Nutr 63(32):11425–11447
35. Desvita H, Faisal M, Mahidin M, Suhendrayatna S (2020) Edible coating for beef preservation from chitosan combined with liquid smoke. Int J Technol 11(4):817–829
36. Seyfzadeh M, Motalebi A, Kakoolaki S, Gholipour H (2013) Chemical, microbiological and sensory evaluation of gutted kilka coated with whey protein based edible film incorporated with sodium alginate during frozen storage. Iran J Fish Sci 12(1):140–153
37. Kandasamy S, Yoo J, Yun J, Kang H-B, Seol K-H, Kim H-W, Ham J-S (2021) Application of whey protein-based edible films and coatings

in food industries: an updated overview. Coatings 11(9):1056

38. Chiralt A, González-Martínez C, Vargas M, Atarés L (2018) Edible films and coatings from proteins. In: Proteins in food processing. Elsevier, pp 477–500
39. Picchio ML, Linck YG, Monti GA, Gugliotta LM, Minari RJ, Alvarez Igarzabal CI (2018) Casein films crosslinked by tannic acid for food packaging applications. Food Hydrocoll 84: 424–434
40. Mohamed A, Ramaswamy HS (2022) Characterization of caseinate–carboxymethyl chitosan-based edible films formulated with and without transglutaminase enzyme. J Compos Sci 6(7):216
41. Farhan A, Hani NM (2017) Characterization of edible packaging films based on semi-refined kappa-carrageenan plasticized with glycerol and sorbitol. Food Hydrocoll 64:48–58
42. Ramos M, Valdés A, Beltran A, Garrigós MC (2016) Gelatin-based films and coatings for food packaging applications. Coatings 6(4):41
43. Lin J, Pan D, Sun Y, Ou C, Wang Y, Cao J (2019) The modification of gelatin films: based on various cross-linking mechanism of glutaraldehyde at acidic and alkaline conditions. Food Sci Nutr 7(12):4140–4146
44. Jridi M, Mora L, Souissi N, Aristoy M-C, Nasri M, Toldrá F (2018) Effects of active gelatin coated with henna (L. inermis) extract on beef meat quality during chilled storage. Food Control 84:238–245
45. González A, Barrera GN, Galimberti PI, Ribotta PD, Igarzabal CIA (2019) Development of edible films prepared by soy protein and the galactomannan fraction extracted from Gleditsia triacanthos (Fabaceae) seed. Food Hydrocoll 97:105227
46. Kim S, Sessa D, Lawton J (2004) Characterization of zein modified with a mild cross-linking agent. Ind Crop Prod 20(3):291–300
47. Khwaldia K, Perez C, Banon S, Desobry S, Hardy J (2004) Milk proteins for edible films and coatings. Crit Rev Food Sci Nutr 44(4): 239–251
48. Wang Q, Padua GW (2005) Properties of zein films coated with drying oils. J Agric Food Chem 53(9):3444–3448
49. Chen H, Wang J, Cheng Y, Wang C, Liu H, Bian H, Pan Y, Sun J, Han W (2019) Application of protein-based films and coatings for food packaging: a review. Polymers 11(12): 2039
50. Punia Bangar S, Chaudhary V, Thakur N, Kajla P, Kumar M, Trif M (2021) Natural antimicrobials as additives for edible food packaging applications: a review. Foods 10(10): 2282
51. Wulandari D, Erwanto Y, Pranoto Y, Rusman R (2017) The properties of edible film derived from bovine split hide gelatin with isolated soy protein using various levels of glycerol in the presence of transglutaminase. Buletin Peternakan 41(3):319–327
52. Pashova S (2023) Application of plant waxes in edible coatings. Coatings 13(5):911
53. Saucedo-Pompa S, Jasso-Cantu D, Ventura-Sobrevilla J, Sáenz-Galindo A, Rodríguez-Herrera R, Aguilar CN (2007) Effect of candelilla wax with natural antioxidants on the shelf life quality of fresh-cut fruits. J Food Qual 30(5):823–836
54. De Pilli T (2020) Development of a vegetable oil and egg proteins edible film to replace preservatives and primary packaging of sweet baked goods. Food Control 114:107273
55. Nair SS, Trafiałek J, Kolanowski W (2023) Edible packaging: a technological update for the sustainable future of the food industry. Appl Sci 13(14):8234
56. Sahraee S, Milani JM, Regenstein JM, Kafil HS (2019) Protection of foods against oxidative deterioration using edible films and coatings: a review. Food Biosci 32:100451
57. Mellinas C, Valdés A, Ramos M, Burgos N, Garrigós MdC, Jiménez A (2016) Active edible films: current state and future trends. J Appl Polym Sci 133(2)
58. Bourtoom T (2009) Edible protein films: properties enhancement. Int Food Res J 16(1):1–9
59. Pérez-Vergara LD, Cifuentes MT, Franco AP, Pérez-Cervera CE, Andrade-Pizarro RD (2020) Development and characterization of edible films based on native cassava starch, beeswax, and propolis. NFS J 21:39–49
60. Senturk Parreidt T, Müller K, Schmid M (2018) Alginate-based edible films and coatings for food packaging applications. Foods 7(10):170
61. Barbosa CH, Andrade MA, Vilarinho F, Fernando AL, Silva AS (2021) Active edible packaging. Encyclopedia 1(2):360–370
62. Castro FVR, Andrade MA, Sanches Silva A, Vaz MF, Vilarinho F (2019) The contribution of a whey protein film incorporated with green tea extract to minimize the lipid oxidation of Salmon (Salmo salar L.). Foods 8(8):327
63. Souza V-G-L, Rodrigues P-F, Duarte M-P, Fernando A-L (2018) Antioxidant migration studies in chitosan films incorporated with plant extracts. J Renew Mater 6(5):548–558

64. Cai L, Wang Y (2021) Physicochemical and antioxidant properties based on fish sarcoplasmic protein/chitosan composite films containing ginger essential oil nanoemulsion. Food Bioprocess Technol 14(1):151–163
65. Qin Y, Xu F, Yuan L, Hu H, Yao X, Liu J (2020) Comparison of the physical and functional properties of starch/polyvinyl alcohol films containing anthocyanins and/or betacyanins. Int J Biol Macromol 163:898–909
66. Sholichah E, Nugroho P, Purwono B (2018) Preparation and characterization of active film made from arrowroot starch/PVA film and isolated quercetin from shallot (Allium cepa L. var, aggregatum). In: AIP conference proceedings. AIP Publishing
67. Abdollahzadeh E, Nematollahi A, Hosseini H (2021) Composition of antimicrobial edible films and methods for assessing their antimicrobial activity: a review. Trends Food Sci Technol 110:291–303
68. Lucera A, Costa C, Conte A, Del Nobile MA (2012) Food applications of natural antimicrobial compounds. Front Microbiol 3:287
69. Al-Hashimi AG, Ammar AB, Lakshmanan G, Cacciola F, Lakhssassi N (2020) Development of a millet starch edible film containing clove essential oil. Foods 9(2):184
70. Fox PF, Kelly AL (2006) Indigenous enzymes in milk: overview and historical aspects—Part 1. Int Dairy J 16(6):500–516
71. Leśnierowski G, Yang T (2021) Lysozyme and its modified forms: a critical appraisal of selected properties and potential. Trends Food Sci Technol 107:333–342
72. Kumar N, Neeraj (2019) Polysaccharide-based component and their relevance in edible film/coating: a review. Nutr Food Sci 49:793–823
73. Sothornvit R, Krochta JM (2005) 23—Plasticizers in edible films and coatings. In: Han JH (ed) Innovations in food packaging. Academic, London, pp 403–433
74. Moura-Alves M, Esteves A, Ciríaco M, Silva JA, Saraiva C (2023) Antimicrobial and antioxidant edible films and coatings in the shelf-life improvement of chicken meat. Foods 12(12): 2308
75. Tharanathan RN (2003) Biodegradable films and composite coatings: past, present and future. Trends Food Sci Technol 14(3):71–78
76. Prakoso FAH, Indiarto R, Utama GL (2023) Edible film casting techniques and materials and their utilization for meat-based product packaging. Polymers 15(13):2800
77. Wang Q, Chen W, Zhu W, McClements DJ, Liu X, Liu F (2022) A review of multilayer and composite films and coatings for active biodegradable packaging. npj Sci Food 6(1):18
78. Ajesh Kumar V, Hasan M, Mangaraj S, Pravitha M, Verma DK, Srivastav PP (2022) Trends in edible packaging films and its prospective future in food: a review. Appl Food Res 2(1):100118
79. Costa C, Conte A, Del Nobile MA (2014) Effective preservation techniques to prolong the shelf life of ready-to-eat oysters. J Sci Food Agric 94(13):2661–2667
80. Faustini M, Louis B, Albouy PA, Kuemmel M, Grosso D (2010) Preparation of sol–gel films by dip-coating in extreme conditions. J Phys Chem C 114(17):7637–7645
81. Raghav PK, Agarwal N, Saini M (2016) Edible coating of fruits and vegetables: a review. Education 1(2):188–204
82. Zarubica A (2015) Modified nanostructured titania based thin films in photocatalysis: kinetic and mechanistic approach. React Kinet Mech Catal 115:159–174
83. Atieno L, Owino W, Ateka EM, Ambuko J (2019) Influence of coating application methods on the postharvest quality of cassava. Int J Food Sci 2019(1):2148914
84. Andrade RD, Skurtys O, Osorio FA (2012) Atomizing spray systems for application of edible coatings. Compr Rev Food Sci Food Saf 11(3):323–337
85. Aayush K, McClements DJ, Sharma S, Sharma R, Singh GP, Sharma K, Oberoi K (2022) Innovations in the development and application of edible coatings for fresh and minimally processed Apple. Food Control 141:109188
86. Ribeiro AM, Estevinho BN, Rocha F (2021) Preparation and incorporation of functional ingredients in edible films and coatings. Food Bioprocess Technol 14:209–231
87. Costa MJ, Marques AM, Pastrana LM, Teixeira JA, Sillankorva SM, Cerqueira MA (2018) Physicochemical properties of alginate-based films: effect of ionic crosslinking and mannuronic and guluronic acid ratio. Food Hydrocoll 81:442–448
88. Bizymis A-P, Tzia C (2022) Edible films and coatings: properties for the selection of the components, evolution through composites and nanomaterials, and safety issues. Crit Rev Food Sci Nutr 62(31):8777–8792
89. Valdés A, Ramos M, Beltrán A, Jiménez A, Garrigós MC (2017) State of the art of antimicrobial edible coatings for food packaging applications. Coatings 7(4):56
90. Peretto G, Du W-X, Avena-Bustillos RJ, De J, Berrios J, Sambo P, McHugh TH (2017)

Electrostatic and conventional spraying of alginate-based edible coating with natural antimicrobials for preserving fresh strawberry quality. Food Bioprocess Technol 10:165–174
91. Rahman MA, Heidrick T, Fleck BA (2009) Correlations between the two-phase gas/liquid spray atomization and the Stokes/aerodynamic Weber numbers. J Phys Conf Ser 2009: 012057
92. Debeaufort F, Voilley A (2009) Lipid-based edible films and coatings. In: Edible films and coatings for food applications. Springer, pp 135–168
93. Embuscado ME, Huber KC (2009) Edible films and coatings for food applications. Springer
94. Basu A, De A, Dey S (2013) Techniques of tablet coating: concepts and advancements: a comprehensive review. RRJPPS 2(4):1–6
95. Méndez-Vilas A (2013) Microbial pathogens and strategies for combating them: science, technology and education. Formatex Research Center
96. Sikalo S, Marengo M, Tropea C, Ganic E (2000) Analysis of impact of droplets on horizontal surfaces. In: Thermal Sciences 2000. Proceedings of the international thermal science seminar, vol 1. Begel House Inc.
97. Dhall RK (2013) Advances in edible coatings for fresh fruits and vegetables: a review. Crit Rev Food Sci Nutr 53(5):435–450

Chapter 14

Circular Economy Techniques for Waste Reduction in Food Packaging

Homa Behmadi, Elham Azarpazhooh, Zahra Emam-Djomeh, and Hosahalli S. Ramaswamy

Abstract

This chapter investigates the application of circular economy (CE) techniques to reduce food packaging waste along the supply chain and address environmental challenges caused by traditional linear models. It focuses on innovative and sustainable materials, economic and social considerations, and waste management strategies to reduce environmental pollution, maintain food quality and safety, and minimize food losses and waste. The discussion also covers intelligent and smart packaging, biodegradable and bio-based materials, nanoengineered material (e.g., Carbon dots), artificial intelligence (AI) designed new materials, Three-Dimensional (3D) and Four-Dimensional (4D) printed materials, reusable systems, advanced recycling technologies and recent applications of Internet of Things (IoT) and AI in supply chain management, improving recycling processes, and enabling intelligent packaging solutions. The chapter emphasizes the importance of moving from linear to circular systems to increase resource efficiency and waste prevention. The role of the food packaging industry in preserving food quality during storage and global supply chain, its significant contribution to environmental pollutions, particularly through sustainable non-biodegradable materials are discussed. Case studies, applications, barriers to adoption, and strategies for scaling up sustainable practices, consumer behavior, policy supports, technological developments, and economic impacts are also mentioned. The chapter concludes by describing future research trends and the potential for cross-sector collaboration to promote sustainable innovation in the field of food packaging.

Key words Bio-based materials, Circular supply chain models, Eco-friendly food packaging, Sustainable packaging, Waste reduction

1 Introduction

Packaging is critical to the sustainability of food systems as it reduces loss and waste, maintains quality, and extends safety in the supply chain. However, packaging materials are often made from petroleum. Conventional synthetic plastics have a negative impact on the environment as they are not sufficiently recyclable and remain in ecosystems in the long term. The circular economy (CE) offers an efficient alternative to solve this problem by

Tanmay Sarkar (ed.), *Food Waste Valorization: Green Techniques in Sustainable Management*, Methods and Protocols in Food Science, https://doi.org/10.1007/978-1-0716-4646-5_14,
© The Author(s), under exclusive license to Springer Science+Business Media, LLC, part of Springer Nature 2025

encouraging systems to reduce waste, increase resource efficiency, recycle materials, and return them to the production cycle. This chapter investigates CE techniques for waste reduction in food packaging, focusing on innovative materials, sustainable designs, and waste management strategies, while referring to implementation and large-scale industrial challenges and future potential. The food packaging industries encountered dual challenges: ensuring food safety and reducing food losses and waste during storage and the supply chain and in the other hand, reducing environmental pollutions by considering economic and social considerations. Increasing the food losses and waste and decreasing the amount of the common plastic wastes are rational approaches. The worldwide food industry produces about 1.3 billion tons of food waste annually. This amount of waste is estimated at 2.6 trillion US dollars and is responsible for 8–10% of greenhouse gas emissions [1]. Plastic waste remains one of the most difficult and largely unresolved challenging global environmental issues. Recent statistics indicate that the 353 million tons of annual plastic waste generated in 2019 will grow to 1014 million tons by 2060 unless the plastics industry undergoes systemic revolutions [2]. The CE provides a framework to address these problems by emphasizing resource efficiency, waste minimization, and material recycling. Development of innovative new materials such as biodegradable materials, intelligent packaging, nanoengineered material (e.g., Carbon dots), artificial intelligence (AI) designed new materials, Three-Dimensional (3D) and Four-Dimensional (4D) printed materials demonstrate how innovation can bring into line with CE principles. Additionally, reusable packaging systems along with recent applications of Internet of Things (IoT) and AI in optimization of supply chain support sustainability goals and can significantly reduce the negative impact on the environment and decrease carbon footprint of food packaging residues [1, 3–5].

The transition CE models is gradually recognized as a necessity for businesses to maintain competitiveness and environmental responsibility. Table 1 shows the key differences between traditional linear economy methods and CE approaches, emphasizing aspects such as resource use, waste generation, and environmental impacts.

Strategies such as using biodegradable materials, developing closed-loop systems, and implementing multi-criteria decision-making frameworks can lead to reducing package waste and improving environmental sustainability [4, 6, 7]. However, obstacles such as financial constraints, lack of knowledge, and legal challenges prevent widespread industrial adoption of this approach. Eliminating these barriers through stakeholder engagement, technological innovations, and supportive policies is essential to implement and advance the CE in food packaging and achieve long-term sustainability goals [8].

Table 1
Comparison of traditional and circular economy models in food packaging

Aspect	Traditional linear economy	Circular economy
Resource utilization	Relies on virgin materials; high resource extraction	Prioritizes renewable, recycled, or upcycled materials
Waste generation	High waste generation; single-use packaging dominates	Minimizes waste; designs for reuse, recycling, or composting
Environmental impact	Significant carbon footprint and pollution	Reduces environmental impact through closed-loop systems
End-of-life management	Landfill or incineration; limited recycling	Focuses on recovery, recycling, and reintegration into production
Economic model	"Take, make, dispose" linear approach	"Make, use, recycle" circular approach

Sustainable packaging is important for minimizing the environmental footprint of food products. It protects foods during transport and extends their shelf life, thereby reduced food losses and waste [9]. The integration of bioplastics and foams, edible packaging, and intelligent systems, such as sensors that monitor food quality in real-time, represents promising avenues for creating environmentally friendly alternatives that fit perfectly with the principles of a CE [10–12]. Circular packaging solutions are based on the principles of "reduce, reuse and recycle" [1]. By redesigning production processes and using innovative materials and systems, the food industry can transit from traditional linear models to circular models. This change will promote a more sustainable and resilient future.

2 Materials

Sustainable materials are getting a lot of attention as eco-friendly alternatives to fossil-based packages. With growing concerns such as global warming, higher carbon emissions, and non-biodegradable waste, researchers are focusing on creating safe, biodegradable, and environmentally friendly materials [13]. Table 2 shows some of the most common and novel materials offering benefits for sustainable food packaging.

2.1 Biodegradable Materials

Biodegradable materials break down into natural components under specific conditions, reducing long-term environmental pollution. Examples include:

2.1.1 Polylactic Acid (PLA)

Polylactic Acid (PLA) is a biodegradable polymer made from renewable resources like corn starch or sugarcane, offering a sustainable alternative to petroleum-based plastics for packaging. It

Table 2
Overview of circular packaging materials: types, examples, and benefits

Material type	Description	Examples	Benefits
Biodegradable materials	Break down into natural components under specific conditions	PLA, PHAs, starch-based plastics, cellulose-based materials	Reduces long-term pollution, compostable
Compostable materials	Fully decompose in composting systems	Cellulose, hemp, bamboo	Returns nutrients to soil, reduces landfill waste
Recycled materials	Incorporate recycled content to reduce virgin material demand	rPET, HDPE, recycled glass and paper	Promotes resource efficiency, reduces landfill waste
Edible packaging	Integrates food-grade materials into packaging solutions	Seaweed-based wraps, casein films	Zero waste, biodegradable alternatives for food preservation
Bio-based plastics and foams	Derived from renewable resources	Sugarcane, corn starch	Reduces fossil fuel reliance, improves biodegradability
Microbial-derived biomaterials	Uses microorganisms to produce biomaterials	Lactic acid bacteria, *E. coli*, *S. cerevisiae*, *Corynebacterium*, *Bacillus*, *Pseudomonas*	Produces postbiotics, bioplastics, and colorants; sustainable for food and packaging
Other biodegradable polymers	Specialized biodegradable materials	Polyglycolic acid, chitosan	Excellent biodegradability, antimicrobial properties

boasts favorable properties such as tensile strength and flexibility, essential for packaging, and can biodegrade under industrial composting conditions, reducing its environmental impact compared to traditional plastics that linger in landfills for centuries [14, 15]. However, PLA faces challenges like higher production costs and brittleness, limiting its use. Recent advancements, such as blending PLA with natural fillers or agricultural waste, have improved its mechanical properties and cost-effectiveness, enhancing its potential as a sustainable packaging material [16, 17].

2.1.2 Polyhydroxyalkanoates (PHAs)

Polyhydroxyalkanoates (PHAs) are biodegradable polymers made through microbial fermentation of renewable resources like plant oils and sugars. Known for their excellent biodegradability and biocompatibility [18], PHAs are ideal for packaging, decomposing in both industrial and home composting environments, reducing their environmental impact compared to petroleum-based plastics [19, 20]. They also offer strong mechanical properties, such as flexibility and tensile strength, essential for packaging [21]. However, high production costs hinder their widespread use. Recent studies focus on optimizing production and enhancing PHA properties by blending with biopolymers or natural additives [22, 23]. As demand for sustainable packaging grows, PHAs develop as a promising eco-friendly alternative, perfectly in line with both environmental and consumer priorities.

2.1.3 Starch-Based Materials

Starch-based plastics, derived from renewable sources like corn, rice, and cassava, are gaining attention as biodegradable packaging materials. They offer benefits such as low density, biodegradability, and the ability to blend with other materials to improve mechanical strength [24–26]. Their production process—mixing, heating, and casting—is simple and scalable [27]. Recent advancements have improved their tensile strength and moisture resistance, making them suitable for food packaging [10, 28]. These materials biodegrade through microbial action, reducing environmental impact compared to petroleum-based plastics [29]. As demand for sustainable packaging grows, starch-based plastics are a promising solution for reducing waste and supporting a circular economy [30].

2.1.4 Cellulose-Based Materials

Cellulose-based materials, derived from plant fibers, are gaining traction as biodegradable packaging solutions due to their renewable origin and environmental benefits [31–34]. Innovations like nanocellulose, particularly cellulose nanofibrils and cellulose nanocrystals [35] and cellulose-based microcellular foam have improved their mechanical properties, density, and barrier performance, making them ideal for food packaging that requires moisture and gas control [10, 36]. For example, cellulose films infused with oils improve food preservation, extending shelf life [37]. These materials can also be customized for UV protection and antimicrobial properties, expanding their use in packaging [38]. By addressing plastic waste and meeting consumer demand for eco-friendly products [39], cellulose-based materials are important in sustainable packaging solutions.

2.1.5 Protein-Based Materials

Protein-based materials are increasingly used in packaging due to their sustainability and biodegradability. Recent advancements show the use of protein-derived bionanocomposites, particularly from soy protein and gelatin, which enhance mechanical strength and barrier properties through nanofillers like montmorillonites [40, 41]. Whey protein has also gained attention for its application in edible films and coatings, offering excellent mechanical properties and the ability to incorporate active ingredients such as antimicrobials [42]. Despite their benefits, problems remain regarding mechanical strength and water resistance, necessitating further research to optimize these materials for broader applications [43].

2.2 Compostable Materials

Compostable materials, derived from renewable resources such as cellulose, hemp, and bamboo, are designed to fully decompose in both industrial and home composting systems. These materials contribute to a circular economy by returning nutrients to the soil and significantly reducing landfill waste [44, 45]. The biodegradation process of compostable materials is accelerated by microorganisms, which break down the organic matter into natural components, thus enriching the soil and promoting plant growth [46].

Switching to compostable materials is a great way to support sustainability efforts, by minimizing environmental impact and promoting resource efficiency. Cellulose-based products can be produced from agricultural residues, which not only use waste but also reduce reliance on fossil fuels [47]. Additionally, the incorporation of natural fibers like hemp and bamboo improves the mechanical properties of these materials while ensuring they remain biodegradable [48]. As consumer awareness of environmental issues grows, the demand for compostable packaging solutions is expected to rise, that supporting sustainable practices in various industries.

2.3 Recycled and Recyclable Materials

Incorporating recycled content into packaging reduces the demand for virgin materials and diverts waste from landfills. Commonly used materials include:

2.3.1 Recycled Polyethylene Terephthalate (rPET)

This durable, eco-friendly material is made from recycled PET—the same lightweight thermoplastic polymer you'll find in everyday items like water bottles. It combines sustainability with strong mechanical performance, giving new life to a commonly used plastics [49]. The recycling process not only conserves resources but also reduces energy consumption compared to producing virgin PET. Recent advancements in recycling technologies, such as biocatalytic methods, have improved the efficiency of rPET production, allowing for the recovery of monomers that can be reused in manufacturing [50]. The rPET represents a vital step toward a circular economy, addressing both waste management and resource conservation challenges [51].

2.3.2 High-Density Polyethylene (HDPE)

High-Density Polyethylene (HDPE) is a widely used thermoplastic known for its strength, durability, and chemical resistance. It is produced through the low-pressure polymerization of ethylene, resulting in a material that is cost-effective and versatile for various applications, including packaging, construction, and medical devices [52]. HDPE's properties, such as low moisture absorption and high tensile strength, make it suitable for demanding environments although its susceptibility to UV degradation limits outdoor use [53]. Recent advancements have focused on enhancing HDPE's performance through composite materials, such as incorporating titanium dioxide to improve UV resistance [54]. Additionally, HDPE can be recycled, contributing to sustainability efforts in the plastics industry, and is increasingly being used in innovative applications, including wood-plastic composites for construction [55]. HDPE's unique properties and adaptability make it an important material in both consumer and industrial sectors.

2.3.3 Recycled Glass

Recycled glass is infinitely recyclable, maintaining its quality without degradation. It is non-toxic, impermeable, and chemically inert, making it ideal for food and beverage packaging. Using recycled glass reduces the need for virgin materials like sand and soda ash and lowers energy consumption compared to producing new glass [56]. However, its heavy weight increases transportation emissions, and the melting process requires significant energy. Despite these challenges, glass recycling is necessary for circular economy strategies, as it minimizes waste and conserves resources [57].

2.3.4 Recycled Paper

Recycled paper, made from post-consumer waste, is biodegradable and a sustainable alternative to virgin paper. Producing recycled paper uses less energy and water than virgin paper production. These papers are often laminated or coated for food safety and for improving water and oil resistance. They can be recycled multiple times, which can reduce deforestation and waste [58]. However, most laminates and coatings rely on nondegradable, nonrenewable materials, complicating recycling and increasing environmental burdens [59]. To address this, using of the biodegradable and renewable materials like chitosan and Pickering emulsions are investigated as promising solution for creating biodegradable, recyclable, and eco-friendly coatings with improved water and oil resistance [60].

2.4 Edible Packaging

Edible packaging represents a promising shift toward environmentally friendly alternatives in food preservation and packaging. Edible packaging is a zero-waste innovative and sustainable solution that uses food-grade materials to package food products, effectively integrating packaging with the food itself. This type of packaging can be made from various biopolymers, including polysaccharides, proteins, and lipids, and are used to create barriers against moisture and oxygen, thereby enhancing the shelf life and safety of food products [61]. Examples of edible packaging include:

2.4.1 Seaweed-Based Wraps

These wraps are made from seaweeds, which provide a biodegradable and edible alternative for food preservation. They help in reducing plastic waste and offer a unique flavor profile that can complement certain foods. Their natural properties make them suitable for wrapping sandwiches or sushi, appealing to environmentally conscious consumers [62].

2.4.2 Casein Films

Casein films are derived from milk proteins, casein-based films, which offer functional benefits such as moisture control and are effective in extending the shelf life of various foods. By using casein, a by-product of cheese production, this packaging solution also contributes to reducing overall packaging waste [62].

2.4.3 Gelatin and Starch Films

Edible films made from gelatin or starch are commonly used for wrapping fruits and vegetables. These films provide a protective barrier which improves the visual appeal of the produce. Their ability to maintain freshness makes them a practical choice for consumers and retailers [63].

2.4.4 Edible Muffin Liners

This innovative application of edible packaging has gained popularity among consumers. Edible muffin liners allow for easy baking and serving while eliminating the need for disposable paper liners. They contribute to reducing single-use plastics in the food industry and offer a fun, interactive way to enjoy baked goods [64].

2.5 Bio-Based Plastics

Bio-based plastics which are usually derived from renewable sources such as sugarcane and corn starch are increasingly recognized as sustainable alternatives to conventional petroleum-based plastics. These materials reduce reliance on fossil fuels and improve biodegradability [1, 18]. Bio-polyethylene (Bio-PE) and bio-polypropylene (Bio-PP) exemplify this trend, as they are chemically similar to their fossil fuel counterparts but sourced from renewable materials, offering a more sustainable option without compromising performance [65]. The development of these new bioplastics is important for mitigating plastic pollution and promoting a CE, as they can be designed to be biodegradable or recyclable, thus contributing to a more sustainable future [66].

2.6 Biomaterials Derived from Microbial Sources

Microbial biomanufacturing is an innovative approach that uses and controls wild-type or genetically engineered microorganisms to produce advanced biomaterials with various applications in the field of the agriculture, food, and medicine. By handling microorganisms as factories, they can synthesize high-value intracellular and extracellular biomaterials. Key microbial strains for this purpose are: lactic acid bacteria, *Escherichia coli*, *Saccharomyces cerevisiae*, *Corynebacterium* spp., *Bacillus* spp., and *Pseudomonas* spp. These microorganisms are used to produce a variety of biomaterials, including postbiotics, biopolymers (like bioplastics and foams), and microbial colorants. Researchers are focused on the potential uses of these bioactive molecules in sustainable, microbial-assisted strategies for food processing, packaging, and preservation, showing their growing importance in addressing industrial and environmental challenges [67–70].

2.7 Other Biodegradable Polymers

Biodegradable polymers have functional groups such as esters and anhydrides, in their main chain that are prone to hydrolysis.

2.7.1 Polyglycolic Acid (PGA)

Polyglycolic Acid (PGA) is a hydro-biodegradable polymer possesses widespread interest due to its excellent barrier properties and high mechanical strength. PGA can be an ideal choice for food packaging and replacing traditional plastics. PGA is eco-friendly and breaks down into glycolic acid, which is a natural metabolite. However, the high cost of production, sensitivity to moisture, and low toughness are the main challenges. Blending with other polymers, nanocomposites and developing new processing technologies are some suggested solutions [71, 72]. PGA also used for several biomedical applications such as sutures and implants [73].

2.7.2 Chitosan

Derived from crustacean shells, chitosan is valued for its antimicrobial properties and biodegradability in food packaging and biomedical applications [74, 75].

3 Methods

The implementation of CE techniques in food packaging requires systemic and multidisciplinary approaches that integrate sustainable design, efficient production, and effective waste management strategies. Key methods include designing for recyclability and sustainability, implementing reusable packaging systems, leveraging industrial composting and advanced recycling technologies, and adopting intelligent and active packaging solutions. Suitable waste management practices and collaboration across supply chains more improve the effectiveness of these strategies. Together, these approaches contribute to reducing waste, minimizing environmental impact, and promoting the circularity of food packaging systems. More details of these key methods are given below:

3.1 Design for Sustainability and Recyclability

Designing packaging with a focus on resource efficiency, recyclability, and end-of-life management is critical for reducing environmental impact. One key principle is material selection, which involves prioritizing mono-materials to simplify recycling processes and ensure compatibility with existing recycling infrastructures [76, 77]. Another important aspect is modular design, which focuses on creating packaging that can be easily disassembled, particularly for multi-component systems, to improve recyclability. Clear and standardized labeling is also essential, as it encourages consumer participation in recycling programs and ensures materials are sorted correctly. Additionally, incorporating eco-design principles and utilizing life cycle assessment (LCA) tools helps evaluate the environmental impacts of packaging throughout its life cycle, enabling more informed decision-making and minimizing resource use [56].

3.2 Reusable Packaging Systems

Reusable packaging systems help reduce waste by allowing a single packaging unit to be used multiple times. One example is deposit-return schemes, where consumers return used packaging for cleaning and reuse, significantly cutting down on single-use packaging waste. Another approach is refill stations, which encourage the use of refillable containers for products like beverages and dry goods, providing scalable solutions to minimize plastic waste.

3.3 Industrial Composting Systems

Biodegradable packaging materials, such as polylactic acid (PLA) and polyhydroxyalkanoates (PHAs), require specialized composting facilities to break down efficiently. Industrial composting systems maintain controlled temperature and moisture conditions, enabling the decomposition of these materials within weeks [78].

3.4 Advanced Recycling Technologies

Advanced recycling methods overcome the limitations of traditional mechanical recycling by breaking down complex polymers into reusable monomers. Key technologies include chemical recycling and enzymatic degradation. Chemical recycling employs techniques such as pyrolysis and depolymerization to break down plastics into their monomers, enabling high-quality recycling and resource recovery. Enzymatic degradation, an emerging approach, uses enzymes to break down polymers, opening new possibilities for processing materials that are challenging to recycle through mechanical means.

3.5 Smart Packaging

Figure 1 illustrates cutting-edge packaging solutions, from antimicrobial additives and gas-absorbing materials to real-time freshness sensors and RFID tracking, designed to reduce waste, improve food safety, and advance sustainability goals through intelligent, CE-supportive technologies. Smart packaging, including intelligent and active packaging technologies offer significant environmental benefits while also presenting certain challenges.

3.5.1 Intelligent Packaging Systems

Intelligent packaging systems integrate advanced technologies such as sensors, indicators, and data carriers to monitor and communicate the condition of food products and their environment. These systems improve food safety, extend shelf life, and reduce waste, particularly for perishable goods like meat, fish, and fresh produce [11, 79, 80]. While intelligent packaging offers significant environmental benefits, it also presents challenges (Fig. 2). On the positive side, it reduces food waste by extending shelf life and minimizing spoilage, which lowers the associated environmental impacts [81, 82]. Additionally, it improves resource efficiency by

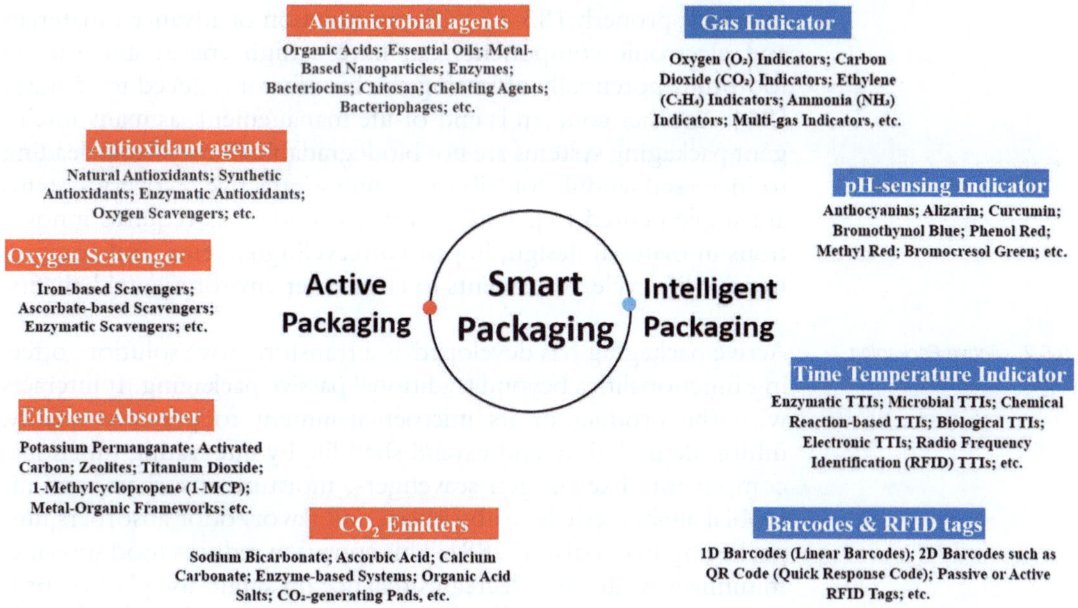

Fig. 1 Smart packaging systems

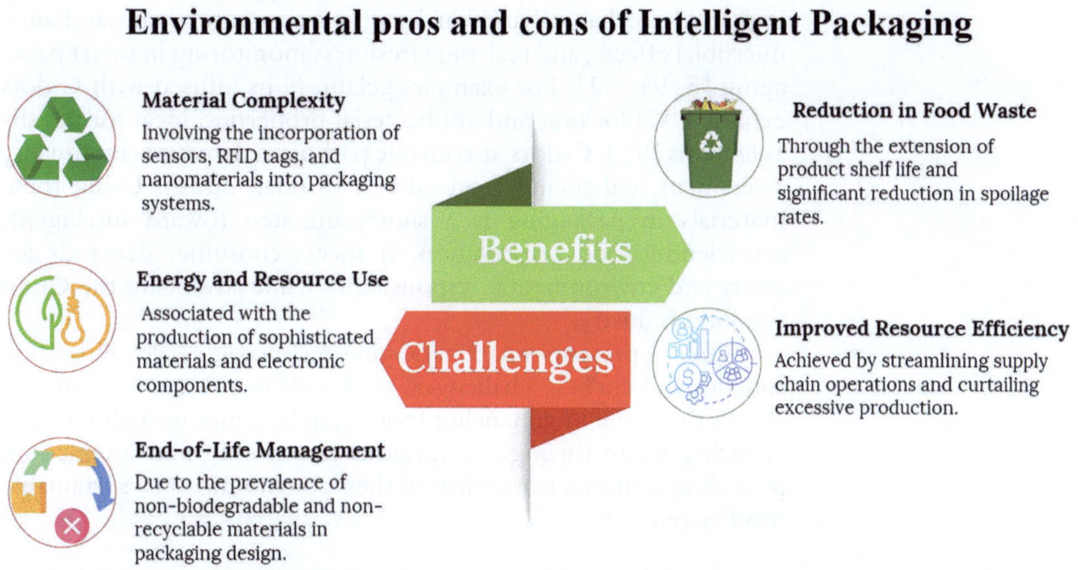

Fig. 2 Environmental pros and cons of intelligent packaging technologies

optimizing supply chains and reducing overproduction, thereby decreasing the overall carbon footprint of food systems. However, challenges arise from the material complexity of these systems, as they often incorporate sensors, RFID tags, and nanomaterials that complicate recycling processes and may increase waste if not managed properly [83–85]. The production of advanced materials and electronic components can have a high energy and carbon footprint, potentially offsetting the benefits of reduced food waste [86]. Another concern is end-of-life management, as many intelligent packaging systems are not biodegradable or recyclable, leading to increased landfill contributions unless effective recovery systems are implemented [9]. Balancing these trade-offs requires innovations in material design, improved recycling systems, and comprehensive life cycle assessments to ensure net environmental benefits.

3.5.2 Active Packaging

Active packaging has developed as a transformative solution, offering functionalities beyond traditional passive packaging. It interacts with the product or its microenvironment to preserve quality, inhibit degradation, and extend shelf life by integrating functional components like oxygen scavengers, moisture absorbers, antimicrobial agents, ethylene absorbers, and flavor/odor absorbers into packaging materials [87, 88]. This approach reduces food spoilage, minimizes waste, and decreases reliance on single-use plastics, promoting sustainability [89]. Recent innovations include biodegradable materials and natural additives, such as essential oils, which improve eco-friendliness [90]. Carbon dots (C-dots) are gaining attention for their multifunctional applications, including improved mechanical and barrier properties, antioxidant and antimicrobial effects, and real-time freshness monitoring in smart packaging [5, 91, 92]. For example, gelatin films infused with C-dots exhibit UV-blocking and antibacterial properties, ideal for perishable items [92]. C-dots also enable pathogen detection and quality assessment, enhancing food safety (Mondal, 2024). Using these materials in packaging is a significant step toward intelligent, eco-friendly food preservation. It meets consumer demands for safety and environmental responsibility while advancing the CE in the food industry.

Table 3 presents smart packaging technologies, detailing their functions, benefits, challenges, and real-world applications. It shows their role in enhancing food safety, extending shelf life, and reducing waste through continuous monitoring and interaction, providing a succinct overview of their contributions to sustainable food systems.

3.6 Nano-engineering, 3D and 4D Printed, and AI-Designed Materials

More recent technologies such as nanoengineering, 3D and 4D printing technologies, and AI designed materials enable the creation of customized packaging solutions customized to specific food products, enhancing functionality and sustainability. 3D printing allows for intricate designs that can incorporate smart features, such as sensors monitoring freshness and quality. This capability reduces waste by ensuring food remains safe for consumption longer while utilizing biocompatible and non-toxic materials [28, 93, 94].

Table 3
Overview of smart packaging technologies

Technology type	Description	Functions	Benefits	Challenges	Examples
Time-temperature indicators (TTIs)	Labels that change color based on time and temperature exposure to indicate freshness	Monitor product exposure to temperature over time	Visual indication of product quality and safety	Limited shelf life of indicators; may not be accurate in all conditions	Freshness labels on refrigerated food products
Freshness sensors	Sensors that detect volatile compounds emitted by spoiled food	Spoilage detection based on gas emissions	Early warning of spoilage, reducing food waste	Requires calibration and may not work for all food types	Sensors integrated into meat packaging
Oxygen scavengers	Materials that absorb oxygen from the packaging environment to extend shelf life	Prevent oxidation and maintain freshness	Prolongs shelf life of oxygen-sensitive products	May alter product chemistry; limited effectiveness in high oxygen environments	Packaged snacks and dried fruits
Moisture control packaging	Packaging that regulates moisture levels to prevent spoilage and maintain quality	Absorbs excess moisture or releases moisture as needed	Helps maintain best product texture and prevents mold growth	Requires precise control; may add costs to packaging	Bakery items and dehydrated foods
RFID (radio-frequency identification)	Tags that use electromagnetic fields to automatically identify and track products	Supply chain tracking and inventory management	Improves traceability and reduces food waste through better inventory control	Higher implementation costs and complexity	RFID tags in large retail operations
Active packaging	Packaging that interacts with the food product to improve its shelf life, such as antimicrobial agents	Release preservatives or absorb contaminants	Reduces microbial growth, extending shelf life	Regulatory approval needed for certain chemicals; potential consumer concerns	Modified atmosphere packaging for meats
Smart labels	Labels with printed electronic circuits that can display real-time information about the product	Display temperature, freshness status, and nutritional information	Provides consumers with immediate information about product safety and quality	Higher production costs and technology limitations	Smart labels on ready-to-eat meals

(continued)

Table 3
(continued)

Technology type	Description	Functions	Benefits	Challenges	Examples
Blockchain-enabled packaging	Packaging that incorporates blockchain technology for transparency in the supply chain	Traceability from farm to fork	Increases consumer trust and improves food safety	Implementation complexity; requires industry-wide cooperation	Traceability systems for organic produce
Colorimetric indicators	Chemical indicators that change color in response to specific conditions (e.g., pH, temperature)	Indicate spoilage or quality changes visually	Simple visual cues for consumers regarding product safety	May not provide precise information; depends on consumer interpretation	pH indicators in dairy products
IoT-enabled packaging	Packaging integrated with IoT devices for real-time monitoring and data collection	Monitor environmental conditions and send alerts	Provides data for supply chain optimization and reduces waste	Requires reliable connectivity and data management systems	

4D printing introduces the dimension of time, allowing packaging to change shape or properties in response to environmental stimuli like temperature or humidity. This adaptability improves food preservation and minimizes spoilage, thereby reducing food waste. The integration of artificial intelligence (AI) in these processes optimizes material selection and design, leading to more efficient production methods and improved packaging performance. AI also accelerates the development of intelligent packaging systems that communicate with consumers and supply chain stakeholders, enhancing sustainability efforts [95, 96].

3.7 Waste Management and Collaboration in the Supply Chain

Effective waste management in the supply chain hinges on well-organized systems for collection, sorting, and recycling to ensure packaging materials are recovered and reintegrated into production processes. A key strategy is the development of closed-loop recycling systems, which strengthen recycling infrastructure and allow materials to be reused repeatedly without compromising quality. Additionally, the implementation of supportive public policies and consumer education campaigns can significantly improve waste management efficiency and encouraging greater consumer participation [97].

Collaboration among all stakeholders in the supply chain—including producers, retailers, and consumers—is essential for optimizing packaging solutions and advancing waste management practices. By working together, these stakeholders can embed CE principles throughout the packaging life cycle, ensuring sustainability at every stage.

3.8 Challenges in Scalability of Biodegradable Materials for Food Packaging

Scaling biodegradable materials for food packaging presents several key challenges. A major issue is cost, as materials like polylactic acid (PLA) and starch-based films are often more expensive to produce than conventional plastics, hindering widespread adoption [98, 99]. Additionally, many biodegradable materials fall short in mechanical and barrier properties, such as strength, flexibility, and resistance to oxygen and moisture, often requiring costly and complex additives to meet food preservation needs [100, 101]. Processing and stability are also concerns, as these materials can be sensitive to heat and humidity, affecting their performance during manufacturing and storage [102].

Another complicating scalability is the lack of adequate composting and recycling infrastructure, which can lead to contamination of traditional recycling streams and undermine their environmental benefits [99]. Consumer awareness and acceptance are equally critical, as misconceptions and limited understanding can reduce demand and slow market growth [103]. Finally, regulatory hurdles, including varying standards across regions, create additional barriers for manufacturers and suppliers [104]. Overcoming these challenges is vital for integrating biodegradable materials into mainstream food packaging.

3.9 Case Studies of Circular Packaging Solutions

Recent case studies illustrate the practical application of CE principles in food packaging. Organizations like the Ellen MacArthur Foundation and Waste and Resources Action Programme (WRAP) have documented numerous examples of companies embedding circular design into their processes, demonstrating innovations in reusable and recyclable packaging materials [105, 106]. Below are some examples of successful implementations of CE techniques in food packaging:

3.9.1 Tetra Pak's Circular Supply Chains

Tetra Pak has developed circular supply chain models in China and Brazil, focusing on creating packaging recovery ecosystems adapted to local market conditions. This approach serves as a model for CE practices in emerging economies, demonstrating how localized solutions can improve sustainability [107].

3.9.2 Reusable Food Packaging

Reusable packaging systems for products such as eggs, bread, and dry foods are gaining popularity among environmentally conscious consumers. These systems significantly reduce single-use packaging waste, promoting a shift toward sustainable consumption [108].

3.9.3 Agro-Food Waste Valorization

Agro-food processing waste is being transformed into bio-based and biodegradable packaging materials. This approach not only reduces waste but also provides sustainable alternatives to synthetic plastics, contributing to a CE by utilizing by-products effectively [62].

3.9.4 Biopolymers from Fish Industry Waste

The fish industry is increasingly utilizing waste materials to produce biopolymers for food packaging. This approach not only addresses waste management issues but also provides sustainable packaging solutions, such as collagen and chitosan, which can improve food preservation [74].

3.9.5 Antimicrobial Packaging from Cheese Whey

Cheese whey is used to produce antimicrobial packaging materials, enhancing food preservation while utilizing waste streams [109].

3.9.6 Ground Coffee Packaging Optimization

A CE framework was applied to optimize ground coffee packaging using multi-objective analysis. The study identified multilayer bags as the most sustainable option, effectively balancing economic and environmental impacts [110].

3.10 SWOT Analysis

Figure 3 presents an SWOT analysis for the implementation of sustainable CE practices in food packaging. The analysis outlines the strengths, weaknesses, opportunities, and threats associated with this approach, which are discussed in detail below.

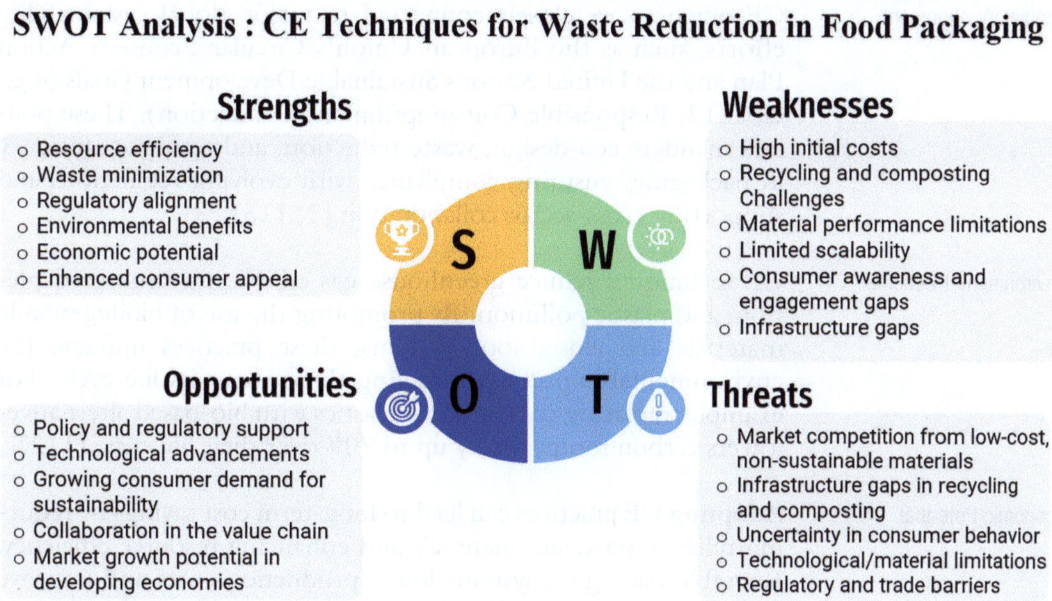

Fig. 3 SWOT Analysis for implementing sustainable CE in food packaging

3.10.1 Strengths

The strengths of implementing CE practices in food packaging lie in their ability to support environmental, economic, and regulatory goals while encouraging innovation and consumer engagement.

Resource Efficiency

CE techniques prioritize the use of renewable, recyclable, and biodegradable materials, such as bio-based plastics (e.g., polylactic acid [PLA] and polyhydroxyalkanoates [PHA]), reducing dependency on finite fossil fuels. This approach minimizes resource extraction, energy consumption, and carbon footprints. Advanced methods like life cycle assessment (LCA) optimize material use during design, while chemical recycling enables the recovery of high-quality materials from previously unrecyclable waste, more enhancing resource efficiency [3, 111]. Closed-loop systems, such as PET reuse, help minimize virgin material extraction and energy consumption.

Waste Minimization

By well-designing packaging for reuse, recycling, or composting, CE techniques significantly reduce landfills and incinerators waste. For example, closed-loop recycling systems allow materials like PET to be reused multiple times without significant quality loss. Biodegradable and compostable materials, such as PLA and PHA, degrade into non-toxic components under appropriate conditions, reducing plastic waste in landfills and marine environments [4, 112]. Innovations in edible packaging and smart sensors more reduce food spoilage, indirectly lowering packaging waste [11].

Regulatory Alignment

CE practices are closely connected to major global sustainability efforts, such as the European Union's Circular Economy Action Plan and the United Nations Sustainable Development Goals (e.g., Goal 12: Responsible Consumption and Production). These policies mandate eco-design, waste reduction, and resource efficiency in packaging, ensuring compliance with evolving regulations and supporting cross-sector collaboration [1, 113].

Environmental Benefits

CE techniques reduce greenhouse gas emissions, resource depletion, and plastic pollution. By promoting the use of biodegradable materials and closed-loop systems, these practices mitigate the environmental impact of packaging throughout its life cycle. For example, replacing conventional plastics with bio-based alternatives lowers carbon footprints by up to 70% over their life cycle [112].

Economic Potential

Adopting CE practices can lead to long-term cost savings by reducing reliance on virgin materials and enhancing resource efficiency. Reusable packaging systems lower production costs and improve material availability, supporting economic resilience and scalability. Industries leveraging recycled materials also benefit from reduced production costs and increased market competitiveness [113]. Public-private partnerships and subsidies further offset initial investments, particularly in developing economies [114].

Improved Consumer Appeal

As consumer awareness of environmental issues grows, demand for sustainable packaging increases. Companies adopting CE principles can use this trend to improve brand loyalty and market competitiveness. Brands adopting CE principles report increased loyalty, with 82% of consumers willing to pay premiums for eco-friendly options [115]. Figure 4 shows the share of respondents saying that eco-friendly packaging is an important aspect of sustainable consumption [116]. Eco-friendly packaging conveys corporate responsibility, influencing purchasing decisions, particularly among environmentally conscious consumers [114, 117]. Clear labeling and campaigns amplify this trend, bridging the intention-action gap [118].

3.10.2 Weaknesses

While CE practices offer transformative potential, several weaknesses delay industrial acceptance, especially in developing countries.

High Initial Costs

Transitioning to CE requires substantial upfront investments in R&D, infrastructure, and material production. Biodegradable plastics like PLA cost 20–50% more than conventional plastics. These financial barriers can deter small- and medium-sized enterprises from adopting circular practices [111, 112, 119].

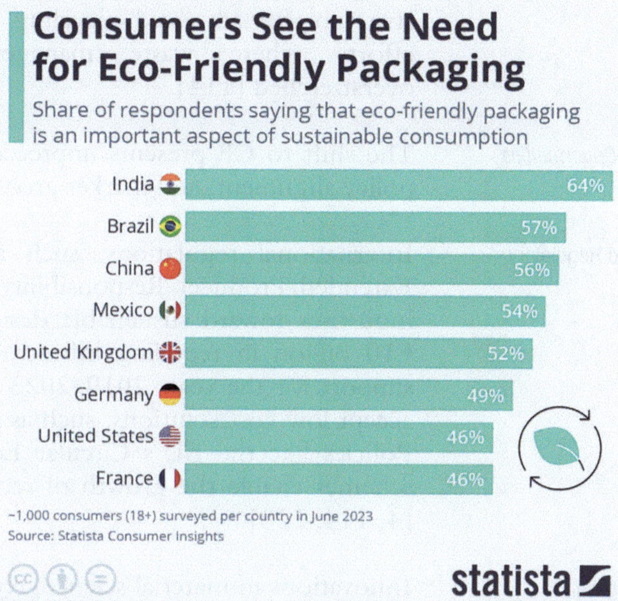

Fig. 4 Importance of eco-friendly packaging in sustainable consumption: Share of respondents [116]

Recycling and Composting Challenges Multilayered or composite packaging is difficult to recycle due to the complexity of separating its components. Biodegradable plastics often require specific industrial composting conditions (e.g., high temperatures and controlled humidity), which are unavailable in many regions. Only 14% of global plastics are recycled, with compostables often landfilled due to inadequate facilities [80, 114].

Material Performance Limitations Biodegradable and compostable materials often lack mechanical strength, permanency, and barrier properties requirements for high moisture or oxygen-sensitive foods, risking food safety and shelf life. These limitations make it challenging to replace conventional plastics with sustainable alternatives in some cases, potentially increasing food waste and counteracting the goals of the CE [112].

Limited Scalability Emerging technologies and most of the innovative materials like edible packaging or nanomaterial are still in the research or pilot phase. This limits applicability in the broader market and delays expected effect on waste reduction [119].

Consumer Engagement Gaps CE techniques depend seriously on consumer participation, such as returning reusable packaging, properly sorting recyclable materials, or disposing of compostable packaging correctly. However, low consumer awareness or accepting often leads to incorrect disposal, making these systems ineffective [111, 117].

Infrastructure Gaps

Many regions, primarily in developing countries, lack the infrastructure for advanced recycling or composting, hindering CE efforts where waste management systems are already overstretched [114].

3.10.3 Opportunities

The shift to CE presents unprecedented chances for innovation, policy alignment, and market growth.

Policy and Regulatory Support

International regulations, such as single-use plastic ban and Extended Producer Responsibility (EPR) programs, are pushing industries toward sustainable design. The EU's commitment of €10 billion in recycling infrastructure indicates real and active support for the years 2019–2023 [120]. Developing nations can accept low-cost solutions, such as agricultural biomass packaging. Policies like the EU's Circular Economy Action Plan and EPR schemes enable the growth of recyclable and reusable packaging [4, 114, 120].

Technological Advancements

Innovations in material science, recycling technologies, and smart packaging offer significant opportunities to improve CE practices, mainly AI- and IoT-based methods. AI-powered waste sorting systems improve recycling efficiency by accurately identifying and separating materials, supporting CE goals [121, 122]. Chemical recycling (e.g., pyrolysis) and smart packaging (e.g., freshness sensors) address material limitations and food waste [11, 112, 123].

Consumer Demand for Sustainability

Growing consumer knowledge of environmental concerns is fueling demand for sustainable packaging, with studies showing a favorability for eco-friendly options, mainly in the food sector. Brands embracing CE principles can capitalize on this trend to stand out and build loyalty [3, 117]. Fortunately, Gen Z's strong inclination toward sustainable packaging, driven by their environmental consciousness and support for waste-reducing practices, indicates a promising shift toward a more sustainable future [124, 125].

Cross-Value Chain Collaboration

Partnerships among packaging manufacturers, food producers, retailers, and waste management companies can speed up CE adoption. Retailer-recycling firm collaborations can improve collection systems for reusable and recyclable packaging. Public-private partnerships are also vital for developing infrastructure and driving innovation in sustainable packaging [114].

Market Growth Potential in Developing Economies

Developing countries, facing infrastructure gaps, offer important opportunities for introducing CE practices. Businesses can adopt low-cost, eco-friendly packaging solutions, such as biodegradable materials derived from locally available sources like agricultural biomass. By addressing local needs and constraints, companies can contribute to sustainable development while creating new revenue streams and building systems from the ground up [112].

3.10.4 Threats

The Challenge of Competing with Low-Cost, Non-sustainable Materials

Conventional plastics derived from fossil fuels remain significantly cheaper and more available than biodegradable or recycled alternatives. This cost difference creates a major barrier to the widespread adoption of CE techniques, especially in price-sensitive markets and developing countries where affordability is prioritized. Without significant subsidies or policy interventions, sustainable packaging options may struggle to compete with conventional plastics [1, 114].

Infrastructure Gaps in Recycling and Composting

Many regions, generally in developing countries, lack proper recycling and composting infrastructure. Even in developed nations, inconsistent waste systems hinder CE efforts. Industrial composting facilities and advanced recycling technologies for complex materials, like multilayer films, are often unavailable, leading to improper disposal and reduced effectiveness [80, 111, 112].

Uncertainty in Consumer Behavior

The success of CE techniques relies heavily on consumer participation, such as returning reusable packaging, properly sorting waste, and disposing of compostable materials correctly. However, inconsistent consumer behavior, driven by limited awareness or motivation, poses a significant challenge. Mislabeling and lack of clear education more intensify confusion, resulting in improper disposal and contamination of waste streams. This uncertainty undermines the effectiveness of circular systems [117, 119].

Technological and Material Limitations

While advancements in biodegradable and recyclable materials are promising, their current limitations pose threats to adoption. Some biodegradable materials require highly specific environmental conditions to decompose, and their failure to degrade in natural environments could lead to public uncertainty about their efficacy. Additionally, chemical recycling processes, though effective, are not yet widely available due to high costs and practical complexity, limiting their practical application [111, 112].

Regulatory and Trade Barriers

Regulatory inconsistencies across regions create challenges for companies operating in large-scale markets. Differing standards for biodegradable materials or recycling protocols may lead to inefficiencies and higher compliance costs. Trade barriers or import restrictions on certain materials may limit the accessibility of sustainable packaging options, slowing the transition to circular systems [114].

3.11 Future Research Directions

The CE in food packaging represents a significant shift toward minimizing waste and maximizing resource regeneration. This transition is driven by innovative technologies and sustainable materials that improve packaging efficiency.

Smart and intelligent packaging, which uses IoT and sensors, can monitor food conditions in real-time. This technology can improve product safety and reduce waste by providing vital data on food quality throughout the supply chain. Future efforts should aim to better integrate these systems into existing processes for improved traceability [35, 126].

The development of biodegradable and bio-based materials is essential to reducing dependence on traditional plastics. Enhancing the properties and biodegradability of materials like polylactic acid (PLA) will ensure they can reintegrate effectively into natural ecosystems [22].

Nanoengineered materials, such as carbon dots, show promise in improving packaging performance with better barrier properties and antimicrobial functions [5, 91]. Using AI in material design can support smarter, more sustainable packaging solutions.

New technologies such as 3D and 4D printing are also involved in this development. These techniques allow for the creation of customized, biodegradable packaging designed to specific food products, thus minimizing waste. 3D printing can produce compostable or even edible packaging, helping to reduce reliance on conventional plastics. In contrast, 4D printing enables packaging to adapt to environmental factors like moisture and temperature, potentially extending food shelf life. This adaptability fits well with the core ideas of with CE principles, promoting resource efficiency and sustainability [127, 128].

To more reduce single-use waste, the focus on durable and reusable packaging systems is vital. Research should emphasize efficient designs that encourage consumer participation in sustainable practices.

Advancements in recycling technologies are vital for achieving a closed-loop system in food packaging. Research into chemical recycling methods and AI-improved sorting processes can significantly improve material recovery [129].

Industry 4.0 (I4.0) technologies have been recognized as key enablers of CE. Integrating IoT, AI, Blockchain, and human-robot can improve recycling practices, optimization of supply chain and also material recovery. Using real-time data analytics can empower businesses to make informed decisions that support sustainability [130].

To sum up, the future of food packaging within a CE is promising, driven by innovations in smart technologies, sustainable materials, and efficient recycling systems. Collaborative efforts among researchers, industry stakeholders, and policymakers will be essential in advancing these innovations and creating a sustainable packaging ecosystem.

4 Notes

This chapter offers an in-depth study of how CE principles can be effectively applied to food packaging to reduce waste and improve sustainability. It emphasizes the serious need to transition from traditional linear models of production and consumption, which often result in significant environmental degradation, to circular systems that focus on resource efficiency and waste minimization.

The chapter begins by defining the CE basis, which promotes the continuous use of resources through strategies such as recycling, reuse, and the development of biodegradable materials. Implementing CE practices can significantly improve sustainability outcomes, as these approaches encourage the reintegration of materials into production cycles, which is essential for addressing the environmental challenges posed by conventional packaging materials, particularly those derived from petroleum origin [131].

Innovative materials designed to be fit perfectly with CE principles. Biodegradable options, such as Polylactic Acid (PLA) and Polyhydroxyalkanoates (PHAs), are viable alternatives to traditional plastics. They not only reduce long-term environmental impacts but also maintain the necessary safety and quality standards for food packaging [15]. The chapter also discusses the mechanical properties and biodegradability of these materials.

The integration of intelligent packaging technologies is another key topic in this chapter. Intelligent packaging systems, which use sensors and indicators to monitor food quality, can significantly extend shelf life and reduce waste. These new technologies have a high potential to addressing food spoilage, a significant contributor to food waste by providing real-time data on product conditions [11].

Effective waste management practices are essential for the successful implementation of CE techniques. This chapter emphasizes the importance of proper recycling and composting systems [112], which can effectively divert waste from landfills and promote the recovery of valuable materials, thereby supporting the overall goals of sustainability and resource regeneration.

Consumer behavior is also a significant factor influencing the transition to CE practices in food packaging. The chapter discusses the growing consumer demand for sustainable packaging solutions [125]. Understanding consumer attitudes and preferences is very important for encouraging greater adoption of eco-friendly packaging options, which can motivate market trends and encourage manufacturers to innovate.

Supportive policy and regulatory frameworks improve the adoption of CE principles. The European Union's CE Action Plan serves as a prominent example of how regulatory efforts can drive sustainability initiatives in packaging. This plan emphasizes the need for eco-design, waste reduction, and resource efficiency [120].

Collaboration among manufacturers, retailers, and consumers is essential for advancing sustainable practices in food packaging. The chapter emphasizes the significance of public-private partnerships in developing infrastructure and driving innovation. Collaborative efforts can improve recycling systems and promote the implementation of CE strategies across different sectors [114]. Table 4 outlines key aspects of CE techniques in food packaging, presenting materials, technologies, and practices in this field, while indicating their benefits, challenges, and exemplative cases to aid readers in understanding their multidimensional applications and implications.

In conclusion, this chapter gives a detailed investigation of how CE techniques can change food packaging systems. It focuses on the key areas such as innovative materials, intelligent technologies, waste management, consumer engagement, supportive policies, and collaboration in the supply chain. The chapter summarizes a roadmap for enhancing sustainability in food packaging.

Future research should focus on several key areas to advance the application of CE principles in food packaging. Empirical studies assessing the effectiveness of various CE practices in real-world settings are essential. This research could provide valuable data on the sustainability impacts of specific materials and technologies, helping to identify best practices.

A deeper study of consumer behavior is necessary to understand the motivations and barriers influencing purchasing decisions regarding sustainable packaging. Investigating the regulatory situation will also be important, identifying gaps and inconsistencies that postpone CE adoption and suggesting reforms to support sustainable initiatives.

Finally, investigating the potential of new technologies, such as AI and IoT, in enhancing waste management and recycling processes could lead to innovative solutions that improve the efficiency of recycling systems. By focusing these research gaps, scholars and practitioners can contribute to the ongoing development of sustainable food packaging solutions that effectively minimize waste and promote a CE.

Table 4
Overview of circular economy (CE) techniques in food packaging

Aspect	Description	Benefits	Challenges	Examples
Biodegradable materials	Materials that decompose naturally in the environment	Reduces long-term waste and environmental impact	Limited industrial composting facilities	PLA, PHAs, starch-based plastics
Recyclable materials	Materials that can be collected, processed, and reused	Conserves raw materials and reduces energy consumption	Contamination in recycling streams can hinder effectiveness	PET, HDPE, glass
Intelligent packaging	Packaging with embedded sensors to monitor freshness and quality	Extends product shelf life and reduces food waste	Higher production costs and technology integration challenges	Smart labels, freshness indicators
Reusable packaging	Packaging designed for multiple uses before disposal	Reduces overall packaging waste and material consumption	Requires infrastructure for collection and cleaning	Glass jars, refillable containers
Composting systems	Systems designed to process organic waste into compost	Diverts waste from landfills and enriches soil	Requires consumer participation and awareness	Municipal composting programs
Supply chain innovations	Strategies that optimize logistics and reduce waste throughout the supply chain	Increases efficiency and reduces costs associated with waste	Requires collaboration and data sharing among stakeholders	Just-in-time delivery, AI-driven inventory management
Consumer engagement	Strategies to educate and motivate consumers to choose sustainable packaging options	Increases demand for sustainable products, driving market change	Overcoming consumer inertia and misinformation	Campaigns promoting eco-friendly packaging
Government policies	Regulations and incentives to promote CE practices in packaging	Creates a supportive framework for sustainability initiatives	Regulatory complexity and variability across regions	EU CE action plan, plastic bans
Life cycle assessment (LCA)	A systematic approach to evaluate the environmental impacts of packaging throughout its life cycle	Identifies areas for improvement and helps inform design decisions	Data availability and variability in methodologies	LCA studies for different packaging materials

(continued)

Table 4 (continued)

Aspect	Description	Benefits	Challenges	Examples
Public-private partnerships	Collaborations between government and businesses to improve recycling and sustainability efforts	Influences resources and expertise for greater impact	Coordination challenges and differing objectives	Industry coalitions for recycling initiatives

References

1. Caner C, Yüceer M, Harte B (2024) Trends in sustainability and innovative food packaging materials: an overview. Akademik Gıda:65–77. https://doi.org/10.24323/akademik-gida.1554476
2. Law KL, Narayan R (2021) Reducing environmental plastic pollution by designing polymer materials for managed end-of-life. Nat Rev Mater 7:104–116. https://doi.org/10.1038/s41578-021-00382-0
3. Guillard V, Gaucel S, Fornaciari C et al (2018) The next generation of sustainable food packaging to preserve our environment in a circular economy context. Front Nutr 5:121. https://doi.org/10.3389/fnut.2018.00121
4. Zambujal-Oliveira J, Fernandes C (2024) The contribution of sustainable packaging to the circular food supply chain. Packag Technol Sci 37:443–456. https://doi.org/10.1002/pts.2802
5. Zhang P, Gao S, Xu L, Xu C (2025) Application of biomass carbon dots in food packaging. Environ Sci Pollut Res 32:1710–1733. https://doi.org/10.1007/s11356-024-35818-x
6. Munro T (2023) Introduction to sustainability. MacEwan Open Books
7. Sah AK, Hong Y-M (2024) Circular economy implementation in an organization: a case study of the Taiwan sugar corporation. Sustainability 16:7865. https://doi.org/10.3390/su16177865
8. Toni M (2023) Conceptualization of circular-economy and sustainability at the business level. Circular economy and sustainable development. IJERM 1:81–89. https://doi.org/10.59762/ijerm205275791220231205140635
9. Verghese K, Lewis H, Lockrey S, Williams H (2013) The role of packaging in minimising food waste in the supply chain of the future. CHEP Australia
10. Afzia N, Prasannavenkadesan V, Katiyar V, Ghosh T (2025) Biodegradable foams for food packaging. In: Sustainable materials for food packaging and preservation. Elsevier, pp 301–316
11. Heising JK, Claassen GDH, Dekker M (2017) Options for reducing food waste by quality-controlled logistics using intelligent packaging along the supply chain. Food Addit Contam Part A 34:1672–1680. https://doi.org/10.1080/19440049.2017.1315776
12. Ikechukwu C (2024) Addressing food security and environmental concerns through bioplastic innovation in foodpackaging. GJESS 1:53–68. https://doi.org/10.69798/49264347
13. Swargiary B, Patel S, Chutia H et al (2025) Sustainable materials in food packaging and preservation: an introduction. In: Sustainable materials for food packaging and preservation. Elsevier, pp 1–25
14. Li G, Zhao M, Xu F et al (2020) Synthesis and biological application of polylactic acid. Molecules 25:5023. https://doi.org/10.3390/molecules25215023
15. Ncube LK, Ude AU, Ogunmuyiwa EN et al (2020) Environmental impact of food packaging materials: a review of contemporary development from conventional plastics to polylactic acid basedmaterials. Materials 13:4994. https://doi.org/10.3390/ma13214994
16. Dorgan JR, Braun B, Wegner JR, Knauss DM (2006) Poly(lactic acids): a brief review. In: Degradable polymers and materials. Chapter 7, pp 102–125
17. Kawale A, Shukla S, Thiruvengadam V, Bansod A (2025) Development of silver nanoparticles infiltrated PLA films for food packaging. Plast Rubber Compos Macromol Eng 2025:14658011241304381. https://doi.org/10.1177/14658011241304381
18. He X, Jin Y, Tang L, Huang R (2023) A brief review and perspective on the functional

biodegradable films for food packaging. Polymer Sci Peer Rev J 5(2)

19. Panou A, Karabagias I (2023) Biodegradable packaging materials for foods preservation: sources, advantages, limitations, and future perspectives. Coatings 13:1176. https://doi.org/10.3390/coatings13071176
20. Zhong ZW, Song B, Huang CX (2009) Environmental impacts of three polyhydroxyalkanoate (PHA) manufacturing processes. Mater Manuf Process 24:519–523. https://doi.org/10.1080/10426910902740120
21. Khodaei D, Álvarez C, Mullen AM (2021) Biodegradable packaging materials from animal processing co-products and wastes: an overview. Polymers 13:2561. https://doi.org/10.3390/polym13152561
22. Amin U, Khan MU, Majeed Y et al (2021) Potentials of polysaccharides, lipids and proteins in biodegradable food packaging applications. Int J Biol Macromol 183:2184–2198. https://doi.org/10.1016/j.ijbiomac.2021.05.182
23. Song JH, Murphy RJ, Narayan R, Davies GBH (2009) Biodegradable and compostable alternatives to conventional plastics. Philos Trans R Soc B 364:2127–2139. https://doi.org/10.1098/rstb.2008.0289
24. Marichelvam MK, Jawaid M, Asim M (2019) Corn and rice starch-based bio-plastics as alternative packaging materials. Fibers 7:32. https://doi.org/10.3390/fib7040032
25. Sriroth K, Sangseethong K (2006) Biodegradable plastics from cassava starch. Acta Hortic:145–152. https://doi.org/10.17660/ActaHortic.2006.703.16
26. Subramani R, Mustafa MA, Ghadir GK et al (2024) Exploring the use of biodegradable polymer materials in sustainable 3D printing. Appl Chem Eng 7:3870. https://doi.org/10.59429/ace.v7i2.3870
27. Kamsiati E, Herawati H, Purwani EY (2017) Potensi pengembangan plastik biodegradable berbasis pati sagu dan ubikayu di Indonesia/ The development potential of sago and cassava starch-based biodegradable plastic in Indonesia. J Litbang Pert 36:67. https://doi.org/10.21082/jp3.v36n2.2017.p67-76
28. Singh S, Gaur S, Sharma N (2024) Starchy films as a sustainable alternative in food industry: current research and applications. Starch Stärke 76:2300078. https://doi.org/10.1002/star.202300078
29. Teseme BW (2020) Review on the manufacturing of biodegradable plastic packaging film from root and tuber starches. Nano 8:1. https://doi.org/10.11648/j.nano.20200801.11
30. Sinha S (2024) An overview of biopolymer-derived packaging material. Polym Renew Resour 15:193–209. https://doi.org/10.1177/20412479241226884
31. Ai B, Zheng L, Li W et al (2021) Biodegradable cellulose film prepared from banana pseudo-stem using an ionic liquid for mango preservation. Front Plant Sci 12:625878. https://doi.org/10.3389/fpls.2021.625878
32. Badii F, Maftoonazad N, Behmadi H (2010) Preparation of cellulose-based edible films and investigating some of their physical and mechanical properties and their application to extend the shelf life of horticultural crops. Agricultural Engineering Research Institute of Iran (AERI)
33. Sadeghipour M, Badii F, Behmadi H, Bazyar B (2012) The effect of methyl cellulose based active edible coatings on the storage life of tomato. Food Sci Technol 9:89–98
34. Shanmugam K, Chandrasekar N, Balaji R (2024) Nanocellulose as sustainable bio-nanomaterial for packaging and biomedical applications. Sci Social Res 6:57–67. https://doi.org/10.26689/ssr.v6i2.6057
35. Bai L, Huan S, Zhu Y et al (2021) Recent advances in food emulsions and engineering foodstuffs using plant-based nanocelluloses. Annu Rev Food Sci Technol 12:383–406. https://doi.org/10.1146/annurev-food-061920-123242
36. Vilarinho F, Sanches Silva A, Vaz MF, Farinha JP (2018) Nanocellulose in green food packaging. Crit Rev Food Sci Nutr 58:1526–1537. https://doi.org/10.1080/10408398.2016.1270254
37. Chen G, Wang K, Chen P et al (2024) Fully biodegradable packaging films for fresh food storage based on oil-infused bacterial cellulose. Adv Sci 11:2400826. https://doi.org/10.1002/advs.202400826
38. Xia G, Zhou Q, Xu Z et al (2021) Cellulose-based films with ultraviolet shielding performance prepared directly from waste corrugated pulp. Polymers 13:3359. https://doi.org/10.3390/polym13193359
39. Su Y, Yang B, Liu J et al (2018) Prospects for replacement of some plastics in packaging with lignocellulose materials: a brief review. Bioresources 13:4550–4576. https://doi.org/10.15376/biores.13.2.Su
40. Dehghan Shoar Z, Badii F, Behmadi H (2009) Effect of glycerol addition on the mechanical and physical properties of soy protein isolates film. Iran J Food Sci Technol 6:1–10
41. Zubair M, Ullah A (2020) Recent advances in protein derived bionanocomposites for food

packaging applications. Crit Rev Food Sci Nutr 60:406–434. https://doi.org/10.1080/10408398.2018.1534800

42. Kandasamy S, Yoo J, Yun J et al (2021) Application of whey protein-based edible films and coatings in food industries: an updated overview. Coatings 11:1056. https://doi.org/10.3390/coatings11091056
43. Purewal SS, Kaur A, Bangar SP et al (2023) Protein-based films and coatings: an innovative approach. Coatings 14:32. https://doi.org/10.3390/coatings14010032
44. Ashik KP, Sharma RS (2015) A review on mechanical properties of natural fiber reinforced hybrid polymer composites. JMMCE 03:420–426. https://doi.org/10.4236/jmmce.2015.35044
45. Yildizhan Ş, Çalik A, Özcanli M, Serin H (2018) Bio-composite materials: a short review of recent trends, mechanical and chemical properties, and applications. Eur Mech Sci 2:83–91. https://doi.org/10.26701/ems.369005
46. Korjenic A, Teichmann F (2024) Building with renewable materials. Automatisierungstechnik 72:679–686. https://doi.org/10.1515/auto-2024-0048
47. Ma Y, Wei C, Wang Z et al (2024) Precise chemical regulation of polar groups to enhance the charge transfer density of cellulosic triboelectric textiles. J Mater Chem A 12:17702–17713. https://doi.org/10.1039/D4TA02816C
48. Lawson L, Degenstein LM, Bates B et al (2022) Cellulose textiles from hemp biomass: opportunities and challenges. Sustainability 14:15337. https://doi.org/10.3390/su142215337
49. Ibrahim ID, Hamam Y, Sadiku ER et al (2022) Need for sustainable packaging: an overview. Polymers 14:4430. https://doi.org/10.3390/polym14204430
50. Zimmermann W (2020) Biocatalytic recycling of polyethylene terephthalate plastic. Phil Trans R Soc A 378:20190273. https://doi.org/10.1098/rsta.2019.0273
51. Cusano I, Campagnolo L, Aurilia M et al (2023) Rheology of recycled PET. Materials 16:3358. https://doi.org/10.3390/ma16093358
52. Daramola OO, Jepkoech R, Aramide FO, Adediran AA (2024) High-density polyethylene matrix hybrid composites: a review of their characterization, processing techniques, and promising applications. Mater Tech 112:501. https://doi.org/10.1051/mattech/2024024
53. Craciun DC, Peter I (2024) Review of the polymeric material composite, optimization for industrial application. AMSET 21:47–51. https://doi.org/10.62838/amset-2024-0008
54. Abeysinghe S, Gunasekara C, Bandara C et al (2021) Engineering performance of concrete incorporated with recycled high-density polyethylene (HDPE)—a systematic review. Polymers 13:1885. https://doi.org/10.3390/polym13111885
55. Ribeiro LS, Stolz CM, Amario M et al (2023) Use of post-consumer plastics in the production of wood-plastic composites for building components: a systematic review. Energies 16:6549. https://doi.org/10.3390/en16186549
56. Mendes AC, Pedersen GA (2021) Perspectives on sustainable food packaging: is bio-based plastics a solution? Trends Food Sci Technol 112:839–846. https://doi.org/10.1016/j.tifs.2021.03.049
57. Wojnarowska M, Muradin M, Paiano A, Ingrao C (2025) Recycled glass bottles for craft-beer packaging: how to make them sustainable? An environmental impact assessment from the combined accounting of cullet content and transport distance. Resources 14:23. https://doi.org/10.3390/resources14020023
58. Koutsoyiannis D, Kundzewicz ZW (2009) Editorial—recycling paper vs recycling papers. Hydrol Sci J 54:3–4. https://doi.org/10.1623/hysj.54.1.3
59. Elkholy HM, Hamdani SS, Alghaysh MO et al (2025) Design of carboxylic acid-functionalized poly(Butylene Adipate-*co*-Terephthalate) for recyclable and biodegradable zero-waste paper packaging. Adv Sustain Syst 9:2400621. https://doi.org/10.1002/adsu.202400621
60. Peng Y, Guo J, Gao Y et al (2024) Facile preparation of pickering emulsion-based coatings for hydrophobic, oil-resistant, biodegradable green paper-based packaging. Food Hydrocoll 149:109624. https://doi.org/10.1016/j.foodhyd.2023.109624
61. Zhao Y, Li B, Li C et al (2021) Comprehensive review of polysaccharide-based materials in edible packaging: a sustainable approach. Foods 10:1845. https://doi.org/10.3390/foods10081845
62. Carnaval LDSC, Jaiswal AK, Jaiswal S (2024) Agro-food waste valorization for sustainable bio-based packaging. J Compos Sci 8:41. https://doi.org/10.3390/jcs8020041

63. Mihalca V, Kerezsi AD, Weber A et al (2021) Protein-based films and coatings for food industry applications. Polymers 13:769. https://doi.org/10.3390/polym13050769
64. Taylor S, Colonna A, Jung J et al (2024) Consumer perception and acceptance of edible packaging for various food products. J Food Sci 89:2423–2437. https://doi.org/10.1111/1750-3841.16992
65. Soo XYD, Muiruri JK, Wu W et al (2024) Bio-polyethylene and polyethylene biocomposites: an alternative toward a sustainable future. Macromol Rapid Commun 45: 2400064. https://doi.org/10.1002/marc.202400064
66. Rizvi SMH (2024) Development of sustainable bio-based polymers as alternatives to petrochemical plastics. Valley Int J Digit Lib 12: 107–124
67. İncili GK, Razavi R, Hayaloğlu AA et al (2025) Microbial derived biomaterials: fabrication, processing, and food application. In: Sustainable materials for food packaging and preservation. Elsevier, pp 55–84
68. Santos MRD, Durval IJB, Medeiros ADMD et al (2024) Biotechnology in food packaging using bacterial cellulose. Foods 13:3327. https://doi.org/10.3390/foods13203327
69. Suvarna V, Nair A, Mallya R et al (2022) Antimicrobial nanomaterials for food packaging. Antibiotics 11:729. https://doi.org/10.3390/antibiotics11060729
70. Tan C, Han F, Zhang S et al (2021) Novel bio-based materials and applications in antimicrobial food packaging: recent advances and future trends. IJMS 22:9663. https://doi.org/10.3390/ijms22189663
71. Abdul Malek NS, Omar H, Rosman N et al (2025) Tailoring synergistic polylactic acid–based nanocomposites for sustainable antimicrobial fruit packaging. Packag Technol Sci 38:163–192. https://doi.org/10.1002/pts.2864
72. Peng C, Lyu M, Guo P et al (2025) Toughening biodegradable poly(glycolic acid) with balanced mechanical properties by biobased poly(butylene 2,5-furanoate). J Polym Environ 33:2161. https://doi.org/10.1007/s10924-025-03511-w
73. Gunja NJ, Athanasiou KA (2006) Biodegradable materials in arthroscopy. Sports Med Arthrosc Rev 14:112–119. https://doi.org/10.1097/00132585-200609000-00002
74. Lionetto F, Esposito Corcione C (2021) Recent applications of biopolymers derived from fish industry waste in food packaging. Polymers 13:2337. https://doi.org/10.3390/polym13142337
75. Mahdizadeh Barzoki Z, Emam-Djomeh Z, Mortazavian E et al (2018) Determination of diffusion coefficient for released nanoparticles from developed gelatin/chitosan bilayered buccal films. Int J Biol Macromol 112:1005–1013. https://doi.org/10.1016/j.ijbiomac.2018.01.215
76. Wever R, Vogtländer J (2013) Eco-efficient value creation: an alternative perspective on packaging and sustainability. Packag Technol Sci 26:229–248. https://doi.org/10.1002/pts.1978
77. World Economic Forum, Ellen MacArthur Foundation and McKinsey & Company (2016) The new plastics economy: rethinking the future of plastics
78. Manea EE, Bumbac C, Dinu LR et al (2024) Composting as a sustainable solution for organic solid waste management: current practices and potential improvements. Sustain For 16:6329. https://doi.org/10.3390/su16156329
79. Beshai H, Sarabha G, Rathi P et al (2020) Freshness monitoring of packaged vegetables. Appl Sci 10:7937. https://doi.org/10.3390/app10217937
80. Poças F, Do Céu Selbourne M (2023) Drivers, advances, and significance of measures for effective circular food packaging. Front Sustain Food Syst 7:1140295. https://doi.org/10.3389/fsufs.2023.1140295
81. Han J, Ruiz-Garcia L, Qian J, Yang X (2018) Food packaging: a comprehensive review and future trends. Compr Rev Food Sci Food Saf 17:860–877. https://doi.org/10.1111/1541-4337.12343
82. Janjarasskul T, Suppakul P (2018) Active and intelligent packaging: the indication of quality and safety. Crit Rev Food Sci Nutr 58:808–831. https://doi.org/10.1080/10408398.2016.1225278
83. Du L, Huang X, Li Z et al (2025) Application of smart packaging in fruit and vegetable preservation: a review. Foods 14:447. https://doi.org/10.3390/foods14030447
84. Kumari S, Soni S, Sharma A et al (2024) Layered double hydroxides based composite materials and their applications in food packaging. Appl Clay Sci 247:107216. https://doi.org/10.1016/j.clay.2023.107216
85. Lisboa HM, Pasquali MB, Dos Anjos AI et al (2024) Innovative and sustainable food preservation techniques: enhancing food quality, safety, and environmental sustainability.

Sustainability 16:8223. https://doi.org/10.3390/su16188223
86. Li X, Liu D, Pu Y, Zhong Y (2023) Recent advance of intelligent packaging aided by artificial intelligence for monitoring food freshness. Foods 12:2976. https://doi.org/10.3390/foods12152976
87. Kadirvel V, Palanisamy Y, Ganesan ND (2025) Active packaging system—an overview of recent advances for enhanced food quality and safety. Packag Technol Sci 38:145–162. https://doi.org/10.1002/pts.2863
88. Priyanka P, Rani R, Yadav R et al (2024) Role of active packaging for food freshness and quality maintenance. IJEAB 9:023–030. https://doi.org/10.22161/ijeab.94.4
89. Omerović N, Djisalov M, Živojević K et al (2021) Antimicrobial nanoparticles and biodegradable polymer composites for active food packaging applications. Compr Rev Food Sci Food Saf 20:2428–2454. https://doi.org/10.1111/1541-4337.12727
90. Carpena M, Nuñez-Estevez B, Soria-Lopez A et al (2021) Essential oils and their application on active packaging systems: a review. Resources 10:7. https://doi.org/10.3390/resources10010007
91. Moradi M, Molaei R, Kousheh SA et al (2023) Carbon dots synthesized from microorganisms and food by-products: active and smart food packaging applications. Crit Rev Food Sci Nutr 63:1943–1959. https://doi.org/10.1080/10408398.2021.2015283
92. Wagh RV, Ezati P, Khan A et al (2024) *Vaccinium corymbosum* –derived carbon dots and anthocyanin-infused gelatin multifunctional films for smart packaging applications. Packag Technol Sci 37:1049–1064. https://doi.org/10.1002/pts.2839
93. Choi J, Kwon O-C, Jo W et al (2015) 4D printing technology: a review. 3D Print Addit Manuf 2:159–167. https://doi.org/10.1089/3dp.2015.0039
94. Sharifi Malvajerdi S, Aboutorabi S, Shahnazi A et al (2023) HVHC-ESD-induced oxygen vacancies: an insight into the phenomena of interfacial interactions of nanostructure oxygen vacancy sites with oxygen ion-containing organic compounds. ACS Appl Mater Interfaces 15:48785–48799. https://doi.org/10.1021/acsami.3c10017
95. Adeyi TA, Jaiyeola CO, Oresegun OI et al (2025) AI and nanotechnology integration in the development of biodegradable food packaging materials. Int J Sci Res Arch 14:726–736. https://doi.org/10.30574/ijsra.2025.14.1.0119
96. Chen S, Brahma S, Mackay J et al (2020) The role of smart packaging system in food supply chain. J Food Sci 85:517–525. https://doi.org/10.1111/1750-3841.15046
97. Franklin Associates (2004) Life cycle inventory of packaging options for shipment of retail mail-order soft goods. Franklin Associates
98. Perera KY, Jaiswal AK, Jaiswal S (2023) Biopolymer-based sustainable food packaging materials: challenges, solutions, and applications. Foods 12:2422. https://doi.org/10.3390/foods12122422
99. Robertson G (2008) State-of-the-art biobased food packaging materials. In: Environmentally compatible food packaging. Elsevier, pp 3–28
100. Jafarzadeh S, Forough M, Kouzegaran VJ et al (2023) Improving the functionality of biodegradable food packaging materials via porous nanomaterials. Compr Rev Food Sci Food Saf 22:2850–2886. https://doi.org/10.1111/1541-4337.13164
101. Siddiqui SA, Sundarsingh A, Bahmid NA et al (2023) A critical review on biodegradable food packaging for meat: materials, sustainability, regulations, and perspectives in the EU. Compr Rev Food Sci Food Saf 22:4147–4185. https://doi.org/10.1111/1541-4337.13202
102. Baghi F, Gharsallaoui A, Dumas E, Ghnimi S (2022) Advancements in biodegradable active films for food packaging: effects of nano/microcapsule incorporation. Foods 11:760. https://doi.org/10.3390/foods11050760
103. Kalina S, Kapilan R, Wickramasinghe I, Navaratne SB (2024) Potential use of plant leaves and sheath as food packaging materials in tackling plastic pollution: a review. Ceylon J Sci 53:21–37. https://doi.org/10.4038/cjs.v53i1.8145
104. Koirala P, Sagar NA, Thuanthong A et al (2025) Revolutionizing seafood packaging: advancements in biopolymer smart nanopackaging for extended shelf-life and quality assurance. Food Res Int 203:115826. https://doi.org/10.1016/j.foodres.2025.115826
105. Ellen MacArthur Foundation (2021) Case studies and examples of circular economy in action. https://www.ellenmacarthurfoundation.org/topics/circular-economy-introduction/examples
106. WRAP Case studies. https://www.wrap.ngo/what-we-do/case-studies
107. Batista L, Gong Y, Pereira S et al (2019) Circular supply chains in emerging

economies—a comparative study of packaging recovery ecosystems in China and Brazil. Int J Prod Res 57:7248–7268. https://doi.org/10.1080/00207543.2018.1558295

108. Noëth E, Van Opstal W, Du Bois E (2024) Introducing reusable food packaging: customer preferences and design implications for successful market entry. Bus Strat Env 33:6507–6532. https://doi.org/10.1002/bse.3820
109. Rollini M, Musatti A, Cavicchioli D et al (2020) From cheese whey permeate to Sakacin-A/bacterial cellulose nanocrystal conjugates for antimicrobial food packaging applications: a circular economy case study. Sci Rep 10:21358. https://doi.org/10.1038/s41598-020-78430-y
110. Munoz-Briones PA, Munguía-Lópeza ADC, Sánchez-Riveraa KL et al (2024) Optimal design of food packaging considering waste management technologies to achieve circular economy. Syst Control Trans 3:820–828
111. Hopewell J, Dvorak R, Kosior E (2009) Plastics recycling: challenges and opportunities. Philos Trans R Soc B 364:2115–2126. https://doi.org/10.1098/rstb.2008.0311
112. Dilkes-Hoffman LS, Pratt S, Lant PA, Laycock B (2019) The role of biodegradable plastic in solving plastic solid waste accumulation. In: Plastics to energy. Elsevier, pp 469–505
113. Ellen MacArthur Foundation (2017) The new plastics economy: rethinking the future of plastics & catalysing action. https://www.ellenmacarthurfoundation.org/the-new-plastics-economy-rethinking-the-future-of-plastics-and-catalysing
114. Kaza S, Yao L, Bhada-Tata P, Van Woerden F (2018) What a waste 2.0: a global snapshot of solid waste management to 2050. World Bank Publications
115. PR Newswire (2023) New data reveals consumers increasingly choose products in sustainable packaging globally, despite rising prices. https://www.prnewswire.com/news-releases/new-data-reveals-consumers-increasingly-choose-products-in-sustainable-packaging-globally-despite-rising-prices-301804273.html. Accessed 27 Jan 2025
116. Richter F (2024) Consumers see the need for eco-friendly packaging
117. Van Loon P, Van Wassenhove LN (2020) Transition to the circular economy: the story of four case companies. Int J Prod Res 58:3415–3422. https://doi.org/10.1080/00207543.2020.1748907
118. Oluyemi AS, Adelabu OS, Oladumiye EB (2024) Moderation of consumers' confusion in the atypicality of packaging design through the use of e-labeling towards sustainable consumption. In: Proceedings—The twelfth international symposium GRID 2024. University of Novi Sad Faculty of Technical Sciences Department of Graphic Engineering and Design
119. Matheus JRV, Dalsasso RR, Rebelatto EA et al (2023) Biopolymers as green-based food packaging materials: a focus on modified and unmodified starch-based films. Compr Rev Food Sci Food Saf 22:1148–1183. https://doi.org/10.1111/1541-4337.13107
120. European Commission (2020) Circular economy action plan. https://environment.ec.europa.eu/strategy/circular-economy-action-plan_en
121. Lakhouit A (2025) Revolutionizing urban solid waste management with AI and IoT: a review of smart solutions for waste collection, sorting, and recycling. Results Eng 25:104018. https://doi.org/10.1016/j.rineng.2025.104018
122. Maniatis P (2025) Artificial intelligence in waste sorting: advancing recycling processes in Greece through AI-driven solutions. IJCCR 23:01–10. https://doi.org/10.31579/2690-4861/697
123. Abu-Bakar H, Charnley F, Hopkinson P, Morasae EK (2024) Towards a typological framework for circular economy roadmaps: a comprehensive analysis of global adoption strategies. J Clean Prod 434:140066. https://doi.org/10.1016/j.jclepro.2023.140066
124. Halicka E, Kaczorowska J, Rejman K, Plichta M (2025) Investigating the consumer choices of Gen Z: a sustainable food system perspective. Nutrients 17:591. https://doi.org/10.3390/nu17030591
125. Lisboa A, Vitorino L, Antunes R (2022) Gen Zers' intention to purchase products with sustainable packaging: an alternative perspective to the attitude-behaviour gap. J Mark Manag 38:967–992. https://doi.org/10.1080/0267257X.2022.2083214
126. Gonzalez-Amarillo C, Cardenas-Garcia C, Mendoza-Moreno M et al (2021) Blockchain-IoT sensor (BIoTS): a solution to IoT-ecosystems security issues. Sensors 21:4388. https://doi.org/10.3390/s21134388
127. Olawade DB, Wada OZ, Ige AO (2024) Advances and recent trends in plant-based

materials and edible films: a mini-review. Front Chem 12:1441650. https://doi.org/10.3389/fchem.2024.1441650

128. Singh P, Pandey VK, Tripathi A et al (2024) 4D food printing technology: structural changes to culinary art and beyond. J Food Process Eng 47:e14535. https://doi.org/10.1111/jfpe.14535

129. Rujnić-Sokele M, Pilipović A (2017) Challenges and opportunities of biodegradable plastics: a mini review. Waste Manag Res 35:132–140. https://doi.org/10.1177/0734242X16683272

130. Ciano MP, Peron M, Panza L, Pozzi R (2025) Industry 4.0 technologies in support of circular economy: a 10R-based integration framework. Comput Ind Eng 201:110867. https://doi.org/10.1016/j.cie.2025.110867

131. Geissdoerfer M, Morioka SN, De Carvalho MM, Evans S (2018) Business models and supply chains for the circular economy. J Clean Prod 190:712–721. https://doi.org/10.1016/j.jclepro.2018.04.159

Chapter 15

Green Valorization Protocols for Marine Food Waste: Collagen and Chitin Recovery

Sibojyoti Baksi, Pritha Pal, and Sibashish Baksi

Abstract

The growing need for sustainable and environmentally friendly alternatives in a variety of sectors has emphasized the importance of effective technologies for extracting important biomolecules such as chitin and collagen from marine food waste. Chitin, a natural polymer found abundantly in crab shells, and collagen, a protein found in marine by-products such as fish skins, heads, and frames, are both widely sought after for their many uses in the pharmaceutical, cosmetic, and food sectors. Conventional extraction processes for chitin and collagen, which sometimes use harsh chemicals and need lengthy processing periods, raise concerns about their environmental effect and scalability. Recent advancements in green extraction technologies, such as deep eutectic solvents (DES), natural deep eutectic solvents (NADES), ionic liquids (ILs), supercritical fluid extraction (SFE), enzyme-assisted extraction (EAE), and non-thermal methods such as ultrasound, have received attention for their lower toxicity, improved biodegradability, and potential for increased yields. Collagen derived from marine sources such as fish, jellyfish, and sea cucumbers has gained popularity due to its high output and uses in a variety of industries; however, traditional techniques pose obstacles. Green extraction techniques, such as enzyme-assisted and ultrasonic extractions, are developing as viable alternatives, with lower toxicity, faster processing times, and lesser environmental effect. These techniques help to improve the sustainability and efficiency of chitin and collagen extraction procedures, respectively. Furthermore, incorporating these technologies into marine biorefinery models provides the opportunity for zero-waste processing and contributes to a circular blue economy, which aligns with global sustainability goals. This study examines both traditional and developing green extraction strategies for chitin and collagen, assessing their benefits, limits, and future prospects, with an emphasis on optimizing procedures for large-scale production and the sustainable use of marine food waste.

Key words Fermentation, Marine food waste, Extraction, Collagen, Chitin

1 Introduction

The global food production is increasing significantly to meet growing consumer demand. As a crucial part of the food industry, the seafood sector delivers diverse marine products to consumers worldwide. Marine seafood includes different types of finfish, such as pelagic species, anchoveta, pollock, tuna, herring, mackerel, and whiting. Additionally, shellfish encompass crustaceans like shrimp,

Tanmay Sarkar (ed.), *Food Waste Valorization: Green Techniques in Sustainable Management*, Methods and Protocols in Food Science, https://doi.org/10.1007/978-1-0716-4646-5_15,
© The Author(s), under exclusive license to Springer Science+Business Media, LLC, part of Springer Nature 2025

krill, crab, and lobster, as well as mollusks, which include bivalves (such as mussels, oysters), cephalopods (like squid and cuttlefish), and gastropods (including abalone and snails). In 2018, global seafood production reached 178.5 million metric tons (MT), with capture fisheries contributing 96.4 MT. Forecasts suggest that by 2030, the consumption of fishery products for human use will increase to 204 MT [1]. Not all harvested species are fully utilized for consumption in the seafood industry. A considerable portion of the catch remains underutilized or unused due to unappealing color, small size, or high-fat content. As a result, a significant amount of the harvest is classified as by-catch—often comprising multiple species, discarded or inadequately utilized as food [2].

According to Gustavsson et al. [3], approximately 35% of global fish catches are lost or wasted, with by-catch accounting for 9–15% of these losses. Centralized preprocessing activities generate substantial amounts of solid waste and wastewater, such as beheading, de-shelling, skinning, gutting, fin and scale removal, filleting, and washing. Waste can make up as much as 50% of whole shellfish like shrimp, krill, and crab in terms of wet weight. Specifically, shrimp waste consists of approximately 70% head and 30% shell [4]. This vast industry produces an estimated 6–8 million tons of crustacean shell waste annually, including materials from crabs, lobsters, and shrimp [5]. Argentine red shrimp is a widely consumed shellfish, with its industrial processing generating approximately 18,000 metric tons of shell waste annually [6].

Due to its low economic value and high disposal costs, most seafood waste is either discarded in landfills or dumped into the ocean. This practice leads to various environmental concerns, including strong odors from decomposition, decreased oxygen levels in seawater, the burial or smothering of marine life, excessive plant and algae growth caused by high organic matter content (eutrophication), and alterations in water salinity, temperature, and pH levels [7].

Achieving food sustainability requires efficiently utilizing resources to maximize benefits, including economic feasibility. Ensuring sustainable seafood processing involves addressing key challenges such as minimizing environmental pollution, conserving water, and preventing nutrient losses [8]. To mitigate environmental hazards and nutrient depletion in the seafood industry, strategies such as selective trawling to reduce by-catch and proper waste and effluent treatment can be implemented. Seafood processing effluents are characterized by total suspended solids (TSS), fats, oils, grease (FOG), pigments, and minerals. TSS consists of protein-based substances, including myofibrillar proteins, collagen, gelatin, enzymes, soluble peptides, and amino acids. Analyzing the composition of over 40 types of seafood processing waste revealed that, on average, these discards contain 60% protein, 19% fat, and 22% ash [9]. This suggests that processing effluents, solid discards, and by-catch contribute to significant nutrient losses.

Table 1
Uses and valuation of seafood by-products through various traditional extraction techniques

Animal residues	Source	Products	Extraction conditions	References
Shell	White shrimps	Chitin	Demineralization: 1 M hydrochloric acid at a 1:10 (w/v) ratio; deproteinization: crude enzyme to substrate at 1:10 (w/v) ratio	[10]
	Shrimp	Chitosan	Acidic extraction: 2–4% HCl at 30–70 °C; alkaline extraction: 2–4% NaOH at 50 °C	[11]
Scale	Katamba fish (*L. lentjam*)	Chitosan	Acid treatment involves 10 mL of HCl. The final step includes an alkaline treatment with 10 mL of NaOH	[12]
	Sardine (*S. pilchardus*)	Chitosan	40% NaOH with a solid-to-solvent ratio of 1:10	[13]
	Tilapia and Grey mullet fish	Collagen	Treated with 0.1 N NaOH, neutralized, and then with 0.5 M acetic acid	[14]
	Fish	Gelatine	Acidic treatment with 2 M lactic acid at a 1:10 (w/w) ratio. Neutralized scales are then boiled at a 1:2 (w/w) ratio, and dehydrated	[15]
Head	Snakehead fish	Albumin	Boiling at 70 °C for 30 min	[16]
	Australian rock lobster (*J. edwardsii*)	Protein	Ultrasonically extraction for 5 min; pH of 13; water to LHBs ratio is maintained at 8 mL/g	[17]
	P. mucilaginosus TKU032	Protease	2 N of HCl solution	[18]
Skin	Grouper fish	Collagen	0.1 M NaOH, followed by neutralization, and treated with acetic acid at a 1:10 skin-to-solution ratio. Precipitation, and finally, dialysis in 0.5 M acetic acid at a 1:10 ratio	[19]
	Java barb (*B. gonionotus*)	Collagen	Alkaline extraction: 0.1 M NaOH (1:10 w/v); acidic treatment: 0.5 M acetic acid at 4 °C under (A) 1:10 (w/v), (B) 1:20 (w/v), (C) 1:10 (w/v), and (D) 1:20 (w/v)	[20]

One possible approach to addressing the economic, environmental, and social challenges associated with seafood residues is to redefine "waste" by promoting sustainable seafood production and the valorization of marine biomass.

Marine waste processing is not a new strategy; as Table 1 summarizes, several chemical techniques have been used for millennia.

Nevertheless, these techniques frequently present serious difficulties, such as high processing costs, chemical toxicity, and the discharge of hazardous waste into the environment. In the end, these problems may pose a bigger risk to the ecosystem than the actual marine garbage.

Green chemistry, also referred to as sustainable chemistry, emerged in the 1990s as an eco-friendly alternative to traditional valorization methods. Its primary goal is to design chemical products and processes that minimize or eliminate the use and production of hazardous substances. This approach applies throughout a chemical product's life cycle, from design and manufacturing to usage and disposal. Conventional reagents are often reactive, persistent, or toxic, with many organic solvents posing health risks. Additionally, traditional processes typically rely on high-energy inputs derived from fossil fuels and may use catalysts that include toxic elements, sometimes requiring extreme heat or pressure. In contrast, green chemistry promotes using inert, recyclable, and sustainable solvents, low-energy chemical reactions, and environmentally friendly catalysts such as microorganisms and enzymes, which are cost-effective and stable. Innovative green processing further supports the safer transformation of seafood by-products into valuable products. These methods operate with higher efficiency, potentially lower costs, and reduced environmental impact, contributing to a more sustainable and eco-conscious economy [21].

By adopting an inclusive and systemic approach, marine biomass waste can be repurposed as a valuable feedstock for producing industrially significant chemicals [4]. A variety of valuable products, including collagen, albumin, fish oil, chitin, chitosan, ink, protease, calcium carbonate, hydroxyapatite nanoparticles, and gelatin, can be derived from seafood by-products such as skin, head, liver, shell, bones, and scales through various suitable extraction methods. Crustacean shell waste contains essential components such as chitin, minerals, proteins, and pigments, which can be transformed into high-value products [16, 22, 23].

Collagen is a fundamental extracellular protein in animals, playing a key role in providing mechanical protection to tissues and organs. With increasing demand and expanding applications in dental surgery, tissue engineering, bone grafting, and the food and pharmaceutical industries, collagen extraction has been growing significantly. Marine vertebrates and invertebrates, such as cuttlefish, jellyfish, sea anemones, octopuses, sea urchins, starfish, prawns, and squid, serve as essential sources of collagen [24, 25]. These marine-derived collagens are among the safest options, often yielding higher amounts than other sources. Various extraction techniques classify collagen into different types, including acid-soluble collagen (ASC), salt-soluble collagen (SSC), ultrasound-assisted collagen (UAC), and pepsin-soluble collagen (PSC). The yield and physicochemical properties of the extracted collagen vary depending on the specific extraction method used [26].

Chitin, the second most abundant natural polysaccharide after cellulose, has gained attention as a valuable raw material with significant environmental, economic, and technical benefits. Its primary application lies in its conversion into chitosan, its most well-known derivative. However, chitin has limited industrial applications due to its insolubility in common solvents. In contrast, chitosan is soluble, biocompatible, and non-toxic, making it highly versatile for various uses [27]. Industrial extraction of chitin from shells involves the removal of minerals, proteins, and pigments through a well-established, solution-based process [22]. Chitin can then undergo deacetylation to produce chitosan, a water-soluble derivative. Both chitin and chitosan have a wide range of applications across various industries, including food, cosmetics, water treatment, agriculture, and biomedicine. This versatility makes chitin the most economically valuable product derived from crustacean shells [23].

This chapter presents a comprehensive approach to green extraction methods for chitin and collagen from marine food waste, incorporating in-depth molecular techniques.

2 Materials

The raw materials are first crushed in order to remove chitin and collagen from marine food waste, such as shrimp, crab, and fish processing wastes. Skin, scales, bones, and fins are the main components of fish waste. Deproteinization, a critical step in chitin extraction, is accomplished with the use of commercial bacterial strains that produce proteases, including *Lactobacillus* species, *Pseudomonas aeruginosa* K-187, and *Bacillus subtilis.* Demineralization is then carried out using commercial organic acid-producing bacteria, such as commercial alcalase or *Lactobacillus pentosus* 4023. Fish scales and skin are the primary sources of fibril-forming Type I collagen, while fish cartilage is the primary source of Type II collagen. Collagen extraction has made extensive use of fish waste from species like horse mackerel (*Trachurus japonicus*), bullhead shark (*Heterodontus japonicus*), chub mackerel (*Scomber japonicus*), yellow sea bream (*Dentex tumifrons*), Japanese sea bass (*Lateolabrax japonicus*), skipjack tuna (*Katsuwonus pelamis*), and ayu (*Plecoglossus altivelis*). These fish residues are mainly obtained from commercial fish markets. Collagen is often extracted using commercial pepsin, which is mostly obtained from the stomach mucosa of pigs. To increase yield, other enzymes such as papain, trypsin, and collagenases have been added. Numerous species, including plants, fungi, viruses, insects, and vertebrates, include aspartic proteases, a unique family of acidic proteases that work at low temperatures and have been investigated as a pepsin substitute. At different phases of extraction and purification, chemical agents including HCl, NaOH, and NaCl are needed.

3 Methods

3.1 Green Extraction Methods for Chitin: Mechanisms and Current State

3.1.1 Biological Extraction

It has been acknowledged that using microorganisms to handle aqua and agricultural industrial wastes is a viable approach. The use of microorganisms to treat chitinous waste has also drawn a lot of interest from the maritime industry. It was discovered that the more environmentally friendly way of employing proteolytic fermentation to carry out the deproteinization process was a sustainable one [28]. For instance, *Paenibacillus mucilaginosus* was utilized to produce protease from shrimp heads [18]. According to Zhang et al. [29], microbial fermentation has also been shown to produce large yields of chitin and chitosan, up to 21.35% and 13.11%, respectively. Better chitin production was demonstrated by the invention of three-step microbial fermentation techniques employing *Rhizopus japonicus*, *L. plantarum*, and *S. marcescens* [29]. *Lactobacillus acidophilus* and single-step fermentation procedures have been carried out to obtain chitin from shrimp waste. Shrimp waste was fermented using the protease enzyme from *Brevibacillus parabrevis*, which demonstrated efficient deproteinization rates. Using crab solid waste and a fermentation process with lactic acid bacteria (*L. plantarum*), high-quality chitin was recovered [28].

Additionally, *Pseudomonas aeruginosa*, *S. marcescens*, and *Bacillus pumilus* were used to remove chitin from shrimp waste in order to identify the bacterial strains that produce protease. *P. aeruginosa* had the greatest rates of demineralization (78.46%) and deproteinization (74.76%) among them. Chitin was extracted from *Litopenaeus vannamei* shell wastes using a protease generated from *Alcaligenes faecalis* and lactic acid-producing (*Bacillus coagulans*) bacterial strains [30]. In another investigation, chitin was extracted from shrimp shell waste using the protease-producing strain *P. aeruginosa* [31]. Different marine sources and their associated bacteria are presented in Table 2, along with yield percentages, to provide insights into the biological extraction of chitin.

When compared to alkali extraction, the proteolytic enzymes generated by the bacterial communities often showed higher yields, making it one of the most promising biological extraction methods.

3.1.2 Enzyme-Assisted Extraction

In response to increasing safety and environmental concerns, the functionalization of natural polymers through enzymatic methods has been actively explored as a viable alternative to toxic chemical approaches that pose environmental risks and lack specificity. Enzyme-Assisted Extraction (EAE) demonstrates greater specificity, rapid action, and a reduction in the consumption of energy, chemicals, and water relative to traditional processes [37]. This process facilitates the recovery of high-value-added products such as chitin, pigments, and peptides [38]. Furthermore, enzyme

Table 2
Microbial fermentation-based biological extraction of chitin and chitosan

Sources of marine waste	Microbial strains and/or proteolytic enzymes	Carbon source	Yield (%)	References
Prawn shell waste	*Lactobacillus plantarum* ATCC 14917 and *Bacillus subtilis* ATCC 6051	Red grape pomace	Chitin: 0.061	[32]
Shrimp head waste	*Paenibacillus mucilaginosus* TKU032	Fishery by-products	Chitin: 20.67	[18]
Lobster shell waste	*L. plantarum*, *Bacillus megaterium* NH21, or *Serratia marcescens* db11	Glucose	Chitin: 82.56	[33]
Shrimp shells (*Penaeus vannamei*)	*S. marcescens* B742, *L. plantarum* ATCC 8014, and *Rhizopus japonicus* M193	Glucose	Chitin: 74.67 Chitosan: 63.42	[29]
Crab (*Allopetrolisthes punctatus*)	*L. plantarum* sp. 47	Sucrose	Chitin: 6.9	[34]
Shrimp wastes	*Lactobacillus acidophilus* FTDC3871	Glucose	57.7	[32]
Shrimp shell powder	*Paenibacillus jamilae* BAT1	Glucose	24.5	[35]
Shrimp shells	*B. subtilis* and *Acetobacter pasteurianus*	Glucose	18	[36]
Shrimp waste	*Penaeus merguiensis*, *Pseudomonas aeruginosa*	Glucose	Chitin: 47 Chitosan: 87	[31]

application mitigates the irregular deacetylation and molecular weight reduction induced by acid and alkali treatments [39]. Chitosan was produced enzymatically using specific enzymes, including chitinases and chitosanases, as well as nonspecific enzymes such as carbohydrases and proteases [40]. Table 3 represents enzyme-based extraction of chitin and chitosan from different marine food waste.

The EAE employs a sequential method combined with mild alkali treatment for the extraction of chitin from shell waste. Shrimp shell waste underwent deproteinization using alcalase, a proteolytic enzyme, followed by demineralization with a 10% HCl solution at 20 °C for 30 min, conducted at 55 °C and a pH of 8.5 [47]. A suspension of demineralized shells in water (1:2, w/v) underwent protein hydrolysis at 55 °C and a pH of 8.5, utilizing an enzyme concentration of 20 AU/kg of protein. This process produced pure chitin with a reduced presence of residual small peptides and amino acids directly associated with chitin molecules. The degree of hydrolysis (DH) was approximately 4.4% residual protein at 30% DH. In a similar manner, the proteolytic enzyme papain achieved a deproteinization rate of 73.1% in shell waste over a 72-h digestion process, while the chemical method resulted in a 98%

Table 3
Chitin and chitosan extracted from natural resources using enzymes

Sources	Experimental conditions	Enzymes used	Recovery efficiency (%)	References
Shrimp waste (*Metapenaeus monoceros*)	pH: 8.0, temp: 90 °C enzyme: substrate-20 (w/w), time: 20 min	Protease (from *Bacillus cereus* SV1)	88	[41]
Shrimp shell (*Litopenaeus vannamei*)	pH: 6.0, temp: 45 °C, enzyme: substrate—7.5–100:25–200 U/g, time: 1–7 h	Chitinase and protease A and B	88.9	[42]
Shrimp shell (*Penaeus vannamei*)	pH 6.2, within a concentration of 0.25% (w/v)	Papain	76	[43]
Shrimp shell (*L. vannamei*)	pH: 7.7 and 7.5, temp: 45 ° C, time: 4 h	Trypsin and ficin	32.12	[44]
Crab shell	pH: 5.60, temp: 50 ° C, time: 24 h	Cellulase (from *Aspergillus niger*)	55 Chitooligosaccharides	[45]
Mealworm's cuticles (*Tenebrio molitor*)	pH: 8.5; time: 10 min	Alcalase (from *Bacillus licheniformis*)	Chitin: 31.9% Chitosan: 45.1%	[46]

deproteinization [48]. Chitin produced via the enzymatic method exhibited a higher degree of acetylation (19.4%) and viscosity compared to that obtained through the chemical method (17.2%). A chitinase was derived from *Lecanicillium fungicola*, resulting in the production of limited crude enzymes (endochitinase and *N*-acetylhexosaminidase) via submerged fermentation (SF) with the addition of chitin to generate chitin oligosaccharides [49]. To enhance the solubility of the substrates, α- and β-chitins underwent alkali treatment, resulting in degrees of deacetylation of 55% and 50%, respectively. Subsequently, chitin hydrolysis conducted with crude enzyme utilizing endochitinase (at 40 °C and a pH of 5) and *N*-acetylhexosaminidase resulted in yields of 619 and 355 U/mg of protein, respectively. The partially deacetylated α-chitin served as a substrate for *L. fungicola* chitinases in slightly acidified reaction media, resulting in a yield of chitin oligosaccharides that was up to 16-fold greater than that obtained from crystalline α-chitin. The maximum concentrations of chitin oligosaccharides obtained from α- and β-deacetylated chitins were 2.77 mmol/L and 4.44 mmol/L, respectively. Partial deacetylation and the application of acidified reaction media markedly enhanced enzyme efficiency regarding chitin oligosaccharide yield and processing duration.

Chitooligosaccharides can be derived from chitosan via hydrolysis using cellulase sourced from *A. niger* [45]. In this process, the chitosan solution was adjusted to a pH of 5.6 using NaOH, after which 1% cellulase was added. The mixture was then incubated at 50 °C for 24 h to recover chitooligosaccharides. This method demonstrated that cellulase was utilized to hydrolyze chitosan, yielding chitooligosaccharides with a degree of polymerization ranging from 3% to 11%. However, a degree of polymerization between 6% and 8% resulted in a superior yield. This method proved to be more cost-effective compared to specific enzymes like chitinase, chitosanase, and lysozyme. Protease derived from *Bacillus cereus* SV1 [41] demonstrated 88.8% deproteinization of shrimp waste at an enzyme to substrate ratio of 20. Demineralization was fully accomplished within 6 h at room temperature using 1.25 M HCl, resulting in a residual calcium content of 0.01% in the chitin. The Nuclear Magnetic Resonance (NMR) study indicates that the structure of chitin derived from the EAE method is comparable to that obtained through the alkaline hydrolysis process and commercial α-chitin. Shrimp shells were hydrolyzed for chitin extraction using both chemical methods (acids and alkalis) and a papain enzymatic method [43]. The chemical method did not modify the morphology of chitin following hydrolysis, maintaining a high degree of N-acetylation and crystallinity. Nonetheless, the enzymatic method utilizing papain compromised the structural integrity of chitin within the tissue. This occurs due to papain's broad selectivity, which concurrently hydrolyzes proteins and deacetylates chitin, resulting in chitosan that requires additional purification steps, presenting a potential drawback. Chemical purification is effective; however, by-products remain unrecoverable. In contrast, papain hydrolysis produces partially purified chitosan while allowing for the isolation of amino acids, which is significant for applications in the food industry. Furthermore, food-grade commercial enzymes including alcalase, flavourzyme, lysozyme, papain, and trypsin VI demonstrated protease activity in the deproteinization of dried shrimp heads [50]. Alcalase (pH of 8.0) emerged as the most effective commercial enzyme for the deproteinization of shrimp heads, as indicated by the increased signal intensity in the carbohydrate fingerprint region (950–1200/cm) observed through infrared spectroscopy. Chitins underwent treatment with alcalase and were demineralized using lactic acid, accompanied by microwave irradiation at 121 °C for 30 min. The integration of enzymatic and physicochemical treatments facilitated the recovery of chitin from dried shrimp heads in an environmentally sustainable manner.

In a similar study, protease-assisted extraction of chitin from shrimp shells under demineralization conditions at 25 °C for 20 min, utilizing a shells to lactic acid ratio of 1:1.1 w/w and a shells to acetic acid ratio of 1:1.2 w/w, achieved maximum demineralization rates of 98.64% for lactic acid and 97.57% for acetic acid

[51]. A protein removal efficiency of 91.10% was achieved using protease from *Streptomyces griseus*, with an enzyme-substrate ratio of 55 U/g, a pH of 7.0, and an incubation time of 3 h. The particle size range of 50–25 μm was identified as the most significant factor influencing this efficiency. The protease-assisted extraction of chitin demonstrated a reduced crystallinity percentage and an increased degree of acetylation, along with improved solubility properties and diminished depolymerization when compared to chitin obtained through chemical extraction methods. Utilizing alcalase, esperase, and neutrase for deproteinization, followed by alkaline treatment for deacetylation, achieved the highest deacetylation levels (>93%) at NaOH concentrations of 61.0–63.7% and hydrolysis durations of 14.9–16.4 h from the by-products of *Illex argentinus pen* [52]. The molecular weight of the chitosan produced in the experimental designs ranged from 143 to 339 kDa. The study's results confirmed the efficacy of alcalase and esperase in deproteinizing squid pens, serving as an alternative to traditional chemical proteolysis. Lopes et al. (2018) utilized alkali and enzymatic hydrolysis on the crustacean *Munida* spp. Chemical hydrolysis was conducted at a regulated temperature of 110 °C using 4 M NaOH. Hydrolysis was conducted enzymatically using alcalase, with protein removal executed at a controlled temperature of 55 °C and a pH of 8. During the demineralization stage, the cake underwent washing with 1 M and 4 M HCl for enzymatic and chemical processes, respectively, maintaining a solid/liquid ratio of 1/5 (w/v) at room temperature. The efficiency of the demineralization step is lower for the enzymatic process, recorded at 19%. Moreover, the enzymatic process failed to eliminate all proteins, as approximately 5–10% of residual protein typically remains associated with the isolated chitin. Alkaline treatment at elevated temperatures is a more effective method for protein removal from shells; however, it results in the depolymerization and deacetylation of chitin [53]. Deng and coworkers employed two recombinant aspartic proteases for protein hydrolysis and a recombinant chitinase for chitin hydrolysis. The deproteinized shell residues underwent treatment with chitinase at 45 °C and a pH of 6.0, with shaking at 150 rpm. The chitin recovery rate (CRR) increased with higher chitinase concentrations and extended incubation times. When the enzyme concentration (50 U/g) and reaction time (6 h) were increased, there was only a minimal enhancement in CRR. Consequently, the parameters of 50 U/g and 6 h were employed for chitin hydrolysis. Following hydrolysis by chitinase, the weight reduction was quantified at 16.8% of the shell mass. A chitin conversion rate of 88.9% was determined based on weight reduction, suggesting that the majority of chitin was transformed into an oligomeric form. The process achieved protein recovery rates of 91.4% and chitin recovery rates of 88.9%. Amino acids in peptides and the N-acetyl group of chitin oligomers were preserved in their natural states [28].

3.1.3 Ultrasonic-Assisted Extraction (UAE)

The cavitation effect of ultrasound (US) enhances the solubility of proteins linked to chitin, likely resulting from the depolymerization of macromolecules, the dissociation of covalent bonds within polymer chains, and the dispersion of aggregates [54]. High-intensity ultrasound signals at 750 W power and an operating frequency of 20 kHz ± 50 Hz improved the efficiency of chitin extraction, thereby reducing extraction time and eliminating the need for elevated temperatures [55]. The pronounced dipole–dipole interactions within the biopolymer chains influence the polymer's crystallinity, thereby significantly impeding chitinase-mediated hydrolysis of chitin. To address this shortfall, the effects of sonication and steam explosion (SE) are regarded as effective pretreatments. This was applied to the chitin oligosaccharide using partially purified chitinases derived from *Lecanicillium lecanii* [56]. Sonication, in conjunction with solvent extraction from a high-pressure reactor following the heterogeneous deacetylation of chitin, resulted in a notable reduction in both the crystallinity indices and the average degree of acetylation in the chitin samples. Although sonication resulted in minor depolymerization, SE treatment significantly affected amino-based hydrogen bonding in the samples. Hydrogen bonding decreased following SE treatment of the samples. However, SE did not decrease sugars at the onset of the reactions, indicating that SE is chemically less destructive than sonication. Consequently, enhanced enzymatic hydrolysis without depolymerization, characterized by a minimal degree of acetylation, can be accomplished through pretreatment with SE. Furthermore, a comparison was made between US irradiation and thermochemical extraction for the conversion of α-chitin to chitosan. The reaction medium undergoes high-intensity ultrasound irradiation under nonisothermal conditions [57]. The study investigated the impact of (1) heat application in the reaction medium (100–105 ° C) and (2) high-intensity ultrasound (45 and 60 min) on chitosan recovery. The α-chitin suspension underwent nonisothermal ultrasound-assisted N-deacetylation for extended durations, resulting in higher chitosan yields (99.4% for 60 min and 95.3% for 45 min) and lower degrees of deacetylation (29.0% for 60 min and 31.0% for 45 min) [57]. An 84.4% degree of deacetylation (DDA) of β-chitosan was achieved at 120 min. However, the longer reaction time (>120 min) resulted in a low DDA and high degradation of β-chitosan into monomers or soluble chitosan by the dissociation of β-chitosan's β (1 → 4) glycosidic bonds [58]. The dimensions of the US probe on the surface with constant reaction time affect the volume of β-chitin, with approximately 88% of β-chitin being converted into chitosan during the initial step [59]. Mild conditions, devoid of an inert gas atmosphere or additives, along with extended reaction time, were determined to be effective. Furthermore, the application of US irradiation at regular intervals (7.5 min/period and non-sonication for 5 min/period) would enhance the yield of chitosan by 100%, achieving the lowest degree of acetylation (DA) of 22.1% [57]. Furthermore, the extremely high pressures (>20 MPa), temperatures (>5000 K), and cooling rates (>10^6 K/s) resulting from US irradiation due to the collapse of sonochemical cavities via acoustic cavitation generally influence the chemical reactions [60].

3.1.4 Microwave-Assisted Extraction (MAE)

Microwave technology finds extensive application in the food and chemical industries. Numerous reports indicate that microwave heating significantly accelerates chemical reactions, improves reaction yield and enhances product purity and properties in comparison to conventional heating [28]. It offers numerous advantages, such as rapid heating, energy efficiency, low processing costs, precise process control, and selective heating [61]. Microwave heating operates through two primary mechanisms: (1) dipolar polarization and (2) ionic conduction [62]. The thermal effects arise from inverted heat transfer, a uniform microwave field within the sample, and the selective absorption of radiation by polar substances [63]. Microwave irradiation efficiently transfers energy directly and rapidly to the biomass substrate and catalyst, enhancing reaction efficiency [64]. Microwave heating has been employed in the extraction of polysaccharides, such as chitosan and cellulose, while also decreasing the reliance on chemicals [28].

Various microwave operating conditions, such as reaction time, solvent concentration, and solid-to-liquid ratio, influence the degree of deacetylation and the molecular mass of the resulting chitosan. Optimizing process conditions is essential for enhancing the reaction rate and minimizing production costs [65]. MAE achieves 82% deacetylation of chitosan in 24 min, whereas a conventional heating method necessitates 6–7 h to attain a similar level of deacetylation (81%) [66]. Both the conventional heating method and MAE produced products with comparable structures, morphologies, and chemical compositions. Chitosan obtained through microwave heating demonstrated a higher molecular weight and comparable crystallinity to commercial chitosan, achieved in a shorter heating duration than the conventional method. Chitosan (2 g) combined with HCl underwent hydrolysis via microwaves and alkali to produce glucosamine, with microwave exposure times ranging from 5 to 20 min, in contrast to conventional hydrolysis conducted on a hot plate at 90 °C for 90 min. Microwaves markedly decreased both the heating temperature and duration. FTIR analysis indicated that the absorption bands of the synthesized glucosamine hydrochloride conformed to those of the standard compound [67]. Increasing microwave irradiation to 130 W for 15 min led to significant deproteinization (11.46%) and a reduced ash content (5.4%) at 700 °C for 2 h [68]. The demineralization reaction facilitated by microwave heating is significantly more effective than conventional heating, owing to the molecular interactions with microwaves that produce uniform heating. Zhang et al. [69] optimized the conditions for microwave-assisted heating treatment to yield low-molecular-weight chitosan with a high degree of deacetylation (DDA). According to the model, microwave conditions involving 80 s of heating time, a 1.5% concentration of H_2O_2, and a solid-to-liquid ratio of 1:40 resulted in the highest degree of deacetylation (90.58%) and the lowest molecular weight (124.25 kDa) of chitosan. The antibacterial activity of chitosan with a low molecular weight and a degree of deacetylation of 90.58% exceeded that of chitosan with a high molecular weight and a degree of deacetylation of 52.14%. Furthermore, certain studies indicated that elevated power in MAE systems necessitates a low alkali concentration for the extraction process [69]. Machałowski et al. [70] presented an effective method for isolating chitin from Theraphosidae spiders using MAE. The lipids and waxes were extracted from the cuticle surface using a chloroform:ethanol (2:1) mixture, subjected to microwave irradiation for 1 min at 750 W and 2450 MHz. The cuticle was subsequently isolated and rinsed with absolute ethanol. In the second step, the sample was treated with 2.5 M NaOH solution and irradiated with microwaves for 3 min at 95 °C, resulting in an orange solution. Subsequently, depigmentation was conducted under microwave irradiation for 2 min utilizing 30% H_2O_2. Finally, the isolated transparent and soft chitin scaffold was washed with distilled water to neutral pH and stored at 4 °C [70].

3.1.5 Subcritical Water Extraction (SWE)

Subcritical water treatment presents significant potential for producing raw materials for new products from waste biomass while simultaneously reducing waste stream volumes. This approach presents multiple benefits, such as accelerated reaction rates and the substitution of acids and bases with a more environmentally friendly solvent.

Subcritical water hydrolysis typically occurs at temperatures between 100 and 374 °C, maintained under adequate pressure (Pc of 22.1 MPa) to ensure water remains in a liquid state [71, 72].

Water and CO_2 are predominantly utilized in subcritical reactions for the processing of agricultural biomass residue, owing to their environmentally friendly properties and greater availability. Pressurized subcritical water, at temperatures ranging from 100 to 374 °C, functions as an exceptional solvent for diverse cellulosic and *proteinaceous biomasses*. At elevated subcritical water temperatures, chitin undergoes hydrolysis, leading to the decomposition of oligosaccharides and the recovery of compounds including amino acids and fatty acids [73, 74]. This process facilitates alterations in chitin morphology to improve enzymatic digestion and aids in the removal of proteins from crab shells, in addition to enabling the extraction of mineral elements such as calcium and citric acid [28].

Table 4
Chitin and chitosan extraction from natural sources using subcritical water

Sources	Extraction conditions		Recovery efficiency (%)	References
	Temperature (°C)	Time		
Shrimp cephalothorax	260	30 min	Chitin: 82.3	[75]
Crab shell	350	7 min	*N,N'*-Diacetylchitobiose: 8	[76]
Swimming crab (*P. trituberculatus*)	170	30 min	Chitosan: 12.2	[77]
Chitin-glucan complexes (*Aspergillus niger*)	300	2.0–50 s	50.8 (overall)	[78]

Recent investigations have employed subcritical treatment for chitosan extraction and the same is represented in Table 4.

Ordoñez et al. [78] extracted chitin and chitin-glucan complexes from the mycelium of *A. niger* using subcritical water treatment. A continuous apparatus was developed for operation at 300 °C, with a pressure of at least 29 MPa and a mycelium residence time ranging from 2.0 to 50 s. The reaction of *A. niger* biomass in subcritical water resulted in complexes with chitin concentrations ranging from 19.7 to 54.2 wt%, molecular weights from 1.7 to 155.2 kDa, and product yields between 22 and 68 wt%. Preliminary exploratory tests indicate the potential to isolate products with chitin concentrations exceeding 80 wt% by conducting the process at temperatures above 230 °C and reaction times ranging from 13.5 to 35 s. The controllable percentage of chitin suggests that these products may be applicable across various domains [78]. Espíndola-Cortés et al. [75] extracted calcareous chitin from shrimp through subcritical water treatment, achieving a deproteinization rate of 96%. The Taguchi statistical model was employed to optimize experimental conditions. The treatments resulted in an increase in the crystalline domain size of the chitin fibers. The maximum α-chitin content of 82.2 wt% was achieved with a 0.17 chitin:dH_2O (wt/wt) ratio following 30 min of treatment at 260 °C. Chemical, structural, and morphological characterizations confirmed the complete removal of protein from the chitin matrix under subcritical conditions within a brief reaction time [75].

3.2 Green Extraction Methods for Collagen: Mechanisms and Current State

3.2.1 Extraction of Enzyme-Soluble Collagen

The principal approach for collagen extraction is enzyme-soluble collagen (ESC), which is a green biotechnology that has the potential to minimize solvent consumption and waste formation by speeding the breakdown reaction without the need of organic solvents and harmful chemicals to manufacture replacements [79]. Adding enzymes breaks down the telopeptides of the triple helix, dissolves collagen peptides, and enhances extraction yield. Yu et al. found that employing optimal circumstances such as a pepsin concentration of 1389 U/g, hydrolysis duration of 8.67 h, and a solid-to-liquid ratio of 1.57, collagen could be extracted with a maximum yield of 84.85%. Pepsin was used to extract collagen from bighead carp tissues, including skin, bones, scales, fins, and swim bladders [80, 81].

Similar levels of collagen were produced by the skin (60.3%) and swim bladder (59%), which were noticeably more than those from the fins (5.1%), scales (2.7%), and bones (2.9%). The kind of fish, its environment, and the particular pretreatment and extraction technique employed may all affect the collagen output. In a recent study, Lee et al. investigated how the enzymatic extraction method (1% Alcalase) affected the physicochemical and antioxidant properties of collagen that was isolated from Alaska pollock skin. Jia et al. used a variety of enzymes, such as Neutrase, Flavourzyme, Alcalase, trypsin, Protamex, papain, and pepsin, to create an enzyme hydrolysate from Alaska pollock skin in order to identify the best enzymes and hydrolysis conditions [82–84].

Collagen hydrolysate derived from Alaska pollock skin using various enzyme extraction techniques has been shown in two experiments to possess strong antioxidant qualities. Further research on the antioxidant properties of scallop protein hydrolysates (SPHs) made by enzymatic hydrolysis using dispase, pepsin, and alkalase was carried out by Wang et al. Because of its exceptional solubility and high concentration of amino acids (35.25%), which have antioxidant qualities, the SPH hydrolyzed by alcalase shows a notable ability to scavenge free radicals [79, 85].

Extraction of pepsin-soluble collagen from eggshell membrane revealed that pepsin concentration is one of the most critical elements in collagen content. The content of pepsin linked highly with hydrolysis duration in terms of improving extraction yield. Increasing the concentration of pepsin from 15 to 50 U/mg significantly improved collagen recovery from eggshell membrane. Increasing the quantity of pepsin to 50–75 U/mg resulted in a steady rise that reached a maximum value after 48 h of hydrolysis, with no further influence. With enough reaction time, hydrolysis in the telopeptide region was expected to be more evident. As a result, the addition of pepsin may accelerate the release of the collagen triple helix into the resultant slurry as compared to collagen extracted without it. Collagen extracted from the skin of gigantic croaker (*Nibea japonica*) showed improved extraction efficiency as the amount of pepsin increased [80, 86].

Papain is another enzyme that has been shown to aid in the extraction of acidic collagen. The primary driving force behind the search for papain, a plant-based protease, was the halal concern around pepsin, which is commonly derived from pig sources and employed in collagen extraction. Papain efficiently generated a considerable yield (22.79%) of collagen from the skin of yellowfin tuna (*Thunnus albacares*) at 7000 U/mg. 20 kU/g papain increased the extraction efficiency of collagen from barramundi (*Lates calcarifer*) skin by up to five times when compared to acid-soluble collagen [87, 88]. Additionally, papain-aided extraction produced results that were equivalent to pepsin-aided extraction when the same quantity of enzyme was used. When papain treatment was optimized for collagen extraction from chicken feet, the highest yield (32.16%) was obtained at 30 °C for 28 h of contact time. Low-molecular-weight components were present together with scattered molecular weight bands in the collagen from chicken foot treated with papain. This was in line with the collagen that was extracted from barramundi skin using papain, which showed similar protein patterns (37–150 kDa). Chuaychan et al. claim that the enzyme's separation of the telopeptide segments was the reason for the occurrence of peptides of lower sizes [89, 90].

3.2.2 Ultrasound-Assisted Extraction

Collagen may be extracted more effectively with ultrasound, which is a promising technique. The technique by which ultrasonic treatment enhances extraction yields is related to cavitation effects, which occur when alternating mechanical waves induce microbubbles to form and grow in the liquid medium to a sufficiently negative threshold pressure. The bursting of cavitation bubbles causes macroturbulence, high velocity interparticle collisions, and disruption in biomass particles, hence enhancing mass transfer during solid-liquid extraction [91, 92]. The higher extraction yield caused by cell rupture is due to liquid jets of solvent resulting from asymmetric microbubble collapse. Cavitation caused by ultrasonic pressure waves fragments the collagen structure inside the tissue matrix, leading in better collagen extraction (Refer Fig. 1). The standard collagen extraction process is time demanding and leaves a substantial number of insoluble collagens in the tissue matrix. Sonication speeds up the extraction process compared to traditional methods by helping to relax the tissue matrix and enabling the acid solution to enter the tissue matrix more effectively by rupturing collagen fibrils [93, 94].

Additionally, because the structure became looser when pepsin was added to the UAE procedure, more collagen was effectively removed. In this instance, pepsin's synchronized cleavage of the telopeptide region and the mechanical action of ultrasound allowed pepsin to more easily penetrate the tissue matrix. The amount of mechanical energy applied and the length of ultrasonic treatment determine how easily collagen may be extracted. To avoid adverse impacts on the resultant collagen's molecular structure, the UAE conditions must be continuously regulated [10, 95].

Collagen extraction from chicken lung rose up to 150 W of ultrasonic power, but the trend reversed when 200 W of ultrasonic power was added. During the extraction procedure, using too much ultrasonic power might break the collagen structure entirely or decrease the collagen's solubility. According to different

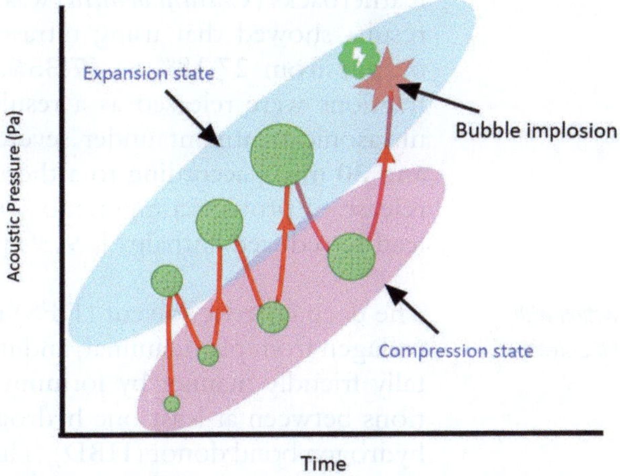

Fig. 1 Schematic depiction showing the mechanism of bubble cavitation. [10]

research, the collagen output from the skin of clown featherbacks (*Chitala ornata*) was proportionate to the ultrasonic amplitude; however, the purity was reduced, as seen by a decrease in the hydroxyproline concentration. As the UAE time rose, the collagen output from the sternal cartilage of chickens also increased. The extended UAE duration of 36 min, however, resulted in a partial disruption of collagen's secondary structure, as seen by a decrease in β-chain components [94, 96, 97].

Similarly, when the UAE time rose up to 30 min, the collagen's triple-helical structure underwent slight changes, according to the Fourier transform infrared (FTIR) spectrum of the skin of clown featherbacks (*Chitala ornata*). However, when UAE and pepsin were combined, the collagen from golden carp (*Probarbus jullieni*) retained its triple-helical structure. It's interesting to note that collagen showed increased heat stability after ultrasonic treatment. This might be because pepsin removed the non-helical telopeptide region, creating a more compact and organized collagen structure. In contrast to inorganic acid (hydrochloric acid), UAE (20 kHz and 500 W for 30 min) with organic acid (acetic acid) produced a threefold greater yield and shown improved functional qualities, according to a comparative analysis of several acid types [94, 95, 98].

Using varying amplitudes (20–80%) and periods (10–30 min), Petcharat et al. investigated the ultrasound-assisted collagen extraction from clown featherback skin. In their investigation, the collagen output rose from 27.18% to 57.35% with an increase in ultrasonication duration ($p > 0.005$) [94]. It was discovered that the ideal ultrasonic extraction parameters for collagen were 20 kHz, 80% amplitude, and 0.1 M acetic solution for 3 h. The impact of ultrasonic therapy on collagen extracted from the skin of clown featherbacks (*Chitala ornata*) was examined by Petcharat et al. The results showed that using ultrasonic therapy raised the collagen output from 27.18% to 57.35%. Low-molecular-weight protein fractions were released as a result of protein cleavage caused by ultrasonic treatment under severe circumstances (80% amplitude and 30 min), according to a thermal transition investigation. The release of protein components with low molecular weight might lead to reduced enthalpy [79, 94].

3.2.3 Extraction with Deep Eutactic Solvent

The deep eutectic solvent (DES) method could be used to extract collagen from plant, animal, and marine sources in an environmentally friendly manner by forming strong hydrogen bond interactions between at least one hydrogen bond acceptor (HBA) and a hydrogen bond donor (HBD). The introduction of water into DES for extraction is a well-known technique that improves mass transfer operations by lowering viscosity [99–101]. Collagen was extracted from Atlantic cod fish by Bisht et al. using aqueous solutions of several DESs, with acetic acid, lactic acid, and propionic acid serving as the HBD and urea (U), betaine (Bet), and cholinium chloride (CC) as the HBA. At the ideal urea: lactic acid (1:2) aqueous solution at a concentration of 0.75 M, the maximum collagen extraction yield was 6%. Using a DES made up of oxalic acid (OA) and choline chloride (CC), Bai et al. also showed that 90% of collagen could be extracted from cod skin. They also looked at how temperature affected the yield of CC:OA extraction. Because of a larger mass transfer coefficient and lower viscosity, they discovered in their investigation that higher temperatures enhanced collagen extraction [102, 103]. Elevated temperatures not only purify extracted collagen but also enhance polymerization above 55 °C, yielding a maximum efficiency of 96.01% at 65 °C. Batista et al. described a green collagen extraction from blue shark skin utilizing a natural deep eutectic solvent made of citric acid, xylitol, and water (1:1:10 molar ratio). The isolated collagen was recognized as type I collagen, with a 2.5-fold greater collagen yield than the typical collagen and a significantly shorter extraction time of 96 to 1 h. The extracted material's biocompatibility was demonstrated by in vitro cytotoxicity research [103].

3.2.4 Supercritical Fluid Extraction

Supercritical fluid extraction (SFE) is a popular green extraction technique that has several advantages over commercial extraction methods, including increased selectivity, improved fractionation capabilities, lower environmental impact, and a higher collagen yield. It can be used to extract high-value compounds from natural products and food with increasing reliability in the heavy metal recovery and drug delivery sectors. The SEF approach depends on the solvent's solvating properties, using pressure and temperature beyond the solvent's critical point to cause significant physical changes that improve the solvent's capabilities [79, 104, 105]. The point at which the substance's critical temperature and critical pressure combine is known as the critical point, according to the pressure-temperature diagram. A homogeneous supercritical fluid (SCF) is produced at the critical point when the characteristics of the gas and liquid phases converge as the material approaches its critical temperature. The behavior of the solvent in this state is in between that of a gas and a liquid, with low viscosity and high diffusivity contributing to higher mobility than a liquid. Furthermore, by adjusting the SFE's temperature and pressure, its solvent properties may be changed, as well as the fluid density. The solvent's capacity to interact with other molecules can be significantly impacted by slight variations in temperature and pressure, which may favor the precipitation of the solutes dissolved in the SCF or the extraction of the target compounds [10, 106]. Mechanism of collagen extraction using supercritical fluid is schematically represented in Fig. 2.

CO_2 is a popular SEF molecule because to its remarkable qualities such as low toxicity, cost-effectiveness, stability, great availability, flammability, environmental acceptability, and gentle working conditions. This approach was employed in a recent study to extract collagen from cod skin, yielding a maximum of 13.8%. The skins were specifically placed in a high-pressure vessel filled with distilled water and compressed to 50 bars with CO_2 gas. The resulting collagen might be employed in the beauty, biomedical, and pharmaceutical industries, providing significant economic benefits [79, 104, 107]. In an investigation, collagen was extracted from *Chondrosia reniformis* sponges using water acidified with CO_2 at different pressures and periods. Even at the mildest SFE conditions (3 h, 10 bar), the isolation was effective. Regardless of the SFE time, the study shows that raising the pressure from 10 to 50 bar at 37 °C decreased the yield. The solid matrix may contract under increased pressure, which might lead to the void fraction and poor

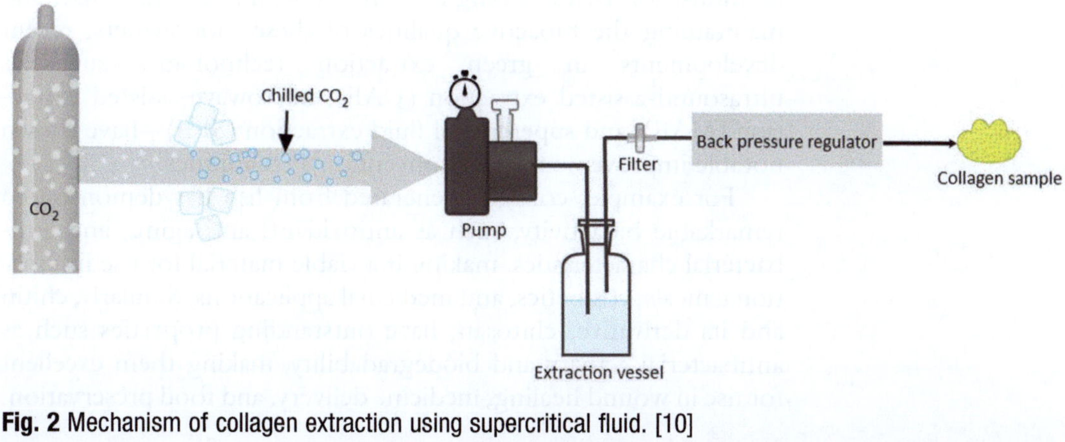

Fig. 2 Mechanism of collagen extraction using supercritical fluid. [10]

extraction results. It's interesting to note that the low CO_2 pressure employed in this investigation can quicken the sponge matrices' swelling process, enabling SCF to enter the cell and promote collagen healing. Since temperature and pressure directly affect an SCF's density, viscosity, and diffusivity, it is essential to comprehend how these two variables relate to one another. For collagen extraction based on SFE, this is still up for debate [108–110].

3.2.5 Extrusion–Hydro-Extraction

The strong bond between collagen and hydroxyapatite in fish scales makes separation difficult. In order to facilitate the release of collagen by water extraction from fish scales, the extrusion–hydro-extraction technique uses extrusion to disrupt the tight bond between collagen and hydroxyapatite. Continuous production, low labor costs, little waste, high yield, ease of use, and a variety of product results are some advantages of this process. Huang and Kuo presented a brand-new extrusion–hydro-extraction technique for removing collagen from the scales of tilapia fish. In comparison to non-extruded scale samples, their results showed that the high pressures attained during extrusion produced collagen yields ranging from 7.5% to 12.3%, which were two to three times greater. Furthermore, the collagens produced using this technology closely matched collagens extracted using traditional techniques in terms of their amino acid profiles and moisture absorption and retention capabilities [111, 112].

4 Notes

Using green extraction methods to extract collagen and chitin from marine food waste is a potential strategy for producing valuable biopolymers and managing garbage in a sustainable manner. Collagen and chitin, which are abundant in marine by-products including fish skins, scales, and crab shells, respectively, have significant uses in the culinary, pharmaceutical, biomedical, and environmental industries. By increasing efficiency, lowering chemical use, and maintaining the bioactive qualities of these biopolymers, recent developments in green extraction technologies—such as ultrasound-assisted extraction (UAE), microwave-assisted extraction (MAE), and supercritical fluid extraction (SFE)—have shown notable improvements over conventional techniques.

For example, collagen generated from fish has demonstrated remarkable bioactivity, such as antioxidant, anti-aging, and antibacterial characteristics, making it a viable material for use in functional meals, cosmetics, and medicinal applications. Similarly, chitin and its derivative, chitosan, have outstanding properties such as antibacterial activity and biodegradability, making them excellent for use in wound healing, medicine delivery, and food preservation.

While these green extraction strategies show promise, there are still hurdles to their optimization and large-scale deployment. Despite breakthroughs, extraction techniques sometimes require additional modification to improve efficiency and cost-effectiveness. Low denaturation and melting temperatures continue to hinder collagen's applicability in high-temperature situations such as biomedical devices and food packaging. Research into structural changes and collagen supply from heat-resistant marine animals might help to address these challenges.

Similarly, existing techniques of chitin extraction are energy-intensive, and while emerging technologies like as pulsed electric fields (PEF) and SFE show promise, they too require additional exploration to increase scalability and operational costs. Furthermore, the limited thermal stability of marine-derived chitin limits its commercial uses, particularly high-temperature procedures. To address these constraints, further research into thermally resistant forms of chitin, either by alternate extraction methods or sourcing from warmer marine species, would be critical for broadening its potential applications.

To summarize, while green extraction approaches for collagen and chitin from marine food waste offer a sustainable and ecologically friendly alternative to traditional methods, more study and optimization are required. By fine-tuning extraction conditions, improving structural stability, and increasing scalability, these biopolymers can become more economically viable for a wide variety of applications, from biomedical to food sectors. With further innovation and the incorporation of biotechnological developments, the sustainable synthesis of collagen and chitin has the potential to significantly advance the circular bioeconomy, reduce waste, and promote the effective use of marine resources.

References

1. Action S (2020) World fisheries and aquaculture. Food and Agriculture Organization, pp 1–244
2. Venugopal V, Shahidi F (1998) Traditional methods to process underutilized fish species for human consumption. Food Rev Intl 14(1):35–97
3. Gustavsson J et al (2011) Global food losses and food waste. FAO, Rome
4. Yan N, Chen X (2015) Sustainability: don't waste seafood waste. Nature 524(7564): 155–157
5. Mekouar MA, Food and Agriculture Organization of the United Nations (FAO) (2018) Yearbook of international environmental law, vol 29. FAO, pp 448–468
6. Cretton M et al (2021) Lipid fraction from industrial crustacean waste and its potential as a supplement for the feed industry: a case study in Argentine Patagonia. Waste Biomass Valor 12:2311–2319
7. Arvanitoyannis IS, Kassaveti A (2008) Fish industry waste: treatments, environmental impacts, current and potential uses. Int J Food Sci Technol 43(4):726–745
8. López-Pedrouso M et al (2020) Aquaculture and by-products: challenges and opportunities in the use of alternative protein sources and bioactive compounds. Adv Food Nutr Res 92:127–185
9. Islam MS, Khan S, Tanaka M (2004) Waste loading in shrimp and fish processing effluents: potential source of hazards to the coastal and nearshore environments. Mar Pollut Bull 49(1–2):103–110

10. Mohamad Razali UH et al (2023) Improving collagen processing towards a greener approach: current progress. J Chem Technol Biotechnol 98(5):1063–1082
11. Carmen P-C et al (2023) Influence of pre-treatment and drying process of the shrimp (Litopenaeus vannamei) exoskeleton for balanced feed production. Heliyon 9(6): e16712
12. Said A (2018) Sintesis plastik biodegradable berbahan komposit pati sagu-kitosan sisik ikan katamba (Lethrinus lentjam). QUANTUM: Jurnal Inovasi Pendidikan Sains 9(1): 23–30
13. Aboudamia FZ et al (2020) Potential of discarded sardine scales (Sardina pilchardus) as chitosan sources. J Air Waste Manage Assoc 70(11):1186–1197
14. Shalaby M et al (2020) Fish scale collagen preparation, characterization and its application in wound healing. J Polym Environ 28: 166–178
15. Boronat Ò et al (2023) Development of added-value culinary ingredients from fish waste: fish bones and fish scales. Int J Gastron Food Sci 31:100657
16. Xia FLW et al (2024) Turning waste into value: extraction and effective valorization strategies of seafood by-products. Waste Manag Bull 2(3):84–100
17. Nguyen TT et al (2020) Highly efficient recovery of nutritional proteins from Australian Rock Lobster heads (Jasus edwardsii) by integrating ultrasonic extraction and chitosan co-precipitation. Innovative Food Sci Emerg Technol 60:102308
18. Doan CT et al (2020) Microbial conversion of shrimp heads to proteases and chitin as an effective dye adsorbent. Polymers 12(10): 2228
19. Suprayitno E et al (2020) Effect of grouper collagen and cork fish albumin for making wound ointment. Int J Sci Res Publ 10(9): 655
20. Prihanto AA et al (2023) Characterization of collagen from Java barb (Barbonymus gonionotus) skin and its effect on selected properties of yogurt. Canrea J Food Technol Nutr Culinary J:17–30
21. Venugopal V, Sasidharan A, Rustad T (2023) Green chemistry to valorize seafood side streams: an ecofriendly roadmap toward sustainability. J Agric Food Chem 71(46): 17494–17509
22. Chen X, Yang H, Yan N (2016) Shell biorefinery: dream or reality? Chemistry 22(38): 13402–13421
23. Shamshina JL, Berton P, Rogers RD (2019) Advances in functional chitin materials: a review. ACS Sustain Chem Eng 7(7): 6444–6457
24. Krishnan S, Perumal P (2013) Preparation and biomedical characterization of jellyfish (Chrysaora Quinquecirrha) collagen from southeast coast of India. Int J Pharm Pharm Sci 5(3):698–701
25. Sugiura H et al (2009) In vivo biological responses and bioresorption of tilapia scale collagen as a potential biomaterial. J Biomater Sci Polym Ed 20(10):1353–1368
26. Ahmed M, Verma AK, Patel R (2020) Collagen extraction and recent biological activities of collagen peptides derived from sea-food waste: a review. Sustain Chem Pharm 18: 100315
27. Aranaz I et al (2021) Chitosan: an overview of its properties and applications. Polymers 13(19):3256
28. Mohan K et al (2022) Green and eco-friendly approaches for the extraction of chitin and chitosan: a review. Carbohydr Polym 287: 119349
29. Zhang H et al (2017) The preparation and characterization of chitin and chitosan under large-scale submerged fermentation level using shrimp by-products as substrate. Int J Biol Macromol 96:334–339
30. Rakshit S et al (2021) Extraction of chitin from Litopenaeus vannamei shell and its subsequent characterization: an approach of waste valorization through microbial bioprocessing. Bioprocess Biosyst Eng 44(9): 1943–1956
31. Sedaghat F et al (2017) Bioconversion of shrimp waste Penaeus merguiensis using lactic acid fermentation: an alternative procedure for chemical extraction of chitin and chitosan. Int J Biol Macromol 104:883–888
32. Tan JS et al (2020) Chitin extraction from shrimp wastes by single step fermentation with Lactobacillus acidophilus FTDC3871 using response surface methodology. J Food Process Preserv 44(11):e14895
33. Chakravarty J et al (2018) Chitin extraction from lobster shell waste using microbial culture-based methods. Appl Food Biotechnol 5(3):141–154
34. Castro R, Guerrero-Legarreta I, Bórquez R (2018) Chitin extraction from Allopetrolisthes punctatus crab using lactic fermentation. Biotechnol Rep 20:e00287
35. Taser B et al (2021) Preparation of chitosan from waste shrimp shells fermented with

Paenibacillus jamilae BAT1. Int J Biol Macromol 183:1191–1199

36. Zhang Q et al (2021) Establishment of successive co-fermentation by Bacillus subtilis and Acetobacter pasteurianus for extracting chitin from shrimp shells. Carbohydr Polym 258:117720
37. Jegannathan KR, Nielsen PH (2013) Environmental assessment of enzyme use in industrial production—a literature review. J Clean Prod 42:228–240
38. Cahú TB et al (2012) Recovery of protein, chitin, carotenoids and glycosaminoglycans from Pacific white shrimp (Litopenaeus vannamei) processing waste. Process Biochem 47(4):570–577
39. Rass-Hansen J et al (2007) Renewable hydrogen: carbon formation on Ni and Ru catalysts during ethanol steam-reforming. Green Chem 9(9):1016–1021
40. Kim JS, Je YH (2010) A novel biopesticide production: attagel-mediated precipitation of chitinase from Beauveria bassiana SFB-205 supernatant for thermotolerance. Appl Microbiol Biotechnol 87:1639–1648
41. Manni L et al (2010) Extraction and characterization of chitin, chitosan, and protein hydrolysates prepared from shrimp waste by treatment with crude protease from Bacillus cereus SV1. Appl Biochem Biotechnol 162: 345–357
42. Deng J-J et al (2020) Enzymatic conversion and recovery of protein, chitin, and astaxanthin from shrimp shell waste. J Clean Prod 271:122655
43. Gartner C, Peláez CA, López BL (2010) Characterization of chitin and chitosan extracted from shrimp shells by two methods. E-Polymers 10(1):069
44. Marzieh M-N et al (2019) Comparison of the physicochemical and structural characteristics of enzymatic produced chitin and commercial chitin. Int J Biol Macromol 139:270–276
45. Xie Y et al (2009) Preparation of chitooligosaccharides by the enzymatic hydrolysis of chitosan. Polym Degrad Stab 94(10): 1895–1899
46. da Silva Lucas AJ et al (2021) Extraction, physicochemical characterization, and morphological properties of chitin and chitosan from cuticles of edible insects. Food Chem 343:128550
47. Synowiecki J, Al-Khateeb NAAQ (2000) The recovery of protein hydrolysate during enzymatic isolation of chitin from shrimp Crangon crangon processing discards. Food Chem 68(2):147–152
48. Gopalakannan A et al (2000) Application of proteolytic enzyme, papain for the production of chitin and chitosan from shrimp waste. J Mar Biol Assoc India 42:167–172
49. Ramírez-Coutiño L et al (2006) Enzymatic hydrolysis of chitin in the production of oligosaccharides using Lecanicillium fungicola chitinases. Process Biochem 41(5): 1106–1110
50. Valdez-Peña AU et al (2010) Screening of industrial enzymes for deproteinization of shrimp head for chitin recovery. Food Sci Biotechnol 19:553–557
51. Hongkulsup C, Khutoryanskiy VV, Niranjan K (2016) Enzyme assisted extraction of chitin from shrimp shells (Litopenaeus vannamei). J Chem Technol Biotechnol 91(5):1250–1256
52. Vázquez JA et al (2017) Optimization of high purity chitin and chitosan production from Illex argentinus pens by a combination of enzymatic and chemical processes. Carbohydr Polym 174:262–272
53. Lopes C et al (2018) Chitin production from crustacean biomass: sustainability assessment of chemical and enzymatic processes. J Clean Prod 172:4140–4151
54. Suryawanshi N, Ayothiraman S, Eswari JS (2020) Ultrasonication mode for the expedition of extraction process of chitin from the maritime shrimp shell waste. Indian J Biochem Biophys 57(4):431–438
55. Singh A, Benjakul S, Prodpran T (2019) Ultrasound-assisted extraction of chitosan from squid pen: molecular characterization and fat binding capacity. J Food Sci 84(2): 224–234
56. Villa-Lerma G et al (2013) Ultrasonication and steam-explosion as chitin pretreatments for chitin oligosaccharide production by chitinases of Lecanicillium lecanii. Bioresour Technol 146:794–798
57. Birolli WG, de Moura Delezuk JA, Campana-Filho SP (2016) Ultrasound-assisted conversion of alpha-chitin into chitosan. Appl Acoust 103:239–242
58. Huang J et al (2011) Ultrasonic-assisted preparation, characterization and antibacterial activity of β-chitosan from squid pens. Adv Mater Res 236:282–287
59. Fiamingo A et al (2016) Extensively deacetylated high molecular weight chitosan from the multistep ultrasound-assisted deacetylation of beta-chitin. Ultrason Sonochem 32:79–85
60. Yen M-T, Yang J-H, Mau J-L (2009) Physicochemical characterization of chitin and chitosan from crab shells. Carbohydr Polym 75(1):15–21

61. Sumnu G (2001) A review on microwave baking of foods. Int J Food Sci Technol 36(2): 117–127
62. Prajapat AL, Gogate PR (2015) Depolymerization of guar gum solution using different approaches based on ultrasound and microwave irradiations. Chem Eng Process Process Intensif 88:1–9
63. Wasikiewicz JM, Yeates SG (2013) "Green" molecular weight degradation of chitosan using microwave irradiation. Polym Degrad Stab 98(4):863–867
64. Tsubaki S, Azuma J-i (2013) Total fractionation of green tea residue by microwave-assisted alkaline pretreatment and enzymatic hydrolysis. Bioresour Technol 131:485–491
65. Nouri M et al (2016) Improvement of chitosan production from Persian Gulf shrimp waste by response surface methodology. Food Hydrocoll 59:50–58
66. El Knidri H et al (2016) Eco-friendly extraction and characterization of chitin and chitosan from the shrimp shell waste via microwave irradiation. Process Saf Environ Prot 104: 395–405
67. Zaeni A et al (2017) Microwave-assisted hydrolysis of chitosan from shrimp shell waste for glucosammine hydrochlorid production. J Phys Conf Ser 846:012011
68. Apriyanti DT, Susanto H, Rokhati N (2018) Influence of microwave irradiation on extraction of chitosan from shrimp shell waste. Reaktor 18(1):45–50
69. Zhang J et al (2018) Base-free preparation of low molecular weight chitin from crab shell. Carbohydr Polym 190:148–155
70. Machałowski T et al (2019) Spider Chitin. The biomimetic potential and applications of Caribena versicolor tubular chitin. Carbohydr Polym 226:115301
71. Lachos-Perez D et al (2016) Subcritical water hydrolysis of sugarcane bagasse: an approach on solid residues characterization. J Supercrit Fluids 108:69–78
72. Yoshida H et al (2015) Recovery of indium from TFT and CF glasses of LCD wastes using NaOH-enhanced sub-critical water. J Supercrit Fluids 104:40–48
73. Kang K et al (2001) Optimization of amino acids production from waste fish entrails by hydrolysis in sub and supercritical water. Can J Chem Eng 79(1):65–70
74. Yoshida H, Terashima M, Takahashi Y (1999) Production of organic acids and amino acids from fish meat by sub-critical water hydrolysis. Biotechnol Prog 15(6):1090–1094
75. Espíndola-Cortés A et al (2017) Hydroxyapatite crystallization in shrimp cephalothorax wastes during subcritical water treatment for chitin extraction. Carbohydr Polym 172:332–341
76. Osada M et al (2012) Effect of sub- and supercritical water pretreatment on enzymatic degradation of chitin. Carbohydr Polym 88(1):308–312
77. Hao G et al (2021) Physicochemical characteristics of chitosan from swimming crab (Portunus trituberculatus) shells prepared by subcritical water pretreatment. Sci Rep 11(1):1646
78. Ordonñez L, Garciía J, Bolanños G (2013) Producing chitin and chitin-glucan complexes from Aspergillus niger biomass using subcritical water. In: The Iberoamerican conference on supercritical fluids, Cartagena, Colombia
79. Gaikwad S, Kim MJ (2024) Fish by-product collagen extraction using different methods and their application. Mar Drugs 22(2):60
80. Yu F et al (2018) Optimization of extraction conditions and characterization of pepsin-solubilised collagen from skin of giant croaker (Nibea japonica). Mar Drugs 16(1):29
81. Liu D et al (2012) Extraction and characterisation of pepsin-solubilised collagen from fins, scales, skins, bones and swim bladders of bighead carp (Hypophthalmichthys nobilis). Food Chem 133(4):1441–1448
82. Song Z et al (2021) Characterization and comparison of collagen extracted from the skin of the Nile tilapia by fermentation and chemical pretreatment. Food Chem 340: 128139
83. Lee JE, Noh S-K, Kim MJ (2022) Effects of enzymatic- and ultrasound-assisted extraction on physicochemical and antioxidant properties of collagen hydrolysate fractions from Alaska pollack (Theragra chalcogramma) skin. Antioxidants 11(11):2112
84. Jia J et al (2010) Enzymatic hydrolysis of Alaska pollack (Theragra chalcogramma) skin and antioxidant activity of the resulting hydrolysate. J Sci Food Agric 90(4):635–640
85. Wang Z et al (2021) Antioxidant activity and functional properties of Alcalase-hydrolyzed scallop protein hydrolysate and its role in the inhibition of cytotoxicity in vitro. Food Chem 344:128566
86. Nalinanon S et al (2008) Tuna pepsin: characteristics and its use for collagen extraction from the skin of threadfin bream (Nemipterus spp.). J Food Sci 73(5):C413–C419
87. Nurilmala M et al (2019) Characterization of collagen and its hydrolysate from yellowfin tuna Thunnus albacares skin and their potencies as antioxidant and antiglycation agents. Fish Sci 85:591–599

88. Jamilah B et al (2013) Properties of collagen from barramundi (Lates calcarifer) skin. Int Food Res J 20(2)
89. Chuaychan S, Benjakul S, Kishimura H (2015) Characteristics of acid- and pepsin-soluble collagens from scale of seabass (Lates calcarifer). LWT Food Sci Technol 63(1): 71–76
90. Dhakal D et al (2018) Optimization of collagen extraction from chicken feet by papain hydrolysis and synthesis of chicken feet collagen based biopolymeric fibres. Food Biosci 23:23–30
91. Preece KE et al (2017) Intensified soy protein extraction by ultrasound. Chem Eng Process Process Intensif 113:94–101
92. Kadam SU, Tiwari BK, O'Donnell CP (2013) Application of novel extraction technologies for bioactives from marine algae. J Agric Food Chem 61(20):4667–4675
93. Shirsath S, Sonawane S, Gogate P (2012) Intensification of extraction of natural products using ultrasonic irradiations—a review of current status. Chem Eng Process Process Intensif 53:10–23
94. Petcharat T et al (2021) Ultrasound-assisted extraction of collagen from clown featherback (Chitala ornata) skin: yield and molecular characteristics. J Sci Food Agric 101(2): 648–658
95. Ali AMM, Kishimura H, Benjakul S (2018) Extraction efficiency and characteristics of acid and pepsin soluble collagens from the skin of golden carp (Probarbus Jullieni) as affected by ultrasonication. Process Biochem 66:237–244
96. Zou Y et al (2020) Effect of ultrasound power on extraction kinetic model, and physicochemical and structural characteristics of collagen from chicken lung. Food Prod Process Nutr 2:1–12
97. Akram AN, Zhang C (2020) Effect of ultrasonication on the yield, functional and physicochemical characteristics of collagen-II from chicken sternal cartilage. Food Chem 307: 125544
98. Shaik M, Effendi NA, Sarbon N (2021) Functional properties of sharpnose stingray (Dasyatis zugei) skin collagen by ultrasonication extraction as influenced by organic and inorganic acids. Biocatal Agric Biotechnol 35: 102103
99. Passos H, Freire MG, Coutinho JA (2014) Ionic liquid solutions as extractive solvents for value-added compounds from biomass. Green Chem 16(12):4786–4815
100. Ruesgas-Ramón M, Figueroa-Espinoza MC, Durand E (2017) Application of deep eutectic solvents (DES) for phenolic compounds extraction: overview, challenges, and opportunities. J Agric Food Chem 65(18): 3591–3601
101. Vieira V et al (2018) Enhanced extraction of phenolic compounds using choline chloride based deep eutectic solvents from Juglans regia L. Ind Crop Prod 115:261–271
102. Bai C, Wei Q, Ren X (2017) Selective extraction of collagen peptides with high purity from cod skins by deep eutectic solvents. ACS Sustain Chem Eng 5(8):7220–7227
103. Bisht M et al (2021) Uncovering the potential of aqueous solutions of deep eutectic solvents on the extraction and purification of collagen type I from Atlantic codfish (Gadus morhua). Green Chem 23(22):8940–8948
104. Jafari H et al (2020) Fish collagen: extraction, characterization, and applications for biomaterials engineering. Polymers 12(10):2230
105. Sharif K et al (2014) Experimental design of supercritical fluid extraction—a review. J Food Eng 124:105–116
106. Sanchez-Camargo ADP et al (2014) Supercritical fluid extraction
107. Sousa RO et al (2020) Collagen from Atlantic cod (Gadus morhua) skins extracted using CO_2 acidified water with potential application in healthcare. J Polym Res 27(3):73
108. Singh S et al (2021) Supercritical fluid extraction (SCFE) as green extraction technology for high-value metabolites of algae, its potential trends in food and human health. Food Res Int 150:110746
109. Khaw K-Y et al (2017) Solvent supercritical fluid technologies to extract bioactive compounds from natural sources: a review. Molecules 22(7):1186
110. Silva JC et al (2016) Extraction of collagen/gelatin from the marine demosponge Chondrosia reniformis (Nardo, 1847) using water acidified with carbon dioxide—process optimization. Ind Eng Chem Res 55(25): 6922–6930
111. Farooq S et al (2024) A review on marine collagen: sources, extraction methods, colloids properties, and food applications. Collagen Leather 6(1):11
112. Huang C-Y et al (2016) Isolation and characterization of fish scale collagen from tilapia (Oreochromis sp.) by a novel extrusion–hydro-extraction process. Food Chem 190: 997–1006

Chapter 16

Green Biorefineries: Converting Food Waste into Renewable Biofuels

Tuhin James Paul, Khadga Raj Aran, Biplab Debnath, Sumel Ashique, Anas Islam, and Yusuf Asad

Abstract

This chapter explores the potential of food waste as a resource for the sustainable production of green biofuels through advanced biorefinery processes. It begins with an introduction to the environmental and economic challenges posed by food waste, while emphasizing the significance of sustainable biofuel production in addressing these issues. The composition of food waste is examined, alongside various pretreatment methods such as thermal, chemical, and mechanical approaches that prepare the waste for biofuel production. The chapter delves into microbial conversion processes, highlighting the use of yeast and algae in biofuel production, particularly through fermentation and anaerobic digestion. It presents an integrated biorefinery approach, illustrating how combining fermentation, anaerobic digestion, and chemical conversion can optimize biofuel yield. Detailed discussions follow on the production of biofuels like biodiesel, bioethanol, and biogas, exploring production techniques, challenges, and optimization strategies to improve efficiency. Sustainability plays a central role in this chapter, where the environmental impact of food waste biorefining is analyzed, with a focus on circular economy principles. Case studies from different global regions demonstrate the practical application of food waste biorefineries, showcasing their success and scalability. The chapter also considers future trends, including technological advancements, emerging feedstocks, and the research directions that could shape the next generation of biofuels. In conclusion, the chapter synthesizes the key findings from the various biorefinery protocols discussed, offering a comprehensive outlook on the future of food waste valorization and its contribution to sustainable biofuels. This chapter aims to provide insights into the practical and theoretical aspects of biorefining food waste and its transformative potential for the biofuel industry.

Key words Food waste, Biorefinery, Green biofuels, Sustainable biofuel production, Food waste valorization

1 Introduction

Food waste is a major issue that affects both the environment and global resources. Every year, about 931 million tons of food is wasted, which is nearly one-third of all food produced worldwide [1]. To describe the grievous situation of food waste, Table 1

Tanmay Sarkar (ed.), *Food Waste Valorization: Green Techniques in Sustainable Management*,
Methods and Protocols in Food Science, https://doi.org/10.1007/978-1-0716-4646-5_16,
© The Author(s), under exclusive license to Springer Science+Business Media, LLC, part of Springer Nature 2025

Table 1
Statistics from 2020 to 2025 foretelling how every year food waste is increasing

Year	Estimated global food waste (billion tons)	Approx. % of global food production wasted	Primary sectors contributing	Estimated global economic cost (USD billions)	Estimated environmental impact (GHG emissions %)	References
2020	1.3	~33	Post-harvest, processing, retail, and households	~1000 (approx.)	8–10	[18]
2021	1.3	~33	Post-harvest, processing, retail, and households	~1000 (approx.)	8–10	[18]
2022	1.05	~30–33	Post-harvest, retail and household consumption	~950 (approx.)	8–10	
2023	1.1	~30–33	Retail, households, and food service sectors	~950–1000 (approx.)	8–10	
2024	1.1	~30–33	Retail, households, and food service sectors	~950–1000 (approx.)	8–10	
2025*	1.15	~33	Retail, households, and food service sectors	~1000 (approx.)	8–10	

provides a detailed statistics from 2020 to 2025 foretelling how every year increasing food waste is wreaking havoc in the environment. Food waste happens at various stages of the food supply chain, including production, processing, and consumer disposal [2]. As the global population continues to grow, food waste will increase, providing a large, untapped resource for biofuel production [3]. Essential resources such as water, energy, and land are overutilized during the production of food waste which also significantly harms the environment, producing greenhouse gas emissions during the process. Traditional methods of handling food waste, such as landfilling and incineration, are inefficient and harmful [4]. However, food waste also presents a valuable opportunity. If properly managed, it can be turned into useful products, particularly biofuels. Biorefineries are facilities designed to convert biomass, such as food waste, into valuable products like biofuels, bioethanol, and biodiesel [5]. These biorefineries help address both food waste management and the global demand for renewable energy. By converting food waste into biofuels, these processes reduce the need for fossil fuels while also solving waste disposal issues. Food residues such as fruit peels, food scraps, and cereal by-products can be repurposed into biofuels, contributing to environmental sustainability [6]. The food waste such as from fruit and vegetable peelings to cereal by-products offers a wide range of organic materials that can be used to create biofuels [7]. One of the most common methods for converting food waste into biofuels is bioethanol production [8]. Bioethanol is made by fermenting the sugars found in organic materials. Food waste can be a key ingredient in this process. Unlike first-generation biofuels that are made from food crops, second-generation bioethanol uses food waste, which does not compete with food production [8]. Research shows that food waste like pineapple peels, orange rinds, and banana bracts can be turned into bioethanol using processes like Simultaneous Saccharification and Fermentation (SSF) [9]. These processes break down food waste into fermentable sugars, which are then turned into bioethanol, providing a sustainable fuel option. In addition to bioethanol, new technologies have improved the way food waste is processed in biorefineries. Green extraction methods, such as microwave-assisted extraction (MAE) and ultrasound-assisted extraction (UAE), have been used to extract valuable compounds from food waste [10, 11]. These compounds, such as phenolics and flavonoids, are important for biofuel production. MAE and UAE are considered environmentally friendly because they use less energy and avoid harmful chemicals. These methods have been shown to work well for extracting materials from food waste like walnut husks, which are high in lignocellulose [12]. By using these techniques, biorefineries can increase the amount of biofuel produced and make the process more efficient [13]. For biofuel production from food waste to be successful, it must be economically feasible. This means looking at both the costs and the potential profits of turning food waste into biofuels. Recent studies have shown that food waste can be turned into bioethanol or biodiesel at a profit [14]. One promising way to manage food waste and turn it into a cost-effective measure is anaerobic digestion, which breaks down food waste into biogas. However, this process also produces a by-product called digestate, which is rich in nutrients and organic matter; this digestate can be used in biorefineries to produce biofuels like bioethanol. By using food waste digestates in biorefineries, companies can produce biofuels and other valuable products, such as biosurfactants like rhamnolipids, which can be sold to create more revenue [14]. This dual-purpose approach producing both biofuels and value-added products makes food waste-based biofuel production more economically viable. As global energy demands rise, food waste-based biofuels have the potential to help meet this need. Biorefineries that turn food waste into biofuels like bioethanol and biodiesel provide an eco-friendly energy source and reduce waste. The ongoing improvements in fermentation processes, enzyme technologies, and green extraction methods will continue to enhance the efficiency of these biorefineries [15, 16]. These advancements make food waste to biofuel conversion more sustainable and cost-effective, helping it become a larger part of the global energy system. In the coming decade, biorefineries in many parts of the world will base their protocols on integration of advanced fermentation techniques, enzyme applications, and green extraction methods that will further improve the efficiency of food waste-based biofuel production [17]. As such technologies evolve, food waste-based biorefineries will play a crucial role in meeting global energy demands while minimizing environmental impact.

2 Materials

Food waste (FW) biomass for biofuel production can be sourced from various locations, including restaurants, households, marketplaces, and food processing facilities. This waste contains essential organic components such as carbohydrates (including starch and cellulose), lipids, and proteins. Specific microbial strains, like *Saccharomyces cerevisiae* for bioethanol production and *Clostridium* species for butanol synthesis, can be obtained from commercial suppliers or isolated in laboratory environments. Enzymes such as amylase, cellulase, and pectinase play a crucial role in hydrolysis, while chemical agents—including sulfuric acid (H_2SO_4), organic acids, and alkalis like sodium hydroxide—are essential for biomass pretreatment and hydrolysis. In thermochemical conversion processes, catalysts such as zeolites, metal oxides (ZnO, Al_2O_3), and activated carbon may also be utilized.

3 Method

3.1 Food Waste Composition and Pretreatment

The rapid increase in food waste generation alongside the growing demand for sustainable energy solutions has spurred interest in converting food waste into green biofuels through biorefinery processes [19]. This strategy not only addresses environmental issues related to waste management and greenhouse gas emissions but also contributes to energy security by transforming low-value organic waste into high-value renewable fuels. In this section, the various types of food waste are examined in detail, focusing on their chemical and physical characteristics, the implications for pretreatment and conversion protocols, and their overall potential in biofuel production [20]. Understanding these aspects is essential for developing efficient biorefinery processes that can accommodate the heterogeneity of food waste. Broadly, food waste can be categorized into three main groups: household waste, commercial and institutional waste, and industrial or processing waste [21]. Household food waste typically originates from domestic kitchens, where remnants of meals, peelings, and expired products accumulate [22]. Although generated in smaller quantities by individual households, the aggregated volume is substantial. Commercial and institutional waste, produced by restaurants, hotels, cafeterias, and supermarkets, usually consists of large volumes of perishable items and preprepared meals [23]. Industrial or processing waste arises during the manufacturing, processing, and packaging stages, and is often more homogeneous, including by-products such as pomace, bran, and other residues [24]. These categories differ in moisture content, nutrient composition, and the presence of contaminants, each factor playing a critical role in shaping the subsequent pretreatment and conversion processes used in biorefineries [25]. Fruit and vegetable wastes represent one of the most abundant fractions of food waste, these materials are characterized by a high-water content along with significant concentrations of simple sugars, organic acids, vitamins, and fibers [26]. The abundance of readily fermentable sugars in fruit waste renders it an attractive substrate for fermentation-based processes that yield ethanol or biogas. Vegetable wastes, although also rich in fermentable compounds, typically contain a considerable proportion of structural carbohydrates such as cellulose, hemicellulose, and lignin [27, 28]. These components contribute to the physical rigidity of plant tissues but require more intensive pretreatment to render them amenable to enzymatic hydrolysis. Consequently, processing strategies for fruit and vegetable wastes often include dehydration or concentration steps to enhance the yield of fermentable sugars. In addition, mechanical separation combined with chemical pretreatment methods (such as acid or alkaline hydrolysis) is employed to disrupt cell walls and release the sugars bound within. Advances in technologies such as steam explosion and enzymatic hydrolysis have significantly improved the conversion efficiency of these waste streams, highlighting the importance of tailored pretreatment protocols. Cereal and grain wastes, generated predominantly during the processing of cereals in milling, baking, and other food manufacturing industries, provide another valuable resource. These wastes, which include husks, bran, and other residual by-products, are notably high in starch content. Their relatively uniform chemical composition—with high levels of polysaccharides and lower moisture compared to fresh produce—enables more predictable processing outcomes. The starch present in cereal waste can serve as an excellent substrate for bioethanol production through fermentation. However, the residual proteins and lipids that are often present in these wastes can interfere with enzymatic hydrolysis. Optimizing parameters such as pH, temperature, and enzyme concentration is therefore crucial to maximize the yield of fermentable sugars. Additionally, preprocessing techniques like milling and particle size reduction are essential to increase the surface area available for enzymatic action, thereby enhancing overall conversion efficiency. Animal-derived food wastes, including those from meat, dairy, and seafood sources, present a distinct set of challenges due to their complex composition. Meat and dairy wastes are rich in proteins and fats, containing compounds such as collagen and casein that require more sophisticated pretreatment methods to break down. Although these biomolecules offer high energy potential, their complex structures necessitate thermal or enzymatic hydrolysis to yield simpler molecules suitable for fermentation into biofuels. Seafood waste, comprising fish, shellfish, and other marine by-products, similarly contains high levels of proteins along with a significant lipid fraction. However, the high salt content and potential presence of heavy metals or other marine contaminants require additional purification steps prior to conversion. Desalination, deboning, and impurity removal are critical measures for preventing equipment corrosion and ensuring the quality of the final biofuel products. For meat and dairy wastes, thermal hydrolysis or enzymatic treatments are typically employed to decompose proteins into amino acids and peptides, which can subsequently be fermented into biogas or other biofuel intermediates. In the case of seafood waste, the extraction and transesterification of lipids to produce biodiesel is often coupled with anaerobic digestion of the remaining proteinaceous material, provided that inhibitors are adequately managed. Bakery and sugar-rich food wastes offer yet another promising substrate for biofuel production. These wastes, which include discarded bread, pastries, confectioneries, and residues from the soft drink and confectionery industries, are abundant in carbohydrates and simple sugars. Their high fermentable sugar content makes them ideal for microbial fermentation processes that yield ethanol and other biofuel intermediates. Given that these substrates are less fibrous than fruit or vegetable wastes, they typically require minimal pretreatment, reducing processing costs and energy inputs. However, the highly perishable nature of bakery waste demands rapid processing to prevent spoilage and maintain its biofuel potential. Efficient logistics, storage, and transportation protocols are therefore essential to preserve the integrity of these substrates until they are processed in biorefineries. Industrial food wastes, generated as by-products of large-scale food manufacturing and processing operations, encompass a diverse range of materials such as fruit pomace, spent grains, vegetable trimmings, and even wastewater containing dissolved organic compounds. Although these waste streams are heterogeneous, they are often produced under controlled conditions that lend them a relatively consistent quality. This uniformity enables the design of standardized processing protocols and facilitates optimization of conversion technologies. Integrated biorefinery concepts that capitalize on industrial food waste are particularly attractive because they enable the simultaneous production of biofuels and value-added co-products such as organic acids, enzymes, biopolymers, and nutritional supplements. By employing multiple conversion pathways—including hydrolysis, fermentation, and catalytic upgrading—integrated systems can achieve higher efficiencies and improved economic viability. The concept of cascading biomass use, wherein every fraction of the feedstock is valorized, is central to the development of a circular bioeconomy. The effectiveness of converting food waste into biofuels through biorefinery processes is largely determined by several intrinsic factors. One of the most critical parameters is moisture content. High-moisture levels, commonly found in fresh produce and bakery waste, can dilute the concentration of fermentable sugars and impede process efficiency. In contrast, low-moisture wastes such as cereal by-products provide a more concentrated source of carbohydrates. To address this issue, various moisture reduction techniques, including dehydration and pressing, are applied to standardize the feedstock before conversion. The chemical composition of food waste, which includes carbohydrates, proteins, fats, and lignocellulosic materials, dictates the choice of pretreatment and conversion technologies. For instance, high cellulose content in vegetable waste necessitates intensive pretreatment to enhance enzymatic digestibility, whereas starch-rich cereal waste may require only simple hydrolysis to achieve effective conversion. In addition to chemical composition, physical characteristics such as particle size and structure play a significant role in determining the efficiency of bioconversion processes. Mechanical milling or grinding is commonly used to increase the surface area of the feedstock, facilitating better enzymatic access and more uniform processing. Optimizing the particle size distribution can lead to improved reactor performance and higher conversion yields. Another factor influencing the conversion process is the presence of inhibitory compounds. Certain food wastes contain phenolic compounds, organic acids, and other secondary metabolites that can interfere with microbial fermentation or enzymatic activity. The inhibitory effects of these compounds necessitate additional detoxification steps within the pretreatment protocol, such as adsorption techniques or chemical neutralization, to ensure high yields of biofuels. In addressing the inherent challenges posed by the diverse nature of food waste, technological innovations in pretreatment and process integration have become pivotal. Pretreatment is the cornerstone of converting heterogeneous food waste into fermentable substrates. Depending on the type of feedstock, pretreatment may involve physical processes (such as milling or homogenization), chemical methods (including acid or alkaline hydrolysis), or biological techniques (such as enzymatic treatment). The objective is to disrupt the cellular structure of the waste material, releasing bound sugars and enhancing the overall accessibility of the biomass. For instance, steam explosion has been particularly effective in breaking down the lignocellulosic matrices found in vegetable wastes, while enzymatic hydrolysis is well-suited for processing starch-rich cereal by-products. Following pretreatment, fermentation processes are tailored to the specific characteristics of the waste stream. High-moisture wastes are typically directed toward anaerobic digestion to produce biogas, whereas sugar-rich substrates are more amenable to ethanol fermentation. In some instances, co-digestion strategies are employed to combine multiple types of food waste, thereby creating synergistic effects that balance nutrient profiles and improve overall conversion efficiency. For example, blending carbohydrate-rich bakery waste with protein-rich meat waste can enhance microbial performance and stabilize biofuel yields. Moreover, the integration of process analytical technologies (PAT) into biorefinery operations facilitates real-time monitoring and adaptive control of critical parameters such as pH, temperature, and substrate concentration. This dynamic feedback is essential for maintaining optimal conversion efficiency despite the variability inherent in food waste feedstocks. The environmental benefits of utilizing food waste in biorefineries are significant. Diverting organic waste from landfills and incinerators reduces the emission of methane and other greenhouse gases, while the production of renewable biofuels contributes to lowering the overall carbon footprint of the energy sector. The dual benefit of waste management and renewable energy production aligns closely with global efforts to mitigate climate change and promote sustainable development. Economically, the valorization of food waste into high-value products offers the potential to transform low-value waste into revenue-generating resources. By reducing the costs associated with waste disposal and creating new markets for renewable fuels and co-products, food waste-based biorefineries can stimulate local economies and support regional development. The decentralized nature of many biorefinery operations further reduces transportation costs and enhances the feasibility of sourcing feedstock from local communities. Policy and regulatory frameworks also play a crucial role in the successful implementation of food waste biorefineries. Governments and international organizations have increasingly recognized the importance of a circular bioeconomy, and policies that encourage the sustainable utilization of waste have emerged in many regions. Financial incentives, subsidies for renewable energy projects, and stricter regulations on waste disposal can all help accelerate the adoption of biorefinery technologies. In this context, robust life cycle assessments and techno-economic analyses are essential tools for evaluating the environmental and economic benefits of these systems and guiding policy decisions. Despite the significant progress made in recent years, challenges remain in the full-scale conversion of food waste into biofuels. Future research must focus on developing novel pretreatment methods that are both cost-effective and environmentally benign. Innovations in enzyme engineering may lead to more efficient hydrolytic processes, while advances in microbial fermentation could improve the conversion rates of diverse substrates. Additionally, integrated biorefineries that produce multiple products—such as biofuels, biochemicals, and biofertilizers—from a single feedstock are a promising avenue for maximizing resource utilization and achieving economic viability. Improved process monitoring and control strategies, supported by digital technologies and machine learning, may further enhance the stability and efficiency of biorefinery operations by mitigating the effects of feedstock variability. In conclusion, food waste represents a diverse and abundant resource with significant potential for conversion into green biofuels through advanced biorefinery processes. The various types of food waste—including residues from fruits and vegetables, cereal and grain by-products, animal-derived wastes, bakery wastes, and industrial food processing by-products—each present unique challenges and opportunities for biofuel production. A comprehensive understanding of the chemical and physical properties of these waste streams, including moisture content, particle size, and inhibitory compounds, is crucial for designing effective pretreatment and conversion strategies. Technological innovations in pretreatment, fermentation, and process integration are key to overcoming the challenges posed by the heterogeneous nature of food waste, ultimately improving biofuel yields and contributing to environmental sustainability. Furthermore, the economic benefits of transforming low-value waste into renewable energy and value-added products, coupled with supportive policy frameworks, underscore the importance of continued research and development in this field. As society seeks sustainable alternatives to fossil fuels, the bioconversion of food waste into biofuels offers a compelling solution that addresses environmental, economic, and social challenges in a holistic manner.

3.2 Microbial Conversion of Food Waste for Biofuel Production

The conversion of food waste into biofuels via microbial processes offers a promising solution to two pressing global issues: waste management and renewable energy production. This approach exploits the natural metabolic pathways of various microorganisms to transform complex organic materials into energy-rich fuels. Among the most effective strategies are microbial fermentation, anaerobic digestion, and the lipid accumulation capabilities of the oleaginous yeast *Rhodotorula glutinis*, each contributing uniquely to the overall bioconversion process. Microbial fermentation has long been a cornerstone for biofuel production, particularly for generating bioethanol. Food waste, which is typically rich in carbohydrates, undergoes pretreatment to release fermentable sugars. These pretreatments can be physical, chemical, or enzymatic and are designed to break down complex matrices, liberating sugars such as glucose and fructose. Once available, these sugars are metabolized by fermentative microorganisms like *Saccharomyces cerevisiae* to produce bioethanol. An integrated method known as simultaneous saccharification and fermentation (SSF) has been widely adopted to enhance process efficiency by combining hydrolysis and fermentation in a single step. This not only shortens the overall processing time but also minimizes product inhibition, leading to improved yields. Recent advances in genetic engineering have led to the development of robust microbial strains that can tolerate higher concentrations of inhibitors and a broader range of substrates, thereby expanding the scope of fermentable food wastes. In addition to bioethanol, microbial fermentation can also produce other biofuels, including biobutanol and biohydrogen, depending on the specific microorganisms and conditions employed [1]. Anaerobic digestion is another vital microbial process for converting food waste into renewable energy, particularly in the form of biogas. Unlike aerobic processes, anaerobic digestion occurs in the absence of oxygen and relies on a consortium of microorganisms that operate through a series of biochemical stages: hydrolysis, acidogenesis, acetogenesis, and methanogenesis. During the initial hydrolysis stage, complex organic compounds such as proteins, fats, and carbohydrates are depolymerized into simpler monomers. These monomers are then converted into volatile fatty acids and alcohols by acidogenic bacteria. In the acetogenesis phase, these intermediates are further transformed into acetic acid, hydrogen, and carbon dioxide. Finally, methanogenic archaea convert these products into methane-rich biogas. This biogas can be utilized directly for energy generation, while the nutrient-rich digestate produced during the process can serve as a fertilizer, contributing to the circular bioeconomy. One of the primary advantages of anaerobic digestion is its ability to handle heterogeneous waste streams, a critical feature given the variability inherent in food waste. Advances in reactor design and process control, such as real-time monitoring systems, have significantly improved the stability and efficiency of anaerobic digestion even under fluctuating feedstock conditions [2]. In addition to these well-established microbial pathways, the oleaginous yeast *Rhodotorula glutinis* plays an emerging role in the microbial conversion of food waste, particularly for biodiesel production. *Rhodotorula glutinis* is notable for its ability to accumulate high levels of intracellular lipids when grown on carbon sources derived from food waste hydrolysates. Under optimized culture conditions—characterized by a high carbon-to-nitrogen (C/N) ratio and appropriate pH and temperature controls—this yeast can convert simple sugars into lipids that may constitute up to 70% of its dry biomass. These lipids, primarily composed of triacylglycerols, are ideal precursors for biodiesel synthesis. The process typically begins with the pretreatment of food waste to release fermentable sugars, which are then fed into bioreactors cultivating *Rhodotorula glutinis*. In this setting, the yeast not only accumulates lipids but also produces co-products such as carotenoids and antioxidants, which have additional market value. The dual production of biodiesel precursors and high-value compounds can significantly enhance the economic viability of the bioconversion process. Furthermore, ongoing research in metabolic engineering is focused on improving lipid yields and tailoring fatty acid profiles to meet specific biodiesel quality standards [3]. The integration of these microbial processes—fermentation, anaerobic digestion, and yeast-based lipid production—into a unified biorefinery framework represents the future of sustainable biofuel production. Each process complements the others; for instance, the effluent from anaerobic digestion, which is rich in volatile fatty acids, can serve as a substrate for further microbial fermentation. Similarly, the residual biomass from lipid-producing yeast can be repurposed as animal feed or soil amendments, thereby closing the resource loop. Such an integrated approach not only maximizes the overall energy yield but also minimizes waste, ensuring that all components of food waste are effectively utilized. Recent technological advancements are continually enhancing the performance of microbial conversion systems. Innovations in reactor design, process monitoring, and genetic engineering are leading to more stable and efficient operations. For example, continuous fermentation systems and advanced anaerobic digesters equipped with real-time control mechanisms are improving process scalability and economic feasibility. Moreover, the development of genetically modified strains capable of metabolizing a broader range of substrates and withstanding higher concentrations of inhibitors is expanding the applicability of microbial conversion processes to diverse food waste streams. In conclusion, microbial conversion of food waste into biofuels provides a comprehensive strategy for transforming waste into energy, thereby addressing both environmental and energy challenges. Microbial fermentation converts carbohydrates into bioethanol and other valuable alcohols, anaerobic digestion

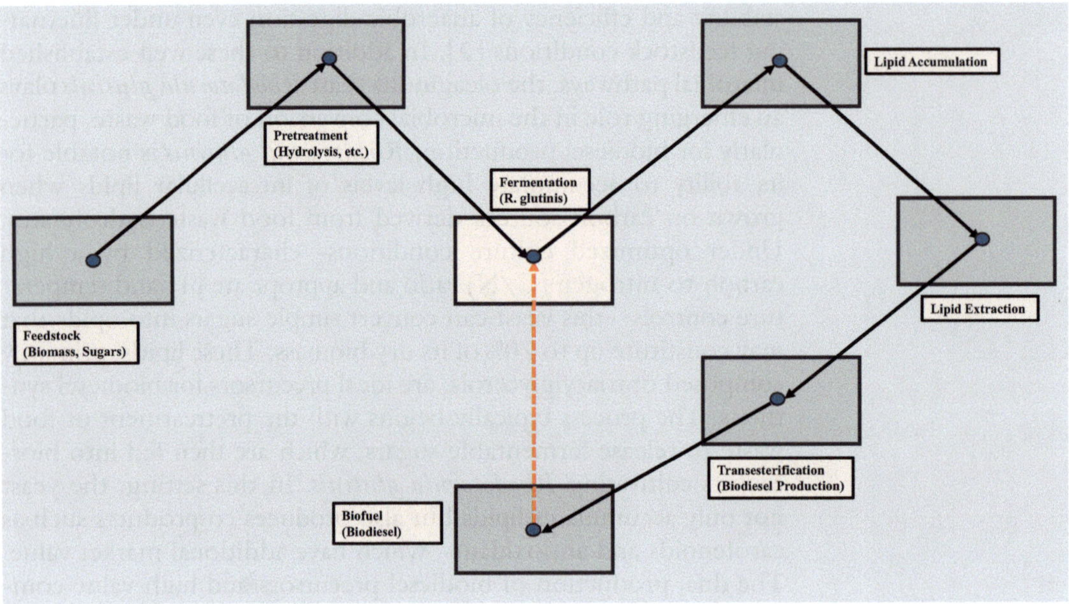

Fig. 1 Schematic representation of the biorefinery process employing *Rhodotorula glutinis* for lipid production and biodiesel synthesis. The diagram delineates the sequential steps—from biomass pretreatment and yeast-mediated fermentation to lipid extraction and transesterification—highlighting the metabolic conversion pathways that enable efficient biofuel production

transforms complex organic matter into methane-rich biogas, and *Rhodotorula glutinis* offers a highly efficient route for lipid accumulation for biodiesel production. Integrating these processes into a unified biorefinery not only enhances overall energy recovery but also promotes a circular bioeconomy by ensuring that every fraction of food waste is valorized. Continued research and technological innovation will be crucial for scaling these processes and unlocking their full potential as part of a sustainable energy future. Figure 1 provides an illustration of the innovative role of *Rhodotorula glutinis* in lipid production within a biorefinery setup. It visually captures the journey from raw biomass through pretreatment, fermentation, and lipid accumulation, culminating in biodiesel production via extraction and transesterification. By leveraging the natural lipid-accumulating properties of this yeast, the process underscores a sustainable and efficient pathway to renewable biofuel.

3.3 Integrated Biorefinery Approach for Green Biofuel Production

The integrated biorefinery approach serves as an innovative method to convert food waste into renewable energy sources. The operational framework achieves optimal unity between multiple processing streams by adopting an approach which embodies resource efficiency and sustainability and process intensification principles.

The system unites diverse conversion methods to recover maximum energy while decreasing waste output and creating multiple biofuel products as well as co-products for the circular bioeconomy framework. The base of food waste consists of heterogeneous material containing carbohydrates with proteins and lipids and lignocellulosic substances before fractional separation produces its elemental components. All fractionated materials receive specific conversion pathways to maximize the value of every stream that reasserts bioethanol and biodiesel along with biogas and bio-based chemical products. A single system which connects different processes generates superior energy efficiency and material efficiency performance. Process intensification occurs through uniting pretreatment with microbial conversion and catalytic upgrading which results in diminished energy waste together with streamlined procedures. The integrated system leads to decreased financial costs and operational costs combined with lower environmental impacts. The main success factor behind this strategy relies on uniting processes that support one another. Operating anaerobic digestion along with fermentation transformation enables product recycling between removal of volatile fatty acids from digestion to microbial fermentation. The collaborative process between both cultivation methods leads to an increased total yield while maintaining a steady operational framework despite feedstock quantity adjustments. The pretreatment phase begins conversion processes by complex organic matrices of food waste through effective breakdown procedures. Different pretreatment methods including thermal hydrolysis as well as chemical and enzymatic hydrolysis and mechanical disruption break down cell structures to release fermentable sugars and target compounds. Pretreatment techniques become more successful when factors such as substrate moisture level and particle size receive proper optimization which enhances substrate accessibility for conversion processes. The pretreated substrate submitted to microbial fermentation results in the production of biofuels including bioethanol, biobutanol, and biohydrogen. The processing time shortens and the inhibitory effects decrease when the single reactor hardware combines the steps of enzymatic hydrolysis and microbial processing through the method of simultaneous saccharification and fermentation (SSF). Through modern genetic engineering scientists developed resistant microorganism strains that show tolerance to elevated inhibitor levels thereby accepting a wider range of raw materials for bioconversion. The process of anaerobic digestion simultaneously operates with fermentation by converting resistant food waste types into biogas containing methane. The chemical breakdown takes place in four consecutive steps called hydrolysis then goes into acidogenesis followed by acetogenesis then finishes with methanogenesis which transforms intricate substances. Raw biogas contains predominantly methane as its main component which operators can use in two ways either for thermal applications and power generation or for biomethane production. Additionally, for feedstock fractions rich in lipids, the cultivation of oleaginous microorganisms such as *Rhodotorula glutinis* facilitates the accumulation of intracellular lipids. When carbon-to-nitrogen ratios reach their maximal levels, the process allows effective biodiesel extraction from lipids followed by their transesterification into biodiesel products. The integrated biofuel system achieves biodiesel diversity while producing high-value co-products made of carotenoids and antioxidants. The integrated biorefinery method enables the complete usage of all refinery streams. The leftover biomass from fermentation services as the basis for making biochemicals and organic acids and developing nutrient supplements from it. Through extensive biomass utilization the principles of circular economy persist as the biorefinery becomes more financially prosperous. Integrated systems excel through their adaptability feature since they must cope with variable food waste patterns both by seasons and regions. The process maintains high conversion performance using adaptive control systems that redesign parameters dynamically while modular architecture allows feedstock adjustments regardless of input variations. The integrated biorefinery model produces maximum energy output from each unit of waste food while it enables revenue production through both biofuel generation and co-product manufacturing. A few major barriers need to be addressed before successfully implementing these processes at industrial scales. Upcoming research needs to develop sophisticated process surveillance methods together with real-time management systems and efficient restrictive compound control techniques. Machine learning and process analytics technologies integrated with digital systems promise to improve integrated biorefineries by enhancing their flexible operations and operational efficiency. The integrated approach uses resource optimization with process intensification and flexible system design techniques to reach maximum energy recovery from food waste in addition to supporting the circular bioeconomy development. Table 2 demonstrates that many benefits can be gained by linking different biorefining process operations together. The integration of fermentation with anaerobic digestion leads to enhanced biofuel production by 18% and improved energy recovery performance up to 27 MJ/kg as well as shorter processing duration by 15% and reduced operational costs by 12%. Pretreatment through enzymatic hydrolysis combined with fermentation process leads to a 22% increase in yield and a 20% decrease in operational time. Quantitative metrics reveal how joint processing of different biorefining methods produces substantial outcomes which increase operational performance while enabling better economic viability. The efficient conversion of mixed food waste into renewable energy depends on these optimization techniques to enable greater industrial and circular bioeconomy achievement.

Table 2
Synergies between different biorefining methods: Quantitative metrics for biofuel yield improvement, energy recovery efficiency, process time reduction, and economic benefits in integrated food waste valorization processes

Biorefining method combination	Feedstock type	Biofuel yield improvement (%)	Energy recovery efficiency (MJ/kg)	Process time reduction (%)	Economic benefit (cost reduction, %)	References
Fermentation + anaerobic digestion	Mixed food waste	18	27	15	12	
Pretreatment (enzymatic hydrolysis) + fermentation	Cereal by-products	22	30	20	15	
Lipid extraction + fermentation	Bakery waste	16	25	10	18	
Microwave-assisted extraction (MAE) + anaerobic digestion	Fruit and vegetable wastes	20	28	**18**	14	
Integrated fermentation + lipid accumulation + anaerobic digestion	Mixed food waste	25	32	**22**	20	

The data highlights in Table 2 receive support from recent studies that demonstrate integrated biorefineries hold enormous potential for sustainable biofuel creation.

3.4 Production of Biofuels from Food Waste

The transformation of food waste into biofuels proves to be a sustainable approach that deals with waste disposal problems together with rising global power requirements. An abundance of carbohydrates along with lipids and proteins exists in the substantial food waste production from homes and industrial sites and commercial venues. The various biochemical and thermochemical processes achieve effective conversion of these food waste components into biodiesel, bioethanol, biogas, and biohydrogen biofuels. The waste transformation process simultaneously prevents large organic waste from filling landfills while reducing greenhouse gas emissions because renewable energy replaces traditional fossil fuels. The primary production method for biodiesel from food waste depends on lipid extraction from waste materials including used cooking oils and bakery residues as well as other fat materials. The extraction of lipids from various food waste components happens by mechanical pressing or solvent extraction methods at the initial stage of the process. Triglycerides that are recovered undergo transesterification through reaction with methanol under catalyst conditions. The conversion produces biodiesel fatty acid methyl esters alongside glycerol as an additional substance. Multiple factors control the transesterification process efficiency including alcohol-to-oil molar ratio and catalyst type and concentration and reaction temperature and reaction time. New advancements in reaction engineering methods alongside catalyst development technologies have led to better conversion rates while requiring less energy thus establishing biodiesel production as both commercially sound and eco-friendly for food waste components. Biomass pretreatment marks the initial step of the process which functions for splitting the complex lignocellulosic biomass structure. The release of fermentable sugar requires physical disruption or thermal processing or chemical hydrolysis treatment of biomass materials. The process of enzyme-based hydrolysis starts after pretreatment by breaking down polysaccharides into their main product which is glucose. Biological fermentation relies on these sugars as the main reactant compound. The microorganism strain *Saccharomyces cerevisiae* along with other selected strains ferment sugars into ethanol and carbon dioxide while operating in anaerobic environments. The fermentation process requires exact optimization of pH temperature together with nutrients and inhibitor controls to achieve maximum ethanol yields. Food waste bioethanol production has received a boost from microbial strain engineering which led to the creation of tolerant microorganisms that work efficiently with high levels of inhibitors while processing a diverse selection of substrates. The organic matter decomposition takes place through the orderly function of microorganisms who accomplish this work without oxygen being present. The anaerobic digestion process includes four specific stages named hydrolysis, acidogenesis, acetogenesis, and methanogenesis. Through hydrolysis the first phase chemically breaks down complete biomolecules of proteins carbohydrates and lipids into basic simpler compounds that dissolve. During acidogenesis microbial organisms convert these compounds into volatile fatty acids together with alcohols. After the transformation of intermediary compounds by acetogenic bacteria, they generate acetic acid together with hydrogen gas and carbon dioxide. Finally, methanogenic archaea convert these products into methane-rich biogas during the methanogenesis phase. Food waste break down efficiency through anaerobic digestion relies directly on operational conditions such as temperature and pH and hydraulic retention time and composition of food materials. Anaerobic digestion generates renewable energy for electricity production or fuel use as well as produces digestate containing valuable nutrients which can function as organic fertilizer to advance the circular bioeconomy framework. The conversion of food waste into biohydrogen shows increasing importance because biohydrogen offers high energetic potential along with environmentally friendly combustion features. Under anaerobic dark fermentation specialized bacteria transform organic substrates into biohydrogen alongside organic acids. Photo-fermentation enables photosynthetic bacteria to use light energy for converting organic acids into hydrogen gas. Research in metabolic engineering and reactor design along with process optimization aims to solve the lower biohydrogen production rates when compared to other biofuels. The integration of biohydrogen production with biogas production leads to mutual beneficial effects because it allows intermediary compounds to be shared between these processes and balances microbial ecosystems to enhance total energy filtration from feedstock materials. A comprehensive biorefinery framework benefits greatly from diverse biofuel production systems because it enhances both resource utilization and sustainability performance. The combination of various conversion processes enables the best possible energy recoveries in addition to the reduction of remaining waste streams. The combined system creates opportunities to generate multiple marketable by-products while giving operators freedom to handle the natural variations seen in food waste streams. Technological advancements regarding pretreatment along with enzymatic hydrolysis and microbial fermentation and catalytic conversion continue to develop process efficiency and yield capacity which brings large-scale commercial biofuel production from food waste within closer reach. The combined bid to produce biodiesel along with bioethanol, biogas, and biohydrogen from food waste represents a progressive approach which delivers major benefits toward sustainable energy security and environmental sustainability and bioeconomic circular

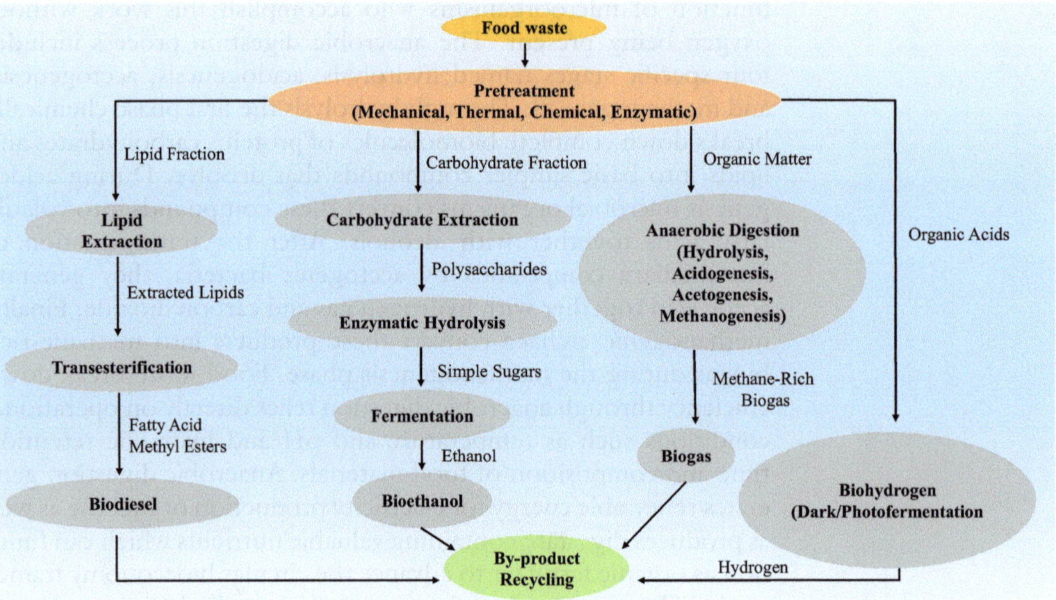

Fig. 2 Schematic overview of an integrated biorefinery approach for converting food waste into multiple biofuels and value-added products. The process begins with mechanical, thermal, chemical, or enzymatic pretreatment, generating lipid, carbohydrate, and organic fractions. Lipids are extracted and trans-esterified to produce biodiesel, while carbohydrate fractions undergo enzymatic hydrolysis and fermentation to yield ethanol. Concurrently, anaerobic digestion of organic matter produces methane-rich biogas or biohydrogen. By-product recycling further enhances overall sustainability and resource efficiency

development. Figure 2 visually encapsulates the multi-pathway conversion of food waste into renewable biofuels. It begins with a pretreatment stage where food waste is fractionated into distinct streams—lipid, carbohydrate, and organic matter fractions. The lipid fraction is directed toward extraction and subsequent transesterification to yield biodiesel, while the carbohydrate fraction is enzymatically hydrolyzed and fermented to produce bioethanol. Simultaneously, a separate pathway processes organic matter through anaerobic digestion to generate biogas, and organic acids are converted into biohydrogen via dark or photo-fermentation. Additionally, the diagram highlights the recycling of by-products, underscoring the circular bioeconomy concept integral to the integrated biorefinery approach.

3.5 Case Studies and Applications

Industrial case studies worldwide have increasingly demonstrated the potential of biorefinery protocols to convert food waste into renewable biofuels while creating valuable co-products, all within a framework of environmental sustainability and circular economy principles. For instance, in the United Kingdom, the start-up Dark Green has been pioneering urban biogas production by employing anaerobic digestion systems to process large volumes of food waste. By capturing methane from organic matter, their approach not only diverts waste from landfills but also produces a renewable energy source capable of powering thousands of homes, thereby significantly reducing greenhouse gas emissions. This example illustrates how integrating cutting-edge microbial processes with robust waste collection networks can transform urban waste management and energy generation. In Scotland, initiatives such as those led by regional councils in partnership with companies like Advetec have turned non-recyclable food waste into solid recovered fuel (SRF). Using advanced digestion reactors, these projects process food waste to produce a fuel that burns with substantially lower CO_2 emissions compared to traditional fossil fuels. This technology supports regulatory efforts aimed at eliminating biodegradable waste from landfills and showcases a scalable model that can be adopted in remote communities with limited waste disposal options. The successful implementation of these technologies underscores the dual benefits of environmental stewardship and improved energy security. The Danish eco-industrial park in Kalundborg stands as a longstanding testament to industrial symbiosis, where companies work together to maximize resource efficiency. Within this network, entities like the Asnaes Power Station supply surplus heat to local industries and residential areas, while by-products such as gypsum are repurposed in construction. Although not exclusively focused on food waste, the principles of resource sharing and waste valorization demonstrated in Kalundborg have inspired similar approaches in biorefinery operations. The Danish model highlights how collaborative efforts between industries can lead to significant economic savings and environmental benefits through integrated process design. North American companies have also played a pivotal role in advancing waste-to-biofuel technologies. Enerkem, based in Montreal, has developed a proprietary process that transforms municipal solid waste—including its organic fraction—into biofuels and renewable chemicals. Their Edmonton facility utilizes a series of thermal, chemical, and catalytic conversion steps to handle heterogeneous waste streams efficiently. The success of Enerkem's technology in processing diverse feedstocks not only reduces dependence on fossil fuels but also exemplifies the industrial scalability of modern biorefinery approaches. In France, TotalEnergies has undertaken a transformative project by converting its traditional fossil fuel refinery at La Mède into a biorefinery capable of producing aviation biofuel from waste cooking oil. Utilizing advanced hydroprocessing and catalytic upgrading techniques, the facility represents a significant step toward integrating renewable energy production within established industrial infrastructures. Despite some controversies surrounding feedstock sustainability, the La Mède project marks an important milestone in repurposing legacy assets to meet contemporary environmental standards and reduce the carbon footprint of the energy sector. In Finland, UPM's biorefinery at Lappeenranta has set a benchmark by processing crude tall oil—a by-product of the pulp and paper industry—into advanced second-generation biodiesel. Employing state-of-the-art processing technologies, the facility not only maximizes biofuel yield but also generates valuable co-products that enhance overall economic feasibility. This model demonstrates how industries with traditional waste streams can effectively harness biorefinery protocols to create renewable energy solutions, thereby supporting broader goals of resource recovery and sustainability. Smaller scale industrial applications further illustrate the versatility of biorefinery processes. In the United Kingdom, Bio-bean has successfully industrialized the recycling of waste coffee grounds into biofuels and biomass pellets. By converting an abundant urban waste stream into a commercially viable energy product, Bio-bean exemplifies how niche feedstocks can be efficiently valorized, contributing to both waste reduction and energy diversification. Similarly, in the United States, Changing World Technologies has developed a thermal conversion process that transforms agricultural residues, such as turkey offal and fats, into renewable diesel. Research initiatives in Australia have also contributed to the industrial application of biorefinery protocols. For example, studies funded by the Australian Renewable Energy Agency are investigating the conversion of sugarcane waste into sustainable aviation fuel. These projects focus on optimizing process parameters to achieve high-quality fuel output while ensuring economic viability. The findings from these studies reinforce the concept that agricultural residues can serve as an effective feedstock for advanced biofuel production, expanding the scope of waste valorization beyond urban and municipal waste streams. Collectively, these case studies highlight that biorefinery protocols offer a robust framework for transforming food and organic waste into renewable energy and value-added products. By leveraging innovative technologies and fostering industrial collaborations, companies like Dark Green, Enerkem, TotalEnergies, UPM, Bio-bean, and Changing World Technologies are setting the stage for a more sustainable energy future. The continued evolution of these approaches is expected to drive further advancements in process efficiency, scalability, and economic performance, ultimately contributing to the global transition toward a circular bioeconomy. Table 3 provides an overview of diverse case studies and applications demonstrating the conversion of food waste into renewable energy across different geographic regions. It categorizes key entities by their location, the type of feedstock or waste utilized, the conversion technology or process employed, and the resultant biofuel or co-product. The table highlights innovative practices—from urban anaerobic digestion to advanced catalytic upgrading that underscore the scalability and adaptability of biorefinery approaches. It serves as a concise reference for understanding how

Table 3
Global case studies and applications of food waste-based biorefineries

Case study/ entity	Country	Feedstock/waste type	Conversion technology/ process	Biofuel/outcome
Dark Green	United Kingdom	Urban food waste	Anaerobic digestion with integrated waste collection networks	Methane-rich biogas powering thousands of homes
Advetec and Regional Councils	Scotland	Non-recyclable food waste	Advanced digestion reactors producing solid recovered fuel (SRF)	Lower CO_2 emission fuel; scalable solution
Enerkem	Canada (North America)	Municipal solid waste (including organic fraction)	Thermal, chemical, and catalytic conversion processes	Biofuels and renewable chemicals
TotalEnergies (La Mède Project)	France	Waste cooking oil	Hydroprocessing and catalytic upgrading	Aviation biofuel
UPM's biorefinery	Finland	Crude tall oil (pulp and paper by-product)	State-of-the-art processing technologies	Second-generation biodiesel and value-added co-products
Bio-bean	United Kingdom	Waste coffee grounds	Industrial recycling and conversion processes	Biofuels and biomass pellets
Changing World Technologies	The USA	Agricultural residues (Turkey offal, fats)	Thermal conversion process	Renewable diesel
Australian Research Initiative	Australia	Sugarcane waste	Emerging conversion process research	Potential sustainable aviation fuel

integrated waste valorization strategies contribute to environmental sustainability and economic viability in the global energy landscape.

4 Notes

Food waste (FW) presents a significant global environmental challenge, necessitating urgent and sustainable management solutions. Extensive efforts are being made to address FW through ecologically and economically viable approaches. This chapter explores the characteristics and composition of FW, along with the diverse types of biofuels that can be derived from it. Furthermore, various technologies and optimization processes for converting FW into bioenergy are thoroughly analyzed. The discussion also emphasizes the economic and environmental factors influencing the costs of refining FW into valuable products. Gaining insight into these aspects enables stakeholders to develop more efficient and sustainable strategies for converting FW into renewable energy, reducing its environmental footprint while fostering economic growth.

References

1. Pandey A (2021) Food wastage: causes, impacts and solutions. Sci Herit J 5:17–20
2. Lahiri A, Daniel S, Kanthapazham R, Vanaraj R, Thambidurai A, Peter LS (2023) A critical review on food waste management for the production of materials and biofuel. J Hazard Mater Adv 10:100266
3. Bagherzadeh M, Inamura M, Jeong H (2014) Food waste along the food chain. OECD, Paris
4. MacRae R, Siu A, Kohn M, Matsubuchi-Shaw M, McCallum D, Cervantes TH et al (2016) Making better use of what we have: strategies to minimize food waste and resource inefficiency in Canada. Can Food Stud 3(2): 145–215
5. Tropea A (2022) Biofuels production and processing technology. MDPI, p 319
6. Ebikade E, Athaley A, Fisher B, Yang K, Wu C, Ierapetritou MG et al (2020) The future is garbage: repurposing of food waste to an integrated biorefinery. ACS Sustain Chem Eng 8(22):8124–8136
7. Zeng J, Zeng H, Wang Z (2022) Review on technology of making biofuel from food waste. Int J Energy Res 46(8):10301–10319
8. Konti A, Kekos D, Mamma D (2020) Life cycle analysis of the bioethanol production from food waste—a review. Energies 13(19):5206
9. Akita H, Shibata S, Komoriya T, Kamei S, Asamoto H, Matsumoto M (2024) Simultaneous saccharification and fermentation for isobutanol production from banana peel. Fermentation 10(3):161
10. Gupta Y, Barrett B, Vlachos DG (2024) Understanding microwave-assisted extraction of phenolic compounds from diverse food waste feedstocks. Chem Eng Process Process Intensification 203:109870
11. Alpural A, Dincoglu B, Imamoglu E (2024) Ultrasound-assisted extraction vs. microwave-assisted extraction for sustainable development goals: selecting the ideal lipid extraction and fatty acid profile. Chem Eng Process Process Intensification 205:110035
12. Álvarez C, Duque A, Sánchez-Monedero A, González EJ, González-Miquel M, Cañadas R (2024) Exploring recent advances in lignocellulosic biomass waste delignification through the combined use of eutectic solvents and intensification techniques. Processes 12(11):2514
13. Momen E, Satari B (2024) Green biorefinery of walnut husk: phenolic extraction and ethanol production. Appl Food Res 4(2):100500
14. Patria RD, Wong JW, Johnravindar D, Uisan K, Kumar R, Kaur G (2021) Food waste digestate-based biorefinery approach for rhamnolipids production: a techno-economic analysis. Sustain Chem 2(2):237–253
15. Saini JK, Kaur A, Mathur A (2022) Strategies to enhance enzymatic hydrolysis of lignocellulosic biomass for biorefinery applications: a review. Bioresour Technol 360:127517
16. Sawatdeenarunat C, Nguyen D, Surendra K, Shrestha S, Rajendran K, Oechsner H et al (2016) Anaerobic biorefinery: current status, challenges, and opportunities. Bioresour Technol 215:304–313
17. Mohammad REA, Abdullahi SSA, Muhammed HA, Musa H, Habibu S, Jagaba AH et al (2024) Recent technical and non-technical biorefinery development barriers and potential solutions for a sustainable environment: a mini review. Case Stud Chem Environ Eng 9: 100586
18. Balan IM, Trasca TI, Brad I, Belc N, Tulcan C, Radoi BP et al (2023) Transitioning to zero hunger. In: Approaches to limiting food loss and food waste. MDPI, pp 215–244
19. Nanda S, Vo D-VN, Sarangi PK (2020) Biorefinery of alternative resources: targeting green fuels and platform chemicals. Springer
20. Economou F, Voukkali I, Papamichael I, Phinikettou V, Loizia P, Naddeo V et al (2024) Turning food loss and food waste into Watts: a review of food waste as an energy source. Energies 17(13):3191
21. Dou Z, Toth JD (2021) Global primary data on consumer food waste: rate and characteristics—a review. Resour Conserv Recycl 168:105332
22. Silvennoinen K, Nisonen S, Katajajuuri J-M (2022) Food waste amount, type, and climate

impact in urban and suburban regions in Finnish households. J Clean Prod 378:134430

23. Armington WR, Babbitt CW, Chen RB (2020) Variability in commercial and institutional food waste generation and implications for sustainable management systems. Resour Conserv Recycl 155:104622
24. Kosseva MR, Webb C (2020) Food industry wastes: assessment and recuperation of commodities. Academic
25. Roka K (2022) Environmental and social impacts of food waste. In: Responsible consumption and production. Springer, pp 216–227
26. Hingston ST, Noseworthy TJ (2020) On the epidemic of food waste: idealized prototypes and the aversion to misshapen fruits and vegetables. Food Qual Prefer 86:103999
27. Chen Y, Pinegar L, Immonen J, Powell KM (2023) Conversion of food waste to renewable energy: a techno-economic and environmental assessment. J Clean Prod 385:135741
28. Wu S, Wang Q, Fang M, Wu D, Cui D, Pan S et al (2023) Hydrothermal carbonization of food waste for sustainable biofuel production: advancements, challenges, and future prospects. Sci Total Environ 897:165327

Correction to: Food Waste Valorization

Tanmay Sarkar

Correction to:
Chapters 4 and 12 in: Tanmay Sarkar (ed.), ***Food Waste Valorization: Green Techniques in Sustainable Management*****, Methods and Protocols in Food Science,**
https://doi.org/10.1007/978-1-0716-4646-5

The original version of this book was inadvertently published with errors. The title of Chap. 4 and the author list of Chap. 12 have been updated.

The updated version of these chapters can be found at
https://doi.org/10.1007/978-1-0716-4646-5_4
https://doi.org/10.1007/978-1-0716-4646-5_12

Tanmay Sarkar (ed.), *Food Waste Valorization: Green Techniques in Sustainable Management*,
Methods and Protocols in Food Science, https://doi.org/10.1007/978-1-0716-4646-5_17,
© The Author(s), under exclusive license to Springer Science+Business Media, LLC, part of Springer Nature 2025

Index

A

B

Tanmay Sarkar (ed.), *Food Waste Valorization: Green Techniques in Sustainable Management*, Methods and Protocols in Food Science, https://doi.org/10.1007/978-1-0716-4646-5,
© The Editor(s) (if applicable) and The Author(s), under exclusive license to Springer Science+Business Media, LLC, part of Springer Nature 2025

C

D

E

F

G

H

MIX
Papier aus verantwortungsvollen Quellen
Paper from responsible sources
FSC® C105338

If you have any concerns about our products,
you can contact us on
ProductSafety@springernature.com

In case Publisher is established outside the EU,
the EU authorized representative is:
Springer Nature Customer Service Center GmbH
Europaplatz 3, 69115 Heidelberg, Germany

Printed by Libri Plureos GmbH
in Hamburg, Germany

